T0348567

CONCISE ENCYCLOPEDIA OF
HISTORY OF ENERGY

Elsevier Science and Technology Homepage - http://www.elsevierdirect.com
Consult the Elsevier homepage for full catalogue information on all books, major reference works, journals, electronic products and services.

Elsevier Titles of Related Interest

C. CLEVELAND ET AL.
Encyclopedia of Energy (2004)
ISBN-13: 978-0-12-176480-7

T. DRENNEN & J. ROSTHAL
Pathways to a Hydrogen Future (2007)
ISBN-13: 978-0-08-046734-4

I. DINCER & M. ROSEN
Energy (2007)
ISBN-13: 978-0-08-044529-8

E. HUSSEIN
Radiation Mechanics (2007)
ISBN-13: 978-0-08-045053-7

I. PÁZSIT & L. PÁL
Neutron Fluctuations (2007)
ISBN-13: 978-0-08-045064-3

CONCISE ENCYCLOPEDIA OF
HISTORY OF ENERGY

Editor

C. CLEVELAND
Boston University, Boston, MA, USA

ELSEVIER

Amsterdam—Boston—Heidelberg—London—New York—Oxford
Paris—San Diego—San Francisco—Singapore—Sydney—Tokyo

Elsevier Inc., 525 B Street, Suite 1900, San Diego, CA 92101-4495, USA

First edition 2009

A catalogue record for this book is available from the British Library

ISBN 978-0-12-375117-1

Transferred to Digital Printing, 2012

Printed and bound by CPI Group (UK) Ltd, Croydon, CR0 4YY
Transferred to Digital Print 2012

CONTENTS

Contents

O

P

Q

T

W

EDITOR'S PREFACE

Human history can be told in terms of the history of energy. The discovery of fire, the domestication of animals, the discovery of fossil fuels, the electrification of cities, and advances in nuclear physics were all pivot points in human history. The importance of energy in human affairs has continued to grow since the publication of the *Encyclopedia of Energy* by Elsevier Science in 2004. The critical connection between energy and climate change was a highlight of the 2007 Assessment Report by the Intergovernmental Panel on Climate Change (IPCC). Record energy prices contributed to a global recession, and investment in "green energy" was a catalyst for economic recovery. Developing nations continue to struggle with meeting the energy needs of population and economic growth in a sustainable and equitable manner. This *Concise Encyclopedia of the History of Energy* pulls together the essential articles related to energy history from the *Encyclopedia of Energy*. Key articles have been updated, including the History of Electricity Use, the History of Natural Gas, the History of OPEC Market Behavior, the History of the Oil Industry, and the History of the Prices of Energy. Readers should find the *Concise Encyclopedia* a valuable reference update to the always-evolving field of energy.

Cutler J. Cleveland
Boston, MA USA
August 2009

ALPHABETICAL LIST OF ARTICLES

C

Coal Industry, History of

Glossary

coal   A black or brown rock that burns; a solid combustible rock formed by the lithification of plant remains; a metamorphosed sedimentary rock consisting of organic components; a solid fossil hydrocarbon with a H/C ratio of less than 1, most commonly around 0.7 (for bituminous coal); a solid fossil fuel; fossilized lithified solar energy.

Coal Age   The historical period when coal was the dominant fuel, from late 18th through middle 20th century; name of a trade journal devoted to the coal industry (ceased publication summer 2003).

Coalbrookdale   A town in England's Black Country, symbol of the Industrial Revolution; site of first iron production using coke produced from coal; center of early steam engine and railroad development; celebrated in paintings by Williams, de Loutherbourg, Turner, and others, damned in poems by Anna Seward and others; part of the Ironbridge Gorge UNESCO World Heritage Site; a British national monument.

coal field   Region in which coal deposits occur.

coal gas   Fuel rich gas produced by partial oxidation (burning) of coal (also producer gas).

coal liquefaction   Method for producing liquid fuels (e.g., gasoline, diesel) from coal.

coal preparation   Treatment of coal to prepare it for any particular use; improve the quality of coal to make it suitable for a particular use by removing impurities, sizing (crushing, screening), and special treatments (e.g., dedusting); upgrading (beneficiation) of coal to a more uniform and consistent fuel (or chemical feedstock) of a particular size range, with specified ash, sulfur, and moisture content.

Industrial Revolution   Controversial term to design the period when modern industrial society developed, initially applied primarily to Britain, later to other countries as well. From about 1760 to about 1850.

International Energy Agency (IEA)   Autonomous body within OECD (Organisation for Economic Co-operation and Development) to implement an international energy program. Publishes reports on the coal industry, many of which are posted on its Web page.

water gas   Gas produced by the interaction of steam on hot coke, used for lighting (primarily during the 19th through early 20th century) and as fuel (well into 20th century).

In a narrow sense, the coal industry can be considered as the coal mining industry, including coal preparation. More broadly it includes major users, such as steam generators for electric power, coking for steel production, and, historically, coal as a feedstock for chemicals and as transportation fuel, especially steam locomotives and ships, as well as the wide range of applications of steam engines. The transportation of coal, given its bulk and low value, can be a major cost of its use. Coal exploration and geology are part of the front end of the coal mining cycle.

1. Introduction

Coal has been mined for centuries. Until shortly before the Industrial Revolution, its use was local, and it did not make a significant contribution to the overall energy consumption of the world. Coal use increased rapidly in the two centuries before the Industrial Revolution and has continued to grow ever since, with occasional temporary down dips. Coal was the dominant fuel during the 19th century and the first half of the 20th century. The rise of modern society is intimately intertwined with the growth of the coal industry. Developments driven by coal with a major impact on technological progress include the steam engine, the railroad, and the steamship, dominant influences on the formation of modern society. For over a century, coal was the major energy source for the world. Its relative contribution declined over the second half of the 20th century. In absolute terms its contribution continues to increase.

The environmental disadvantages of coal have been recognized for centuries. Efforts to reduce these disadvantages accelerated over the last third of the 20th century. Because coal remains the largest and most readily available energy source, its use is likely to continue, if not increase, but will have to be supported by improved environmental control.

2. Pre- and Early History

Where coal was mined and used initially is not clear. Secondary sources vary as to when coal use may have started in China, from as early as 1500 BC to as late as well after 1000 AD. It has been stated that a Chinese coal industry existed by 300 AD, that coal then was used to heat buildings and to smelt metals, and that coal had become the leading fuel in China by the 1000s. Marco Polo reported its widespread use in China in the 13th century. Coal may have been used by mammoth hunters in eastern central Europe.

The Greeks knew coal, but as a geological curiosity rather than as a useful mineral. Aristotle mentions coal, and the context implies he refers to "mineral" or "earth" coal. Some mine historians suggest he referred to brown coal, known in Greece and nearby areas. Theophrastus, Aristotle's pupil and collaborator, used the term ανθραζ(anthrax), root of anthracite. Although Theophrastus reported its use by smiths, the very brief note devoted to coal, compared to the many pages dealing with charcoal, suggest, as does archaeological evidence, that it was a minor fuel. The term is used with ambiguous meanings, but at least one was that of a solid fossil fuel. Theophrastus describes spontaneous combustion, still a problem for coal storage and transportation.

Coal was used in South Wales during the Bronze Age. The Romans used coal in Britain, in multiple locations, and in significant quantities. After the Romans left, no coal was used until well into the second millennium. Lignite and especially peat, geological precursors to coal, were used earlier and on a larger scale in northern and western Europe. Pliny, in the first century AD, mentions the use of earth fuel by inhabitants of Gaul near the Rhine mouth, to heat their food and themselves. All indications are that he describes the use of peat in an area currently part of The Netherlands, where peat still was used as a fuel nearly twenty centuries later. Romans observed coal burning near St. Étienne, later a major French coal mining center. The Romans carved jet, a hard black highly polishable brown coal, into jewelry. Jet was used as a gemstone in Central Europe no later than 10000 BC.

The recorded history of coal mining in India dates from the late 18th century. Place and river names in the Bengal-Bihar region suggest that coal may have been used, or at least that its presence was recognized, in ancient times.

Coal use was rare, even in most parts of the world where it was readily accessible in surface outcrops, until well into the second millennium, at which time its use became widespread, be it on a small scale.

The Hopis mined coal at the southern edge of Black Mesa, in northern Arizona, from about the 13th through the 17th century AD. Most was used for house fuel, some for firing pottery. Coal was mined by a primitive form of strip mining, conceptually remarkably similar to modern strip mining. At least a few underground outcrop mines were pursued beyond the edge of the last (deepest) strip, a practice reminiscent of modern auger mining. Gob stowing was used to prevent or control overburden collapse.

3. Middle Age and Renaissance

Coal and iron ore formed the basis of the steel industry in Liège in present Belgium. Coal mining rights were granted in Liege in 1195 (or 1198?) by the Prince Bishop. At about the same time, the Holyrood Abbey in Scotland was granted coal mining rights. A typical medieval situation involved abbeys or cloisters driving technology. In France, a 13th-century real estate document established property limits defined by a coal quarry. The earliest reliable written documentation in Germany appears to be a 1302 real estate transaction that included rights to mine coal. The transaction covered land near Dortmund in the heart of the Ruhr.

Coal use started before it was documented in writing. Religious orders tended to keep and preserve written materials more reliably than others. It seems reasonable to postulate that coal use began no later than the 12th century in several West European countries. Documented evidence remains anecdotal for several more centuries but indicates that coal use increased steadily from the 12th century on. England led in coal production until late in the 19th century. Contributing to this sustained leadership were that wood (and hence charcoal) shortages developed earlier and more acutely there than in other countries, there was ready access to shipping by water (sea and navigable rivers), and large coal deposits were present close to the surface.

By the 13th century, London imported significant amounts of coal, primarily sea-coal, shipped from Newcastle-upon-Tyne. The use of coal grew steadily, even though it was controversial, for what now would be called environmental impact reasons. Smoke, soot, sulfurous odors, and health concerns made it undesirable. From the 13th century on, ordinances were passed to control, reduce, or prevent its use. None of these succeeded, presumably because the only alternatives, wood and charcoal, had become too expensive, if available at all. Critical for the acceptance of coal was the development of chimneys and improved fire places. By the early 1600s, sea-coal was the general fuel in the city. The population of the city grew from 50,000 in 1500 to more than 500,000 by 1700, the coal imported from less than 10,000 tons per year to more than 500,000 tons.

Early coal use was stimulated by industrial applications: salt production (by brine evaporation), lime burning (for cement for building construction), metal working, brewing, and lesser uses. By the late Middle Ages, Newcastle exported coal to Flanders, Holland, France, and Scotland.

By the early 16th century, coal was mined in several regions in France. In Saint Étienne, long the dominant coal producer, coal was the common household fuel, and the city was surrounded by metals and weapons manufacturing based on its coal. Legal and transportation constraints were major factors in the slower development of coal on the European continent compared to England. In the latter, surface ownership also gave subsurface coal ownership. In the former, the state owned subsurface minerals. Notwithstanding royal incentives, France found it difficult to promulgate coal development on typically small real estate properties under complex and frequently changing government regulations. Only late in the 17th century did domestic coal become competitive with English coal in Paris, as a result both of the digging of new canals and of the imposition of stiff tariffs on imported coal. Coal mining remained artisanal, with primitive exploitation "technology" of shallow outcrops, notably in comparison with the by this time highly developed underground metal mining.

The earliest coal mining proceeded by simple strip mining: the overburden was stripped off the coal and the coal dug out. As the thickness of the overburden increased, underground mines were dug into the sides of hills, the development of drift mines. Mining uphill, strongly preferred, allowed free water drainage and facilitated coal haulage. Some workings in down-dipping seams were drained by driving excavations below the mined seams.

Shafts were sunk to deeper coal formations. When coal was reached, it was mined out around the shaft, resulting in typical bell pits. Coal was carried out in baskets, on ladders, or pulled out on a rope. Where necessary, shafts were lined with timber.

In larger mines, once the coal was intersected by the shaft, "headings" were driven in several directions. From these, small coal faces were developed, typically about 10 ft wide, usually at right angles from the heading, and pillars, blocks of coal, were left in between these so-called bords, the mined-out sections. Pillar widths were selected to prevent roof, pillar, and overburden failure, which could endanger people and could lead to loss or abandonment of coal. Each working face was assigned to a hewer, who mined the coal, primarily with a pick. The hewer first undercut the coal face: he cut a groove along the floor as deep as possible. He then broke out the overhanging coal with picks, wedges, and hammers. Lump coal was hand loaded into baskets, usually by the hewer's helper. The baskets were pushed, dragged, or carried to and sometimes up the shaft.

This haulage commonly was performed by women and children, usually the family of the hewer. The hewer operated as an independent contractor. He was paid according to the amount of coal he and his family delivered to the surface. Once established as a hewer, after multiple years of apprenticeship, the hewer was among the elite of workers in his community.

In deeper or better equipped mines, the coal baskets or corves were hoisted up the shaft using a windlass, later horse-driven gins. These also were used for hoisting water filled baskets or buckets, to dewater wet mines. By the end of the 17th century, shaft depths reached 90 ft, occasionally deeper. Encountering water to the extent of having to abandon pits became frequent. Improving methods to cope with water was a major preoccupation for mine operators.

Ventilation also posed major challenges. Larger mines sank at least two shafts, primarily to facilitate operations. It was recognized that this greatly improved air flow. On occasion shafts were sunk specifically to improve ventilation. During the 17th century, the use of fires, and sometimes chimneys, was introduced to enhance air updraft by heating the air at the bottom of one shaft. In "fiery" mines, firemen, crawling along the floor, heavily dressed in wetted down cloths, ignited and exploded gas accumulations with a lighted pole. This practice was well established in the Liège basin by the middle of the 16th century. As mines grew deeper, and production increased, explosions became more frequent, on occasion killing tens of people.

In North America, coal was reported on Cape Breton in 1672. Some coal was mined for the French garrison on the island, some was shipped to New England before 1700. Coal was found by Joliet, Marquette, and Father Hennepin, in 1673, along the Illinois river. It is possible that some was used at that time by the Algonquins in this area.

4. Precursors to the Industrial Revolution

The best sun we have is made of Newcastle coal.

—*Horace Walpole, 1768*

Whether or not one subscribes to the (controversial) concept of a first industrial revolution, the use of coal increased significantly from about the middle of the 16th century. A prime cause was the shortage of firewood, acute in England, noticeable in France. In Britain coal production increased from less than 3 million tons in 1700 to more than 15 million tons in 1800 (and then doubled again to more than 30 million tons per year by 1830). Driving were the growing industrial uses of coal, in particular the technology that made possible the smelting and forging of iron with coal. Steam power, developed to dewater mines, created a voracious demand for coal.

Land positions and legal statutes facilitated the development of coal. The frequent closeness of iron ore, limestone, and coal deposits stimulated the iron industry. Canals provided the essential low cost water transportation for inland coal fields. Tramways and turnpikes fed canal transport and led to railroads and highways.

Major technical advances were made in mining technologies: dewatering, haulage, and ventilation, challenges to mining that continue to this day and presumably always will.

Thomas Savery's *Miners Friend*, patented in 1698, promised the use of steam for a mine dewatering pump. Thomas Newcomen's steam-driven pump fulfilled the promise. The first Newcomen steam pump, or "fire-engine" or "Invention to Raise Water by Fire," was installed at a coal mine in 1712. Although expensive and extraordinarily inefficient by modern standards or even by comparison with Watt's steam engine of 60 years later, 78 Newcomen engines operated by 1733. The first Boulton and Watt engine, based on Watt's patent, was installed at a coal mine in 1776. By the end of the century, Boulton and Watt engines also found increasing use for shaft hoisting. For most of the century, horse power had been the dominant shaft hoisting method.

Underground haulage improved. Wheels were mounted on the baskets previously dragged as or on sledges. Planks were placed on the floor and were replaced by wooden and eventually iron rails—forerunners of the railroad. Horses replaced boys for pulling, at least in those mines where haulage-ways were large enough.

Dewatering became practical to great depths. Haulage improved greatly. Coping with explosive gases proved difficult. Fires at shaft bottoms remained the dominant method to induce airflow. Numerous ingenious devices were invented to course air along working faces, as air coursing was pursued vigorously. But explosions remained a major and highly visible hazard, resulting in an increasing cost in lives.

The physical coal winning—hewer with pick—remained unchanged. Improvements in production were associated with better lay-outs of the workings. Most mines continued to operate the room and pillar method, leaving over half the coal behind. An exception was the development of the longwall method. Here, a single long face was mined out entirely. As the coal was mined, the roof was supported with wooden props. The gob, the mined out area, was back-filled with small waste coal. This method allowed a nominal 100% recovery or extraction but resulted in surface subsidence. It now has become, in a highly mechanized form, the dominant underground coal mining method in the United States, Australia, and Europe.

The European continent lagged far behind Britain in all aspects of coal industry development: production, technology, transportation, use, and legal framework. By 1789, France produced somewhat more than 1 million tons per year, about one-tenth of that produced in Britain. The Anzin Coal Company, formed after considerable government prodding to try to increase coal production, mined 30% of all French coal. France now recognized the growing threat posed by British industrialization.

Although by 1800 some 158 coal mines operated in the Ruhr, production that year barely exceeded 250,000 tons. The first steam engine at a coal mine in the region became operational only in 1801.

In North America, wood remained widely available at low cost, as well as water power in the more developed areas (e.g., New England). There was no need to use coal, and little was used. Major deposits were discovered and would prove extremely significant later on.

Coal was known to exist near Richmond, Virginia, by 1701, and some was mined by 1750. Some was sold to Philadelphia, New York and Boston by 1789. In 1742, coal was discovered in the Kanawha Valley, West Virginia. The Pocahontas seam, a major coal source for many decades, was mapped in Kentucky and West Virginia in 1750. An extensive description of coal along the Ohio River was made in 1751. George Washington visited a coal mine on the Ohio River in 1770. Thomas Jefferson recorded coal occurrences in the Appalachian mountains.

5. Industrial Revolution

We may well call it black diamonds. Every basket is power and civilization. For coal is a portable climate. It carries the heat of the tropics to Labrador and the polar circle; and it is the means of transporting itself whithersoever it is wanted.

—Emerson, **Conduct of Life**, 1860

Coal fueled the Industrial Revolution, arguably the most important development in history. It transformed the world. While the term "Industrial Revolution" is controversial, it remains helpful to identify that period when fundamental changes developed in the basic economic structure, initially in Great Britain, shortly thereafter on the European and North American continents, and eventually throughout much of the world. The transition to the predominant use of fossil fuels, initially coal, was a major aspect and driving force of industrialization.

In Britain, the Industrial Revolution usually is considered the period from about 1760 through about 1840. In Belgium it started about 1830; in France, the United States, and Germany it started within the following decades. Other West European countries followed well before the end of the

19th century. Russia and Japan initiated their industrial revolutions by the end of the century.

The reasons why Britain took the lead are complex and involve all aspects of society. Readily accessible coal and iron ore in close proximity to each other was a major factor. Mine dewatering needs led to the development of the steam engine and mine haulage to railroads, two core technologies of the Industrial Revolution.

Belgium had coal and iron ore of good quality in close proximity and the necessary navigable river system. In France and Germany, coal and iron ore were far apart, and in France they were not of a quality that could be used for steel production until late in the century, when its metallurgy was understood. France and Germany lacked the water transportation necessary for the development of a heavy industry (Fig. 1). Steam locomotives and steam ships permanently altered travel and perceptions of space and time. Both were major users of coal and required that coal be provided at refueling points all around the world, starting worldwide coal trade and transportation. For many decades, Britain dominated this coal shipping and trading. Coal bunkering ports were major way stations toward building the British Empire.

Steam converted the Mississippi into the commercial artery it still is, although it is now diesel fueled. Railroads opened up the Midwest, the West, and later Canada. As the railroads expanded, so did the coal mines feeding them. Anthracite built eastern Pennsylvania, bituminous coal built Pittsburgh.

Coal gas and water gas dramatically, often traumatically, increased available lighting and changed living habits. Initially used in manufacturing plants, to allow around the clock work, gas light slowly conquered streets and homes and turned Paris into the "city of light."

Social, political, and cultural revolutions accompanied the Industrial Revolution. Working and living conditions of the lower classes became a societal issue and concern and industrial societies accumulated the wealth needed to address such concerns. Dickens, Marx, Engels, and many others identified social problems and proposed solutions either within or outside the existing political structure.

Coal mining was marked by difficult labor relations from the beginning and continues to be so in most parts of the world where it is a significant economic activity. Responsibility of the government in the social arena found its expression in multiple studies and reports and numerous labor laws. A focal point was child and female labor. The age at which children were allowed to start working underground was gradually raised. In Belgium, women miners were more adamant and more successful in their opposition to being banned from underground and worked there legally until late in the 19th century (Fig. 2).

Although science and the scientific approach were well developed by this time, they had little influence

Figure 1
The Coal Wagon. Théôdore Géricault (1821). Transportation cost always has been a major item in coal use. Road haulage was and is extremely expensive. A significant advantage for early British industrial development was its ready access to sea, river, and canal haulage of coal. Courtesy of the Fogg Art Museum, Harvard University Art Museums Bequest of Grenville L. Winthrop. Gray Collection of Engravings Fund. © President and Fellows of Harvard College. Used with permission.

Figure 2
The Bearers of Burden. Vincent Van Gogh (1881). The first bearer carries a safety lamp, suggesting she came from underground, which still was legal in the Belgian Borinage coal mining region when Van Gogh worked there in the 1870s. Collection Kröller-Müller Museum, Otterlo, The Netherlands. Photograph and reproduction permission courtesy of the Kröller-Müller Foundation.

on the early technological developments. The lack of geological understanding made the search for deeper coal uncertain and expensive. Geological sciences grew rapidly toward the end of this period and would soon make major contributions to finding coal and understanding the complexities of coal—even though its origin and formation remained a matter of controversy for decades. While the official geological community had little interaction with coal mining, William Smith, author of the first geological map of Britain, a masterpiece of geological mapping, was involved in coal mines, although his prime engineering work was for canals, mainly coal shipping canals. "Strata Smith" was a founder of stratigraphy because of his recognition of the possibility to use fossils to correlate strata. The first paper published by James Hutton, the founder of modern geology, was a coal classification. Hutton devised a method to allow customs agents and shippers to agree on a coal classification to set customs fees. While Hutton recognized the usefulness of the vast amount of geological information disclosed by coal mining, his later interest was strictly scientific. He did present a theory for coal formation based on metamorphosis of plant materials. While coal mining influenced the early development of geology, its impact was less than that of metal mining.

During the 1840s, the leading British geologist Charles Lyell visited several coal mining areas in North America. He proposed igneous activity combined with structural disturbance as the mechanism of formation of the anthracites in northeastern Pennsylvania, then the booming heart of the U.S. Industrial Revolution. He used the creep of mine roof and floor as an analog for the slow large deformations of rock formations.

As in Britain, much early U.S. geological mapping was for surveys for canals built during the beginning of the U.S. Industrial Revolution. Because much of the initial geological mapping was utilitarian, including looking for building stone and metal ore, at the time when the coal industry was developing, finding and mapping coal was a major objective of early geologists. This included the First Geological Survey of Pennsylvania, and the mapping of the upper Midwest by David Dale Owen, first state geologist of Indiana.

Although the connection between coal and geology may seem obvious, and it was far less so then than now, even less obvious might be relations between Industrial Revolution technology and fundamental geology. Yet Hutton, good friend of James Watt, was influenced by the *modus operandi* of the steam engine in his understanding of the uplifting of mountains.

6

Among the members of the Oysters Club in Edinburgh that included James Watt and James Hutton was Adam Smith. *The Wealth of Nations* addresses many strengths and weaknesses of the coal industry, while laying the theoretical economic basis for the free market economy that allowed it and the Industrial Revolution to thrive. Adam Smith identified the need for a coal deposit to be large enough to make its development economically feasible, the need for it to be located in the right place (accessible to markets, i.e., on good roads or water-carriage), and the need for it to provide fuel cheaper than wood (i.e., to be price competitive). Also, as Smith stated, "the most fertile coal-mine regulates the price of coals at all other mines in its neighbourhood." He discussed the appropriate purchase price and income of a coal mine and the economic differences between coal and metal mines—in sum, the basics of mineral and fuel economics.

The French engineer-scientist Coulomb reported on steam engines based on his observations during a trip to Britain and pushed French industrialists toward the use of coal. It was recognized that steam engines remained extremely inefficient. Scientific investigations of the performance of the engines evolved toward the science of thermodynamics. In 1824, Carnot published a theoretical analysis of the heat engine, a founding publication of thermodynamics, although the term was introduced only 25 years later by Lord Kelvin in his *Account of Carnot's Theory*. From 1850 on there was considerable interaction between the development of the steam engine and of thermodynamics, as demonstrated by the involvement in both of Rankine, one of the Scottish engineers-scientists who drove the technological developments of the Industrial Revolution.

6. 19th Century

The 19th century was the century of King Coal. Coal production grew dramatically in most countries that became industrialized, and coal use grew in all of them. While Britain continued to lead in coal production, the growth rate in the United States was so fast that by the end of the century its production exceeded Britain's (Table I). In 1900, coal exports accounted for 13.3% of total British export value. Coal trading was greatly liberated during the first half of the century. Direct sales between coal producers and consumers (e.g., gas producers) became legal. In this free and open domestic trading environment coal flourished. Internationally, the Empire encouraged and protected its domestic industry, supporting domestic production for export. Capital was needed to develop the mines—for example, to sink shafts, construct surface and underground operating plant (e.g., pumps, hoisting engines, fans), build loading, processing, and transportation

facilities. Coal mine ownership slowly shifted from private and partnerships to stock-issuing public corporations. Employment in British coal mining grew from about 60,000 to 80,000 in 1800 to nearly 800,000 in 1900.

In 1800, the United States produced barely over a 100,000 tons of soft coal and almost certainly much less anthracite. Although many Pennsylvania anthracite outcrops were well known by then and were within hauling distance from Philadelphia, Boston, and New York City, no satisfactory method for burning anthracite had been developed.

The American Industrial Revolution started around 1820, in parallel with the growth of the anthracite industry in northeastern Pennsylvania. Canals were built to provide transportation but were soon superseded by railroads. Anthracite had been promoted and used somewhat as a domestic fuel late in the 18th century. It did not become accepted until efficient fire grates became available, well into the 19th century, and until the price of wood fuel had risen significantly, particularly in the large cities. Following intense development of burning technologies, anthracite became the fuel of choice for public buildings and was used in steam engines (stationary, e.g., for coal mine pumping, manufacturing, and mobile: locomotives and boats) and iron production. The commercialization of improved and easier to use stoves made anthracite the home heating fuel of choice for well over the next century.

Early mining of anthracite was easy, as it was exposed in many outcrops. Many small operators could start mining it with little or no capital, manpower, or knowledge required. Once underground, the geological complexity of the intensely folded and faulted deposits required operating knowledge that led to consolidation in the industry. The drift mines and shallow shaft mines (rarely deeper than 30 ft) started in the 1810s were followed in the 1830s by slope mines. Breast and pillar mining was common: coal pillars were left in between the breasts (faces, rooms) that were mined out. The pillars carried the weight of the overburden. Horses and mules pulled cars running on rails. Black powder was used to shoot the coal (break out the coal). Wooden props (timber posts) provided safety by holding up the immediate roof.

Contract mining was standard. Each miner worked a breast, assisted by one or two helpers who loaded the coal and installed timber. The miner was paid on a piece rate and paid his helpers. The miner worked as an independent, with minimal supervision. Death and injury rates were high, predominantly from roof falls, not from the explosions that occurred all too often. By the end of the century the fatality rate in U.S. coal mining approached 1000 per year.

Over the course of the 19th century, numerous safety lamps were developed, typically designed with a protective wire mesh screen and glass to prevent the

Table 1
Annual Coal Production for Some Countries, in Millions of Tonnes[a] (Mt) per Year.

Year	World	United States	United Kingdom	Belgium	Germany	France	Japan	China	Australia	South Africa	Indonesia	Poland[c]	Russia, USSR/FSU[d]	Colombia	India
1700			2.5												
1800		0.1	>10 (157)		1	<0.1									
1815			22.6		<1.3	1									
1830		0.9	22.8 (30.9)	2.3	1.8	0.9									
1850		6.4	69.5	5.8	4.2	5									
1860	135	13.2	80.0	9.6	12.3	8.3									
1870									0.9			8	0.4		0.4
1871		42.5	117.4	13.8	24	13.2									
1880	332	64.9	147.0	16.9	47	19.4	0.9		1.6			14	2		1
1890	512	143.2	181.6	20.4	70.2	26.1	2.6			<0.1					
1900	700	244.2	225.2	23.5	109.3	33.4	7.4	<1	6.5	0.9	0.2	31	10		6.2
1910	1160	454.6	264.4	25.5	151.1	38.3	15.7	4.2							
1913	1341	517.2	287.4	22.9	277.4[b]	40.9	21.3	14.0	12.6	8.9	0.5	36	32.2	<0.1	16.5
1920	1320	597.1	233.2	22.4	131.4	24.3	30.5	19.5	13.2	10.4	1.1	32	7.6		18.2
1930	1414	487.1	262.1	27.4	142.7	53.9	33.4	26.4	11.5	12.2	1.7	34.0	35.2		21.9
1932	1124	326.1	212.1	21.4	104.7	46.3	28.1	28.0	11.3	9.9		28.8	53.7		22.0
1940	1497	462.0	227.9	25.6	240.1	39.3	57.3	46.5	11.8	17.2	2.0	77.1	148.7	0.5	29.9
1950	1508	509.4	220	27.3	113.8	52.5	38.5	41.1	16.5	26.5	0.8	78.0	185.2	1.0	32.8
1952	1496	484.2	228.5	30.4	143.7	52.4	43.4	66.6	19.4	28.1	1.0	84.4	215.0	1.0	36.9
1960	1991	391.5	197.8	22.5	148.0	56.0	57.5	397.2	21.9	38.2	0.7	104.4	374.9	2.6	52.6
1970	2208	550.4	147.1	11.4	118.0	37.8	40.9	354.0	45.4	54.6	0.2	140.1	205.7	2.75	73.7
1980	2810	710.2	130.1	8.0	94.5	20.2	18.0	620.2	72.4	115.1	0.3	193.1	553.0	4.2	113.9
1990	3566	853.6	94.4	2.4	76.6	11.2	8.3	1050.7	158.8	174.8	10.5	147.7	527.7	21.4	211.7
2000	3639	899.1	32.0	0.4	37.4	4.4	3.1	1171.1	238.1	225.3	78.6	102.2	321.6	37.1	309.9

[a] 1 tonne = 1000 kg = 2204.6 lb = 1.102 (short) tons

[b] Production for pre-World War I territory, sum of hard coal (lignite)

[c] There are considerable differences in data, especially before 1920, depending on territory considered—that is, on how changed boundaries (one of which has intersected, in different ways, Upper Silesia, the major coal producing area) affected coal production. Given is an exceedingly simplified summary of very approximate production data for the area currently (2003) Poland.

[d] FSU = Former Soviet Union (data from 1980 through 2000). Data from Coal Information (2001), "International Energy Agency," Paris, France (2001); "Energy Information Adminstration," U.S. Department of Energy, Washington, D.C; "Minerals Yearbook, U.S. Department of the Interior," Washington, D.C., (1918), (1923), (1934), (1950); "Historical Statistics of the United States to 1957: A Statistical Abstract Supplement," Washington, D.C. (1960); B. R. Mitchell, "International Historical Statistics, Africa, Asia & Oceania," 2nd Rev. ed., Stockton Press, New York, NY (1992); B. R. Mitchell, "International Historical Statistics, Europe (1750-1988)," 3rd ed. Stockton Press, New York, NY (1992); B. R. Mitchell, "International Historical Statistics, Africa, Asia & Oceania," 2nd Rev. ed., Stockton Press, New York, NY (1995); K. Takahashi (1969), "The Rise and Development of Japan's Modern Economy," The Jiji Tsushinsha (The Jiji Press, Ltd.), Tokyo; A. L. Dunham (1955), "The Industrial Revolution in France (1815-1848)," Exposition Press, New York; B. R. Mitchell (1962), "Abstract of British Historical Statistics," Cambridge at the University Press; W. W. Lockwood (1954), "The Economic Development of Japan," Princeton University Press, Princeton, NJ; A. S. Milward and S. B. Saul (1973), "The Economic Development of Continental Europe," Rowman and Littlefield, Totowa, NJ; W. Ashworth (1975), "A Short History of the International Economy Since 1850," Longman, London; M. Gillet (1973), "Les Charbonnages du Nord de la France au XIX Siècle," Mouton, Paris; R. Church (1986), "The History of the British Coal Industry, Vol. 3 (1830-1913): Victorian Pre-eminence," Clarendon Press, Oxford; V Muthesius (1943), "Ruhrkohle (1893-1943)," Essener Verlagsanstalt, Essen, Germany; B. R. V Mitchell (1980), "European Historical Statistics (1750-1975)," Facts on File, New York; M. W Flinn (1984), "The History of the British Coal Industry, Vol. 2 (1700-1830): The Industrial Revolution," Clarendon Press, Oxford; J. A. Hodgkins (1961), "Soviet Power," Prentice-Hall, Englewood Cliffs, NJ; H. N. Eavenson (1935), "Coal Through the Ages," AIME, New York; S. H. Schurr and B. C. Netschert (1960), "Energy in the American Economy (1850-1975)," The Johns Hopkins Press, Baltimore; "A History of Technology," T. I. Williams, Ed. (1978), Clarendon Press, Oxford; "COAL, Clarendon Mining in Art (1680-1980)," Arts Council of Great Britain, London; T. Wright, Growth of the modern chinese coal industry, "Modern China," Vol. 7, No. 3, July 1981, 317-350; R. L. Gordon (1970), "The Evolution of Energy Policy in Western Europe," Praeger, New York; J. S. Furnivall (1939) (1967), "Netherlands, India, A Study of Plural Economy," Cambridge at the University Press; D. Kumar, Ed. (1983) "The Cambridge Economic History of India," Cambridge University Press; Z. Kalix, L. M. Fraser, and R. I. Rawson (1966), "Australian Mineral Industry: Production and Trade (1842-1964)," Bulletin No. 81, Bureau of Mineral Resources, Geology and Geophysics, Commonwealth of Australia Canberra; P. Mathias and M. M. Postan, Eds., "The Cambridge Economic History of Europe," Vol. VII, Part I, Cambridge University Press, Cambridge (1978); N. J. G. Pounds, "The Spread of Mining in the Coal Basin of Upper Silesia and Southern Moravia," Annals of the Association of American Geographers, Vol. 48, No. 2, 149-163 (1958). Most data prior to 1900 must be considered as subject to large uncertainties and to significant differences between different sources.

flame from igniting an explosion. The Davy lamp is the most famous, although it was not the one most widely used. Whether these lamps improved safety, (i.e., reduced accidents) or whether they simply allowed miners to work in more gaseous (i.e., more dangerous) conditions remains controversial.

Bituminous or soft coal mines operated with room and pillar mining, similar to the anthracite mines, but predominantly in flat or nearly flat beds—not the steeply dipping beds of the anthracite region. In 1871, virtually all coal was still undercut by hand and shot with black powder. Coal preparation was almost nonexistent. Animal haulage was universal.

Major efforts were started to mechanize coal mining. Cutter machines, designed to replace the manual undercutting, were patented, but it took several more decades before they performed well and became accepted. Late in the century, electric locomotives were introduced. They quickly gained widespread acceptance, replacing animal haulage. In 1871, most mines that used artificial ventilation—and many did not—used furnaces. By the end of the century, mechanical fans dominated. They were driven by steam engines, as were the pumps that dewatered wet mines.

By the end of the century, a pattern of difficult labor relations was well established in coal fields around the world. The first strike in the Pennsylvania anthracite region took place in 1849. Labor actions and organizations started in Scotland, England, and Wales in the 18th century. Even though by the early 19th century miners' wages were substantially higher than those in manufacturing, the relentless demand for labor, driven by the rapidly increasing demand for coal, facilitated the growth of labor movements, notwithstanding the dominant political power of the coal producers. The high fatality and injury rate gave impetus to a labor emphasis on improving safety.

The 1892 first national miners strike in Britain stopped work from July through November. Two men were killed in riots. In the 1890s the UMWA (United Mine Workers of America) was trying to organize. General strikes were called in 1894 and 1897. In 1898, an agreement was reached between the UMWA and the main operators in Illinois, Ohio, Indiana, and western Pennsylvania. Missing from this list are the anthracite region of northeastern Pennsylvania and West Virginia.

By the end of the century, coal had broadened its consumption base. In 1900, the world produced 28 million tons of steel, the United States, 11.4 million tons. Coking coal for steel had become a major coal consumer. To improve steel quality, steel producers tightened specifications on coke and thereby on the source coal.

Coal gas was introduced early in the century, and had become the major light source in both large and small cities by midcentury. Electric light was introduced by the end of the century. Electric power generation became a major coal user only by the middle of the 20th century, however. Both the conversion to gas light (from candles) and the later one to electric light (from gas) required adjustments on the part of the users. For both changes a major complaint was the excessive brightness of the new lights. (Initially electric light bulbs for domestic use were about 25 W.)

The heavy chemical industry received a major boost when it was discovered that coal tar, a waste by-product of coke and gas production, was an excellent feedstock for chemicals, notably organic dyes. Discovered in Britain, shortly after mid-century, Germany dominated the industry by the end of the century and did so until the first world war. The industry became a textbook example of the application of research and development (R&D) for the advancement of science, technology, and industry.

An important step during the 19th century was the development of coal classifications. Coal is complex and variable. It can be classified from many points of view. Geologically, coal classification is desirable to bring order in a chaotic confusion of materials (macerals) with little resemblance to the mineralogy and petrography of other rocks. Chemical classification is complicated by the fact that the material is a highly variable, heterogeneous mixture of complex hydrocarbons. From the users and the producers point of view, the buyer and the seller, some agreement needs to be reached as to what is the quality of the delivered and the received product. Depending on the use, "quality" refers to many different factors. It includes calorific value (i.e., how much heat it generates), moisture content, and chemical composition (e.g., how much carbon or hydrogen it contains). Impurities are particularly important (e.g., ash and sulfur content and the behavior of the coal during and after burning or coking).

From 1800 to 1889, world production of coal increased from 11.6 to 485 million tons. In 1899, coal provided 90.3% of the primary energy in the United States. It was indeed the century of King Coal.

7. 20th Century

Well into the second half of the 20th century, coal remained the world's major primary energy source. Throughout most of the century, the relative importance of coal declined—that is, as a fraction of total energy production coal decreased. Worldwide coal use increased steadily, but the major production centers shifted. The coal industry changed fundamentally.

Two world wars changed the world and the coal industry. During both wars most industrial countries pushed their steel production, and hence their coal production, as high as possible. Due to the wartime destructions, both wars were followed by severe coal shortages. In response, coal-producing nations pushed hard to increase coal production. In both cases, the demand peak was reached quickly and subsided quickly. A large overcapacity developed on both occasions, resulting in steep drops in coal prices, in major production cutbacks, and in severe social and business dislocations. After the second world war, two major changes impacted coal: railroads converted from steam to diesel and heating of houses and buildings switched to fuel oil and natural gas. (Diesel replaced coal in shipping after World War I.) In the United States, railroads burned 110 million tons of coal in 1946, 2 million tons in 1960. Retail deliveries, primarily for domestic and commercial building heating, dropped from 99 million tons in 1946 to less than 9 million tons in 1972. One hundred fifty years of anthracite mining in northeastern Pennsylvania was ending.

A remarkable aspect of the coal user industry is the growth in efficiency in using coal. Railroads reduced coal consumption per 1000 gross ton-miles from 178 lbs in 1917 to 117 lbs in 1937. One kWh of electrical power consumed 3.5 lbs of coal in 1917, 1.4 lbs in 1937. A ton of pig iron required 2,900 lbs of coking coal in 1936, down from 3500 lbs in 1917. Improvements in use efficiency continued until late into the century. Coal consumption per kWh of electric power dropped from 3.2 lbs in 1920 to 1.2 lbs in 1950, and to 0.8 lbs in the 1960s. After that efficiency decreased somewhat due to inefficiencies required to comply with environmental regulations. Super efficient steam generating and using technologies in the 1990s again improved efficiencies. Even more dramatic was the efficiency improvement in steel production (i.e., the reduction in coal needed to produce steel). Concurrent with the loss of traditional markets came the growth in the use of coal for generating electrical power, the basis for the steadily increasing demand for coal over the later decades of the century and for the foreseeable future.

Major shifts took place in worldwide production patterns. Britain dropped from first place to a minor producer. Early in the century, the United States became the largest coal producer in the world and maintained that position except for a few years near the very end of the century when the People's Republic of China became the largest producer. Most West European countries and Japan reached their peak production in the 1950s, after which their production declined steeply. In the much later industrialized eastern European countries, in Russia (then the Soviet Union), and in South Korea, the peak was reached much later, typically in the late 1980s.

In Australia, India, and South Africa, coal production increased over most of the century, with major growth in the last few decades. The large production growth in China shows a complex past, with major disruptions during the 1940s (World War II) and the 1960s (cultural revolution). The recent entries among the top coal producers, Colombia and Indonesia, grew primarily during the 1980s and 1990s.

Worldwide production patterns have changed in response to major transportation developments. Large bulk carrier ships reduced the cost of shipping coal across oceans. While some international coal trading existed for centuries (e.g., from Newcastle to Flanders, Paris, and Berlin and later from Britain to Singapore, Cape Horn, and Peru), only during the second half of the 20th century did a competitive market develop in which overseas imports affect domestic production worldwide. Imports from the United States contributed to coal mine closures in Western Europe and Japan during the 1950s. Imports from Australia, South Africa, Canada, Poland, and the United States contributed to the demise of coal mining in Japan, South Korea, and most of Western Europe.

Inland, unit trains haul coal at competitive cost over large distances: Wyoming coal competes in Midwestern and even East Coast utility markets. It became feasible to haul Utah, Colorado, and Alberta coking coal to West Coast ports and ship it to Japan and South Korea.

Coal mining reinvented itself over the 20th century. A coal hewer from 1800 would readily recognize a coal production face of 1900. A coal miner from 1900 would not have a clue as to what was going on at a coal face in 2000.

The most obvious and highly visible change is the move from underground to surface mining (Fig. 3). Large earthmoving equipment makes it possible to expose deeper coal seams by removing the overburden. Although large-scale surface mining of coal started early in the century, by 1940 only 50 million tons per year was surface mined, barely over 10% of the total U.S. production. Not until the 1970s did surface production surpass underground production. By 2000, two-thirds of U.S. coal production was surface mined.

Room and pillar mining dominated underground U.S. coal mining until very late in the century. Early mechanization included mechanical undercutting and loading. Conventional mining, in which the coal is drilled and blasted with explosives, decreased steadily over the second half of the century and was negligible by the end of the century. Continuous mining grew, from its introduction in the late 1940s (Fig. 4), until very late in the century, when it was overtaken by longwall mining (Fig. 5). Modern mechanized longwall mining, in which the coal is broken out mechanically over the entire face, was developed in Germany and Britain by the middle of the century. Geological conditions made room and pillar mining impractical or even impossible in many European deposits. In the last two decades of the century, American (and Australian) underground coal mines adopted longwalling, and greatly increased its

Figure 3
Early mechanized surface coal mining. A 1920s vintage P&H shovel loading in a coal wagon pulled by three horses. Photograph courtesy of P&H Mining Equipment, A Joy Global Inc. Company, Milwaukee, WI. Used with permission.

Figure 4
Early attempt at mechanized underground coal mining. The Jeffrey 34 F Coal Cutter, or Konnerth miner, introduced in the early 1950s. The machine was designed to mechanize in a combined unit the most demanding tasks of manual coal mining: breaking and loading coal. Photograph and reproduction courtesy of Ohio Historical Society, Columbus, OH. Used with permission.

Figure 5
A major, highly successful advance in mechanizing underground coal mining: replacing manual undercutting by mechanical cutting. Jeffrey 24-B Longwall Cutter. Photograph and reproduction courtesy of Ohio Historical Society, Columbus, OH. Used with permission.

productivity. In conjunction with the increased production arose a serious safety problem: coal is being mined so fast that methane gas is liberated at a rate difficult to control safely with ventilation systems.

Technological advances depend on equipment manufacturers. Surface mining equipment size peaked in the 1960s and 1970s. The largest mobile land-based machine ever built was the *Captain*, a Marion 6360 stripping shovel that weighed 15,000 tons. *Big Muskie*, the largest dragline ever built, swung a 220 cu yd bucket on a 310 ft boom. The demise of the stripping shovel came about because coal seams sufficiently close to the surface yet deep enough to warrant a stripping shovel were mined out. While the stripping shovel was exceedingly productive and efficient, its large cost required that it operate in a deposit that could guarantee a mine life of at least 10 to 20 years. Capital cost for a stripping shovel was markedly higher than for a dragline.

Large draglines shipped during the 1980s were mostly in the 60 to 80 cu yard bucket size range. A few larger machines (120 to 140 cu yd) were build in the 1990s. By the end of the century, the conventional mine shovel reached a bucket size approaching that of all but the largest stripping shovels ever built.

Worldwide research was conducted in support of the coal industry. The U.S. Bureau of Mines was established in 1910 and abolished in 1996. Its mission changed, but it always conducted health and safety research. The Bureau tested electrical equipment for underground coal mines, permissible explosives, designed to minimize the chances of initiating a gas or dust explosion, and improved ground control. The Bureau produced educational materials for health and safety training. In 1941, Congress authorized Bureau inspectors to enter mines. In 1947, approval was granted for a federal mine health and safety code. The 1969 Coal Mine Health and Safety Act removed the regulatory authority from the Bureau, and transferred it to the Mine Safety and Health Administration (MSHA).

Organizations similar to the Bureau were established in most countries that produced coal. In Britain, the Safety in Mines Research and Testing Board focused on explosions, electrical equipment, and health, the latter particularly with regard to dust control. In West Germany, the Steinkohlenbergbau-verein was known for its authoritative work in ground control, especially for longwalls. CERCHAR in France, INICHAR in Belgium, and CANMET in Canada studied coal mine health and safety. In the Soviet Union and the People's Republic of China, highly regarded institutes ran under the auspices of their respective National Academy of Science.

Over the course of the 20th century, the classification of coal took on ever more importance, resulting in a proliferation of classification methods. Early in the century, when international coal trade was not common and user quality specifications less comprehensive, national and regional classifications were developed. As international coal trade grew, over the second half of the century the need arose for classification schemes that could be applied worldwide.

In situ coal gasification has been demonstrated and could be developed if economics made it attractive. Conceptually simple, a controlled burning is started in a coal seam to produce gas containing CO, H_2, CH_4, and higher order hydrocarbons. The complexity of the fuel, the variability of the deposits, and the potential environmental impacts complicate implementation.

> *You can't dig coal with bayonets.*
> —*John L. Lewis, president, UMWA, 1956*

> *You can't mine coal without machine guns.*
> —*Richard B. Mellon, American industrialist*

Difficult labor relations plagued coal mining through much of the century in most free economy countries that mined coal. In many parts of the world, coal miners formed the most militant labor unions. The West Virginia mine wars, lasting for most of the first three decades of the century, were among the most prolonged, violent, and bitter labor disputes in U.S. history. In Britain, the number of labor days lost to coal mine strikes far exceeded comparative numbers for other industries. Intense violent labor actions, frequently involving political objectives, have recurred throughout much of the century in Britain, France, Germany, Belgium, Australia, Canada, and Japan. During the last few decades of the century, strikes in Western Europe, Poland, Japan, and Canada were driven largely by mine closure issues. Strikes in Russia, the Ukraine, and Australia dealt primarily with living and working conditions. Coal miners in Poland, Serbia, and Rumania were leaders, or at least followers, in strikes with primarily political objectives. The last major strikes in Britain also had a strong political component, although pit closure concerns were the root cause. In the United States, the last two decades of the century were remarkably quiet on the labor front, especially compared to the 1970s.

The structure of the coal mining industry changed significantly over the course of the 20th century. In the United States during the 1930s, many family-owned coal mining businesses were taken over by corporations. Even so, the historical pattern of coal mining by a large number of small producers continued until late in the century. Production concentration remained low compared to other industries

and showed an erratic pattern until late in the century. In 1950, the largest producer mined 4.8% of the total, in 1970 11.4%, in 1980 7.2%. In 1950, the largest eight producers mined 19.4% of the total; in 1970, 41%; in 1980, 29.5%. High prices during the energy crisis of the 1970s facilitated entry of small independents. The top 50 companies produced 45.2% of the total in 1950, 68.3% in 1970, 66.3% in 1980, confirming the significant reduction of the small producers.

During the 1970s, oil and chemical companies took over a significant number of coal companies because they believed widely made claims during that decade of an impending depletion of oil and gas reserves. As the hydrocarbon glut of the 1980s and 1990s progressed, most of these coal subsidiaries were spun off and operated again as independent coal producers.

Toward the end of the century, there was significant consolidation of large coal producers, domestically and internationally. Even so, the industry remained characterized by a relatively large number of major producers. In 2000, the 10 largest private companies controlled barely over 23% of the world production. In the United States the largest producer, Peabody, mined 16% of U.S. coal, the second largest one, Arch Coal, 11%. Coal remained highly competitive, domestically and internationally.

8. The Future

Coal resources are the largest known primary energy resource. Supplies will last for centuries, even if use grows at a moderate rate. Reserves are widely distributed in politically stable areas. The many uses of coal, from electrical power generation to the production of gaseous or liquid fuels and the use as petrochemical feedstock, make it likely that this versatile hydrocarbon will remain a major raw material for the foreseeable future.

The mining and especially the use of coal will become more complicated and hence the energy produced more expensive. Coal mining, coal transportation, and coal burning have been subjected to ever more stringent regulations. This trend toward tighter regulations will continue. A major environmental factor that will affect the future of coal is the growing concern about global warming. While technologies such as CO_2 capture and sequestration are being researched, restrictions on CO_2 releases will add significantly to the cost of producing energy, in particular electricity, from coal.

Predictions for the near future suggest a modest, steady increase in coal production. The main competition in the next few decades will be from natural gas. Natural gas reserves have risen steadily over the past 30 years, in parallel with the increased demand and use, and hence the increased interest in exploration for gas. Natural gas is preferred because it is richer in hydrogen, poorer in carbon, and hence the combustion products contain more steam rather than CO_2. If, in the somewhat more distant future, the predictions of a reduction in supply of natural gas and oil were to come through—and for over a century such predictions have proved "premature" —coal might once again become the dominant fossil fuel. The rise in demand for electric power seems likely to continue in most of the world to reach reasonable living standards and in the developed world for such needs as electric and fuel cell-driven vehicles and continued growth in computers and electronics in general.

To make coal acceptable in the future, steps need to be taken at all phases of the coal life cycle, from production through end use. A major focus in the production cycle is minimizing methane release associated with mining. Methane (CH_4) is a greenhouse gas. It also is a main cause of coal mine explosions. Great strides have been made in capturing methane prior to and during mining. In gassy seams, it now is collected as a fuel. In less gassy seams, especially in less technologically sophisticated mines, it remains uneconomical and impractical to control methane releases. Extensive research is in progress to reduce methane releases caused by coal mining.

Other environmental problems associated with mining coal include acid mine drainage, burning of abandoned mines and waste piles, subsidence, spoil pile stability issues, and mine site restoration and reclamation. Technological remedies exist, but their implementation may need societal decisions for regulatory requirements.

Coal preparation is critical for improving environmental acceptability of coal. Super clean coal preparation is feasible. Technically, virtually any impurity can be removed from coal, including mercury, which has drawn a great deal of attention over the last few years. Coal transportation, particularly in ocean going vessels, has modest environmental impacts, certainly compared to oil.

Coal users carry the heaviest burden to assure that coal remains an acceptable fuel. Enormous progress has been made in reducing sulfur and nitrogen oxide emissions. Capturing and sequestering CO_2 will pose a major challenge to the producers of electric power.

More efficient coal utilization contributes to the reduction in power plant emissions. Modern power plants run at efficiencies of about 37%. During the 1990s, power plants have come on stream that run at over 40%. It is likely that 50% can be achieved by 2010. Increasing efficiency from 37 to 50% reduces by one-third the coal burned to generate electricity and reduces by one-third gas (and other) emissions.

Coal has been attacked for environmental reasons for over seven centuries. With ups and downs, its use has grown over those seven centuries because of its desirable characteristics: low cost, wide availability, ease of transport and use. It will be interesting to see whether it can maintain its position as a major energy source for another seven centuries or whether more desirable alternatives will indeed be developed.

9. Coal and Culture

Given the pervasive influence of coal and in particular of its uses on the fundamental transformations of societies, especially during the 19th century, it is not surprising to see coal reflected in cultural contexts. Pervasive was the sense of progress associated with industrial development, the perception of dreadful social problems associated with the industrial progress, and the disintegration of an older world.

Zola's *Germinal* remains the major novel rooted in coal mining, a classic in which have been read different, contradictory meanings from revolutionary to bourgeois conservative. D. H. Lawrence grew up in a coal mining town, and it and its collieries pervade several of his masterpieces. Again, these incorporate deep ambiguities with respect to coal mining: admiration for the male camaraderie in the pits, the daily dealing with a hostile dangerous environment, the solidarity in the mining community, and the stifling constraints of it all. Similar ambiguities are found in Orwell's *The Road to Wigan Pier* and in the reactions to it. It was published by the Left Book Club. These publishers inserted an apologetic introduction, justifying why they published this book, frequently so critical of the left. The book includes a widely quoted description of underground mining practices in Britain in the 1930s, as well as sympathetic descriptions of life in the mining communities—not deemed sufficiently negative by many on the left.

Richard Llewellyn's popular *How Green Was My Valley* introduces a small coal mining community in South Wales through a voluptuous description of rich miner's meals. (Orwell, as described in his personal diary, published long after *Wigan Pier*, similarly was struck by the rich miners meals.)

Lawrence's descriptions of coal mine towns in the early 20th century gained wide distribution through films made of his works, several of which were successful commercially. The critically acclaimed and commercially successful movie adaptation by John Ford of *How Green Was My Valley* introduced a wide audience to a coal mining community in South Wales. While justifiably criticized as sentimental, the book and film offer a sympathetic view of a community that often has felt looked down upon, examples of nostalgic flashback descriptions of mining communities often found in the regional literature of mining districts.

Joseph Conrad, in his 1902 story "Youth," fictionalized, although barely, the loss at sea, to spontaneous combustion, of a barque hauling coal from

Newcastle to Bangkok. The self-ignition of coal, particularly during long hauls, was a major problem during the late 19th century and continues to pose headaches for those who haul or store coal over extended periods of time. "Youth" shows that Conrad, an experienced coal hauling sailor, was thoroughly familiar with spontaneous combustion.

Upton Sinclair, a leading American muckraker of the early 20th century, based *King Coal* on the long and bitter 1913–1914 Colorado coal strikes. Neither a critical nor a commercial success (the sequel *The Coal War* was not published until over half a century later and then decidedly for academic purposes), it describes the recurring problem that has plagued coal forever: difficult labor relations.

Van Gogh sketched and painted female coal bearers during his stay in the Borinage, the heart of early coal mining in Belgium. Also originating in the Borinage were the sculptures by Constantin Meunier, including many of miners. The tour of his work through Eastern and Midwestern cities (1913–1914) brought to the American public a visual art that recognized the contribution, hardships, and strengths of blue collar workers.

More visible than coal mining, visual celebrations of the Industrial Revolution focused on the offspring of coal mining and its customers, railroads and the steam locomotive. Among the better known ones is *Rain, Steam and Speed*, by J. W. Turner, the painter of the British industrial landscape, immortalized in *Keelmen Heaving in Coals by Night, Coalbrookdale by Night*, and his paintings of steamships.

The opening up of the American landscape and space, or the intrusion into the landscape, was depicted by Cole's 1843 *River in the Catskills*, his *The Oxbow*, Melrose's *Westward the Star of Empire Takes Its Way*, Bierstadt's *Donner Lake*, and especially George Inness' *The Lackawanna Valley*, Henry's *The First Railway on the Mohawk and Hudson Road*, and Durand's *Progress*. During the 19th century, Currier & Ives' prints popularized progress and penetration of the wilderness thanks to the railroads and the steam engine. Walt Whitman's *Ode to a Locomotive* summarized the widely held view that a continent was being conquered and opened up thanks to the railroad and its most visible emblem, the locomotive's steam.

Nikolai Kasatkin, a leading Russian realist and one of the Itinerants (peredvizhniki—wanderers, travelers), lived several months in the Donetsk coal basin at a time (late 1800s) when Russian coal production was increasing rapidly. One of his most famous paintings, *Miners Changing Shift* (Fig. 6), is often referenced but rarely reproduced. His *Poor People Collecting Coal in an Abandoned Pit* and *Woman Miner*, both far more lyrical than the subject might suggest, also are far more readily accessible. Closer to socialist realism is Deineka's *Before Descending into the Mine*. One of the better known Soviet railroad celebrations is *Transport Returns to Normal* by Boris Yakovlev.

Figure 6
Coal Miners: Shift Change. Nikolai A. Kasatkin (1895). Photograph and reproduction courtesy of Tretyakov State Gallery, Moscow.

Charles Sheeler and Edward Hopper continue the American tradition of painting railroads in land- and cityscapes. American primitive John Kane, born in Scotland, includes coal barges on the river in his exuberant *Monongahela River Valley* and *Industry's Increase*. Probably the only one among these artists who worked as a coal miner ("the best work I knew and enjoyed"), Kane published his autobiography *Sky Hooks*, one of the sunnier recollections among the many written by people who grew up in mining communities. More representative of the bleak conditions during the 1930s may be Ben Shahn's sad and haunting *Miners' Wives*.

In France, Théodore Géricault painted the horse haulage of coal, common well into the 19th century, in *Coal Wagon*. Manet, Monet, Caillebotte, Seurat, and Pissarro recorded their impressions of railroads, railroad stations, steamboats, and their impact on landscape and society. The responses ranged from reluctant acceptance, at best, by Pissarro, to the enthusiastic celebrations by Monet of railroads, steam engines, and in particular the Gare St. Lazare, the first railroad station in France and Paris. Monet's *Men Unloading Coal*, a rather darker picture than most Monets, illustrates the manual unloading of coal from river barges late in the 19th century.

Further Reading

Berkowitz N 1997 *Fossil Hydrocarbons: Chemistry and Technology*. Academic Press, San Diego, CA

Bryan A M Sir. 1975 *The Evolution of Health and Safety in Mines*. Ashire, London

Gregory C E 2001 *A Concise History of Mining*. A. A. Balkema, Lisse, The Netherlands

Jones A V Tarkenter R P 1992 *Electrical Technology in Mining: The Dawn of a New Age*. P. Peregrinus, London, in association with the Science Museum

Peirce W S 1996 *Economics of the Energy Industries*. 2nd ed. Praeger, Westport, CT, London

Pietrobono J T (ed.) 1985 *Coal Mining: A PETEX Primer*. Petroleum Extension Service, The University of Texas at Austin, Austin, TX, in cooperation with National Coal Association, Washington, DC

Shepherd R 1993 *Ancient Mining*. Elsevier Applied Science, London and New York

Stefanko R 1983 *Coal Mining Technology*. Society of Mining Engineers of AIME, New York

Thesing W B (ed.) 2000 *Caverns of Night: Coal Mines in Art, Literature, and Film*. University of South Carolina Press, Columbia, SC

Trinder B (ed.) 1992 *The Blackwell Encyclopedia of Industrial Archaeology*. Blackwell, Oxford, UK

Jaak J. K. Daemen

Coal Mining in Appalachia, History of

Glossary

Appalachian Regional Commission   Federal agency established in 1965 "to serve the needs of 21,000,000 people residing in one of the most economically distressed regions of the country."

black lung   The common name for coal workers' pneumoconiosis, a respiratory disease caused by the inhalation of coal dust.

longwall mining   A method of mining in which long sections of coal—up to 1000 feet across—are mined and deposited directly onto a conveyor system with the help of shields that support the roof and advance the longwall rock sections.

mountaintop removal   A surface mining technique in which the top of a mountain is blasted away to expose a seam of coal; debris is then deposited in a valley fill.

rock dusting   White, powdered limestone is sprayed on the roofs, bottoms, and ribs of a mine to mitigate the explosive qualities of coal dust.

roof bolting   A process in which holes are drilled into the roof of a mine; long bolts coated with glue are then screwed into the holes to support the roof.

scrip   Nonlegal tender issued to workers in place of cash; generally, redeemable only at the company-owned store.

timbering   The use of wood for constructing a roof support in an underground mine.

tipple   An elevated structure, located near a mine entrance, that receives coal from mine cars or conveyors and from which coal is dumped or "tipped," washed, screened, and then loaded into railroad cars or trucks.

welfare capitalism   An industrial strategy in which company flexibility and benevolence are exercised in an attempt to maximize production and promote social control of the workforce.

yellow-dog contracts   Agreements that prohibit employees from joining or supporting labor unions.

Appalachia—as defined by the United States Appalachian Regional Commission (ARC)—is a sprawling region encompassing approximately 400 counties in 13 eastern states: New York, Pennsylvania, Ohio, Maryland, West Virginia, Virginia, Kentucky, Tennessee, North Carolina, South Carolina, Georgia, Alabama, and Mississippi. Rich in natural resources, such as timber, coal, oil, iron ore, gold, and copper, and located within relatively easy reach of major industrial centers along the Atlantic seaboard, this expansive area supplied the raw materials that fueled America's Industrial Revolution. With vast deposits of bituminous ("soft") coal underlying approximately 72,000 square miles from Pennsylvania to Alabama, and an additional 484 square miles of anthracite ("hard") coal concentrated in northeastern Pennsylvania, exploitation of this one resource uniquely shaped the region.

1. Introduction

Opening the Appalachian coalfields was no easy feat. Nor was it achieved overnight. For aspiring "captains of industry," there were great financial risks involved. Substantial capital investment was required to survey and purchase coal lands, to construct mines and provide for the needs of a large workforce, and to establish a reliable and cost-effective means of delivering this bulky and relatively low-value good to market. For many, the road to success led instead to financial ruin. For those charged with the task of extracting "black diamonds" from the earth, there were great personal risks involved. Mining is and always has been hazardous work. Historically, periodic layoffs and the threat of black lung added to the worries of the miner and his family. In the final analysis, it must be concluded that coal mining in Appalachia has had as profound an impact on the region's inhabitants as it has had on the region's forest and water resources. Although much diminished, King Coal's influence is still discernible today—both in the region's historic mining districts and in those areas where underground and surface mining are still being carried out (Fig. 1).

2. Historical Overview

In his opening address to the White House Conference on Conservation in 1908, President Theodore Roosevelt reminded those in attendance that coal's ascendancy was a relatively recent occurrence: "In [George] Washington's time anthracite coal was known only as a useless black stone; and the great fields of bituminous coal were undiscovered. As steam was unknown, the use of coal for power production was undreamed of. Water was practically the only source of power, save the labor of men and animals; and this power was used only in the most primitive fashion." Over the course of the next century, however, coal's utility would be demonstrated and its value to industry affirmed. By 1900, the year before Roosevelt took office, coal accounted for nearly 70% of the national energy market in the United States. In short, it had become America's fuel of choice and Appalachia was the nation's primary producer.

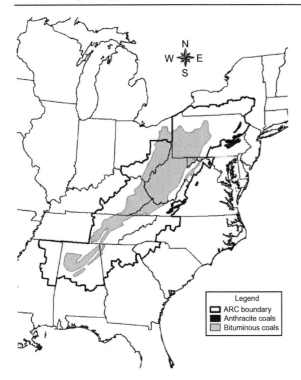

Figure 1
Location of mineral resources in Appalachia. ARC,
Appalachian Regional Commission. Adapted from
Drake (2000).

2.1 Fueling the Industrial Revolution

During the early years of the 19th century, coal was
used primarily to fire blacksmith forges and as a
domestic heating fuel. In ensuing years, it was valued
as a fuel for salt and iron furnaces, brick and pottery
kilns, and steam engines. Although annual coal
production increased significantly between 1800 and
1860—from a mere 100,000 tons to 20,000,000
tons—it was the post-Civil War years that first witnessed truly explosive growth in coal production and
consumption in the United States. In 1885, the annual production figure had climbed to 110,000,000
tons. Stimulated by growth in domestic consumption,
as well as increased use as a fuel for steamship travel,
coal was now being used to stoke the furnaces of an
emerging steel industry and to provide the motive
power for steam locomotives. By 1900, production
topped 243,000,000 tons, making the United States
the world's leading producer. With respect to this
tremendous growth, historian Duane A. Smith notes:
"Without mining—from coal to iron to gold—the
United States could not have emerged as a world
power by the turn of the century, nor could it have
successfully launched its international career of the
twentieth century."

It would be difficult to overstate the importance of
Appalachian coal in propelling America into the industrial age. Prior to World War II, virtually all coal
mined in the United States came from the Appalachian fields. In 1925, for example, the Appalachian
region produced 92% of the total amount of coal
mined in the United States, with Pennsylvania alone
accounting for approximately one-third of this output. Between 1880 and 1930, Pennsylvania consistently ranked as the leading producer of both
bituminous and anthracite coal. By the mid-1920s,
however, it had become apparent that all was not
well with the coal industry in Appalachia.

2.2 The Decline of King Coal

Several years in advance of the stock market crash of
1929, the coal industry in general and miners in particular felt the first tremors of the Great Depression.
In Appalachia, overexpansion stimulated by World
War I, interstate competition with midwestern states,
and competition between "union" and "nonunion"
mines led to market gluts, falling prices, spiraling
unemployment, and labor unrest. Although mining
had weathered economic downturns in the past, most
notably in the mid-1890s, nothing compared with the
depression that beset the industry during the 1920s
and 1930s. West Virginia, which lost 33,000 coal jobs
as production fell from 146 million tons in 1927 to 83.3
million tons in 1932, was hit particularly hard.

In subsequent years, the rising popularity of new
energy sources—petroleum, natural gas, and electricity—challenged coal's dominance of the domestic
heating market. Although World War II provided a
boost to the region's coal-based economy, it was only
temporary. When railroads completed the conversion
from coal to diesel after the war, the industry received
another blow, contributing further to the industry's
recent history of boom and bust. The coal operators'
response was to accelerate the process of mechanization begun in earnest during the 1930s and to opt for
surface mining over underground mining wherever
feasible. Economic instability coupled with widespread utilization of new labor-saving machinery,
typified most noticeably by the introduction of enormous electric-powered shovels to surface operations,
forced many miners and their families to leave the
region altogether in the post-World War II era.

Since the 1960s, the only major growth market for
coal has been to supply electric power producers and
for export abroad. With respect to the former, coal
remains an important player. Today 52% of U.S.
electricity is generated by coal-fired plants. However,
increased production from western states starting in
the late 1950s, but gaining momentum especially
during the 1980s and 1990s, has further eroded
Appalachia's share of coal markets. Compliance with
environmental laws since the 1970s, including the

Clean Air Act and its amendments, the Clean Water Act, and the Surface Mining Control and Reclamation Act (SMCRA), not to mention various state-level regulations, has also contributed to a rise in coal production costs. By 1999, more than half of the coal mined in the United States was coming from states located west of the Mississippi River, and far and away the largest single producer was Wyoming, not Pennsylvania. Nevertheless, coal mining has managed to remain viable in some portions of the Appalachian region due in large part to controversial mining techniques such as mountaintop removal.

3. Breaking Ground

Given the increase in demand for coal spurred on by industrialization, the existence of coal so near to the Atlantic seaboard aroused early interest among local boosters, land speculators, foreign investors, and others wishing to purchase mineral lands and mine them for profit. Before large-scale coal-mining operations could commence, however, certain obstacles had to be surmounted. First, coal lands had to be surveyed and evaluated, and the rights to the coal purchased or leased. Equally important, a reliable and cost-effective means of transporting the coal to eastern and Great Lakes markets, as well as to the steel mills of the North and South, had to be financed and developed. As historian Crandall A. Shifflett has pointed out: "All of the land and coal was worthless... if it could not be gotten out of its mountain redoubts. There was no use to open mines if the coal could not be transported to distant markets."

3.1 Railroads Usher in the Industrial Era

As early as the 1820s, coal mining had become an important, albeit seasonal, activity in some remote locations. During the first decade of the 19th century, for example, small quantities of coal, along with lumber and a variety of agricultural goods, were sent down the Potomac River via flatboat to Georgetown. On arrival, the large rafts were dismantled and their crews walked home. Such trade could only take place during the spring season when floodwaters raised the level of the river. Without significant investment in transportation improvements, it was clear that coal could not be mined profitably on a large scale.

Although coal and coke (fuel derived from coal) were slow to replace wood and charcoal in the manufacture of iron prior to the Civil War, coal found an early use in blacksmith shops and as a domestic heating fuel. After about 1840, coal gradually began to replace charcoal in the furnaces of the burgeoning iron industry. By this time, it was also being recognized as an important source of power for steam engines. After the Civil War, it would play an indispensable role in the manufacture of steel.

Initially, it was the anthracite fields of northeastern Pennsylvania that attracted great attention. Once problems associated with the burning of hard coal were resolved, production rose sharply. Between 1830 and 1860, production of Pennsylvania anthracite, much of it destined for the Northeast, increased more than 40-fold. As Andrew Roy put it in 1903, "the 'black rocks' of the mountains, heretofore deemed useless, at once rose in public estimation." Miners' villages sprouted wherever new mines were opened as English, Welsh, Irish, Germans, and, in later years, eastern and southern Europeans entered the region in search of high wages and a new beginning in America.

Before America's Industrial Revolution could "take off," another revolution—this one in the transportation sector—was needed. As the country grew in geographic extent, government officials became increasingly aware that improvements in communication and trade depended largely on the development of new, more efficient modes of travel. In 1808, Albert Gallatin, Secretary of the Treasury under Thomas Jefferson, envisioned a system of roads and canals crisscrossing the country and financed by the federal government. Although the Gallatin Plan was never carried out, great changes were in the works. The spectacular success of New York's Erie Canal served as a catalyst. Indeed, the completion of the Erie Canal in 1825 ushered in a period of canal building in the United States as eastern cities competed with one another in a race to expand their hinterlands and capture market share. Government officials and businessmen in Pennsylvania, Maryland, Ohio, and Virginia soon unveiled plans for canal systems they hoped would allow them to remain competitive with New York. In the end, none of the canal systems to the south could match the financial success of the Erie Canal and its branch connections. Difficult terrain and competition from a new form of transportation, the railroad, ensured that the reign of the canal would be relatively short (Fig. 2).

On the eve of the Civil War, large parts of northern Appalachia had established outlets for coal via rivers, canals, and railroads. In addition to the anthracite fields of eastern Pennsylvania, bituminous coal from the great "Pittsburgh" seam in western Pennsylvania, western Maryland, northern West Virginia, and Ohio was now being tapped. In contrast, central and southern Appalachia remained relatively unaffected by the transportation revolution that had seized the north. After the Civil War, an already impressive network of railroads in the Appalachian north was expanded. As these railroads, and to a lesser extent canals, penetrated the mountains, operations were expanded in western Maryland, western Pennsylvania, northern West Virginia, and southeastern Ohio, contributing to the growth of the iron industry and, especially after the introduction of the Bessemer hot-blast furnace in the 1860s, the nascent steel industry.

Figure 2
Rodgers locomotive on turntable at Corning, Ohio,
ca. 1910. Reprinted from the Buhla Collection, Ohio
University, Archives and Special Collections, with
permission.

Following the steady progress of the railroads, new
fields were opened up in southern West Virginia,
northeastern Tennessee, and northern Alabama, and
finally, by the 1890s, southwestern Virginia and
eastern Kentucky. As was the case with Pennsylvania's anthracite fields, the opening of mines in these
areas was accompanied by a flurry of coal town
construction and an influx of immigrant labor. As
historian Richard Drake reminds us, in addition to
carrying a great deal of coal, rail lines such as the
Baltimore and Ohio, Chesapeake and Ohio, Norfolk
and Western, Louisville and Nashville, and the
Southern Railway, to name but a few, figured
prominently in the growth of major industrial centers
concentrated in the northern and southern extremities of the region. Of particular importance was
the development of Pittsburgh and its industrial
satellites—Johnstown, Altoona, Morgantown, and
Fairmont—and the "A.B.C." triangle of the Deep
South—Atlanta, Birmingham, and Chattanooga.

3.2 Clearing the Way

Long before the new transportation lines reached
their destinations, however, land speculators were
clearing the way for this initial wave of industrialization. During the first half of the 19th century,
corporations interested in mining coal and iron ore
and cutting timber, and backed by capital from
centers of commerce such as New York, Baltimore,
Boston, and London, dispatched geologists and engineers into the mountains to measure the size and
evaluate the quality of coal and iron ore deposits.
When favorable reports were returned, they began
accumulating property rights, a process that was
often mediated by land speculators. Some of these
speculators and businessmen were outsiders who

entered the region as representatives of mining
interests headquartered in the east. Others were local
residents who recognized the value of coal and accumulated property rights based on the assumption
that the arrival of the railroad would cause real estate
values to soar. Still others were opportunistic politicians—men like Henry Gassaway Davis, Johnson
Camden, and Clarence Watson—who used their
power and influence in the chambers of the U.S.
Senate and in various state capitols to finance railroads, purchase coal lands, and amass personal fortunes. Invariably the result was the same. By the time
the railroad arrived, the coalfields were controlled for
the most part by outside interests eager to begin the
mining process.

One of the best known of these early entrepreneurs was Pike County, Kentucky native John
C. C. Mayo. During the 1880s and 1890s, Mayo traveled throughout eastern Kentucky purchasing land
or mineral rights from local farmers. Using the infamous "Broadform Deed," which permitted the
mountain farmer to retain rights to the surface while
authorizing the owner of the mineral rights to use any
means necessary to extract the minerals, Mayo acquired thousands of acres of land that he eventually
sold to companies such as the giant Consolidation
Coal Company. As late as the 1980s, coal companies
in Kentucky could count on the courts to accept the
binding nature of Broadform Deeds signed in the late
19th and early 20th centuries. However, rather than
extract the coal using underground mining techniques,
as was the practice during the 19th and early 20th
centuries, companies were now employing surface
mining techniques. Essentially, the Broadform Deeds
permitted coal operators to destroy the surface to
reach the mineral deposits. The stability of many rural
communities was sorely tested. In 1988, the citizens of
Kentucky passed an amendment to the state constitution requiring coal companies to obtain permission
from surface owners before strip mining land. Five
years later, the Kentucky Court of Appeals accepted
the constitutionality of this amendment.

4. Life in the Company Town

A vexing problem that confronted coal operators in
the early stages of mine development was a shortage
of labor. Because exploitable seams of coal were
often found in sparsely populated areas, operators
were forced to recruit and import labor. Although
some coal operators used prison labor, as in northern
Alabama, or drew from the pool of displaced
mountain farmers, by far the greater number chose to
lure white labor from Europe or, in the years leading
up to World War I, to enlist black migrants from the
South. Thus, in addition to mine development and
railroad construction, coal companies had to provide
housing and other amenities for newly arrived miners
and their families.

These company towns, as they came to be known, played an especially important role in central and southern Appalachia during the 19th and early 20th centuries. In eastern Kentucky, southern West Virginia, and southwestern Virginia during the 1920s, for example, an estimated two-thirds to three-fourths of all miners and their families resided in such communities. In southern West Virginia, the figure approached 98%. Beginning in the 1930s, the company town's usefulness began to fade. A number of factors contributed to this decline, including the diffusion of the automobile, which greatly enhanced worker mobility; the introduction of labor-saving devices, which made workers redundant; and, later, the adoption of surface-mining techniques, which further reduced the need for a large labor force. No longer profitable, by the 1950s and 1960s most mining firms had sold or otherwise disposed of company housing.

4.1 A Distinctive Feature of the Mining Landscape

There can be little doubt that the company town is one of the most distinctive features associated with Appalachia's historic mining districts. Although some scholars have argued that company towns were a necessity, at least at the outset, and that many companies treated their workers with respect and fairness, others contend that these carefully planned and controlled settlements allowed companies to exert an objectionable level of power over miners and other residents. Considering the extent to which some companies dominated local law enforcement, monitored educational and religious institutions, curtailed personal liberties, and influenced worker behavior, their power was pervasive. As Shifflett reminds us, however, some miners' families, especially the first generation to leave hardscrabble farms in rural Appalachia, may have favored life in the company town, especially the more durably constructed "model" towns: "Contrary to the views that the coal mines and company controlled towns led to social fragmentation, disaffection, and alienation of the workforce, many miners and their families found life in them to be a great improvement over the past. Miners viewed the company town, not in comparison to some idyllic world of freedom, independence and harmony, but against the backdrop of small farms on rocky hillsides of Tennessee, Kentucky, and Virginia, or a sharecropper's life in the deep South states."

Company towns in Appalachia generally shared several distinguishing features (Fig. 3). First, the town bore the stamp of the company in every conceivable way. Everything in the town, from the houses and the company store to the schools, churches, hospitals, and the recreational facilities, if such "frills" even existed, was provided by the company. Second, housing generally conformed to a standard design. In general, one worker's house was identical

Figure 3
Railroad station, Fleming, Kentucky, ca. November 1915. Reprinted from the Smithsonian Institution, National Museum of American History, with permission.

to the next with respect to style and construction. A third trait was economy of construction. Companies typically limited their investment in towns because they knew they were not likely to evolve into permanent settlements. Renting houses to workers ensured that even a substantial investment in housing would be recovered quickly. Because companies generally preferred to hire married men with families over single men, believing they were less inclined to quit their jobs, single- and two-family dwellings were far more common in the coalfields than were large boarding houses. A fourth feature was that the company houses and other structures were usually laid out in a standard gridlike pattern, with long, narrow lots lined up along the railroad tracks. Finally, there was an outward manifestation of socioeconomic stratification and segregation according to racial and ethnic group expressed in the landscape. With respect to the former, sharp distinctions could be found in the size and quality of housing occupied by miners and company officials. Regarding the latter, historian Ronald Lewis notes that some companies sought a "judicious mix" of whites, blacks, and immigrants as a means of maintaining order and fending off the advances of union organizers. Thus it was not unusual to find distinct "neighborhoods"—an immigrant town, a Negro town, and an "American" town—existing side by side in the same town, each with its own separate facilities and social and cultural life.

One of the most controversial features of the typical coal mining town was the company store. Indeed, it represented both the best and worst of company town life. Often doubling as a meeting hall, lodge, and recreation facility, while also housing the company offices, the store carried a great variety of goods, including food and clothing, mining supplies, tools, and durable household goods. It also served as an important informal gathering place for both miners and managers alike. Given the remote location of many towns, the company store was often the only commercial retail establishment to which the mining population had regular access. In some cases, coal companies prohibited the establishment of other stores; in other cases, they penalized miners for making purchases in neighboring towns. Customer loyalty was achieved through the issuance of company scrip. Paying wages at least partially in scrip ensured that a portion of a worker's paycheck would be funneled back to the company. Although company officials claimed that such an arrangement protected families, critics have argued that companies charged higher prices in an attempt to keep the cost of coal production down. It is not surprising, then, that charges of debt peonage, coercion, price gouging, and differential pricing were leveled at the company store.

Some companies encouraged the planting of vegetable and flower gardens to improve the appearance of towns and to supplement food supplies. With the men at work in the mines, primary responsibility for the planting and care of the gardens rested on the shoulders of women and children. While they contributed significantly to the miners' larder, the gardens may have served another purpose. By encouraging miners to grow much of their own food, company officials could justify paying lower wages. Such a strategy enabled companies located at a considerable distance from major coal markets to keep the price of coal low and remain competitive with companies located nearer to centers of industry and population.

4.2 Culturally Diverse Communities

Coal towns in Appalachia often exhibited a high degree of cultural diversity. During the first half of the 19th century, immigrants to the minefields typically came from England, Wales, Scotland, Ireland, and Germany. By the end of the century, large numbers were being recruited from southern and eastern Europe, a trend that continued well into the first two decades of the 20th century (Fig. 4). When immigration from Europe plummeted during the war years, the door was opened wide for large numbers of African-Americans to enter the mines. Discouraged by low wages in the mines of northern Alabama or in search of seasonal work when there was a lull in activities on the family farm, this group made up a significant proportion of the workforce in places such as southern West Virginia and eastern Kentucky from approximately 1910 to the 1930s.

Cultural heterogeneity in the company town manifested itself in many and varied ways. It was not

Figure 4
Miners in front of Oakdale or Jumbo Mine No. 311, Ohio, 1887–1907. Reprinted from the Buhla Collection, Ohio University, Archives and Special Collections, with permission.

uncommon to find safety instructions posted at the mine opening written in several languages. Nor was it uncommon to find a diversity of religions represented in a single company town. Although some have argued that building churches and paying ministers' salaries allowed company officials to exercise even greater control over residents of the company town, others, citing low attendance, have downplayed the importance of religion in these communities.

Although some coal operators looked on their creations with a certain measure of pride, believing them superior to anything one might find in outlying areas, living and working conditions, in fact, varied considerably from one location to the next. The time and money an operator was willing to invest in the development of a town depended on several factors, including the projected life of the mine, the number of houses that needed to be built, and the amount of capital the company had at its disposal. Towns constructed of durable materials and offering a range of amenities were clearly meant to last. Ramshackle housing, on the other hand, was a sure sign that a company's investment in a mine site was limited. According to contemporary theories of corporate paternalism and welfare capitalism, well-built towns equipped with electricity and indoor plumbing, and offering a wide range of amenities, such as churches, recreation buildings, meeting halls, ballparks, and stores, attracted a more dependable and loyal breed of miner—one who, in the end, would reject the temptation to join a union. Because the company maintained ownership of housing, the threat of eviction was never far from the mind of the miner. In the years following the First World War, companies fought off the advances of the unions by equating union membership with communist sympathy or radicalism. Companies employing a high number of foreign-born workers kept their workforce in line by equating loyalty to the company with patriotism and "Americanism." Eager to prove they were good American citizens and avoid persecution, recent immigrants were often reluctant to participate in union activities.

5. Working in a Coal Mine

For the most part, coal miners during the 19th and early 20th centuries worked side by side in a dark, damp, and often dangerous environment, regardless of their cultural or ethnic background (Fig. 5). Given the stressful nature of their work and the dangers they faced, it was absolutely essential that safety and teamwork take precedence over all other matters in the mines. Dust and gas accumulations, roof falls, and heavy equipment malfunctions and accidents were just a few of the hazards miners dealt with on a daily basis. Under such conditions, even the most safety-conscious miner could be caught off guard. If a miner was fortunate enough to avoid a serious lost-time injury over the course of his career, black lung, a condition caused by the inhalation of coal dust, could cut his life short. Aboveground, miners went their separate ways. As Robert Armstead recently detailed in an autobiographical account, segregation and Jim Crow were a fact of life when the shift was over.

The coal mining industry in Appalachia has witnessed a great many changes since the days when miners toiled underground with picks and shovels and hand-loaded coal into wagons for transport to the surface. Although many of these changes have had a positive effect on the lives of miners and their families, e.g., improved safety, wage increases, and

Figure 5
Miner loading car at Mine No. 26, West Virginia. Reprinted from the Smithsonian Institution, National Museum of American History, with permission.

retirement benefits, to name just a few, others, such as the introduction of labor-saving equipment, have had the effect of putting miners out of work. Indeed, the effect that mechanization had on miners and mining communities during the middle years of the 20th century was far reaching.

5.1 Social Relations among Miners

Although relations between whites, blacks, and European immigrants were sometimes strained, especially when blacks or immigrants were brought in as strikebreakers, antagonism appears to have been the exception rather than the rule, at least on the job. According to Ronald Eller, "[a] relatively high degree of harmony existed between the races at a personal level. Working side by side in the mines, the men came to depend upon each other for their own safety, and the lack of major differences in housing, pay, and living conditions mitigated caste feelings and gave rise to a common consciousness of class." If integration typified relations in the mines, segregation characterized relations on the surface. When the workday was over, blacks and whites showered in separate facilities (if such facilities were, in fact, available) and then walked or drove home to their own neighborhoods. Foreign-born miners and their families were often subjected to similar, albeit more subtle, forms of discrimination.

Although small numbers of women worked in small "country bank" mines, they were generally discouraged from engaging in such work. In some states, such as Ohio, they were prohibited by law from coal mining work unless they actually owned the mine. Their absence is frequently attributed to the superstitious belief that a woman's presence in the mines was sure to bring bad luck. Starting in the early 1970s, small numbers of women were hired to join the ranks of male coal miners. Hiring peaked in 1978 and then tailed off again during the 1980s. The historical record seems to support geographer Richard Francaviglia's statement on the matter: "mining has traditionally been 'man's work', and no amount of neutralized language can…conceal that fact."

5.2 Mechanization

Coal mining was one of the last major industries in America to mechanize production. From the beginning of the 19th century through the first two decades of the 20th century, the manner in which coal was extracted from underground mines and brought to the surface changed very little. Before the advent of the mechanical loading machine in the 1920s, miners typically broke coal from the mine face using picks and wedges and shoveled it by hand into coal cars. Loaded cars were then pushed to the surface by miners or pulled by horses or other animals. By the end of the 19th century, miners were "undercutting" the seam and using drills and explosives to increase productivity. Considering that a miner's pay was directly linked to the amount of coal he mined and the number of cars he filled, new methods of mining that saved labor and increased productivity were quickly adopted.

As demand for coal increased over time, companies sought to overcome bottlenecks in the production process. When overexpansion and competition began to cut into profits, and unionization forced the cost of coal production to go up, the drive to mechanize underground mining gained momentum. The logical starting point was to replace the hand-loading system with one that emphasized the use of mechanical loading machines. From the 1930s to the 1970s, the introduction of mechanical loaders, cutting machines, continuous miners, and longwall mining changed forever the way coal would be mined.

Another aspect of mining transformed by mechanization was in the area of haulage. During the pick and hand-loading era, miners shoveled coal into wagons and pushed them to the surface. Drift mines were generally constructed so that they pierced the hillside at a slight upward angle to allow for drainage and to facilitate the transport of loaded cars to the tipple. As mines grew in size and the distance to the tipple increased, companies turned to animal power as the principal means of haulage. When larger coal cars came on the scene, mechanical and electrical haulage, including conveyors, became increasingly necessary.

From the vantage point of the coal operator, mechanization offered several advantages. It allowed for more easily loaded coal, reduced the need for explosives, and lowered timbering costs. Most important, mechanization permitted greater amounts of coal to be mined. Mechanization had its drawbacks, however, at least from the perspective of the miner, for the introduction of laborsaving equipment greatly reduced the need for a large workforce. The introduction of the mechanical coal-loading machine, for instance, reduced the need for coal miners by approximately 30% industrywide between 1930 and 1950. According to Ronald Lewis, black miners bore the brunt of layoffs during this period because they were disproportionately employed as hand loaders. The continuous miner, which combined cutting, drilling, blasting, and loading functions in one machine, had a similar impact. After the introduction of the continuous miner, the number of miners working in West Virginia was cut by more than half between 1950 and 1960. The impact of mechanization also shows up clearly in production and employment figures for Ohio. Between 1950 and 1970, coal production in this state climbed steadily despite a 90% decrease in the number of underground mines and an 83% decline in the size of the labor force. Starting in the 1970s, longwall mining cut even further into employee rolls.

Mechanization presented miners with other problems as well. In mining's early days, miners walked to and from the mine face and relied on natural ventilation to prevent the buildup of mine gas. As mines expanded in size, miners had to walk greater distances to get to work, and problems with ventilation and illumination arose. To facilitate the movement of men, electric trolleys were introduced to the mines. Dust and gas problems were alleviated somewhat by the installation of mechanical fans. Meanwhile, illumination was improved when safety lamps and electric lights were substituted for candles. While solving some problems, these and other solutions contributed to others. In the words of one authority on the matter, now workers could be "crushed, run over, squeezed between cars, or electrocuted on contact with bare trolley wires." In addition, some of the new machinery generated sparks and produced tremendous amounts of ignitable coal dust.

5.3 Health and Safety

Coal mining has always been and continues to be a hazardous occupation. Indeed, fatalities and serious nonfatal injuries occur still. The risks miners face today, however, pale in comparison to the perils miners faced prior to World War II. Accidents resulting from roof falls, ignition of mine gas and coal dust, the handling of mechanical haulage equipment, and the operation of electric-powered equipment of all types contributed to a high accident rate in the United States. Increased demand, fierce competition, advanced technology, negligence on the part of miners, and a poorly trained immigrant workforce have all been blamed for the high accident and fatality rates. Coal operators must share a portion of the blame as well. Opposed to unions and wary of burdensome safety regulations, operators often stood in the way of meaningful safety reform.

Initially, enactment of mine safety legislation rested with the states. In 1870, Pennsylvania passed the first substantive mine safety law. Over the next 25 years, several states passed mine inspection laws of their own. By the mid-20th century, 29 states had mine safety laws on the books. Unfortunately, these laws were often poorly enforced. Although the U.S. Bureau of Mines had been created in 1910, it was primarily an "information-gathering" agency. Authority to inspect underground mines was not conferred until 1941. Enforcement powers were not granted until 1953. Continued high fatality rates eventually sparked an effort to pass comprehensive federal legislation. Proponents of federal legislation had to wait until 1969 before such measures were enacted.

By far, the greatest number of deaths was caused by roof falls. Nevertheless, it was the ignition of methane gas and coal dust that captured the attention of the news media. Wherever miners used candles or open-flame safety lamps for illumination and black powder to break coal from the face, and where the mine was deep, "gassy," and poorly ventilated, conditions were ripe for an explosion. During the 10-year period from 1891 to 1900, 38 major explosions rocked American mines, resulting in 1006 deaths. Some of the worst explosions occurred during the first decade of the 20th century, including the worst one in American history, the dreadful Monongah mine disaster. On December 6, 1907, 362 miners employed by the Fairmont Coal Company of West Virginia were killed when the company's No. 6 and No. 8 mines blew up. Previously, Andrew Roy had noted that "more miners are annually killed by explosions in West Virginia, man for man employed, or ton for ton mined, than in any coal producing State in the Union or any nation in the world." It was clear that much needed to be done to improve mine safety in the United States.

Discovering the cause of mine explosions was of paramount importance to officials at the fledgling U.S. Bureau of Mines. Several causes were identified. First, as mines grew deeper, providing fresh air to miners and dispersing mine gas became more of a challenge. Given that miners typically used candles or open-flame safety lamps to light their way, this was a particularly serious problem. There was also a preference on the part of the miners to use greater amounts of explosives to "shoot off the solid," that is, to blast without first undermining the coal. As with the open flame, explosives were another source of ignition. Finally, there was the widely held belief that coal dust alone could not set off an explosion. Disproving this belief was particularly important (one that Europeans had confirmed by a much earlier date) given the increasing amounts of dust being produced by new mining machinery. The adoption of new technologies eventually reduced the risk of explosion. Improved ventilation, electric cap lamps, rock dusting, safer mining machinery, first-aid and mine rescue training, and passage of legislation permitting federal inspection of mines combined to improve conditions considerably. Although the frequency of mine explosions diminished considerably after the 1930s, the Farmington, West Virginia disaster of 1968, in which 78 miners were killed, served as a reminder that mine explosions were still a potential threat. With respect to roof falls, more effective supervision, systematic timbering, and roof bolting would, in time, reduce the frequency of these accidents.

5.4 Coal Mine Unionism

No history of coal mining in Appalachia would be complete without mentioning organized labor. Given the conditions that coal miners often had to work under, it is not surprising that they were among the first workers in the United States to organize. Given the power and control coal companies wielded throughout the better part of the 19th and early 20th centuries, it is also not surprising that coal operators fought the unions at every turn. Proponents of corporate paternalism and welfare capitalism believed that company benevolence could effectively stave off the unions, but others believed that coercion and intimidation would more effectively produce the desired results.

Although coal mine unionism in Appalachia can be traced to the early 1840s, the modern era of unionism began in 1890 when two groups, the National Progressive Union of Miners and Mine Laborers and the National Assembly Number 135 of Miners of the Knights of Labor, joined forces in Columbus, Ohio to form the United Mine Workers of America (UMWA). After winning recognition from Central Competitive Field Operators in 1897, the UMWA set out to stake a middle ground between ineffectual "company" unions and more radical unions, such as the National Miners Union. Under the able leadership of John Mitchell, the UMWA won strikes and concessions from the coal operators and built membership. Withdrawing from what they saw as a temporary truce with union organizers during the war years, coal operators sought to roll back the gains the UMWA had made during the first two decades of the 20th century. Citing a 1917 U.S. Supreme Court decision in which yellow-dog contracts forbidding union membership were ruled legal, coal operators set out to reclaim lost ground. An attempt by new union leader, John L. Lewis, to organize the southern coalfields during the middle of the decade was crushed. With demand for coal down and nonunion mines undercutting the cost of production in union mines, UMWA membership plummeted during the 1920s.

The 1930s, and more specifically President Franklin D. Roosevelt's New Deal legislation, breathed new life into the beleaguered union. Passage of the National Industrial Recovery Act (1933), the Bituminous Coal Code, or "Appalachian Agreement" (1933), the National Labor Relations, or "Wagner," Act (1935), and the first and second Bituminous Coal Conservation, or "Guffey," acts (1936 and 1937) shifted the advantage to the unions once again. Particularly important was the fact that the union had finally won the right to be recognized as a collective bargaining agent. Among the union's "victories" at this time was approval of an 8-hour workday, a 40-hour workweek, and higher wages. A minimum work age of 17 was also set, ending, in theory at least, the industry's long-standing practice of utilizing child labor. In addition, miners were no longer forced to use company scrip, to shop only at the company store, or to rent a company house, all mechanisms by which some companies attempted to keep up with the rising cost of production.

The gains of the 1930s and 1940s were replaced by the uncertainties of the 1950s and 1960s. The terms of the Love–Lewis agreement, signed in 1950 by

Bituminous Coal Operators Association president George Love and UMWA president Lewis, reflect the ever-changing fortunes of the miners and their union. Although the agreement provided a high wage and improved health benefits for miners, it prevented the union from opposing any form of mechanization. The adoption of new technologies, including auguring and stripping, resulted in layoffs and a decline in membership. Although certain gains were made, e.g., portal-to-portal pay, improved insurance and pension benefits, and passage of black lung legislation, the 1970s and 1980s saw many workers return to nonunion status.

Sometimes, relations between management and labor took a violent turn. Students of Appalachia know well the price that was paid in human life in places such as Matewan, West Virginia, "bloody" Harlan County, and at Blair Mountain, where 3000 UMWA marchers clashed on the battlefield with an army of West Virginia State Police, company mine guards, and assorted others representing the interests of the coal operators. In more recent times, violence has flared up in places such as Brookside, Kentucky. In 1969, charges of fraud and the murder of a UMWA presidential candidate tarnished the reputation of the union.

Although unions, the UMWA in particular, can take justifiable pride in improving the lot of miners, some scholars have argued that the higher labor costs associated with these victories ended up increasing the cost of coal production and accelerating the move toward mechanization. Thus, an impressive and hard-earned package of benefits was eventually passed down to an increasingly smaller pool of beneficiaries.

Further Reading

Armstead R 2002 *Black Days, Black Dust: The Memories of an African American Coal Miner*. The University of Tennessee Press, Knoxville

Corbin D 1981 *Life, Work, and Rebellion in the Coal Fields: The Southern West Virginia Miners, 1880–1922*. University of Illinois Press, Urbana

Crowell D 1995 *History of the Coal-Mining Industry in Ohio*. Department of Natural Resources, Division of Geological Survey, Columbus, Ohio

Drake R 2000 *A History of Appalachia*. The University Press of Kentucky, Lexington

Eller R 1982 *Miners, Millhands, and Mountaineers: Industrialization of the Appalachian South, 1880–1930*. The University of Tennessee Press, Knoxville

Francaviglia R 1991 *Hard Places: Reading the Landscape of America's Historic Mining Districts*. University of Iowa Press, Iowa City

Gaventa J 1980 *Power and Powerlessness, Quiescence and Rebellion in an Appalachian Valley*. University of Illinois Press, Urbana

Harvey K 1969 *The Best-Dressed Miners: Life and Labor in the Maryland Coal Region, 1835–1910*. Cornell University, Ithaca, New York

Hennen J 1996 *The Americanization of West Virginia: Creating a Modern Industrial State, 1916–1925*. University Press of Kentucky, Lexington

Lewis R 1987 *Black Coal Miners in America: Race, Class, and Community Conflict, 1780–1980*. University Press of Kentucky, Lexington

Roy A 1903 *A History of the Coal Miners of the United States: From the Development of the Mines to the Close of the Anthracite Strike of 1902 Including a Brief Sketch of British Miners*. J. L. Trauger Printing Co., Columbus, Ohio

Salstrom P 1994 *Appalachia's Path to Dependency: Rethinking a Region's Economic History, 1730–1940*. University Press of Kentucky, Lexington

Shifflett C 1991 *Coal Towns: Life, Work, and Culture in Company Towns of Southern Appalachia, 1880–1960*. The University of Tennessee Press, Knoxville

Thomas J 1998 *An Appalachian New Deal: West Virginia in the Great Depression*. University Press of Kentucky, Lexington

Williams J 1976 *West Virginia and the Captains of Industry*. West Virginia University Library, Morgantown

Geoffrey L. Buckley

Conservation Measures for Energy, History of

Glossary

end-use efficiency Substituting technological sophistication for energy consumption through (1) obtaining higher efficiency in energy production and utilization and (2) accommodating behavior to maximize personal welfare in response to changing prices of competing goods and services.

energy conservation The wise and thoughtful use of energy; changing technology and policy to reduce the demand for energy without corresponding reductions in living standards.

energy intensity The amount of energy consumed to produce a given economic product or service; often measured as the ratio of the energy consumption (E) of a society to its economic output (gross domestic product, GDP), measured in dollars of constant purchasing power (the E/GDP ratio).

energy services The "ends," or amenities, to which energy is the "means," e.g., space conditioning, lighting, transportation, communication, and industrial processes.

externalities The environmental, national security, human health, and other social costs of providing energy services.

least-cost strategy A strategy for providing individual and institutional energy consumers with all of the energy services, or amenities, they require or want at the least possible cost. It includes the internal economic costs of energy services (fuel, capital, and other operating costs) as well as external costs.

The history of energy conservation reflects the influence of technology and policy on moderating the growth of demand for energy. It traces the evolution of concepts of conservation from curtailment of energy use to least-cost strategies through end-use efficiency. Through understanding the past efforts to address the need for energy conservation, the future potential of the role for energy conservation in the global economy is better understood.

1. Introduction

Just as beauty is in the eye of the beholder, energy conservation means different things to different people. Concerned about deforestation in Pennsylvania, Benjamin Franklin took heed of his own advice ("a penny saved is a penny earned") by inventing the Franklin stove. In his *Account of the New Invented Fireplaces*, published in 1744, Franklin said "As therefore so much of the comfort and conveniency of our lives for great a part of the year depend on this article of fire; since fuel is become so expensive, and (as the country is more clear'd and settle'd) will of course grow scarcer and dearer; any new proposals for saving the wood, and for lessening the charge and augmenting the benefit of fire, by some particular method of making and managing it, may at least be thought worth consideration." The Franklin stove dramatically improved the efficiency of space heating. In Franklin's time, however, others viewed energy consumption as a signal of economic progress and activity. This view was epitomized by company letterheads picturing a factory with chimneys proudly billowing clouds of smoke!

Viewed as a matter of how much energy it takes to run an economy—often referred to as a society's energy intensity—energy conservation has been with us since the middle of the 20th century. Between 1949 and 2000, the amount of energy required to produce a (1996 constant) dollar of economic output fell 49%. Much of this improvement occurred by the combined effects of improved technology (including increased use of higher quality fuels in new technologies, such as high-efficiency gas turbines) and perceptions of higher energy costs after the oil shocks of the 1970s, when energy intensity began to fall at an average rate of 2% per year. (Fig. 1).

Although largely oblivious to long-term improvements in energy intensity, consumers undoubtedly felt the oil shocks, and came to equate energy conservation with waiting in long lines to purchase gasoline. For many people, even to the present day, energy conservation connotes curtailment or denial of the services energy provides, e.g., turning down the

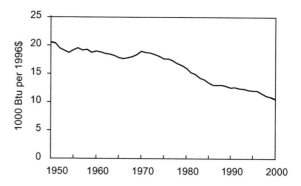

Figure 1
Energy use per dollar of gross domestic product. From the U.S. Energy Information Administration, *Annual Review of Energy 2000*.

heat on a cold winter's day, turning off the lights, or driving lighter cars at slower speeds. This viewpoint came into sharp focus one night during the 1973 oil embargo, when one author (Gibbons), director of the newly established Federal Office of Energy Conservation, fielded a call from an irate Texan who proclaimed that "America didn't *conserve* its way to greatness; it *produced* its way to greatness!"

For purposes of this article, we accept and adapt the term "wise and thoughtful use" for energy conservation. This term implies that, when technically feasible and economically sound, a society will use technology and/or adopt social policy to reduce energy demand in providing energy services. It also implies that society's effort to bring energy supply and energy demand into balance will take into full account the total costs involved. Market prices for energy are generally less than its total social cost due to various production subsidies and nonmarket costs, or externalities, such as environmental impacts and security requirements. This approach has been labeled a "least-total cost" strategy for optimizing the investment trade-off between an increment of supply versus an increment of conservation.

2. Decade of Crisis: 1970s

Events of the late 1960s highlighted many of the external costs of energy production and consumption. Many of the worsening problems of air and water pollution were directly attributable to energy use. The growth in the power of the Organization of the Petroleum Exporting Countries (OPEC) and increasing tensions in the Middle East contributed to a growing wariness of import dependence. High projected electric demand growth exacerbated worries not only about coal and the environment, but also about the sheer amount of capital investment that was implied. Gas curtailments began to show up along with spot shortages of heating oil and gasoline. And, surprising to many in the energy field, there were growing doubts about the nuclear option.

2.1 Rethinking Energy Demand

Those forces converged to trigger work on improved and alternative energy sources as well as on the dynamics and technologies of demand. Public support for analysis of conservation strategies came at first not from the "energy agencies" but from the National Science Foundation (NSF). In 1970, Oak Ridge National Laboratory (ORNL) researchers (who were well-sensitized to the vexing problems of coal and fission) seized on the opportunity provided by the NSF to investigate demand dynamics and technologies. The owner/operator of ORNL, the Atomic Energy Commission (AEC), was not keen on

the idea but accepted the new direction for research as "work for others"!

It quickly became apparent that despite declining energy prices in previous years, energy efficiency had been slowly improving. Technology relevant to efficient use was advancing so impressively that more efficiency was economically attractive in virtually every sector of consumption. Further examination at ORNL and elsewhere showed that such improvements would accelerate more rapidly if energy prices were to rise and/or if consumers became better informed. For example, it was demonstrated that cost-effective technical improvements in refrigerators could double the energy efficiency of household refrigerators for a very modest price increment (Fig. 2). Clearly, major environmental benefits would also accrue from more conservation.

The environmental concerns of the 1960s prompted the first serious efforts to project future energy demand. Scenario analyses, prepared by many different authors throughout the 1970s, produced a wide range of estimates, helping to shape a diverse range of views on the importance of conservation. Eleven forecasts of consumption made during the 1970s turned out to be remarkably consistent but ludicrously high, even for 1985 (ranging from 88 to 116 quads—the actual number was 75 quads). Prior to the oil embargo of 1973, most projections of U.S. energy consumption for the year 2000 fell in the range of 130–175 quads. The actual number turned out to be about 100 quads. Some of the difference can be attributed to lower than anticipated economic growth, but the dominant difference is due to the unexpected role of conservation.

Limits to Growth, published in 1972, was the first analysis to capture the public's attention. This computerized extrapolation of then-current consumption trends described a disastrous disjuncture of the growth curves of population and food supply and of resource use, including energy, and pollution. One scenario climaxed in a collapse mode, the sudden and drastic decline in the human population as a consequence of famine and environmental poisoning. This apocryphal report correctly made the point that exponential growth in finite systems is unsustainable, but its macroeconomic model failed because it did not consider responses to resource prices or advances in technologies.

A Time to Choose: America's Energy Future, a group of scenarios commissioned by the Ford Foundation and published in 1974, became notable, and controversial, for its "zero energy growth" (ZEG) scenario. This ZEG scenario clearly was meant to show that energy growth could technically be decoupled from economic growth. ZEG called for a leveling off in energy demand at about 100 quads per year shortly after 1985. In addition to those economically feasible measures taken by 1985, reduced demand would be effected by means of energy

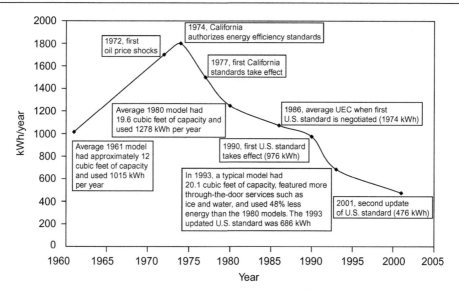

Figure 2
Changes in energy demand of refrigerators in the United States since 1961. UEC, unit energy consumption (kWh/year). From the Lawrence Berkeley National Laboratory.

consumption taxes and other constraints on use. Taxes would increase the price of fuels so that substitution of energy-saving capital investments could be made economical while simultaneously encouraging shifts to less energy-intensive services. *A Time to Choose* also included a "technical fix" scenario that estimated 1985 consumption at 91 quads by coupling a microeconomic model with analysis of the engineering potential for conservation as a response to increased energy price and government policies. In 1976, Amory Lovins published a far more controversial and attention-grabbing scenario called "Energy Strategy: The Road Not Taken" in the journal *Foreign Affairs*. Lovins advocated a "soft energy path" based on three components: (1) greatly increased efficiency in the use of energy, (2) rapid deployment of "soft" technologies (diverse, decentralized, renewable energy sources), and (3) the transitional use of fossil fuels. His assertion that the "road less traveled" could bring America to greatly reduced energy demand (less than 50 quads per year in 2025) is both venerated and despised to this day.

These private sector analyses competed with scenarios developed or commissioned by government agencies for the attention of energy policy makers throughout the 1970s. *Guidelines for Growth of the Electric Power Industry*, produced by the Federal Power Commission (FPC) in 1970, was a classic study of energy demand based largely on an extrapolation of past trends. The FPC had credibility because of its 1964 success in predicting 1970 electricity demand, but its forecast of a doubling of

energy demand from 1970 to 1990 was dramatically far from the mark. The Atomic Energy Commission made more accurate estimates in *The Nation's Energy Future*, published in 1973, but it mistakenly forecast U.S. independence from imported oil by 1980—no doubt caused or at least influenced by President Nixon, who was determined to achieve energy independence before the end of his term.

In 1975, the U.S. National Academy of Sciences established a Committee on Nuclear and Alternative Energy Strategies (CONAES). The CONAES Panel on Demand and Conservation published a pathbreaking paper in the journal *Science* in 1978: "U.S. Energy Demand: Some Low Energy Futures." The CONAES scenarios not only incorporated different assumptions for gross domestic product (GDP) and population growth, but also, for the first time, incorporated varied assumptions about energy prices. The lowest CONAES estimate, based on a 4% annual real price increase coupled with strong non-market conservation incentives and regulations, came in at only 58 quads for 2010. The highest estimate, based on a 2% annual real price increase coupled with a rapidly growing economy, came in at 136 quads for 2010. This study clearly illustrated the importance in scenario analysis of assumptions about everything from income, labor force, and the rate of household formation, to worker productivity, energy productivity, energy price, income, and energy substitutability in scenario analysis. It also highlighted the importance of capital stock turnover in energy futures, in that the largest technical opportunities for increased efficiency derive from new stock, where state-of-the-art technology can be most effectively employed. The CONAES project had a major influence on all of the energy modeling work that followed.

2.2 Oil Shocks

In October 1973, the United States helped Israel beat back the Yom Kippur invasion by Egypt and Syria. OPEC (dominated by Arab oil producers) retaliated with an oil embargo on the United States and a 70% oil price increase on West European allies of the United States. Homeowners were asked to turn down their thermostats and gas stations were requested to limit sales to 10 gallons per customer. Illumination and speed limits were lowered, and the traffic law allowing a right turn on a red light came into use. Gasoline lines, conspiracy theories, and recession followed in rapid succession. As Adlai Stevenson once observed, "Americans never seem to see the handwriting on the wall until their backs are up against it." We perceived the wall behind our backs and responded accordingly—for a brief spell. American humor also surfaced. Some wonderful cartoons depicted innovative ways to save energy—for example, by sleeping with a chicken, which has a normal body temperature of 107°F, and public interest highway signs appeared proclaiming "Don't be fuelish!"

OPEC lifted the embargo in 1974, but political unrest in the Middle East again precipitated a major oil price increase in 1979. When anti-Western Islamic fundamentalists gained control of Iran, oil production in that nation dropped off dramatically and U.S. crude oil prices tripled. These price increases also led to skyrocketing inflation and recession, and rekindled interest in energy efficiency.

2.3 Policy Responses

The need to curtail energy consumption in response to U.S. vulnerability to oil supply disruptions dominated energy policy in the 1970s. In the summer of 1973, President Nixon, anticipating possible shortages of heating oil in the coming winter, and summer gasoline shortfalls, established the Office of Energy Conservation to coordinate efforts to cut energy consumption in federal agencies by about 5%, raise public attention to the need for increased efficiency, encourage private industry to give greater attention to saving energy, and devote more federal research and development to efficiency. After the first oil shock, President Nixon urged voluntary rationing and pushed for a commitment to U.S. energy independence by 1985. President Carter emphasized the importance of conservation in the effort to achieve energy independence, but appeared to equate conservation with wearing a sweater in a chilly White

House. Early actions by our political leaders certainly contributed to a negative association of conservation with sacrifice. But they also contributed to an atmosphere conducive to serious public and private sector actions to incorporate efficient end-use technologies in the U.S. economy. Congress enacted legislation to create standards for Corporate Average Fuel Economy (CAFE) and building energy performance. They also created a cabinet-level Department of Energy (DOE) by merging the functions of preexisting agencies scattered throughout government. In response to an analysis by the congressional Office of Technology Assessment, Congress explicitly included energy conservation in the DOE's new mandate.

3. Decade of No- and Low-cost Options: 1980s

The combination of the oil price shocks with increasing public concern about the environmental consequences of energy use prompted policy makers to take a serious look at policy tools for decreasing energy demand growth during the 1980s. Faced with higher oil prices, industrial and commercial oil consumers also sought out cost-effective investments in energy conservation during this decade. A plethora of no- and low-cost options were ripe for the picking, and progressive industries did just that, with great success at the bottom line.

3.1 Policy Tools

At the beginning of the 1980s, energy analysts converged on a set of principles for a comprehensive energy policy that had conservation at its core. These principles included the following ideas:

1. Consider the production and use of energy as means to certain ends, not as goals in and of themselves. Remember always that, given time and the capital for adjustment, energy is a largely substitutable input in the provision of most goods and services.

2. Application of technical ingenuity and institutional innovation can greatly facilitate energy options.

3. Energy decisions, like other investment decisions, should be made using clear signals of comparative total long-run costs, marginal costs, and cost trends.

4. It is important to correct distorted or inadequate market signals with policy instruments; otherwise, external costs can be ignored and resources can be squandered. This correction includes internalizing in energy price and/or regulation, to the extent possible, the national security, human health, and environmental costs attributable to energy.

5. Investment in both energy supply and utilization research and development is an appropriate

activity for both the public and private sectors, because costs and benefits accrue to both sectors.

6. There are other, generally more productive, ways (for example, assistance with insulation or lighting retrofits, fuel funds, or even refundable tax credits) to assist underprivileged citizens with their energy needs, rather than subsidizing energy's price to them.

7. In a world characterized by tightly integrated economies, we need to increase our cognizance of world energy resource conditions and needs with special regard for international security as well as concern for the special needs of poor nations.

By the mid-1980s, the national energy situation had begun to reflect some of these ideas. CAFE standards reached their peak (27.5 miles per gallon) in 1985. Congress created the Strategic Petroleum Reserve (SPR) to respond to energy supply emergencies. Oil and gas price deregulation was virtually complete, but that was only the first step in internalizing the costs of energy consumption. Energy prices still did not reflect the manifold environmental costs of production and use, the costs of defending Middle East oil production and shipping lanes, the costs of purchasing and storing oil to meet emergencies, or the impacts of U.S. competition for scarce oil resources on developing nations. As the price of imported oil decreased, U.S. policy makers were lulled into complacency about the nation's energy strategy. A dearth of energy research and development expenditures symbolized this declining interest. Before the end of the decade, however, Congress passed the National Appliance Energy Conservation Act, which authorized higher appliance efficiency standards and efficiency labeling that have significantly impacted the market for energy-efficient appliances.

3.2 Private Sector Activities

Immediately following the oil shocks, private sector efforts to save energy focused on changing patterns of energy use within the existing infrastructure, such as lowering thermostats. Most actions involved investments in technology, however—either retrofits of existing technology, such as insulating existing homes, or new investments in technology, such as energy-efficient new construction or autos with improved mileage. Later in the 1980s, energy efficiency gains accrued as incidental benefits to other, much larger investments aimed at improving competitiveness of U.S. products in world markets.

All in all, energy efficiency investments in the 1970s and the 1980s turned out to be generally easier and much more cost-effective than finding new sources of energy. Nongovernmental institutions helped to keep the conservation ball of progress rolling, especially in the residential and commercial sectors, with information to increase awareness of cost-effective conservation opportunities, rebates, and prizes. In the first decade after the oil shocks, industry cut energy requirements per unit of output by over 30%. Households cut energy use by 20% and owners and operators of commercial buildings cut energy use per square foot by more than 10%. Transportation energy efficiency steadily improved as CAFE requirements slowly increased fleet average mileage. However, political maneuvering by car companies resulted in excepting "light trucks," including sport utility vehicles (SUVs), from the mileage requirements for passenger cars. This has resulted in a major slow down of improvement in fleet performance.

Some surprising partnerships formed to promote energy efficiency. For example, the Natural Resources Defense Council designed a Golden Carrot Award to encourage production of a superefficient refrigerator. In response, 24 major utilities pooled their resources to create the $30 million prize. Whirlpool won the competition and collected its winnings after manufacturing and selling at least 250,000 units that were at least 25% more efficient than the 1993 efficiency standard required. By the end of the 1980s, energy efficiency improvements lost momentum. One reason was the decline in oil prices: corrected for inflation, the average price of gasoline in 1987 was half that in 1980. Natural gas and electricity prices also fell, and the price-driven impetus for consumers and businesses to conserve was diminished.

3.3 Lessons from Experience

By the end of the decade, conservation advocates were urging policy makers to take several lessons from the experiences of 1970 and 1980s:

1. First and foremost, that energy conservation worked. Nothing contributed more to the improved American energy situation than energy efficiency. By the late 1980s, the United States used little more energy than in 1973, yet it produced 40% more goods and services. According to one estimate, efficiency cut the nation's annual energy bill by $160 billion.

2. Oil and gas price controls are counterproductive. Price increases of the 1970s and 1980s sparked operating changes, technology improvements, and other conservation actions throughout the United States.

3. Technological ingenuity can substitute for energy. A quiet revolution in technology transformed energy use. Numerous efficient products and processes—appliances, lighting products, building techniques, automobiles, motor drives, and manufacturing techniques—were developed and commercialized by private companies. Energy productivity rose as more efficient equipment and processes were incorporated into buildings, vehicles, and factory stock.

4. Complementary policies work best. The most effective policies and programs helped overcome barriers to greater investment in conservation—barriers such as the lack of awareness and low priority among consumers; lack of investment capital; reluctance among some manufacturers to conduct research and to innovate; subsidies for energy production; and the problem of split incentives, exemplified by the building owner or landlord who does not pay the utility bills but does make decisions on insulation and appliance efficiency.

5. The private sector is the primary vehicle for delivering efficiency improvements, but government-industry cooperation is essential.

Many activists advocated adoption of a new national goal for energy intensity in order to regain momentum toward a secure energy future. The United States achieved a 2.7% annual rate of reduction in energy intensity between 1976 and 1986. The DOE was predicting, however, that U.S. energy intensity would decline at less than 1% per year through the rest of the 20th century. This is the rate of improvement now widely assumed as part of "business as usual." Recognizing that economic structural change had also influenced this rapid improvement and that many of the least expensive investments in efficiency had already been made, the present authors and others, including the American Council for an Energy Efficient Economy, nonetheless recommended a goal of reducing the energy intensity of the U.S. economy by at least 2.5% per year into the 21st century. For the United States, even with an apparently disinterested Administration, it is now proposed that energy intensity over the next decade be improved by 18% (roughly the average yearly gain over the past two decades).

4. Decade of Globalization: 1990s

The United States went to war for oil at the beginning of the 1990s. Oil prices initially spiraled upward, temporarily reinvigorating interest in conservation investments. That interest dwindled as quickly as prices fell. Still, after an inauspicious beginning, during the economic boom of the late 1990s, with major investments in new capital stocks, energy intensity dropped at an average rate of over 2% per year.

The role of carbon emissions in global climate change began to dominate the debate over energy in the 1990s. Although U.S. total energy consumption far outstripped any other country's, consumption was projected to grow at much higher rates in the developing countries, and U.S. policy makers shifted their attention to U.S. participation in international cooperation on energy innovation. In 1993, the U.S. government and the U.S. automobile industry forged an unprecedented alliance under the leadership of President Clinton and Vice President Gore. The partnership included seven federal agencies, 19 federal laboratories, and more than 300 automotive suppliers and universities and the United States Council for Automotive Research, the pre-competitive research arm of Ford, DaimlerChrysler, and General Motors. The Partnership for a New Generation of Vehicles (PNGV) supported research and development of technologies to achieve the program's three research goals: (1) to significantly improve international competitiveness in automotive manufacturing by upgrading manufacturing technology; (2) to apply commercially viable innovations resulting from ongoing research to conventional vehicles, especially technologies that improve fuel efficiency and reduce emissions; and (3) to develop advanced technologies for midsized vehicles that deliver up to triple the fuel efficiency of today's cars (equivalent to 80 miles per gallon), without sacrificing affordability, performance, or safety. The research plan and the program's progress were peer-reviewed annually by the National Research Council. PNGV made extraordinary progress toward achieving its aggressive technical goals. In March 2000, PNGV unveiled three concept cars demonstrating the technical feasibility of creating cars capable of getting 80 miles per gallon. All three cars employ some form of hybrid technology that combines a gasoline- or diesel-powered engine with an electric motor to increase fuel economy. The three major automakers also confirmed their commitment to move PNGV technology out of the lab and onto the road by putting vehicles with significant improvements in fuel economy into volume production and into dealers' showrooms. Work continues on technologies that might contribute to the full achievement of goals for the 2004 prototype.

In 1999, the President's Committee of Advisors on Science and Technology (PCAST) issued recommendations for improving international cooperation on energy conservation. Reflecting a growing national consensus, they reported that efficient energy use helps satisfy basic human needs and powers economic development. The industrialized world depends on massive energy flows to power factories, fuel transport, and heat, cool, and light homes, and must grapple with the environmental and security dilemmas these uses cause. These energy services are fundamental to a modern economy. In contrast, many developing-country households are not yet heated or cooled. Some 2 billion people do not yet have access to electric lighting or refrigeration. The Chinese enjoy less than one-tenth as much commercial energy per person as do Americans, and Indians use less than one-thirtieth as much. Raising energy use of the world population to only half that of today's average American would nearly triple world energy demand. Such growth in per capita energy use coupled with an increase by over 50% in world population over the next half-century and no improvement in energy efficiency would together increase global energy use more than four times. Using conventional technologies, energy use of this magnitude would generate almost unimaginable demands on energy resources, capital, and environmental resources—air, water, and land.

Energy-efficient technologies can cost-effectively moderate those energy supply demands. For example, investments in currently available technologies for efficient electric power use could reduce initial capital costs by 10%, life-cycle costs by 24%, and electricity use by nearly 50%, compared to the current mix of technologies in use. Modest investments in efficient end-use technologies lead to larger reductions in the need for capital-intensive electricity generation plants. Conversely, when unnecessary power plants and mines are built, less money is available for factories, schools, and health care.

In the language of economics, there is a large opportunity cost associated with energy inefficiency. It is both an economic and environmental imperative that energy needs be satisfied effectively. Incorporating energy efficiency measures as economies develop can help hold energy demand growth to manageable levels, while reducing total costs, dependence on foreign sources of energy, and impacts on the environment. Ironically, energy is most often wasted where it is most precious. Developing and transition economies, such as in China, the world's second largest energy consumer, use much more energy to produce a ton of industrial material compared to modern market economies. Developing-nation inefficiency reflects both market distortions and under-development. Decades of energy subsidies and market distortions in developing and transition economies have exacerbated energy waste. Energy efficiency requires a technically sophisticated society. Every society in the world, including the United States, has at one time encouraged energy inefficiency by controlling energy prices, erecting utility monopolies, subsidizing loans for power plant development, and ignoring environmental pollution. Many nations, including the United States, continue these wasteful practices. These subsidies and market distortions can seriously delay technological advances, even by decades, compared to best practice. The worldwide shift to more open, competitive markets may reduce some of the distortions on the supply side, but will do little to change the inherent market barriers on the demand side.

China, India, Brazil, Russia, and other developing and transition economies are building homes and factories with outdated technology, which will be used for many decades. In cold northern China and Western Siberia, for example, apartments are built with low thermal integrity and leaky windows and are equipped with appliances half as efficient as those available in the United States. Developing nations

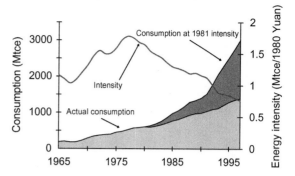

Figure 3
Energy intensity and consumption in China. Mtce, Million tons of coal equivalent. From Advanced International Studies Unit, Battelle Pacific Northwest National Laboratory.

thus tend to build in excessive energy costs, lock out environmental protection, and diminish their own development potential. There are important exceptions. Progress can be both rapid and significant. Over recent decades, China has, through energy-sector reform and pursuit of least-cost paths via energy efficiency opportunities, made unprecedented progress in energy intensity reduction, faster than any developing nation in history. China has held energy demand growth to half of its rate of overall economic growth over the past two decades (Fig. 3). This example demonstrates that economic development can proceed while restraining energy demand growth; indeed, by employing cost-effective energy-efficient technologies, funds are freed for investment in other critical development needs. Energy efficiency can thus be a significant contributor to social and economic development.

5. Decades Ahead

Energy projections indicate differences in U.S. energy demand in 2020 between the high and low scenarios of approximately 25 quads (Fig. 4). The difference between these scenarios is substantially due to differing assumptions about development and deployment of energy-efficient technologies. The differences also reflect the plausible range of key consumption-driving factors that, operating over 20 years, can have a profound influence. These factors include economic growth rate and the (rising) urgency to cut emissions of greenhouse gases in order to slow global climate change.

5.1 Cultural Barriers to Conservation

Sadly, there remains a persistent struggle between advocates of energy supply and conservation advocates.

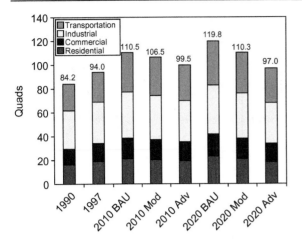

Figure 4
Primary energy consumption by sector. BAU, Business-as-usual model, compared to moderate (Mod) and advanced (Adv) conservation models. From Interlaboratory Working Group (2000).

On the supply side, the energy industry is massive and concentrated, wielding great political influence (e.g., obtaining tax subsidies, allowing unhealthy levels of environmental emissions). On the conservation side, the relevant industry is diffusely spread over the entire economy, and energy efficiency is seldom the primary concern. Furthermore, decisions regarding the energy efficiency of appliances (e.g., heating, ventilation and air-conditioning systems and refrigerators) are frequently made (e.g., by developers, not consumers) on the basis of minimum first cost rather than life-cycle cost, thus generally favoring lower efficiency.

Two streams of progress over the past 50 years have combined to accentuate the attention given to energy conservation. First, steady and ubiquitous advances in technology (such as computers and various special materials) have enabled remarkable improvements (lower cost, higher efficiency) in productivity, including energy. Second, there has been a growing awareness of major direct and indirect externalities associated with energy production and conversion. Although such externalities (e.g., health, ecological stress, climate change, and air and water pollution) are now widely recognized, they still are not generally incorporated into national economic accounts. Instead, various standards, regulations, and policies have been developed or proposed to reflect the external costs by acting as a "shadow price," in order to level the playing field of price signals.

Despite its profound impacts, the conservation revolution has resulted in only a fraction of its ultimate potential. Here are some of the reasons:

1. Lack of leadership. Whereas an energy consumer can make short-term adjustments to save energy (driving slower, driving fewer miles and traveling less, adjusting the thermostat, or turning off lights), these mostly have the unpopular effect (other than saving money) of curtailing the very energy services that we seek. Sadly, such heroic short-term curtailment measures are too often posed as the essence of conservation rather than emergency measures. On national television, President Carter advocated energy conservation as a moral imperative, but delivering his speech while wearing a sweater and sitting by an open fire connoted sacrifice and denial of comfort. Worse yet, President Reagan referred to conservation as being hot in summer and cold in winter! In the short run, curtailment is about all anyone can do, but if we stop there, we lose by far the most impressive opportunities. After the shocks of the 1970s subsided, attention shifted to the more substantial but also more challenging task of capturing longer term opportunities.

2. Time lags. The largest opportunities for energy conservation derive from incorporating more efficient technologies into new capital investments (e.g., cars, buildings, appliances, and industrial plants). These logically accrue at a rate that reflects depreciation of existing capital stock, ranging from years to many decades. Just as for the case of major transitions in energy systems, the major opportunities for conservation inherently require decades to be fully achieved.

3. Discount rates. With a few exceptions, capital purchase decisions tend to favor minimum first cost, rather than minimum life-cycle cost. Generally, the minimum first-cost model corresponds to the least energy-efficient model. Thus the "least-cost" solution is typically not taken by most consumers and this slows the rate at which efficiency is gained.

5.2 Remaining Potential for Conservation of Energy Resources

Liquid fuels, so attractive because they can be stored and transported with relative ease, loom as a global challenge because the main supply (petroleum) is very unevenly distributed geographically and is often concentrated in insecure locations. Technologies for petroleum conservation are steadily improving, from elegant, new methods of exploration and production to highly efficient combustion in motors and turbines. Thus the petroleum resource base, although headed for depletion in this century, can be stretched (conserved) by a variety of technologies. Advances in high-temperature and low-corrosion materials now enable a variety of combustion turbines that can be employed to generate electricity with 60% or better thermodynamic efficiency—double the best-practice several decades ago.

Probably the greatest challenge for liquid fuel conservation for the future is in the transportation sector. Technology, the auto industry, public policies, and consumer preference all play important roles. The industry is most profitable in making large, heavy and powerful cars and SUVs; consumers in the United States presently have a love affair with the image of power, ruggedness, and convenience with little regard for efficiency. And the record shows that U.S. public policy is to keep fuel price below total cost. Advances in automotive efficiency such as hybrid-electric/advanced diesel and fuel cells could help alleviate the situation. For example, over the past decade, work on streamlining, electrical power control systems, high-compression/direct-injection engines, electric drive trains, etc. has resulted in the recent market entry of hybrid cars that are very popular and have about 50% better mileage; in addition, rapid advances in fuel cell technology promise within a decade or so more conservation as well as lower environmental emissions. However, there seems to be little reason to hope for major gains in efficiency without major increases in fuel price or its equivalent shadow price (performance regulation such as CAFE).

Electricity generation is getting very efficient and the technologies that use electricity are also advancing. The world is still electrifying at a rapid pace. Electricity can now be generated from diverse energy sources and with diverse sizes of generators. Source diversity means inherent resilience of the system and capability for disaggregated generation means that "waste heat" from generators can be more readily utilized (cogeneration). Electricity is such a high-quality and versatile energy form that it is likely it will continue to increase in market share. Thus further advances in the efficiency of conversion and use of electricity should find ready markets.

The historic succession of energy sources over the past 150 years has led to progressively higher hydrogen-to-carbon (H/C) ratios for the fuels (e.g., coal, followed by oil, followed by natural gas). Global climate protection argues that we need to move with dispatch toward even higher H/C ratios in the coming decades. This implies a combination of removal (sequestering) of carbon from combustion releases, increased carbon recycle (e.g., renewable fuels from biomass), direct use of solar energy, and increased use of nuclear energy. Under this scenario, hydrogen becomes the most likely candidate for the derived fuel. This underscores the promise of high-efficiency fuel cells for energy conversion to electricity.

The amount of conservation achieved through advancing technology has been outstanding over the past half-century. Many more gains are possible, but we can also discern limits dictated by limits imposed by natural law. Carnot's thermodynamic rule for efficiency of heat engines is one limit. On the other hand, fuel cells give an opportunity to bypass Carnot. Success in this area could effectively enable a doubling of the useful work out of a unit of energy compared to an internal combustion engine, and with basically no environmental emissions. Success beckons but will require many technical advances in special materials and also in providing an acceptable primary energy source—probably hydrogen—all at a much lower production cost than now. The artificially low current price of gasoline constitutes a serious barrier to market entry of the new technologies.

Human ingenuity in the form of technological innovation will continue to provide new options to society. Energy conservation, which derives from sophisticated and elegant technologies, has every chance to continue paying big dividends to technologically capable societies. Conservation enables humanity to receive more "goods" along with fewer "bads." But most things have limits. Ultimately, we must recognize the folly of trying to follow an exponential path of population growth and resource consumption in its present form. We must equip ourselves with new tools, including conservation, and move with dispatch on a journey toward sustainability in the 21st century. There is little time to spare.

Further Reading

Chandler W 2000 *Energy and Environment in the Transition Economies.* Westview Press, Boulder and Oxford

Chandler W, Geller H, Ledbetter M 1988 *Energy Efficiency: A New Agenda.* GW Press, Springfield, Virginia

Demand and Conservation Panel of the Committee on Nuclear and Alternative Energy Systems 1978 U.S. Energy Demand: Some Low Energy Futures (14 April 1978). *Science* **200**, 142–52

Gibbons J 1997 *This Gifted Age: Science and Technology at the Millennium.* AIP Press, Woodbury, New York

Gibbons J, Chandler W 1981 *Energy: The Conservation Revolution.* Plenum Press, New York and London

Gibbons J, Blair P, Gwin H 1989 Strategies for energy use. *Sci. Am.* **261**, 136–43

Interlaboratory Working Group. 2000. *Scenarios for a Clean Energy Future.* Oak Ridge National Laboratory and Lawrence National Laboratory, Oak Ridge, Tennessee, and Berkeley, California.

President's Committee of Advisors on Science and Technology. 1997. *Federal Energy Research and Development for the Challenges of the 21st Century.* Office of Science and Technology Policy, Washington, D.C.

President's Committee of Advisors on Science and Technology. 1999. *Powerful Partnerships: The Federal Role in International Cooperation on Energy Innovation.* Office of Science and Technology Policy, Washington, D.C.

Sant R 1979 *The Least-Cost Energy Strategy: Minimizing Consumer Costs through Competition.* Carnegie-Mellon University Press, Pittsburgh, Pennsylvania

U.S. Congress, Office of Technology Assessment 2004 *OTA Legacy* (includes reports on Conservation). U.S. Govt. Printing Office, Washington, D.C. See also http://www.wws.princeton.edu/ota/

John H. Gibbons and Holly L. Gwin

Conservation of Energy Concept, History of

Glossary

caloric theory   The theory that heat is a substance that flows into or out of a body as it is heated or cooled; this theory was replaced by the notion of the conservation of energy around 1850.

conservation of energy   The notion that in a closed system, energy cannot be created or destroyed but can only be converted into other forms.

energy   The capacity of a body to do work; it comes in various forms that can be converted into each other.

ether   A medium that was assumed to fill space to explain the propagation of light, heat, and electromagnetic waves; it was discarded during the early 20th century.

heat engine   Any device that takes heat and converts it into mechanical work.

mechanical equivalent of heat   A constant that expresses the number of units of heat in terms of a unit of work.

mechanical theory of heat   The theory that heat consists of the motions of the particles that make up a substance.

perpetual motion   The fallacious theory that a machine, producing work, can be operated without any external source of power; it was discussed by philosophers and sold to kings but has been ignored by engineers since ancient times.

vitalism   The theory that life depends on a unique force and cannot be reduced to chemical and physical explanations; this theory has been discarded.

work done   Measured by force multiplied by the distance through which its point of application moves.

The modern principle of the conservation of energy emerged during the middle of the 19th century from a complex of problems across the sciences, from medicine to engineering. It was quickly accepted as a principle of nature and was applied to problems in mechanics and electromagnetism, and then in other sciences, even when it was not the most useful approach. It survived challenges even into the 20th century.

1.   Motion and Mechanics Before 1800

Some ancient philosophers and most engineers assumed that perpetual motion was impossible, but nobody examined the efficiency of machines. The latter did not become central to engineering until the 18th century, after exhaustive metaphysical and methodological debates about motion and mathematics as necessary components of a comprehensive philosophy of nature. Natural philosophers and mathematicians argued over the conditions under which particular conservation laws were valid, and through these debates motion and collisions drew more attention from engineers and mathematicians.

In Rene Descartes's natural philosophy, the universe consisted of matter, whose quantity was measured by size, and motion. Collisions between particles was the mechanism for change, and the total amount of motion in the universe remained constant, although motion could be lost or gained in each collision. Descartes also assumed that the natural motion of bodies was linear—not circular—and was a vector. In collisions, the quantity of motion, size multiplied by velocity, could be transferred from one particle to another. The force of a particle was its size multiplied by the square of its velocity. Although Descartes modeled his philosophical method on geometry, very little mathematics was contained within it. It was left to later generations to put his philosophy of nature into mathematical form.

One of the most successful was Isaac Newton, who began as a follower of Descartes and ended as his scientific and philosophical rival. In his explorations of Descartes's system, Newton became dissatisfied with Cartesian explanations of cohesion and other phenomena. He eventually accepted the reality of the concept of "force," banned by Descartes as "occult," as acting on micro and macro levels. Newton's universe was as empty as Descartes's was full, and God was necessary to keep His vast creation going. In such a universe, conservation laws and collisions were not of central importance. Yet Newton introduced a measure of force, momentum, that became part of the debates on conservation laws in mechanics.

One of the first entries in this debate was that of the Cartesian Christian Huyghens, who examined circular motion and the collisions of perfectly "hard" (perfectly elastic) bodies. He demonstrated that in such collisions, the sum of the magnitude of the bodies multiplied by the square of the velocity was a constant. However, for Huyghens, this expression was mathematically significant but not physically significant.

This pattern was repeated as 18th century continental mathematicians debated the measure of force and the quantities conserved in collisions. Gottfried Leibniz argued that Descartes's measure of force was insufficient. According to Leibniz, a body dropped a certain distance acquired just enough force to return to its original position, and this was a truer measure of force. Descartes had confused motive force with quantity of motion. Force should be measured by the effect it can produce, not by the velocity that a body can impress on another. In perfectly hard body collisions "vis viva," living force ($\frac{1}{2}mv^2$) was conserved. Central to subsequent debates were the experimental results of Giovanni Poleni and Wilhelm s'Gravesande, who measured the depths of the indentations left by metal balls dropped at different speeds into clay targets. If balls of different masses but the same size were dropped, the indentations remained the same. Lighter balls left the same indentations if they were dropped from heights proportional to 1/masses. s'Gravesande's experiments also showed that the force of a body in motion was proportional to v^2, that is, to its vis viva. He also did some elegant experiments in which hollow metal balls were filled with clay to investigate collisions of "soft" bodies. The results only fueled debates about the conditions over which measure of force was appropriate, mv or $\frac{1}{2}mv^2$.

s'Gravesande began his experiment to prove Newton's theories were correct, but the experiments seemed to Cartesians to support their point of view. Despite this evidence, s'Gravesande remained a Newtonian; for him, $\frac{1}{2}mv^2$ remained a number of no physical significance. Mathematicians in these debates over force included Madame de Chatelet, who published a French translation of Newton's *Principia* just 2 years after she had published *Institutiones* to spread Leibniz's ideas among French mathematicians. The subject of vis viva became a part of ongoing debates in the salons and Paris Academie des Sciences over Newtonian or Cartesian natural philosophy and the accompanying politics of each of those institutions.

Nothing was resolved, and mathematicians eventually moved on to other problems. They were inclined to use whatever physical principles seemed the most appropriate in their quest to develop the calculus using mechanical problems of increasing sophistication. Mechanics, or problems within it, became the instrument to display mathematical expertise and brilliance rather than explorations into physical problems. A perfect illustration of this is the work of Jean Louis Lagrange. By considering a four-dimensional space and using calculus, Lagrange reduced solid and fluid dynamics to algebra. He began with the fundamental axiom of virtual velocity for mass particles,

$$\sum m \frac{d\vec{v}}{dt} \cdot \delta\vec{r} = \sum \vec{F} \cdot \delta\vec{r},$$

where m is mass, v is velocity, $\delta\vec{r}$ is virtual displacement, and $\vec{F}$ is the applied force acting on a particle. Lagrange then introduced generalized coordinates, $q_1 \ldots q_n$, and generated the equations,

$$\frac{\partial T}{\partial q_i} - \frac{d}{dt}\left(\frac{\partial T}{\partial \dot{q}_i}\right) - \frac{\partial \Phi}{\partial q_i} = 0,$$

where T is the vis viva and Φ is a function. Thus, he succeeded in making rational mechanics as abstract, general, and algebraic as possible. No diagrams, geometry, or mechanical reasoning graced the pages of his text. It consisted "only of algebraic operations."

2. Caloric Theory and Heat Engines

During the 18th century, French engineers developed entirely different, yet mathematical, approaches to mechanics. Mathematicians had used problems on the construction of bridges, the construction of pillars, and the design of ships to explore the calculus. The education of French engineers began to include some calculus to solve engineering problems, especially in those schools of engineering within branches of the military. Here they developed an independent approach to the understanding of structures and the operations of machines.

Out of this tradition emerged Lazare Carnot, whose 1783 treatise on the efficiency of machines also began with the principle of virtual velocities. From this, he developed a relationship among the moment of activity, force multiplied by distance (work), and vis viva. He concluded that to get the maximum "effect" from the cycles of a machine, it was necessary to avoid inelastic impact in the working of its parts. During the 1820s, his son Sadi turned the same analytical approach toward the new "machine" that was revolutionizing the economy of Great Britain, the steam engine. As Carnot noted, removing this source of power would "ruin her prosperity" and "annihilate that colossal power." By this time, steam engines had been used in Britain for more than a century as pumps for mines. More recently, through the development by James Watt of the separable condenser and a governor, the steam engine became a new driving force for machinery. By the 1830s, the use of steam power and associated machinery was integrated into American and British cultures as a measure of the "civilized" states of societies.

Carnot wanted to establish the conditions under which maximum work could be produced by a heat engine. He stripped the heat engine down to its essential elements and operations. Through a series of examples, he demonstrated that successive expansions and contractions from heating and cooling—not heat alone—produced the motive power. Steam was not unique; any substance that expanded and contracted could be a source of power. Gases generally afforded the most expansion and contraction on heating and cooling. Carnot thought of heating and then cooling as the "reestablishment of equilibrium" in caloric and the substance of heat as the source of motive power. Therefore, he began to search for the general conditions that would produce the maximum motive power, and in principle, this could be applied to any heat engine.

Sadi Carnot used the same approach as his father and other French engineers in analyzing machines by looking at infinitesimal changes. In this case, the temperatures of the heat source and sink differed only slightly from that of the working substance. Carnot examined the changes in temperature and pressure of the working substance in a complete cycle of the engine, again a familiar approach in French engineering. Initially, the body is in contact with the heat source, and its piston rises and its volume expands at a constant temperature. It is then removed from this source, and the expansion continues without any heat input and the temperature of the working substance decreases to that of the heat sink. Placed in contact with the cooler body, the gas is compressed at constant temperature and then is removed and further compressed until the cycle is closed and the working substance attains the temperature of the heat source. The Carnot cycle is then complete. However, Carnot's analysis of its operation was in terms of caloric theory. In the first operation, the caloric is removed from the heat source to the working substance, and motive power is produced as the caloric attains the temperature of the heat sink. Equilibrium is restored as the caloric again reaches the temperature of the heat source. Using the incomplete image of a waterfall used to drive machinery, the caloric is conserved. It has fallen from one temperature to a lower one, and in this fall we can extract motive power just as we do from the gravitational fall of water.

The caloric theory of heat was accepted across Europe by the latter part of the 18th century. Experimental research into heat occupied many chemists across Europe and had an important place in the development of the reformed chemistry of Antoine Lavoisier. "Airs" (gases) were also the object of avid research throughout Europe by many chemists from before the flight of the Montgolfier brothers in 1784 through the first third of the 19th century. Caloric theory drew these two strands of research together. During the 1780s, Lavoisier had measured the specific heats of a number of substances with Pierre Simon de Laplace. These experiments were aspects of Lavoisier's research on airs and heat. In another series of such experiments, Lavoisier studied digestion and animal heat. Using an ice calorimeter, he demonstrated that respiration was a slow form of combustion and predicted heat outputs that agreed with his experimental results. Both Lavoisier and Laplace believed that heat was a material substance that manifested itself in two forms: free caloric (detected by thermometers) and latent caloric. Laplace and Simon Poisson developed caloric theory mathematically by assuming that the caloric, Q, in a gas was a function $Q = f(P, \rho, T)$, where P is the pressure, T is the temperature, and ρ is the density of the gas. Manipulating the partial differential equations in a theory of adiabatic compression,

this general equation led to relationships in accord with known experimental results such as the ratio of the specific heats of gases. Caloric was connected to the notion of work through Laplace's theory. Free heat was vis viva, combined (latent) heat was the loss of vis viva, and heat released was an increase in vis viva. Caloric theory was sophisticated and was taught at the Ecole Polytechnique when Sadi Carnot was a student, although toward the end of his short life he rejected it to explore the idea that heat was the motion of particles.

Carnot used the latest experimental results on the properties of gases to demonstrate his contentions and then discussed the limitations of heat engine design at the time. He concluded that because of gases' great expansion and contraction with heating and cooling, and because of the need for large differences in temperature to gain the most motive power, gases were the only feasible working substances for heat engines. He used current research on gases to demonstrate that any gas might be used.

Although Carnot's small treatise, *Reflections on the Motive Power of Heat*, received a good review from the *Revue Encyclopedique*, only one engineering professor at the Ecole Polytechnique developed Carnot's ideas. Emile Clapyeron gave geometrical and analytical expression to Carnot's analysis, creating the first pressure-volume (PV) diagram of a Carnot cycle. In his analysis, this cycle was reduced to an infinitesimally small parallelogram for the case pursued by Carnot. Again using the results from work on the properties of gases, Clapeyron obtained an expression for the "maximum quantity of action" developed in a cycle from the area within the parallelogram of the heat cycle for a perfect gas heat engine. Again, the analytical expressions were in terms of general expressions for heat, $Q = Q(P, V, T)$, $P = P(V, T)$, and $V = V(P, T)$. In the concluding paragraphs of his memoir, Clapeyron described the mechanical work available from his heat engine in terms of the vis viva the caloric developed in its passage from the boiler to the condenser of the engine. Many of his expressions were independent of any particular theory of the nature of heat and remained valid even as their physical explanation changed during the 1850s.

The works of both Carnot and Clapeyron sank into oblivion. Carnot died young and tragically in an asylum, but his oblivion was partially political. His father, Lazare, had enthusiastically embraced the French Revolution, was important in its early wars, and rose to prominence in Napoleon's regime. He remained loyal to Napoleon until the end and died in exile. The Carnot name (except that of his brother, a prominent politician), was politically tainted even into the 1830s. And however useful caloric might be for engineers, among scientists the caloric theory held less interest during this decade. They extended the wave theory of light to heat, and through it they explained other scientifically pressing problems such as the conduction and radiation of heat.

So deep was Carnot's oblivion that in 1845 William Thomson, a young ambitious Scottish academic, could find no copies of his treatise in Paris. Thomson had recently graduated Second Wrangler and first Smith's Prizeman from Cambridge University and was actively seeking the empty chair in natural philosophy at Glasgow University. The Glasgow that Thomson grew up in was Presbyterian and was the center of heavy industry (especially shipbuilding) and chemical engineering in Great Britain. Although his father was a distinguished professor of mathematics at Glasgow, the younger Thomson needed the support of the rest of the faculty and of the town representatives to ensure his election. He had to cultivate the image of the practical experimenter rather than flaunt his mathematical abilities, hence Thomson's sojourn at the laboratory of Henri Victor Regnault, well known for his empirical studies of steam engines. These were the circumstances that led Thomson to search out Carnot's work. Undaunted, Thomson made full use of Clapyeron's publications in a series of papers (published in 1847 and 1850) extending Clapyeron's analysis. In his first paper he postulated an absolute zero of temperature implied by Clapyeron's analysis and put forward the notion of an absolute scale of temperature. Both emerged from the mathematical forms independent of their physical interpretation. Thomson's brother, James, noted that one of the results of Clapyeron's work was that the freezing point of water decreased with an increase in pressure. William Thomson confirmed this through experiments done at Glasgow. The position of caloric theory seemed to be becoming firmer once again. However, in the middle of this theoretical and experimental triumph, Thomson went to the 1847 annual meeting of the British Association for the Advancement of Science (BAAS), where he heard a paper given by James Joule on experiments whose results contradicted caloric theory.

3. Equivalence of Heat and Work

Joule was the son of a very successful brewer in Manchester, England. He was educated privately and with ambitions to spend his life in science. His first papers dealt with "electromagnetic engines," and in the course of his experiments he focused on what we now call Joulean heating in wires carrying electric currents. This drew attention to the production of heat in both physical and chemical processes. From his experiments and a developing theory, he concluded that an electric generator enables one to convert mechanical power into heat through the electric currents induced into it. Joule also implied that an electromagnetic "engine"

producing mechanical power would also convert heat into mechanical power. By 1843, he presented his first paper to the British Association for the Advancement of Science (BAAS) on the "mechanical value of heat." Joule believed that only the Creator could destroy or create and that the great agents of nature, such as electricity, magnetism, and heat, were indestructible yet transformable into each other. As his experiments progressed, he elaborated his mechanical theory of nature. Important to observers, and to Thomson, was Joule's idea that mechanical work and electricity could be converted into heat and that the ratio between mechanical work and heat could be measured. In a series of papers on the mechanical equivalent of heat presented annually to BAAS and politely received, Joule labored alone until the stark reactions of Thomson in 1847. Some scientists could not believe Joule's contentions because they were based on such small temperature changes. Presumably, their thermometers were not that accurate. However, Thomson understood that Joule was challenging caloric theory head on. Until 1850, Thomson continued publishing papers within a caloric framework and an account of Carnot's theory. He had finally located a copy of the treatise. Clearly, he began to doubt its validity, noting that more work on the experimental foundations of the theory were "urgent," especially claims that heat was a substance. It took Joule 3 years to convince Thomson of the correctness of his insight. Only in 1850 did Thomson concede that heat was convertible into work after he read Rudoph Julius Emmanuel Clausius's papers and those of William McQuorne Rankine on heat. He then reinterpreted some parts of his caloric papers that depended on heat as a substance in terms of the interconversion of heat and work. Also, he rederived and reinterpreted his expression for the maximum amount of work that could be extracted from the cycle of a perfect heat engine and other results and then gave his version of the second law of thermodynamics.

4. Conservation of Energy

Others had come to conclusions similar to those of Joule before Thomson. One of the most notable was Hermann von Helmholtz. Trained as a physician while studying as much mathematics and physics as possible, his doctoral dissertation was a thorough microscopic study of the nervous systems of invertebrates. His mentor was Hermann Mueller, who (along with Helmholtz's fellow students) was determined to erase vitalism from physiology. Their goal was to make physiology an exact experimental science and to base its explanatory structures in physics and chemistry. Helmholtz managed to continue his physiological research while a physician in the Prussian army, where he was dependent on his army postings and his early research on fermentation and putrefaction was cut short by the lack of facilities. He turned to a study of animal heat and the heat generated by muscular action and found that during contraction chemical changes occurred in working muscles that also generated heat. In the midst of this research, Helmholtz published "On the Conservation of Force." This paper displayed the extent of his mechanical vision of life and his commitment to a Kantian foundation for his physics. Central forces bound together the mass points of his theory of matter. These commitments resurfaced during the 1880s with his work on the principle of least action.

With considerable pressure brought to bear on the ministry of education and the Prussian military, Helmholtz was released from his army obligations and entered German academic life. He never practiced medicine again, yet his research over the subsequent two decades began with questions in physiology that expanded into studies on sound, hearing, and vision. His work on nerve impulses in 1848 developed into a study of electrical pulses and then into a study of the problem of induction before growing into a critical overview of the state of the theory of electricity during the 1870s.

The closeness of Helmholtz's work in physics and physiology is evident in his paper on the conservation of force. He had actually stated the principle earlier (in 1845) in a paper reviewing the work of Humphry Davy and Antoine Lavoisier on animal heat. He concluded that if one accepts the "motion theory of heat," mechanical, electrical, and chemical forces not only were equivalent but also could be transformed one into each other. This assumption bound together results already known from chemistry, electricity, and electromagnetism. The structure of this review formed the outline of Helmholtz's 1847 paper. He added a theory of matter that he believed demonstrated the universality of the principle of conservation. Helmholtz began in metaphysical principles from which he extracted specific physical results. In his model, force was either attractive or repulsive. Change in a closed system was measured by changes in vis viva expressed as changes in the "intensity of the force," that is, the potential of the forces that acted between the particles, $\frac{1}{2}mv_1^2 - \frac{1}{2}mv_2^2 = \int_r^R \varphi dr$, where m is the mass of the particle, whose velocity changes from v_1 to v_2, and ϕ is the intensity of the force, whose measure is constructed by looking at changes in v^2. Because both velocities and forces were functions of the coordinates, Helmholtz expressed these last changes as $\frac{1}{2}md(v^2) = Xdx + Ydy + Zdz$, where $X = (x/r)\phi$. He demonstrated that for central forces, vis viva was conserved and, thus, so was the intensity of the force. He then expanded the argument to a closed system of such forces. The unifying concept in the paper was "tensional force" and its intensity ϕ. This allowed Helmholtz to extend his argument to electricity to obtain the "force

equivalent" of electrical processes. In this case, the change in vis viva of two charges moving from distance r to R apart was $\int_1^2 \frac{1}{2} md(v^2) = -\int_r^R \varphi dr$, where he identified ϕ with Gauss's potential function, to which Helmholtz gave physical meaning. However, he did not give this entity a new name. He was reinterpreting physically a well-known approach used when mathematizing a physical problem. Unlike mathematicians, Helmholtz drew important physical conclusions from his equations, including the Joulean heating effect. He also included Franz Neumann's work on electrical induction in his conceptual net. Helmholtz used examples from chemistry and electricity to argue that heat was a measure of the vis viva of thermal motions, not a substance. He used a broad range of experimental results across the sciences to illustrate the plausibility of his contentions.

Helmholtz had difficulty in getting these speculations into print. They eventually appeared as a pamphlet. The thrust of the speculations went against prevailing standards in the emerging field of physics. German physicists had rejected grand speculative visions of nature and its operations for precision work in the laboratory, from which they assumed theoretical images would emerge. The model of this new approach was Wilhelm Weber. In Weber's case, precise, painstaking numerical results, analyzed in detail for possible sources of error, led to a center-of-force model that explained the range of electrical phenomena he had researched experimentally. Neumann was known for his careful experiments on crystals and their optical properties. Helmholtz presented his mechanical model before his analysis of the experimental work of others that ranged across disciplinary boundaries. This approach also dissolved the recently hard-won academic and specialist designation of physics in German academia.

Helmholtz was not the only German claimant to the "discovery" of the principle of conservation of energy. After his graduation in 1840, Julius Robert Mayer, a physician, became a doctor on a ship bound for the East Indies. Mayer noted the deeper color of venous blood in the tropics compared with that in Northern Europe. He interpreted this as more oxygen in the blood, with the body using less oxygen in the tropics because it required less heat. He speculated that muscular force, heat, and the chemical oxidation of food in the body were interconvertible forces. They were not permanent but were transformable into each other. On his return, Mayer learned physics to demonstrate the truth of this conviction and then set about to calculate the mechanical value of a unit of heat. To do this, he looked at the works on the thermal properties of gases and took the difference between the two principal specific heats as the amount of heat required to expand the gas against atmospheric pressure. This heat was transformed into work, and from this he computed the conversion factor between heat and work. Further memoirs followed, extending his use of available experimental results from the chemistry and physics of gases. Mayer even included the solar system. He argued that meteorites falling under gravity built up enormous amounts of mechanical force that yielded considerable amounts of heat on impact. This could account for the sun's seemingly inexhaustible supply of heat.

Mayer also encountered difficulties in publishing his work. His first paper of 1842, "Remarks on the Forces of Inorganic Nature," was published in a chemistry journal rather than a physics journal. As with Helmholtz, the combination of metaphysics and calculations derived from other people's experiments was not acceptable in the physics community in the German states. In addition, Mayer did not have the academic credentials or connections of Helmholtz. In fact, Mayer had to underwrite the publication of his later papers. At the end of the decade, Mayer suffered personal tragedy and in 1850 tried to take his own life. He was confined to a series of mental institutions. However, from 1858 onward, he began to gain recognition for his work from his colleagues, including Helmholtz, and he received the Copley Medal from the Royal Society in 1871.

So far, there were several approaches leading to the statement of the first—and even the second—law of thermodynamics. None of these claimants replaced Carnot's and Clapeyron's work mathematically from a thermodynamic point of view except the work of Clausius. In his 1850 paper, Clausius began from the observation that the conduction of heat is possible only from a hotter to a colder body. This led him logically to replace Carnot's principle, that heat is conserved in the cycle of a perfect heat engine, to that of the equivalence of heat and work. As a consequence, when work is done, heat is consumed in any heat engine. Although he kept many of the mathematical forms developed by Clapeyron, Clausius assumed that the mathematical characteristics of the functions that entered into his equations determined the physical characteristics of his system. In analyzing the Carnot cycle, Clausius chose the perfect gas as his physical system and looked at the state of the gas as it traversed an infinitely small cycle. Following Clapeyron, he constructed the heat added or expended for each leg of the cycle and constructed an expression for the inverse of the mechanical equivalent of heat, A, as the ratio of the heat expended over the work produced. To construct an expression for the work done in completing the cycle, Clausius calculated the expressions for the heat added at each leg in the cycle and constructed

$$\frac{d}{dt}\left(\frac{dQ}{dV}\right) - \frac{d}{dV}\left(\frac{dQ}{dt}\right) = \frac{A \cdot R}{V},$$

where dQ is the heat added, V is the volume of the gas, and R is the gas constant. Arguing mathematically

that Q could not be a function of A and t so long as they were independent of each other, the right-hand side of the equation must be zero. This brought the equation into the form of a complete differential,

$$dQ = dU + A \cdot R\left(\frac{a+t}{V}\right)dV,$$

where $a + t =$ the absolute temperature and $U = U(V, t)$, is the heat necessary for internal work, and depends only on the initial and final states of the gas. The second term on the right-hand side, the work done, depended on the path taken between the initial and final states of the gas. Clausius then introduced $1/T$, where T is the temperature that, as a multiplier of dQ, makes a complete differential of the form $Xdx + Ydy$. He applied this combination of physics and mathematics to the behavior of gases and saturated vapors, with the latter being of great interest due to the engineering importance of steam engines during the mid-19th century. In this analysis, he restated the results, some mathematically equivalent to those of Clapeyron and Thomson, but interpreted them in terms of the equivalence of heat and work. At the end of this first paper, he gave an initial statement of the second law of thermodynamics.

Therefore, there are several different paths and claimants to some statements equivalent to the first law of thermodynamics. During the decades immediately following 1850, many others were championed as "discoverers" of the principle of the conservation of energy among the engineers and scientists working on related problems. These claims came in the wake of a change in language from the equivalence of heat and work to the conservation of energy. Some of the assertions gave rise to bitter complaints, especially the contentions of Peter Guthrie Tait's, whose *Thermodynamics* trumpeted Joule as the "discoverer" of the equivalence of heat and work yet readily recognized Helmholtz's priority in the matter of the first law. Mayer's work was recognized belatedly by Helmholtz and Rankine against Joule's claims. Mayer published an estimate of the mechanical equivalent of heat a year before the appearance of Joule's first value. And so it went. A great deal was at stake in these disagreements, which reflected on industrial prowess as well as on intellectual authority in the growing political and industrial rivalry between Germany and Great Britain. Using hindsight, later historians also tried to claim some of the credit for other Europeans.

5. Energy Physics

A decade before the development of the mechanical meanings of the conservation of energy, Thomson, Clausius, and Helmholtz had extended its reach into electrostatics, thermoelectricity, the theory of electromagnetism, and current electricity. During the 1860s, one of the more important applications of this new approach was the establishment of an absolute system of electrical units. Again this involved national prestige given that the Germans already had their own system based on the measurements of Weber. The BAAS committee included Thomson, James Clerk Maxwell, and the young electrical engineer Fleming Jenkin. Maxwell and Jenkin measured resistance in absolute (mechanical) units and deduced the others from there.

So far, what existed were disparate statements and uses for energy, none of which shared a common language or set of meanings. Thomson and Tait created that language for energy in their 1868 text-book on mechanics, *A Treatise on Natural Philosophy*. The text was directed to the needs of their many engineering students. During the 19th century, numerous new technical terms were coined, many of which derived from ancient Greek, and energy was no exception. Derived from the Greek word ενεργεια, energy was commonly used for people who demonstrated a high level of mental or physical activity. Also, it expressed activity or working, a vigorous mode of expression in writing or speech, a natural power, or the capacity to display some power. More technically, Thomson and Tait used energy as the foundation of their 1868 text. They claimed that dynamics was contained in the "law of energy" that included a term for the energy lost through friction. They introduced and defined kinetic and potential energy and even claimed that conservation of energy was hidden in the work of Newton. William Whewell grumbled about the book's philosophical foundations, but it was translated into German as well as being widely used in Great Britain.

Conservation of energy was becoming the foundation for teaching physics as well as for new approaches in research in the physical sciences. Thomson considered that conservation of energy justified a purely mechanical vision of nature and tried to reduce all phenomena, including electromagnetic phenomena, to purely mechanical terms. In his early papers in electromagnetism, Maxwell used conservation of energy in constructing and analyzing mechanical models as mechanical analogues to electrical and magnetic phenomena. He then abandoned this mechanical scaffolding and established his results on general energetic grounds, including his electromagnetic theory of light. In defense of his approach to electromagnetism, Maxwell criticized Weber's action-at-a-distance approach as being incompatible with the conservation of energy. In 1870, Helmholtz began a systematic critique of Weber's electrical theory and repeated Maxwell's judgment that he located in Weber's use of velocity-dependent forces. Helmholtz had first stated this criticism in his 1847 paper on the conservation of force, and it became a starting point for alternative approaches to

electromagnetism among German physicists. These included Neumann and Bernard Riemann, who developed the notion of a retarded potential. By 1870, this criticism undermined the validity of his work and Weber was forced to answer. He demonstrated that, indeed, his theory was compatible with conservation of energy.

Clausius was the first to try to develop an expression for the laws of thermodynamics in terms of the motions of the molecules of gases. His models were mechanical, whereas the later ones of Maxwell and Ludwig Boltzmann were statistical. Maxwell became more interested in the properties of gases, such as viscosity, heat conduction, and diffusion, relating them to statistical expressions of energy and other molecular properties. Boltzmann developed a statistical expression for the first and second laws of thermodynamics. During the late 19th century, thermodynamics itself was subject to great mathematical manipulation, which in some hands became so remote from physics that it once again became mathematics.

By the time of Maxwell's death in 1879, the conservation of energy had become a principle by which physical theories were judged and developed. This was particularly true in Great Britain, where "energy physics" propelled the theoretical work of most physicists. Much of this work, such as that of Oliver Lodge, was directed toward the development of Maxwell's electromagnetism, although the work was expressed as mechanical models. Lodge also drew up a classification of energy in its varied physical forms, from the level of atoms to that of the ether. He also used an economic image for energy, equating it with "capital." He extolled John Henry Poynting's reworking of Maxwell's *Treatise* in 1884 and saw it as a new form of the law of conservation of energy. In his treatment of the energy of the electromagnetic field, Poynting studied the paths energy took through the ether. He investigated the rate at which energy entered into a region of space, where it was stored in the field or dissipated as heat, and found that they depended only on the value of the electric and magnetic forces at the boundaries of the space. This work was duplicated by Oliver Heaviside, who then set about reworking Maxwell's equations into their still familiar form. The focus was now firmly on the "energetics" of the electromagnetic field.

The emphasis on energy stored in the ether separated British interpretations of Maxwell's work from those on the continent, including that of Heinrich Hertz. However, energy in the German context became the foundation for a new philosophy of and approach to the physical sciences, energetics. Initiated by Ernst Mach, energetics found its most ardent scientific advocate in Wilhelm Ostwald. The scientific outcome of Ostwald's work was the development of physical chemistry. Yet his dismissal of atoms led to clashes with Boltzmann, and during the 1890s his pronouncements drew the criticism of Max Planck. Planck had, since his dissertation during the 1860s, explored the meaning of entropy, carefully avoiding atoms and molecules. He was forced into Boltzmann's camp in his research on black body radiation, essentially a problem of the thermodynamics of electromagnetic radiation. Planck was not alone in understanding the importance of the problem; however, other theorists were tracking the energy, whereas Planck followed the entropy changes in the radiation.

6. Energy in 20th Century Physics

As an editor of the prestigious journal *Annalen der Physik*, Planck encouraged the early work of Albert Einstein, not only his papers in statistical mechanics but also those on the theory of special relativity. Planck then published his own papers on relativistic thermodynamics, even before Einstein's 1907 paper on the loss of energy of a radiating body. Two years before this, Einstein had investigated the dependence of the inertia of body on its energy content. In this case, the mass of the body in the moving system decreased by E/c^2, where E is the energy emitted as a plane wave and c is the velocity of light. Einstein concluded that "the mass of a body is a measure of its energy content." Einstein constructed his relativistic thermodynamics of 1907 largely on the foundation of Planck's work and reached the conclusion that $dQ = dQ_0 \sqrt{(1 - v^2/c^2)}$, where dQ is the heat supplied to the moving system, dQ_0 is that for the system at rest, and v is the velocity of the moving system to that at rest.

Thus, the law of conservation of energy was integrated into relativity theory. However, its passage through the quantum revolution was not as smooth. In 1923, Niels Bohr expressed doubts about its validity in the cases of the interaction of radiation and matter. This was after Arthur Holly Compton explained the increase in the wavelengths of scattered X rays using the conservation of energy and momentum but treating the X rays as particles. The focus of Compton's research, the interaction of radiation and atoms, was shared with many groups, including that around Bohr in Copenhagen, Denmark. Bohr did not accept Einstein's particle theory of light, even with the evidence of the Compton effect. By using the idea of "virtual oscillators" that carried no momentum or energy, Bohr, H. A. Kramers, and John Slater (BKS) hoped to maintain a wave theory of the interaction of radiation and matter. To do so, they had to abandon the idea of causality, the connection between events in distant atoms, and the conservation of energy and momentum in individual interactions. Struggling with the meanings of the "BKS theory" became crucial in

the development of Werner Karl Heisenberg's formulation of quantum mechanics. However, the BKS theory was bitterly opposed by Einstein as well as by Wolfgang Pauli, who, after a brief flirtation with the theory, demonstrated its impossibility using relativistic arguments.

Conservation laws were reinstated at the center of physical theory. Conservation of energy became the criterion for abandoning otherwise promising theoretical approaches during the late 1920s and early 1930s in quantum field and nuclear theory. In the latter case, Pauli posited the existence of a massless particle, the neutrino, to explain beta decay. The neutrino remained elusive until the experiments of Ray Cowan and Fred Reines during the 1950s. Conservation of energy also governed the analysis of the collisions of nuclear particles and the identification of new ones, whether from cosmic rays, scattering, or particle decay experiments. Relativistic considerations were important in understanding the mass defects of nuclei throughout the atomic table. The difference between the measured masses and those calculated from the sum of the masses of their constituent particles was explained as the mass equivalence of the energy necessary to bind those components together.

7. Energy Conservation Beyond Physics

Beyond physics, conservation of energy became a foundation for other sciences, including chemistry and meteorology, although in this case the second law of thermodynamics was more powerful in explaining vertical meteorological phenomena. In the natural sciences, the most dramatic demonstration of the conservation of energy's usefulness was in the study of animal heat and human digestion. The production of heat as synonymous with life can be seen in the aphorisms of Hippocrates during the fifth century BCE. In Galen's physiology, developed during the second century AD, the source of bodily heat was the heart. Heat was carried in the arterial system throughout the body, where it was dissipated through the lungs and the extremities. Aspects of Galen's physiology remained in William Harvey's explanation of the function of the heart. Yet Harvey's experiments, published in 1628 as *Anatomical Experiments on the Motion of the Heart and Blood in Animals*, demonstrated the circulation of the blood and the pumping action of the heart. This mechanical function of the heart became part of Descartes's mechanical philosophy that changed debates over the source of heat in the body. Was it generated in the heart or from friction as the blood flowed through the vessels of the body? Harvey's argument that the lungs dissipated the heat from the body fitted into this mechanical vision of the body.

During the 18th century, Descartes's mechanistic vision of the body was replaced by vitalism based on chemical experiments. This was the vitalism that Helmholtz sought to root out from medicine during the 1840s. The experiments were those performed on "airs" consumed and expended by plants that vitiated or revived the atmosphere. Combined with these experiments were others showing that combustion and respiration made the atmosphere less able to sustain life, usually of birds and small mammals. After the introduction of Lavoisier's reform of chemical nomenclature, which also carried with it a new theory of chemistry and chemical reactions, experiments on animal chemistry became more quantitative and lay within his new theoretical framework. Treadmill studies by Edward Smith in 1857 demonstrated that muscles derived energy, measured by heat output, from foods not containing nitrogen. However, Justus Leibig's idea that bodily energy derived from muscle degradation was not easily deposed. It took the mountain-climbing experiments of Adolph Fick and Johannes Wislicenus in 1866 and the analysis of them by Edward Frankland to finally displace Leibig's theory. Frank-land pointed out that Fick and Wislicenus's results could be interpreted by looking at the energy involved in combustion in a unit of muscle tissue and considering the work output the equivalent of this energy. Conservation of energy was necessary in understanding the dynamics of the human body. Frankland also showed that the energy the body derived from protein equaled the difference between the total heat of combustion and the heat of combustion of the urea expelled, with the urea being the pathway for the body to rid itself of the nitrogen in proteins. Thus, conservation of energy entered the realm of nutrition, biophysics, and biochemistry. In later experimental studies, W. O Atwater measured, by a whole body calorimeter, that the heat given off by the human body equaled the heat released by the metabolism of the food within it. Even when the complexities of the pathways of amino acids and other sources of nitrogen were understood, along with their routes through the digestive system, conservation of energy remained the foundation for interpreting increasingly complex results.

Thus, conservation of energy has proved to be crucial for analyzing the results of experiments across—and in some cases beyond—the scientific spectrum. There have even been experiments—so far unsuccessful—aimed at measuring the mass of the soul on its departure from the body.

Further Reading

Adas M 1989 *Machines as the Measure of Man*. Cornell University Press, Ithaca, NY

Caneva K L 1993 *Robert Mayer and the Conservation of Energy*. Princeton University Press, Princeton, NJ

Cardwell D 1971 *From Watt to Clausius*. Cornell University Press, Ithaca, NY

Carpenter K J 1994 *Protein and Energy: The Study of Changing Ideas in Nutrition*. Cambridge University Press, Cambridge, UK

Cassidy D 1991 *Uncertainty: The Life and Times of Werner Heisenberg*. W H Freeman, San Francisco

Cooper N G, Geoffrey W B (eds.) 1988 *Particle Physics: A Los Alamos Primer*. Cambridge University Press, Cambridge, UK

Einstein A 1905 Does the inertia of a body depend upon its energy content? In: Beck Trans A (ed.) *The Collected Papers of Albert Einstein*, vol. 2. Princeton University Press, Princeton, NJ

Einstein A 1907 On the relativity principle and the conclusions drawn from it. In: Beck Trans A (ed.) *The Collected Papers of Albert Einstein*, vol. 2. Princeton University Press, Princeton, NJ

Everitt C W F 1975 *James Clerk Maxwell: Physicist and Natural Philosopher*. Scribner, New York

Helmholtz H 1853 On the conservation of force. In: Tyndall J, Francis W (eds.) *Scientific Memoirs*. Taylor & Francis, London. (Original work published 1847.)

Smith C 1998 *The Science of Energy: A Cultural History of Energy Physics in Victorian Britain*. University of Chicago Press, Chicago

Smith C, Norton Wise M 1989 *Energy and Empire: A Biographical Study of Lord Kelvin*. Cambridge University Press, Cambridge, UK

Elizabeth Garber

E

Early Industrial World, Energy Flow in

Glossary

agricultural revolution The period prior to England's industrial revolution, when significant improvements in agricultural production were achieved through changes in free labor, land reform, technological innovation, and proto-industrial output.

capitalism A socioeconomic and political system in which private investors, rather than states, control trade and industry for profit.

industry Economic activity concerned with processing raw materials and producing large volumes of consistently manufactured goods, with maximum fuel exploitation.

industrial revolution The period of technological innovation and mechanized development that gathered momentum in 18th-century Britain; fuel shortages and rising population pressures, with nation-states growing in power and competing for energy, led to dramatic improvements in energy technology, manufacture, and agriculture and to massive increases in the volume of international trade.

plague Referred to as the Black Death, or Bubonic Plague, which killed one-third the population of England and Europe from the mid-14th century until around 1500; millions of people perished, and commerce, government, energy production, and industry waned as farms were abandoned and cities emptied.

proto-industry Supplemented agricultural income via the production of handcrafted work; this piecework was especially common in the creation of textiles by spinning, weaving, and sewing.

steel A strong hard metal that is an alloy of iron with carbon and other elements; it is used as a structural material and in manufacturing.

work energy Defined by natural scientist James Joule as $kg\,m^2/s^2$ or a joule (J); it enabled physicists to realize that heat and mechanical energy are convertible.

Population growth and competition for resources have continuously been an impetus for the control and use of energy. In harnessing energy, inventions evolved from previous innovations, ebbing and flowing with changes in population pressures and the availability of resources—necessity shaping history. For millennia, societies depended on resources provided by solar energy (radiation stored as vegetation), wind (a type of solar energy), animal and human energy, water, and fossil fuels. The advent of mechanized labor and its synchronized production demands transformed the lives of workers previously tuned to seasonal energy rhythms. Technological innovations during the early 18th century resulted in enormous growth in the economic wealth and sociopolitical power of nation-states. Sustained growth in Western Europe and Britain spread via exploration and trade. The competition to colonize and control the raw natural resources in the Caribbean and the Americas, in conjunction with imperialism in Indochina, the Mideast, and the Mediterranean, changed the global energy map.

1. Agriculture and Urbanization

Preindustrial economies in Britain and Europe were widespread in scale, dominated by agricultural production, processing, and manufacture, with populations based primarily in rural villages. A range of processed goods were essential to farms and households, although restricted incomes reduced purchasing power. Solar energy, water, regional fuels, and the ebb and flow of the seasons dictated the energy of labor expenditure in workers' lives.

There was little capital accumulation or technological progress during the widespread population destruction of the Black Death in medieval Britain and Europe. In Italy, Spain, France, and Poland, agricultural output per worker diminished, and non-agricultural employment tended to be based in rural areas, corresponding with only minor increases in urban populations. The subsequent recovery of populations after the plague led to the growth of urbanization, and local and national government structures increased their control of energy resources.

The Netherlands, Belgium, and England were distinguished from the rest of Europe by the inter-related factors of rising agricultural productivity, high wages, and urbanization. At the end of the medieval period, rural industry and urban economies expanded. Despite rates of population growth exceeding those elsewhere in Europe, economic development offset population growth, wages were increased or maintained.

By 1750, Britain had the highest portion of population growth in cities, and agricultural output per worker surpassed Belgian and Dutch levels. Urban and rural nonagricultural populations grew more rapidly in England than anywhere else in Europe. The share of the workforce in agriculture dropped by nearly half, while the trend of providing real wages rose.

The wool and cotton textile industry was the first widespread manufacturing enterprise in Britain. When farm laborers began producing piecework for cash, the social and labor organization of agrarian economies changed dramatically. Labor–market interactions and productivity growth in agriculture required a degree of labor mobility that was incompatible with previous agricultural economic models such as serfdom. England's division between landlords and tenant farmers caused reorganization to occur through reductions in farm employment, enclosure, and farm amalgamation. In Europe, where landowners/occupiers were more prevalent, farm amalgamation and enclosure were disruptive to productivity and institutional change was less profitable.

Rural industrialization was a response to population growth combined with increased urban demands for farm products, and sustained urbanization created dependency on labor based in the countryside. Substantial migrations of people to cities, and the spread of industry to the countryside, provided opportunities to escape the restrictions of guilds.

Providing enough farm produce for cities required increasing agricultural output and value per acre and introducing new crops. By the dawn of the 18th century, the industrial and commercial sectors of the British economy contributed at least one-third of the gross national product.

2. Competition for Energy Resources

For millennia, the Mediterranean had been the center of international trade, with trade routes among Europe, Africa, and Asia. After the plague during the 15th and 16th centuries, Mediterranean networks of international trade with Europe declined. European nation-states were not powerful enough to govern beyond their transient borders, and inadequate local resources could not support population growth. This led to each state employing protective measures to stimulate domestic industrial production, self-sufficiency, and competition for resources. During the 16th and 17th centuries, these policies were formalized, extended, and consolidated, severely limiting the growth of trade.

The high costs of inland transportation and restrictive national practices inhibited trade opportunities and led to an economic crisis. As trading opportunities ceased to expand after the plague's decimations, the need for new sources of revenue provided the impetus to search far afield for resources. With the diminishing extent of accessible markets, Europeans explored the Atlantic, Africa, the Caribbean, and the Americas, where raw resources and cheap commodity production offered immense opportunities for trade expansion. During the 15th century, the Portuguese established trading posts along the west coast region of Africa, produced sugar on slave-worked plantations, and traded gold.

The advent of the Atlantic world trading system and its abundant resources extended the production and consumption frontier of Western Europe. Because available transportation of goods was slow and expensive, unit cost of production in the Americas had to be sufficiently low for commodities to bear the cost of trans-Atlantic transportation. Large-scale production in the Americas increasingly depended on coerced labor. Indigenous peoples of the Americas were forced into slavery to mine silver in the Spanish colonies and provision Europeans. Native American land was appropriated, and populations were decimated by an unwelcome import, Old World diseases. Colonizers obtained vast quantities of acreage, becoming land rich beyond the limits of European caste and class systems. The average population density in the Americas was less than one person per square mile during the 17th century.

As indigenous peoples in the colonies suffered disease, displacement, and slavery, the production of commodities for Atlantic commerce became more dependent on slaves from Africa. Labor costs on plantations throughout the Caribbean and the Americas were below subsistence costs, enabling large levels of production and maximum exploitation of human energy resources. From Barbados to the Carolinas and the Mississippi delta, a major source of power were the thousands of enslaved Africans and Native Americans as well as indentured servants from England, Scotland, and Ireland. Plantations prospered due to the high value of cotton in the world market and by producing coffee, sugar, cocoa, tobacco, and rum. As these export commodities increased in volume, prices fell in Europe, with the commodities becoming commonly consumed goods rather than rarified luxuries. The production of raw materials for mass markets contributed greatly to the development and economic homogenization of industry and energy exploitation.

England combined naval power and commercial development to secure choice Atlantic territories and create advantageous treaties to control the colonies' resources. The majority of commodity shares, production, and trade of the slave-based economy of the Atlantic world trading system were primarily British controlled.

The estimated percentage share of these commodities produced by enslaved Africans and indentured servants in the Americas was 50 to 80%. Between 1650 and 1850, the slave-based economy of the Atlantic system transformed international trade, transformed European and British economics, and spurred technological innovation. Slavery failed as an energy source due to the socioeconomic upheavals that eventually prevented and censured its expansion as well as the fact that it could not compete with developing sources of industrial energy such as steam power. Slave labor, rather than free labor, failed as a socioeconomic source of power.

In per capita terms, the exposure of England's economy and society to the development of the Atlantic world market was greater than that of any other country, although Europe profited immensely from participation in the slave-based Atlantic world economy. Newly acquired gold and silver from the Americas bolstered weak European economies. Imported textiles from England were welcomed in the Americas, with these sustained sales creating employment in England's manufacturing regions and stimulating population growth and changes in agrarian social structures. Combined with export demands, this created a ripe environment for transforming the organization and technology of manufacturing in domestic and export industries. This international economic energy propelled the technological industrial revolution.

The transition of intensified industrialization depended on improvements in transportation systems to provide dispersion networks of fuels and raw materials. With the advent of the Victorian empire, improvements in the infrastructure of transportation (e.g., roads, canals, shipping) spread. The construction of railroad systems in Britain, Europe, Asia, Russia, Africa, and the Americas connected, amalgamated, and exploited energy resources in most regions of the world, creating an engine of economic dynamism that itself was a product of mechanized industry.

3. Wood Energy

Woodlands have been an essential resource for constructing homes, vehicles, ships, furniture, tools, and machines as well as for providing heat for households and cooking. Poor management of woodlands caused low productivity, and as timber stands became more remote, logging and transporting timber became less economical. During the 13th century in England, lime kilns devoured hundreds of ancient oaks in Wellington Forest. By the 1540s, the salt industry was forced to search far

afield for the wood used in its processing furnaces. British navel power judged this shortage to be a national security threat. Authorities attempted to impose stiff fines for timber poaching and forbade bakers, brewers, and tile makers from purchasing boatloads of wood.

Severe timber shortages reached national crisis proportions in Britain, impelling the import of expensive timber from the Baltic and North America. Native turf, peat, reeds, gorse, and other plant fuels simply could not provide adequate supplementation. The climate did not cooperate. Between around 1450 and 1850, winters became unusually long and harsh, the Thames river froze, and diseases and cold increased in their killing power. Populations suffered as low birth rates could not keep up with expanding death rates.

In Scotland, woodlands were carefully managed during the early medieval period, and native oak supplied the needs of developing burghs. Norse incursions pushed populations inland, allowing woodland regeneration in coastal areas. During the 15th century, the quality and quantity of native oak dwindled, and alternative sources (e.g., imported oak, conifers) were increasingly exploited. France had abundant internal supplies of timber, so there was no need to invest time, money, and politico–military resources in obtaining timber from elsewhere.

By the close of the 17th century, industry and manufacturing devoured more than a third of all the fuel burned in Britain. The textile industry's immense scale made it a leading consumer of fuels, although fuel cost was not a major constituent in manufacturing textiles, dyeing, calendaring, and bleaching. Energy consumption by industries increased in local and national terms because industry was not uniformly distributed, and this exacerbated and caused severe local and regional scarcity. Local needs concerning brewing, baking, smithing, potting, and lime processing had survived on whatever fuel was available. High temperatures were required to brew ale and beer and to process salt and sugar. The construction industry depended on kilns and furnaces for the production of bricks, tiles, glass, firing pottery, and burning lime. Smelting and working metals, and refining the base compounds of alum, copperas, saltpeter, and starch, were expensive processes.

Even when located within extensive woodlands, levels of consumption by iron forges, glass factories, and lime kilns led to shortages. Fuel needs grew voraciously as industries expanded in scale and became more centralized and urbanized. There was a fine balance between fuel needs in communities and industrial demands. Urban centers consisting of large concentrations of consumers inherently tended to outgrow the capacity of adjacent woodland fuel supplies. Transport costs commensurately inflated as supply lines were extended.

4. Wind and Water Energy

Windmills were developed in Persia by the 9th century BCE. This technological innovation was spurred by the need to mill corn in areas lacking consistent water supplies. Early windmills used an upright shaft, rather than a horizontal one, to hold the blades. This system was housed in a vertical adobe tunnel with flues to catch the wind (similar to a revolving door). The concept of harnessing wind energy via windmills reached Britain and Europe by 1137, where it underwent significant changes, with horizontal (rather than vertical) shafts and horizontal rotation. A form of solar energy, climate-dependent wind power has been unreliable and difficult to accumulate and store.

Roman aqueducts provided a system of fast-flowing water to watermills. By the 1st century BCE, Romans used an efficient horizontal axis where the waterwheel converted rotary motion around a horizontal axis into motion around a vertical axis; water passed under the wheel, and kinetic water energy turned the wheel. Draft animals and water-power were harnessed to turn millstones. Rudimentary primitive watermills used vertical axes, with millstones mounted directly to waterwheel shafts. In mountainous regions, water was channeled through a chute into the wheel. Although power output was minimal ($\sim 300\,W$), this type of watermill was used in Europe until the late medieval period.

During the late 18th and early 19th centuries, waterwheel efficiency was improved. English engineer John Smeaton, expanding on the systems of early Roman waterwheels, found that more energy was generated when water was delivered to the top of the wheel via a chute. An overshot waterwheel used both kinetic and potential water energy and was capable of harnessing 63% more potential energy than was the undershot. The growing demand for energy intensified use of the waterwheel. While toiling in the gold mines of California during the 1870s, a British mining engineer, Lester Pelton, discovered that a waterwheel could be made more powerful if high-pressure jets of water were directed into hemispherical cups placed around its circumference. This waterwheel design, generating 500 horsepower (hp), was used 20 years later in the sodden mines of Alaska.

5. The Transition to Coal Energy

Coal had been widely harvested and exploited as an energy source for thousands of years. In Asia, coal provided little illumination but plenty of heat. Romans in Britain found that stones from black outcrops were flammable, and Roman soldiers and blacksmiths used this coal to provide heat. During the 13th and 14th centuries, a multitude of industries became dependent on a regular supply of coal. In salt

production, coal was used to heat enormous pans of sea water. If woodlands had not been devoted nearly exclusively to the production of timber, their yield could not have met the fuel value generated by that produced by coal mines. It took an annual yield from 2 acres of well-managed woodlands to equal the heat energy of a ton of coal. Timber became more scarce and expensive to procure during the Tudor and Stuart periods. Consequently, the processes of salt boiling, ironworking, brewing, baking, textile manufacturing, and lime burning adopted coal, rather than wood, as a readily available and economical fuel source.

Confronted by the unrelenting scarcity and high price of timber, manufacturers in fuel-intensive industries were further prompted to restrain costs and enhance profits by switching to coal. The problems caused by burning coal in processes previously dependent on wood (e.g., producing glass, bricks, tiles, pottery, alum, copperas, saltpeter, and paper; smelting lead, tin, and copper; refining soap and sugar) swelled the demand for progressive technology. Increased energy efficiency in industrial processes stimulated competition, innovation, and the emergence of fuel economies. Urbanization expanded, and the availability of inexpensive coal enabled agriculture to supplant arboriculture.

During the late 17th century, British coal mining employed 12,000 to 15,000 people, from miners to those involved in the transportation and distribution of coal to the consumer. During this period, coal prices remained virtually constant and were not dependent on revolutionary changes in structure or technology, and mine shaft depth did not increase dramatically (safe ventilation of deep pits did not occur until the 19th century). Conservatively estimated, coal production increased 370% between 1650 and 1680, creating an environment in which profound industrial advances in mining and trade occurred.

The substantial expansion of the textile industry was accommodated within preexisting systems of production, relying on traditional procedures, artisan workshops, and piecework arrangements. Advances were built on the potential that had existed in preindustrial economies, while labor and finance, marketing, transportation, and distribution evolved. Industrial enterprises incorporated factories, warehouses, forges, mills, and furnaces, and small collieries produced thousands of tons of coal per year. By 1750, the British population had nearly doubled, and urbanization and industrialization continued to amalgamate.

During the 1790s, French engineer Philippe Lebon distilled gas from heated wood and concluded that it could provide warmth, conveniently light interiors with gas in glass globes (fuel distributed via small pipes within walls), and inflate balloons. Gregory Watt, son of inventor James Watt, journeyed to

France to investigate these experiments. In Cornwall, England, William Murdock lit a house with gas generated from coal. Concerned with commercial possibilities, he experimented with coal of varying qualities, different types of burners, and devices that could store enough gas to make portable light. In 1798, he returned to the foundry in Soho, England, where he illuminated the main building with coal gas using cast-iron retorts.

By 1816, London had 26 miles of gas mains, and soon there were three rival gas companies. With the availability of kerosene and gas light, the whale oil industry futilely attempted to compete. The advent of gas light enabled Victorian readers to become increasingly literate. During the 1860s, gas utilities spread in England and Europe, while the eastern United States and Canada adopted gas lighting for streets. At the turn of the 20th century, coal gas cooking and radiant heat to warm homes from gas fires became more common. The gas industry was determined to increase efficiency in all of these domestic and industrial heating functions.

Massive petroleum and gas industries resulted from the need for improved lighting during the industrial revolution. Although oil lamps had improved, the illuminating capabilities of vegetable and animal oils were limited. Gas light was superior but was restricted to urban areas. Gas oil, consisting of the fraction intermediate between kerosene and lubricants, was inexpensive and less dangerous than petroleum and also provided a convenient means of enriching the coal gas.

At the turn of the 20th century, when the automobile industry was still rudimentary, crude petroleum and rubber remained insignificant, while coal gas reached its zenith of full technological development as an energy source. Versatile and inexpensive kerosene continued to provide illuminating energy in much of the world, while gas and electricity competed in urban centers. As an illuminant, gas light began to lose momentum, while electrical power grew enormously.

The turning point for the British coal industry, and a defining moment of the industrial revolution, came with the innovation of the steam engine. It was used in processing both coal and metals, and by the close of the 18th century more than 15 million tons of coal were produced annually.

6. Steam Power

In the 6th century BCE, Greek philosophers described the universal elements as consisting of air, earth, fire, and water. It was established that water displaced air and that steam condensed back to liquid and could rotate wheels. During the 17th century, the physical nature of the atmosphere was explored in qualitative rather than quantitative terms; water and

other liquids entering vacuous space was attributed to nature abhorring a vacuum. The persistent problem of mine drainage inspired understanding of atmospheric pressure, leading to the earliest steam engines.

When engineers failed to drain excess water from Italian mines during the 1600s, Galileo was enlisted. In 1644, his pupil, Evangelista, found that atmospheric pressure was equal to that of a column of mercury 30 inches high, corresponding to a column of water 30 feet high, and that atmospheric pressure falls with increasing altitude. Torricelli postulated that a method of repeatedly creating a vacuum would enable atmospheric pressure to be a source of power and could be used in mine drainage. Steam at 100°C equals sea-level atmospheric pressure and so can raise water 30 feet. Tremendous pressure can be generated by superheating water to create high-pressure steam; at 200°C, pressure is 15 times greater than at 100°C, lifting water 450 feet.

The first steam engines were termed atmospheric engines. In 1680, Christian Huygens, a Dutch scientist, designed a system where gunpowder exploding in a cylinder created hot gases that were expelled through relief valves. On cooling, the valves closed and a partial vacuum existed within the cylinder. The gunpowder gas occupied a smaller volume than when heated; thus, atmosphere pressure drove a piston in the cylinder. In 1690, Huygens's assistant, Denis Papin, found that 1 volume of water yields 1300 volumes of steam. A vacuum could be achieved by converting steam back into water via condensation, propelling a piston up and down in a cylinder.

Prolific inventor Thomas Savery improved on this in 1698, designing a steam pump using heat to raise water, generating approximately 1 hp. Steam from a boiler passed through a valve-fitted pipe into a vessel full of water, expelling the water up through a second pipe. When steam filled the vessel, it was condensed by pouring cold water onto the outer surface. This created a partial vacuum. When the vessel was connected with another pipe with water at a lower level, atmospheric pressure forced the water up. Savery's engine pumped water for large buildings and waterwheels, but its maximum lift was inadequate for mine drainage.

Thomas Newcomen, an inventor and blacksmith, adopted Papin's cylinder and piston and built his own version of the steam pump, replacing costly horse-drawn pump engines. Newcomen's direct experience with drainage problems in Cornish tin mines led to his construction of an efficient engine using atmospheric pressure rather than high-pressure steam. The Newcomen system had a boiler produce steam at atmospheric pressure within a cylinder, where a pump rod hung from a beam so as to apply weight to drive a piston. When the cylinder was full of steam and closed by a valve, a jet of cold water condensed the steam and atmospheric pressure forced the pump rod to push the piston down. Generating approximately 5 hp, this engine was a success in draining coal mines, lifting 10 gallons of water 153 feet through a series of pumps. Its primary users were coal mine proprietors who ran their engines on low-grade coal.

With improved control of mine flooding, the depth of mines and the coal industry grew exponentially. Use of the steam engine spread throughout Europe and to the American colonies. Newcomen's design was so reliable that even as other fuel technologies were developed, the last engine of this type worked for more than a century in Yorkshire, England, without serious breakdown (it was finally dismantled in 1934). Steam power was also used to pump water for waterwheels in cotton mills, accommodating the growing energy needs of textile manufacture as well as a multitude of other industries.

Between 1769 and 1776, James Watt discovered that energy was lost in the process of cooling and reheating the Newcomen engine. The addition of a separate condenser cylinder permitted a steady injection of steam into the first cylinder without having to continuously reheat it. In 1774, Watt and Matthew Boulton, an English manufacturer in Soho, became partners, providing Watt with skilled engineers and workers.

An impediment to steam engine design was the shortage of precise cylinders; imprecise boring allowed steam to escape between the cylinder walls and piston. John Wilkinson designed a mill for boring iron cannons that was capable of making precise cylinders for any purpose. His blast furnaces, with air from Watt's engine, produced consistent cylinders. Using this newfound precision in cylinder construction, the rotative Watt engine extracted four times as much energy, a minimum of 10 hp, from coal as the Newcomen model. By the turn of the 19th century, Watt's and Boulton's Soho factory had produced hundreds of engines powering industrial machinery.

Improvements in steam technology continued to be made during the 19th century. In 1802, a Cornish mining engineer, Richard Trevithick, built a small pumping engine with a cylinder 7 inches in diameter, a cast-iron boiler 1.5 inches thick, and steam pressure of 145 pounds per square inch—10 times atmospheric pressure. Two years later, Trevithick built 50 stationary engines and the first successful railway locomotive. By 1844, high-pressure steam engines averaged 68 million foot pounds per bushel of coal, compared with 6 million foot pounds capacity for the Newcomen engine. When Watt's engine was combined with Trevithick's design, a pressure of up to 150 pounds per square inch was achieved, referred to as "compounding." Trevithick's steam engine eliminated the huge rocking beam of Watt's design and became the new standard, suited to factories with limited space. His engines were

unrivaled and remained in a variety of industrial purposes, including pumping, iron works, corn grinding, and sugar milling, until the end of the 19th century.

7. *Steam Engines and Transportation*

Steam-powered engines were crucial in the development of transportation. Roads often were little more than deeply gouged tracks, slowing people, carriages, and carts in an energy-hindering quagmire. The growing demands for bulk fuel beyond what was locally available, with the high cost of transport added to the expense of energy resources, expedited innovation.

During the 1700s, British canals were improved, and tow paths allowed heavily burdened vessels to be pulled by teams of horses. Combining water and steam power, a paddlewheel steamboat ascended a river in 1783 near Lyons, France. William Symington, an English engineer, constructed an atmospheric steam engine to propel a small boat in 1788 using a horizontal double-acting cylinder with a connecting rod driving the paddlewheel's crankshaft. In the United States, Robert Fulton used a Boulton and Watt engine to operate a commercial steamer, and during the 1840s the U.S. Navy adapted this propulsion system in military craft, locating the engine below the waterline for protection.

In 1801, Trevithick built his first steam carriage (later driving one wildly through the streets of London). He combined the steam locomotive with railways, demonstrating its usefulness in conveying heavy goods and passengers. Rails had already been constructed for horses to tow carriages and coal carts more efficiently than on unreliable roads. The development of railway passenger services was a crucial formative factor in the phenomenon of the industrial revolution. Horsepower was replaced by the more powerful steam engine. By 1830, George Stephenson developed a locomotive railway line between Liverpool and Manchester without the need for stationary steam engines or horses in steep areas to supplement power. Railroad carriages were finally capable of hauling tons of coal and hundreds of passengers. Britain exported locomotives to Europe and America, where railway systems were rapidly established. Rail speeds exceeding 60 miles per hour were common, and engine efficiency increased, requiring only 23 pounds of coal per mile, although countries with abundant forests, such as the United States and Canada, still used wood fuel. Improved materials and methods of manufacture, as well as the steam turbine, resulted in higher power/weight ratios and increased fuel economy. During the 1880s, industrial steam power transformed networks of world communication, extending and unifying international economic systems.

8. Petroleum

Coal gas and petroleum energy share the property of having hydrocarbons as their principal constituents; coal gas is rich in the hydrocarbon methane, whereas petroleum consists of a complex mixture of liquid hydrocarbons. The use of petroleum products long antedates the use of coal gas. In Mesopotamia, concentrations of petroleum deposits provided commerce and production. Babylonians called inflammable oil "naphtha." For thousands of years, mixtures containing bitumen were used in roads, ship caulking, floor waterproofing, hard mortar, and medicines. Naphtha was militarily important as a weapon, although Romans derived their bitumen from wood pitch. Aztecs made chewing gum made with bitumen, and during the 16th century oil from seepages in Havana was used to caulk ships. By the 15th and 16th centuries, there was an international trade in petroleum.

Distillation of crude oil revealed that its derivatives were suitable in axle grease, paints and varnishes, dressing leather, lamp fuel, and mineral turpentine. A pliable caulking for ships was made by thinning pitch thinned with turpentine, and hot asphalt pitch and powdered rock were used in flooring and steps. During the 19th century, asphalt mixed with mineral oil was used in pavements and roads, and oil cloth and linoleum became popular and inexpensive household items.

During the early 19th century, deep exploratory drilling for water and salt required sufficiently hard drills and mechanical power. Steam engines provided the energy, and the development of derricks aided in the manipulation of rapid-drilling machines. Prospecting for oil ceased to be dependent on surface seepages when hollow drills enabled drill samples to be removed, revealing the structure of underground formations. The compositions of crude oil varied; thus, methods of oil refinery were important. During the 1860s, the United States had a steady supply of 10 million barrels of petroleum annually. By 1901, Russian crude oil output reached 11 million tons, and petroleum-fueled railways connected world markets.

Kerosene was distilled from petroleum with sulfuric acid and lime, providing inexpensive and plentiful fuel for lamplight. Although gasoline remained a worthless and inflammable by-product of the industry, oil derivatives provided lubricants that became increasingly necessary with innovations in vehicles and machinery, and heavy machine oils were used on railways.

9. Electrical Power

In early Greece, it was found that the *elektron* (amber), when rubbed, acquired the power of attracting low-density materials. The widespread use of electricity for heat, light, and power depended on the development of mechanical methods of generation. During the 17th and 18th centuries, static electricity was found to be distinct from electric currents. Electricity could be positive or negative as charged bodies repelled or attracted each other, distinguishing conductors from nonconductors.

In 1754, John Canton devised an instrument to measure electricity based on the repulsion of like-charged pith suspended by threads, and this was later standardized and redesigned as the gold leaf electroscope. Benjamin Franklin identified the electrical discharge of lightning, leading to the invention of the lightning conductor. In Italy, Allesandro Volta, a natural scientist, found that electricity was derived when alternate metal plates of silver or copper and zinc contacted each other within a solution. The importance of this discovery was that it provided a source of continuous electric current. The original voltaic energy battery gave stimulus to the experimental study of electricity. In a matter of months, laboratories produced electric batteries, converting the energy released in chemical reactions.

In 1820, Danish physicist H. C. Oersted described the magnetic field surrounding a conductor carrying an electric current. The relationship between the strength of a magnetic field and its electric current established that a continuous conductor in a magnetic field causes electric currents to flow. From 1834, rotating coil generators were made commercially in London. The earliest generators produced alternating currents. Converting alternating energy into direct electric currents was resolved via a mechanical commutator and rectangular coils that rotated in a magnetic field. Maximum voltage in each coil was generated in succession, alleviating the irregularities at a given speed of rotation providing constant voltage.

By 1825, electromagnets were used as an alternative to permanent magnets by the founder of the first English electrical journal, William Sturgeon. Electromagnets possessed enough residual magnetism in their iron cores to provide the necessary magnetic field to start output from an electric generator. The electric generator became a self-contained machine that needed only to be rotated to produce electrical energy. The application of a steam engine to rotate the armature generated enough electricity to supply arc lamps for lighthouses, bringing large-scale use of electricity a step closer.

An armature using a continuous winding of copper wire increased the possibilities of using arc lamps for lighting streets. Arc lamps developed after the discovery that a spark struck between two pieces of carbon creates brilliant light. Inventors Thomas A. Edison and Sir Joseph Swan developed filament lamps for domestic use during the 1870s, and an incandescent filament lamp used an electric current to heat its conductor.

The growing needs of industrial light required electric power generators to be substantial in size. In New York, Edison's electric-generating station was in operation in 1882. Electric storage batteries were used for lighting railway carriages and propelling vehicles.

At Niagara Falls during the 1890s, George Westinghouse developed the first large-scale hydroelectric installation, with a capacity of 200,000 hp. Hydroelectric generators demanded a considerable capital outlay for successful operation, so coal-fired steam engines lingered as a favored energy source at the end of the 19th century. During the late 1800s, the electricity industry expanded, output per electric power stations tripled, and coal consumption per unit of electricity fell by nearly half.

Conductors consisting of concentric copper tubes were used with alternating currents because copper had a lower electrostatic capacity and the concentric cable had no inductive effect on neighboring telegraph, telephone, or other electrical installations. The electric telegraph required a reliable battery cell to provide constant voltage that was capable of prolonged output. Attendant construction needs for telegraph poles required the production of enormous quantities of timber, porcelain insulators, and rubber insulation for cables.

Local electrical generation evolved into large centralized distribution utilities, and surplus electricity was sold to local consumers. In 1889, the London Electricity Supply Corporation's power station provided high voltage by operating four 10,000-hp steam engines to fuel 10,000-volt alternators. The economic advantages of central power stations to generate electricity at high voltages for serving large areas brought new practical and economic problems of distribution and, thus, further innovations. By the end of the 19th century, underground distribution systems of electricity had color-coded cables. Each strand was identified in this manner, a major consideration given that the electrical network beneath city streets increased in complexity.

10. Energy Flow and Industrial Momentum

During the early industrial period, increases in scale and diversity of manufacturing and trade corresponded with a shortage of readily available consistent energy, and this pressure induced developments in fuel technology industries. Competition for dwindling resources spurred the expansion of international trade and exploration, and fuel efficiency became more crucial. The fusion between mechanized growth and economic fuel use created the momentum of energy exploitation and technological innovation, and socioeconomic systems became known as the industrial revolution. Transportation systems connected a myriad of worldwide energy sources, and improved communication expedited the process of efficient global fuel exploitation and industry. Global commerce and fuel exploitation decisively affected the flow of materials via worldwide transportation and trade systems. As technological innovations were applied to harvesting energy and the subsequent momentum of labor mechanization, production uniformity and the quantity of fuel used in mass processing vast quantities of goods coalesced into an enormous worldwide energy flow.

Further Reading

Cotterell B, Kamminga J 1990 *Mechanics of Pre-Industrial Technology*. Cambridge University Press, Cambridge, UK

Crone A, Mills C M 2002 Seeing the wood and the trees: Dendrochronological studies in Scotland. *Antiquity* **76**, 788–94

Derry T K, Williams T I 1960 *A Short History of Technology from the Earliest Times to A.D. 1900*. Oxford University Press, Oxford, UK

Fields G 1999 City systems, urban history, and economic modernity: Urbanization and the Transition from agrarian to industrial society. *Berkeley Planning J.* **13**, 102–28

Freese B 2003 *Coal: A Human History*. Perseus Publishing, Cambridge, MA

Harman P, Mitton S (eds.) 2002 *Cambridge Scientific Minds*. Cambridge University Press, Cambridge, UK

Harrison R, Gillespie M, Peuramaki-Brown M (eds.). 2002. *Eureka: The Archaeology of Innovation and Science—Proceedings of the Twenty-Ninth Annual Conference of the Archaeological Association of the University of Calgary*. Archaeological Association of the University of Calgary, Alberta, Canada.

Hatcher J 1993 *The History of the British Coal Industry,* vol. 1: *Before 1700: Towards the Age of Coal*. Clarendon, Oxford, UK

Licht W 1995 *Industrializing America: The Nineteenth Century*. Johns Hopkins University Press, Baltimore, MD

Nef J 1964 *The Conquest of the Material World: Collected Essays*. Chicago University Press, Chicago

Theibault J 1998 Town and countryside, and the proto-industrialization in early modern Europe. *J. Interdisc. Hist.* **29**, 263–72

Toynbee A 1956. *The Industrial Revolution*. Beacon, Boston.

Wallerstein I 1974. *The Modern World-System I: Capitalist Agriculture and the Origins of the European World-Economy in the Sixteenth Century*. Academic Press, San Diego.

Wallerstein I 1980 *The Modern World-System II: Mercantilism an the Consolidation of the European World-Economy 1600–1750*. Academic Press, San Diego

Wallerstein I 1989 *The Modern World-System III: The Second Era of Great Expansion of the Capitalist World-Economy 1730–1840s*. Academic Press, San Diego

Zell M 1994 *Industry in the Countryside: Wealden Society in the Sixteenth Century*. Cambridge University Press, Cambridge, UK

Richard D. Periman

Economic Thought, History of Energy in

Glossary

ancient atomism Natural bodies composed of indivisible units of matter possessing size, shape, weight, and an intrinsic source of motion.

classical economics Early modern theories based on the primacy of production processes.

early idea of energy The principle of activity, force in action, the "go of things."

fire A very active and subtle substance or element used to explain motion and activity in nature and in organisms, including sensitivity in antiquity and in early modern chemistry and physiology.

neoclassical theory The modern exchange approach to economics based on scarcity and marginal utility.

phlogiston A volatile component of combustible substances in 18th century chemistry captured from sunlight by plants and released in vital processes, combustion, and reduction.

work A late-18th-century measure of the effect produced by machines and labor employing natural powers.

Economic theory and its history conveniently divide in two parts: an "early modern" production theory focused on how a society uses resources provided by nature to produce physical goods and services and a "modern" exchange theory centered on the study of the efficient allocation of scarce resources to satisfy the unlimited demands of consumers. The first, pre-classical and classical economics (1650–1850), evolved in a close and ongoing relationship with early modern scientific theories of nature's physical fecundity and organization including mechanics, matter theory, chemistry, physiology, and psychology. Mainstream neoclassical economics (1870 to the present) took its direction from the conceptual and mathematical framework of 19th-century analytical mechanics. Production is treated as an analogue of a spatial (and nondissipative) energy/output gradient, which effectively eliminated any substantive treatment of material and energy resources and physical "converters" from production theory. Neoclassical economics neglects material and energy resources and technologies and is inconsistent with their inclusion. The theory is inherently static and cannot explain physical processes, the dynamics of change in economics, or even how an economic system is coordinated.

1. Introduction

The production approach to economics that emerges in the mid-17th century took its inspiration form a broad range of sciences: Galilean mechanics, ancient ideas of active matter, Harvey's new circulation physiology, the proto-ecological natural histories developed by the scientific academies, and psychological and political theories of sentiment and organization with roots in antiquity. For nearly two centuries, economic ideas of physical production and economic behavior drew on the natural sciences, including physiology and medicine. This early connection between economics and natural philosophy has been little studied by economists but provides an essential key for understanding the evolution and structure of the classical approach.

In the second half of the 19th century, the focus of theoretical inquiry shifted from a production economics to a theory of market exchange. Neoclassical economics was constructed on two related ideas: the universal scarcity of resources and diminishing marginal productivity applied to resources and consumption goods alike. This basic idea, formulated in terms of simple partial derivatives, is that additions of any single input (or good), everything else held constant, yield a stream of positive albeit diminishing additions to output (or psychic benefit). But according to the conservation matter and energy, physical output cannot be increased if material and energy inputs are held constant. The central idea of neoclassical theory, the doctrine of marginal productivity, violates the laws of thermodynamics. Basing economic theory and its regulatory mechanisms on the conceptual structures and mathematics of analytical mechanics and field theory eliminated material and energetic processes (and much else including technology, organization, and psychology) from economic theory. Economics became an autonomous science isolated any real connection to nature or to society.

Our concern is to show how the material and energy ideas of early modern science were used to shape early economic theory and how these ideas were gradually lost. These early connections between science and economics have been hidden for the reason that they were considered heretical and were not generally openly avowed by the scientists and the economic thinkers who took them up to avoid censorship or religious condemnation. Contemporary readers could be assumed to know the coded language of their presentation. Modern investigators face the double difficulty of not knowing the language of the early sciences or the coded messages of radical materialism. They are also trained in economic theories that have severed any connection to nature and therefore lack any conception of the necessary interdependence that exists between material and energy processes in nature and in human economies. The discovery of these connections requires a sustained investigation of the shifting and forgotten labyrinths of early science. Thanks to the breadth of work in the history of science, an archaeology of the material and energetic foundations of early economics is now under way.

The next section surveys the main energy ideas employed in the sciences and economic thought during the first two centuries of early economic theory. Subsequent sections focus on the main economic writers and their very individual connections to the scientific ideas of their time starting with Thomas Hobbes and William Petty in the 17th century. The main developments move back and forth between Britain and France. Classical theory ends in Britain with J. S. Mill in the mid-19th century who moves away from grappling with the fundamental question of the meaning of energy for economic production. A final section briefly sketches the analytical focus that has governed neoclassical theory.

2. Early Energy Ideas

The idea of energy we use is Wicken's suggestion that energy "is the go of things." We are interested in concepts that provided explanations of physical and vital motions in early modern natural philosophy and the influence of these ideas on the formation of early economic theory. Since energy comes in a material form and is not used for itself but to make or do something, we are also concerned with material transformations and physical technologies that convert energy and materials. These components of production are inextricably linked in the economic process.

Material and energy conversions are subject to physical principles: (1) matter and energy are neither created nor destroyed in ordinary processes; (2) energy is degraded in quality as it is used and becomes unavailable for use (useless energy); and (3) material processes also follow physical and chemical rules. Theories of economic production must be consistent with scientific principles including the basic complementarity that exists between materials, energy, and physical converters. If economic principles are not consistent with the sciences of nature, then economic theory must be reformulated. Economics is not an autonomous science.

The first energetic concept we isolate is the material fire, which provided the active principle of matter theory in the natural philosophies of antiquity and early modern era. Fire was a very subtle and active matter used to explain the visible fire of combustion (by a disassociation of the matter of combustible bodies by the penetration of fire atoms), the invisible flame of the vital heat in animal bodies, and sensitivity in plants and animals.

The mechanist philosophies that emerged in the wake of the victory of the new science of mechanics of Kepler and Galileo are generally assumed to have regarded matter as entirely passive. But nature required sources of motion and ideas of active matter remained central in many 17th-century philosophies. The revival of Stoic, Epicurean, and Democritean theories of matter, the close connection between a mechanical explanation of nature and ideas of active matter in antiquity, the Renaissance Aristotelianism taught in the Italian medical schools, and the growing influence of an alchemy joined to atomism all promoted ideas of active matter. Fire, the active principle of matter theory, was a basic doctrine in these theories.

Attempts to construct a unified mechanical philosophy of nature reopened the conflict between doctrines of active matter and religion. Nature, in church doctrine, was passive and God the sole source of motion in the universe. After Galileo, matter theory proved more problematic for natural philosophers than Copernicanism. Descartes canceled plans to publish *Le Monde* with its naturalistic account of the universe and moved to ground his natural philosophy on a metaphysical argument that all motion was due to God. Despite a prohibition against teaching atomism, Gassendi, a Catholic priest and astronomer protected by powerful friends, proposed a version of atomism consistent with the creation. God, according to Gassendi, could have made the world in any number of ways but chose to make it from atoms endowed with an enduring source of motion. His union of atomism with 17th-century atomistic chemistry and with Hippocratic biology provided an attractive package for the development of chemical and physiological ideas and was quickly adopted by physicians in France and by Harvey's disciples in England. His followers had to express their ideas clandestinely. Hobbes developed a entirely naturalistic account of motion but had to parry accusations of antagonists that he held a theory of self-moving matter. Newton fought materialism all his life and denied material powers to "brute matter." But by treating gravity and other active principles as powers of bodies and speculating on material ethers and vegetative powers, he fostered the materialism of the Enlightenment.

Thanks perhaps to the flux in religious regimes, England was fertile soil for energetic and atomistic ideas of matter as is evident in the work of Thomas Harriot and Walter Warner in the late 1500s and after, Bacon's atomistic speculations in the 1620s, and the Cavendish circle at Welbeck in the 1630s (which included Hobbes) who knew Warner's work. Harvey's *De generatione*, which circulated for two decades before publication in 1651, was another source of energetic ideas. His physician disciples in Oxford and London were quick to take up the "Democritean and Epicurean system" associated with Hobbes and Gassendi. The latter had a considerable influence thanks to the flexibility of his system and close connection to chemistry.

The introduction of Gassendi's atomism in England has been credited to Walter Charleton's translations of Gassendi's early works on Epicurus in the 1650s. But the young physician and future economist William Petty provides an earlier source. Petty met and became friends with Gassendi during the mid-1640s in Paris when he served as Hobbes's scribe and tutor in anatomy. The two also took a course in chemistry. We can assume he introduced the ideas of Hobbes and Gassendi to his colleagues, including Robert Boyle, in Hartlib's circle in London, where he learned the chemistry of Cornelius Drebbel; in Oxford in 1649 where he led research meetings in his lodgings, and to his cousin, Nathanial Highmore, who became a Gassendian.

Gassendi's synthesis of atomistic chemistry and Hippocratic physiology provided the basic framework that Harvey's physician disciples employed for four decades to develop the new circulation physiology. They conducted a remarkable series of experiments investigating combustion and the nutrition and respiration of plants and animals using nitre, sulfur, and other chemically active substances that brought them close to modern ideas about combustion and respiration. They also used ideas of fire as the active element of the nutritive, animate, and intellectual souls of living things, concepts Descartes had supposedly banned from biology.

The union of Harvey's circulation physiology and atomistic chemistry was quickly embraced by the scientists of the newly organized Academy of Sciences in Paris who initiated anatomical and chemical studies of plant and animals that were sustained, thanks to Royal patronage, over many decades. Early papers extended the circulation model to the preparation and movement of sap in plants. Long the neglected orphans of science, plants were now recognized as occupying the critical position in the physical and chemical circuits linking living organisms to the soil, air, and sunlight. They were the assemblers of the materials and energetic substances feeding and maintaining the other organisms. Cycles of production and consumption bound all the parts of nature together in conditions of "mutual need." Harvey's circulation physiology extended to larger systems of circulating nutrients and materials provided a new vision of the unity and interdependence of nature and an obvious model for the human economy.

A second and closely related locus of energy ideas in 18th-century scientific and economic thought was the phlogiston chemistry formulated by the German chemist and physician, Georg Stahl, at the beginning of the century. Phlogiston, or "matter of fire," was an imponderable (unweighable) substance that joined other substances in chemical reactions (synthesis) and

was released in chemical decompositions (analysis). Combustible substances such as plants, alcohol, and coal were assumed to have high concentrations of phlogiston. One of Stahl's key contributions was to emphasize the importance of the principle of conservation of matter in the methods of analysis and synthesis of compounds that became a basic tool of chemical research thereafter. Together with the "elective affinity" that different elements of matter revealed for undergoing reactions with other elements, phlogiston chemistry provided the main framework for chemical theorizing until it was supplanted by Lavoisier's oxygen theory. As an early version of potential energy, it continued to provide an energetic explanation of nature's operations into the next century.

Phlogiston chemistry was taught in the medical schools in Montpelier and Edinburgh and employed in conjunction with physiological ideas of sensitive matter. G.-F. Rouelle taught a version of phlogiston chemistry at the Royal Garden under Buffon. His students, some of whom wrote the chemistry articles in Diderot's *Encyclopédie*, rejected mechanistic chemistry in favor of ideas of energetic and sensitive matter. This vision of sensitive molecules informed the theories of psychological sensibility (sentiment) advanced by the scientists of the radical Enlightenment. These ideas informed François Quesnay's *Essay on the Animal Oeconomy* (1847) and were implicit in his doctrine of economic productivity. Adam Smith in turn praised the "philosophical chemistry" of his friends for the theoretical connections it made between the processes of the art of the furnace and the "vital flame" in animals. His associates justified their physiological and medical theories of sensitive substances by referencing the psychological theories of Hume and Smith. Phlogiston chemistry and its ideas of energetic substance were also central to the 18th-century scientists who influenced Thomas Malthus and David Ricardo.

The third basic energy idea to shape economic thinking in the classical period was the new understanding and definition of work done in harnessing and using natural powers. The British engineer, William Smeaton, made the critical contribution in a study of the efficiency of different types of water wheels in converting a given head of waterpower into work. His study, published in 1759, established the superior efficiency of overshot wheels and enabled an increase in effective water power capacity in Britain by a third in a period when water power was critical to economic growth. He defined work as the capacity to lift a given weight a given distance in a given amount of time and made comparisons of the "work done" by horses, humans, and wind, water, and steam engines. Adam Smith adopted his definition of work but unfortunately confined its consideration to human labor, which he made the center of this theory.

The new mechanical theory of work was taken up by the French engineers Charles Coloumb and Lazare Carnot. Coulomb's thrust was to assimilate machines to the work done by humans and to make machines "an economic object, a producer of work." Carnot's great study, published in 1783, was concerned with a theoretical analysis of machines as devices for transmitting motion form one body to another. Carnot was concerned with general principles irrespective of the source of power. He emphasized the notion of power over that of force and machines as agents for the transmission of power. His work received little notice until he was promoted to the top leadership in the new Republic and celebrated for his reorganization of the military as "the organizer of the victory." His contributions to the theory of machines were taken up by J.-B. Say who recognized Smith's neglect of the central role of machines employing natural powers in industrial productivity. Say's construction of a general theory of production based on materials, forces of nature, and machines to multiply labor will receive important elaborations by the mathematician and engineer Charles Babbage, the economist Nassau Senior, and the physicist William Whewell. Their contributions do not, however, receive any sustained attention by subsequent economists. As happens to many good ideas, they were passed over in silence by the leading theorists of the time who followed J. S. Mill's shift of focus from a physical theory of production to economizing, which brought the classical epoch to an end.

3. Early Physiological Economics: Hobbes, Petty, Boisguilbert

Despite his considerable contributions to the psychological and political framework of early economics, Thomas Hobbes is considered to have made little or no contribution to the core ideas of economic theory. He is also widely seen as a mechanistic philosopher who reduced everything to Galileo's new mechanics. Most historians of philosophy and economics have not understood the central importance physiological ideas have in his system, his hierarchical conception of the sciences, and the use he makes of Harveyian physiology to extend the materials provisioning approach Aristotle initiated in his *Politics* but failed to complete. Hobbes in fact makes the first sketch the core ideas that will characterize preclassical and classical theory thereafter.

Like Gassendi, Hobbes constructs a conception of nature and society that proceeds from a basic energetic physics of matter in motion and avoids reductionism by treating each level of activity according to what he took to be the organizing principles and evidence for the topic at hand. In addition to the atomism of Democritus and Galileo,

he knew and used Aristotle's biological and psychological ideas; Harvey's theories of generation, organic sensibility, and circulation; and the economic ideas of Aristotle's *Politics* and the pseudo-Aristotelian *Oeconomia*. This physics and physiology informs the production approach to economics he sets out under the heading of "The nutrition and procreation of commonwealths" in *Leviathan* in 1651. An economy, like living bodies, obtains the materials and "nutriments" it needs to produce a commodious existence from the land and sea by human industry. These material resources are passed through various stages of concoction (heating) and preparation and circulated via the conduits of trade to all parts of the body politic. His chapter also covers the institutions of property, contracts, and so on that were part of the evolutionary sociopolitical framework initiated by Democritus and his successors.

The energetic thrust of Hobbes's ideas is subtle but clear enough. He knew the atomism and active matter theories of Harriot, Warner, Bacon, and Gassendi but adopts the more dynamical approach of Galileo. He had met Galileo in 1635 and reports carrying away a vision of a world filled with motion. Every seemingly solid body was alive with the internal motions (agitations) of its smallest parts. He defines each specie of body by the imperceptible motions of its smallest particles. These motions constituted the specific capacities of each body for the initiating motions (endeavors) triggered by interactions with other bodies (the explosion of gunpowder by a spark, etc.). Hobbes's idea of endeavor as an imperceptible but actual motion is recognized as an influence on the very different ideas of force developed by Leibniz and Newton.

The basic source of motion in nature was the undulating wave-like motions of sunlight. Hobbes did not attempt any explanation of the connection between sunlight, plants, and the food consumed by animals. In his *De homine* (of man) of 1658, he attributes the motions of the heart and muscles to motions of "the invisible atoms ... of a salt or niter" taken into blood via respiration. This idea, long popular among Harvey's and Hobbes's alchemically minded friends, was restated in 1654 by Ralph Bathurst, a supporter of Hobbes and associate of Petty at Oxford.

Hobbes's sketch was fleshed out by the brilliant William Petty (who may have contributed to their formulation in Paris). Petty had studied navigation, geography, physics, mathematics, and medicine in France and Holland, and he was familiar with the work of Hobbes and Gassendi and the scientific ideas of the Puritan reformers in the Hartlib's circle in London. He returned to take his medical degree at Oxford in 1650 where he organized research projects and quickly attained high positions in the university. The siren of political preferment and reform called him to Ireland. Briefly chief physician for the army, he reorganized and completed the Down Survey of Irish lands in breathtaking speed and became one of Ireland's largest landholders. Spending most of his time defending and managing his estates and seeking support for his political and economic projects, his scientific interests shifted towards the socioeconomic topics he called "political arithmetic." His Baconian-inspired works include a succession of acute but scattered theoretical insights central to classical economics, which were never drawn together in a single work.

His first contribution was the elaboration, in a *Treatise on Taxes* in 1662, of the idea of a surplus that the ancient atomists made the basis of social and economic evolution. Petty takes up the surplus in order to explain the source of taxable revenue. The surplus is the difference between the total food produced on the land and the portion that must be reserved to replace the inputs (food, fodder, seed, and equipment) used in production. It is effectively the energy available to support and feed the rest of society, including the artisan and commercial sectors. His proposal was to tax the revenue going to the wealthiest land owners and direct it to projects increasing access to markets, land productivity, and the skills and habits of unemployed workers.

He sketches the implications of the surplus for a theory of competitive profits across sectors, which he terms "neat rents" (not the differential rents earned on more productive or favorably located soils). A "neat" or net rent is the ratio of net earnings from an activity to the cost of carrying it out. He applies this idea to agriculture and extends it to mining and casting coin. Higher returns in one sector, he suggests, lead to a flow of labor and capital goods to that sector, equalizing the rate of return between activities. In addition to this idea of an equal rate of profit (our term), he also developed a concept of differential rent for favorably located lands.

Energy ideas inform a production approach to prices in his *Political Arithmetic*. These are cost-based prices set by the food, materials, and tools needed plus profit. He reduces these in turn to labor, land, and food, given that humans and beasts must "eat, so as to live, labor and generate." The ultimate determinate of prices is thus the "energy" required, directly and indirectly in production. He distinguishes between short-run prices (reflecting day to day movements of market forces) and the "intrinsic" prices determined by production costs. This distinction becomes a basic feature in classical economics via Richard Cantillon and Adam Smith.

The next significant development of a physiological economics was made by Pierre Boisguilbert, the Lieutenant-General of Normandy (active 1695–1714). Boisguilbert was obviously aware of Hobbes's *Leviathan* and Petty's theory of the surplus, which he made the basis of the emergence of the professions in modern society. A deep study of agriculture and its

commerce inform his writings. He was also a close student of the new sciences of nature of the Academy of Sciences in Paris and the work of his younger cousin, Bernard Fontenelle, who was establishing himself as a leading interpreter of the new science. Neither he nor his more radical cousin were Cartesians. Boisguilbert's commitments were to the Stoicism of his uncle (Pierre Corneille) and the Epicureanism of Gassendi and François Bernier. The latter was well known for his travels in the East and critique of Oriental Despotism—the aggrandizement of resources by the few and the absence of property rights and basic civil protections, which the Epicureans regarded as critical incentives for social intercourse, production, and prosperity. Bernier implies a critique of France's aristocratic economy, which Boisguilbert makes explicit.

The work of the Royal Academy of Science provided a new conception of nature linked together in a circuit of chemical exchanges. Plants were the prime producers of the energetic substances maintaining all organisms in "a state of reciprocal and mutual need." Boisguilbert develops similar ideas in economics. Agriculture produces the "basic" goods (grain, wine, and plant and animal tissues) that, like the chemical elements, produce the rest. The fecundity of nature supplies the means that give birth to the 200 or so professions of the state.

The academy's model of the circulation of sap in plants provided Boisguilbert (as Harvey's theory did for Hobbes) a model for economic production and circulation. The plant model makes a clear separation of production sites. Primary materials assembled and processed at one site are perfected at others and move in distinct "circuits." Boisguilbert likewise sees goods produced on the land and in artisan shops moving along distinct circuits between farms and towns. He extends a distinction between the sites of agricultural and manufacturing production to recognize and explain the very different price and quantity behavior between the two sectors.

He recognizes that the strict requirements that nutritional needs impose for maintaining the production and flow of food to consumers requires in turn the reciprocal needs of artisans for markets for their goods and services. Any break in this mutual circulation threatens a widening disequilibrium endangering the health of the whole society. Given the ferocious pursuit of self-interest by traders, equilibrium is maintained on the edge of a sword where a strong movement of supplies or fall in demand can become self-feeding. Such instability is magnified by the sectoral differences in conditions of production and market behavior. The price instability of agricultural markets has disastrous effects for maintaining productive investments and fertility of the soil. In urban industries, workers resist cuts in pay and employers respond to falling demand by laying off workers. The result exacerbates the crisis.

His acute analysis captures an asymmetry in market behavior that continues to perplex modern equilibrium theorists.

Prices obviously play an important role in maintaining economic equilibrium but the analysis goes beyond the sphere of exchange to assume that nature's conserving operations extend to the economy. Nature has the capacity to sense injury to the economy and the power to regulate and to heal itself. His ideas reflect the Stoic vision of natural order and harmony that Gassendi and Bernier extend from animate nature to the universe and the Hippocratic medicine advocated by Gassendian physicians. Good policy can maintain health but when illnesses arrive they must follow a natural course. Nature not the physician heals, thus, "let nature alone." These properties of self-regulation are the laws, God's invisible hand, that maintain order in the world.

Boisguilbert's psychological and social ideas also reflect Gassendi's reinterpretation of Epicurean psychological and ethics. Living organisms are designed with natural capacities for sensing what is good for them and what to avoid. Pleasure and pain are part of God's design to direct the activity of nature's creatures to life, health, and reproduction. This naturalist psychology extends to the direction and regulation of human society where the desire for the goods of the world, property, and security induce individuals to establish collective agreements to respect and protect the property and rights of others. The ruler is bound in turn to protect property and individual rights and to advance the common good, which includes protecting the weak from the depredations of the strong and wicked. Boisguilbert's use of this clandestine current of ideas links him to a lineage of economic thinkers going back to Hobbes and forward to Smith.

4. François Quesnay: The Energetic Premises of Physiocracy

Before Quesnay became an economist, he was a surgeon and physician whose *Physical Essay on the Animal Economy* of 1747 was dedicated to establishing the physical and chemical foundation of the nutrition, generation, and sensitivity of living organisms. For Quesnay, this was the ancient physics revived by Gassendi, the energetic chemistry taught by Rouelle, and an Epicurean psychology extended by the physiological and psychological theories of sentiment of the Montpelier physicians—the chemical and psychological doctrines of Diderot's *Encyclopedia*. Quesnay presents this radical energetic materialism in the form of a critique of ancient ideas, a skeptical strategy designed to avoid censorship (which has also misled historians of economics).

Thanks to his skills as a surgeon and systematic thinker, Quesnay rose from lowly origins to become physician to Madam de Pompadour at the Court of Versailles. The grant of a rural estate by the king turned his attention to the problems of French agriculture. His articles in Diderot's *Encylopedia* on farming in the 1650s and his famous *Economic Table* at the end of the decade include his infamous economic doctrine of the exclusive productivity of agriculture and corresponding sterility of commerce and manufacturing. Developed within the context of an agrarian economy, his argument was that only agriculture is regenerative in the sense of replacing what is consumed. Artisan industry (assuming all inputs come from the land) only transforms and consumes things that have been produced by the farmers. These ideas can be seen to depend on the energetic theory of his physiology that plants are the only producers in nature and the rest of nature (and humans) only exist by consuming what plants have made.

According to the 1747 *Essay*, the cycle of nature's production begins "in the womb of the earth" where elements from the atmosphere and soil are chemically combined in simple compounds and taken up by the plants where they are joined to the "very subtle and active matter" of fire provided by the sun. These organic compounds provide the active energetic substance of fire circulated throughout nature by the animals that eat the plants, the animals that eat animals, and so on. The materials are returned to the earth (recycled), but the subtle energetic matter is lost in the decaying compounds and has to be replaced by a new cycle of primary production.

Two basic criticisms have been leveled at Quesnay. The first is that he mistakenly assumed that only agriculture was capable of producing a surplus (and profit) and that he neglected the obvious importance of manufacturing and growing industrial productivity to human welfare. The criticism is valid. The second criticism is that he failed to apply the conservation of matter to agriculture in the same way he does in artisan manufacturing. Quesnay, however, would agree that production in both sectors obeys the conservation of matter. No output, let alone a surplus of grain, can be produced without the appropriate flow of material inputs from the soil and air. His point is rather that only plants can produce the vital energetic matter. His argument neglects the energy resources and inorganic materials used in industry that do not come from agricultural land. But within the assumption of a renewable resource economy limited to the use of plant biomass for materials and energy, his theory of the exclusive productivity of agriculture holds.

5. Adam Smith: The Work and Powers of Nature and Labor

Adam Smith is popularly regarded as the founder of classical economics, a judgment that owes more to the accessibility of his vision of a self-regulating market economy operating within an institutional framework offering something for everyone than to any fundamental contribution to theory. His laissez-faire vision excites some, his production orientation others. The heart of his production theory is the proposition that increases in the division of labor are the prime driver of productivity growth and the wealth of nations. He employs the new engineering treatment of work and natural powers to set out his theory. What is astounding, however, is the absence of any mention of the work done by natural powers other than labor (and land). This lacunae will have far reaching consequences for the treatment of energy and technology by later writers.

Smith wrote impressive histories of astronomy and ancient physics and reviews of Buffon, Linnaeus, Diderot, Rousseau, and others designed to introduce the ideas of the Enlightenment to his countrymen. He knew Gassendi's work and was good friends with Quesnay. His intimate circle included the philosopher David Hume; the chemist-physicians William Cullen, Joseph Black, and James Hutton (famous as a geologist); and the physicist John Playfair. Smith's early writings and *Theory of Moral Sentiments*, published in 1756, suggest theories of nature's design and operations, organic sensibility, and social evolution at the center of Enlightenment thought.

Smith's references to the system of the world as a great self-regulating machine have been taken to indicate an adherence to Newtonian orthodoxy. This inference is problematic. Smith, as did Hume, placed himself under Newton's mantle, but his ideas of nature's design and principles of operation are at variance with the theistic, antimaterialistic, and neo-Platonic metaphysics of immaterial principles connected with Newtonian "orthodoxy." In his "History of Astronomy," Smith interpreted gravity and inertia as balancing "material principles" that ensure a self-regulating universe. In his "Essay on Ancient Physics," he favorably compares the Stoic system of the world which used the idea of fire as the vital principle forming the world and living things to unify the design and the operation of the world, to the theism of Plato (and Newton), which artificially separated the principles of intelligence that formed and animated the world from those carrying out the ongoing operations of nature.

He also praises the "philosophical chemistry" of his associates for the "connecting principles" it provided for explaining the operation of fire in furnaces and the "invisible fire" or "vital principle" of plants and animals. He noted with regret how little this science was known outside the chemists who practice the "art of the furnace." Indeed, his idea of fire and what he calls the Stoical system was straight out of Bernier, which makes Gassendi a central source for the vision of nature's capacities of operation and self-regulation he both lauds and applies in his own

work. Like Hutton and Playfair, who saw an eternal and self-regulating universe operating by natural laws, Smith appears closer to the deism of the Enlightenment than to Newton's theism.

The chemical ideas of fire and phlogiston continue to inform Smith's idea of the productivity of land in the *Wealth of Nations*. In this regard, he remains close to Quesnay. Agriculture is the source of the energetic matter feeding the human and animal populations. But he employs the new terminology of work and natural powers to displace Quesnay's view that "land is the unique source of wealth" and shift the emphasis of production theory to labor and the factors contributing to labor productivity as the source of wealth. The productivity differences characterizing nations and regions are not related to differences in nature but to the effects of the division of labor, which augments the efficiency of workers. Labor (work) is the primary source of growing wealth, which he contrasts to the "niggardliness of nature."

Smeaton's reports on the conversion of mechanical power by hydraulic and animate engines and definition of work as a universal measure of mechanical effect and value are the obvious source for Smith's ideas. He likely knew these ideas directly (the volumes are in his library) and discussed them with his friend James Watt in the 1760s when they were colleagues at the University of Glasgow. We know that Watt discussed his innovation of a separate condenser for the steam engine with Smith. Smeaton provides the engineer's concern with efficiency in the use of a given quantity of work capacity, the idea of work as a universal measure of effect and value in all times and places, and the notion of work as equivalent to effort or pain exerted over a defined period. All these ideas are prominent in Smith.

While this framework is pregnant with possibilities for a new foundation for production, Smith makes no mention of the work of machines in harnessing nature's resources and powers in his economics. The new ideas of work are confined to labor and the related ideas of natural powers are confined to land and human labor (and working animals). The efficiency concerns of Smeaton's analysis picked up, but there is no discussion of the work done in the economy by natural powers of wind, falling water, or the immense sources of heat from coal, which had replaced wood in the industrial processes powering Britain's industrial growth.

Smith notes in passing the great investments made in "furnaces, forges, and slit mills." He would have known the great Carron ironworks near Glasgow—the first integrated iron works in Britain to employ coke in all its operations, whose blowing engines for the blast furnaces and water wheels and steam pumping engines supplying the mills with a continuous source of motive power were designed by Smeaton. He should also have known that nails and pins were made by power-driven machinery in rolling, slitting, and wire-drawing mills. Yet he illustrates the principle of the division of labor using the example of an artisan pin factory unassisted by machinery.

Modern economists tend to see Smith's division of labor and increases in the efficiency of labor in combination with increasing market size as an ever enlarging spiral of growth. It could not have been in Smith's theory since the original source of natural "powers" converted by labor in his theory continued to be, as it was in Quesnay, the energy fixed by plants. In Book II of the *Wealth of Nations*, he declares agriculture to be more productive than manufacturing because it employs two powers (nature and labor) in contrast to manufacturing which only employs one (labor). But the energy supplied by plants must be divided between workers in agriculture, manufacturing, transport, and a multitude of other uses and remains the fundamental limit on the economy. Hence his emphasis on the division of labor and productive employment as opposed to unproductive servants. His lack of clarity about the role of energy in agriculture and his neglect of industrial sources of power are major reasons for the neglect of energy in economics.

6. J.-B. Say: General Theory of the Work of Nature, Labor, and Machines

The first general application of the new ideas of work in production theory was made by the French economist Jean-Baptiste Say in 1803. Despite strong literary interests, his father sent him to a commercial school in England and then apprenticed him to a powerful French banker. Say left business to work for a newspaper run by Mirabeau (Quesnay's close associate), which led to his appointment as an editor for an important revolutionary journal that included a coverage of science. Aided by his brother, an engineer, he attempted to acquaint himself with "all of science." He knew Lavoisier's chemistry and the contributions of Lazare Carnot and Charles Coloumb to the new science of work. Reading Smith he was immediately critical of Smith's emphasis on labor and neglect of nature and machines. The greatest development of productive forces, he argued, came not from the division of labor but from machinery using the forces of nature. Smith did not understand the true theory of machines in producing wealth.

Say was even more critical of the physiocrats who failed to see that all production, whether it was the production of wheat in a field or the making of a watch in a workshop, did not create new matter but only rearranged or assembled existing matter. All production was subject to the conservation of matter and required the forces of nature. Quesnay, we have seen, would not have disagreed but was

making an argument that only agriculture renewed the energetic properties of materials required for life. Say's contribution was to see that human production was exploiting other sources of power by inventing new machines and that production theory must be concerned with the fundamental contributions of materials and forces of nature in human production. His ideas were clearly shaped by Lavoisier's emphasis on the importance of the conservation of matter in chemical analysis and the energetic ideas employed by French engineers and scientists. He knew the energy mechanics of Lazare Carnot (a revolutionary hero) whose work was reviewed in Say's journal, Charles Coulomb's comparisons of the work of machines, animals, and labor, and most likely Lavoisier's ideas of energy conversions by animals and machines.

Say defined the work of humans and the work of nature as "the continuous action which is exerted for executing an operation of industry." He argues that all material production whether in the field or the factory requires the same basic categories of elements (i.e., materials, forces of nature, and actions of labor, and machines). He compares land to a machine but knows that labor and machines are powerless to produce on their own without materials and a source of power. He understands the importance of production in economics and insists that economic theory must begin with production before taking up the topics of value, distribution, and consumption.

Say provided the first general physical theory of production based on the new theory of work and mechanical energy. Unfortunately, his achievement was obscured by a diffuse presentation (especially in the first edition of the *Treatise*) and his failure to develop the implications of his new approach for a production-based theory of prices. He incorrectly argues that production creates nothing new, only utilities. While production does not create new matter, a watch is more than the gross weight of its parts. It is a new physical entity, and its conditions of production as well as its utility are important for how it is priced or valued. Say's attempt to assimilate his theory to the French scarcity and utility approach to value ignored the interdependence between production and value and contributed to the misunderstanding of his important contribution to classical production theory by the English classicals who failed to take up his production approach and its specific concerns with materials, energy, and machines.

7. Malthus and Ricardo: Limits on Powers of Land

Like Smith, Robert Malthus has been seen as a disciple of Newton in his scientific, economic, and theological ideas. Malthus took his degree in mathematics at Cambridge, which was basically the rational mechanics of Newton and he made good use of his training in formulating regulatory mechanisms for population and price theory. But the natural philosophy and theology he outlined in the first *Essay on Population* in 1798 is neither orthodox nor Newtonian. His ideas of an ongoing creation, the evolution of mind from "a speck of matter," and the forging of mental and moral qualities by the struggle to overcome evil have their sources in the associationist psychology and theology of David Hartley; the scientific and theological ideas of Joseph Priestley, Hartley's leading disciple; and the evolutionary physiology of the physician and biologist, Erasmus Darwin, grandfather of Charles Darwin.

Out of this heady brew, Malthus constructs a decidedly heterodox philosophy and theology which sees nature, mind, and human progress as products of a long struggle against adversity. He makes an important break with ideas of natural harmony by drawing out the implications of Priestley's theological vision of God's imperfect but evolving creation and recognition of limited resources (even God faced limits in what he could do). Given the inherent drive to reproduce and the scarcity of resources, the capacities of the population of each species to outrun its food supply must eventually produce a struggle for existence in nature. Malthus recasts Priestley's progressive theology of struggle and growth into a theodicy of scarcity refracted through the mathematics of exponential growth and the focusing lens of regulatory mechanics. His ideas also reflect the absolute limits of an energetic perspective. Priestley was a phlogiston chemist and co-discoverer of photosynthesis. In a period when the nutrient limits of soils were not understood, the primary limit on agricultural production appeared to be the ability of plants to fix energetic matter as opposed to a shortage of particular (and augmentable) nutrients.

Malthus's preoccupation with the availability of food was reinforced by a theological vision of nature's design. Struggle against limits was necessary for progress. A preoccupation with food and limits, reflecting energy theory and theology, helps explain why Malthus's contributions to areas of production theory outside agriculture were so limited. Despite his acute observations on the central differences between returns and costs in industry (where techniques are replicated and constantly improved) compared to agriculture and an understanding of the energetic basis of production, he does not develop a general theory of the material and energetic transformations specific to industrial production. He certainly knows Say's production theory. It would appear that the idea of constraints and limits, which were rooted in theology and energy theory, carry implications for limits on human progress that explain his reluctance to engage Say's more general theoretical framework with its quite different vision of human progress.

The scientific underpinning of David Ricardo's powerful theorizing has received little attention from historians. What is known of his scientific interests and cast of his thinking suggests that natural philosophy played an important methodological and conceptual role in his economics. He had early interests in chemistry and geology. A convert to Unitarianism, we can expect he was familiar with the philosophical and chemical ideas of Priestley, the intellectual leader of the movement. The fact that Ricardo was a trustee of the London Geological Society, whose directors sought to advance James Hutton's uniformitarian geology, suggests he knew Hutton's powerful energetic arguments about the geological and biological processes forming the earth and his method of dynamic (energy) equilibrium (and his tendency to ignore historical processes). Hutton was one of the leading energy thinkers of his generation and the likely source for Ricardo's similarly powerful focus on dynamic processes.

Energy ideas play a central role in Ricardo's early corn theory of profit, which is set out as a ratio of the surplus of corn to its costs of production (measured in corn). If land (and energy) increase relative to labor, profits rise. Corn continues to play a central role in the dynamics of his labor theory of production and value after Malthus convinced him that capital could not be limited to corn. The relative availability and cost of corn (the energy source running the organic machines of the agricultural and artisan economy) continued to provide the foundation of his theory of relative costs, profits, and rents, and thus the long-run dynamics of the economy.

Although Ricardo noted the merit of Say's division of economic theory into production, distribution, and value, he did not make the explicit treatment of the production principles that Say's plan suggested. Instead, he begins his *Principles* with a chapter on how the values of commodities are determined. His analysis is clearly informed by a vision of production but the physical perspective of contemporary scientific ideas that informs it must be reconstructed from the wording Ricardo employs, a footnote or two, and the tenor of his arguments. We can catch a glimpse of his acute understanding of the contribution of the natural powers employed in industry when he refutes Smith's notion that nature does nothing in manufactures. The passage reflects Say's language but is pure Hutton in its energetic focus:

Are the powers of wind and water which move machinery, and assist navigation, nothing? The pressure of the atmosphere and elasticity of steam which move the most stupendous engines—are they not the gifts of nature? to say nothing of the effects of the matter of heat in softening and melting metals, of the decomposition of the atmosphere in the process of dyeing and fermentation. There is not a manufacture ... in which nature does not give her assistance to man, and give it too, generously and gratuitously.

—Ricardo, *Principles of Political Economy,* [1817], chapter 2, n10

8. *Late Classical Theory: Babbage, Senior, and J. S. Mill*

Until the mid-1830s, the energy ideas entertained by classical economists, apart from Say and Ricardo, were effectively confined to plants and biological engines. Thanks to Babbage's pioneering study of the new industrial "Economy of Machinery and Manufactures" in 1832 and Andrew Ure's study of the factory system in 1835, the importance of coal and steam engines for industrial production and transportation and for Britain's industrial ascendancy were now in the public eye. In reviews of Babbage and Ure, the economist John McCulloch noted the link between Britain's industrial prosperity and the exploitation of coal technologies. He argued that the increase in the quantity of iron made with coal lowered the costs of machinery (and steam engines) leading to more output at lower unit costs and more demand for coal, iron, and machines. He did not, however, attempt to incorporate these ideas into a theory of production.

Babbage's great study of principles of a machine economy distinguished two types of machines, those which produced power and those designed to transmit force and execute work. The first class included water and wind mills. But it was the adoption and continued innovation in steam engines employing coal for fuel that captured Babbage's imagination (the heat-based industries need equal attention as well). Economies of scale and high fixed costs pushed mill owners to continuous operation and opened opportunities for new professions (accountants, etc.) and industries, such as gas lighting. New power sources and technologies created new products and materials and new uses for materials that formerly had little value. The principles of the machine (and coal and iron) economy extended beyond the factory to the larger political economy and society generating new knowledge and new models for administration.

The only economist to apply Babbage's theory of machines to a theory of production inputs and returns was Nassau Senior, whose contributions in this regard, however limited, were completely ignored by his contemporaries. Babbage's suggestion that "practical knowledge" would continue to grow without limit appeared in the third of Senior's "elementary propositions of economic science," which stated that "the powers of labor and the other instruments that produce wealth can be increased indefinitely." Senior's fourth proposition posited the existence of diminishing returns to additional labor employed on the land and increasing returns in manufacturing. While he understands the conservation of matter in manufacturing industry when he states that "no additional labour or machinery can work up a pound of raw cotton into more than a pound of manufactured cotton," his attribution of diminishing returns in agricultural production to the employment of additional labor "on the *same* material" does not make it clear that additional output here requires additional materials from the soil and atmosphere and additional solar energy to produce another bushel of corn or pound of cotton seed.

Senior used Babbage to advance Adam Smith's distinction between fixed and circulating capital (structures and agents versus material flows). He distinguished tools and instruments from materials embodied in production and subdivided machines into those producing and transferring power. He separated natural powers from ordinary materials on the grounds the former are not embodied in output but did not know where to put food, coal, and other natural powers. Instead of distinguishing a subcategory of natural powers or energetic flows along side materials, he lumped food and coal with "fixed capital." Clearly, more insight about the nature of power and the agents and machines applying and using power was needed. His laudable efforts to grapple with a physical classification of production inputs and their implications for production returns were not, unfortunately, taken up by subsequent economic writers. Indeed, J. S. Mill attempts to undermine both Senior's classification efforts and treatment of production returns.

Another extremely important treatment of the ideas of the sources and application of power in human economy was that of the scientist and mathematician William Whewell in 1841, who extended Babbage's classification of machinery to include moving power, trains of mechanism connecting power to work and work done. Whewell also recognized that the engine economy and its sources of power were dissipative and would eventually be exhausted (including the engine of the sun). He concluded, Norton Wise points out, that no balance existed in nature (or economy). In Whewell, the steam economy replaced the equilibrium of the balance with disequilibrium and decay, foreshadowing the second law of thermodynamics. Whewell embraced decay as part of a Christian escatology but would have not been happy that evolutionary processes are the other side of a dissipative and temporal universe.

Despite his friendship with Babbage and adoption of J.-B. Say's ordering of economic theory (Book I of his *Principles* is devoted to production), the socially progressive J. S. Mill made no attempt to apply physical or engineering insights to production theory. He follows Say (and Senior) in classifying production inputs under the headings of labor, natural agents, and capital, but his concern throughout is to emphasize economic aspects of production. He shows no interest in the physical foundations of economic theory. Book I has three chapters on capital and at least two on labor but none on the characteristics of land or natural agents. He isolated "labor and natural objects" as the two "requisites of production."

But in contrast to his elaborate classification of indirect labor, and the separation he makes between materials and powers in nature, he refuses to make any distinction between materials, food, and fuels as circulating capital.

Ignoring Babbage, Senior, and Whewell, Mill argued that "political economists generally include all things used as *immediate* means of production either in the class of implements or materials to avoid a multiplication of classes [and] distinctions of no scientific importance." He made the distinction between materials and machines turn on whether an "instrument of production can only be used once" and declared that a "fleece is destroyed as a fleece" in making a garment just as a fuel is consumed to make heat (which indicates he did not understand the distinction between energy and materials). He responded to the "able reviewer" of the first edition of his work (Senior) who objected to subsuming coals for engines and food for workers under the class of materials by admitting that while his terminology was not in accord with the physical or scientific meaning of material, the distinction between materials and fuels is "almost irrelevant to political economy."

Mill correctly recognized that tools and machinery lack productive powers of their own. But after declaring that the only productive powers are labor and natural agents he denied this status to "food and other materials" (and thus fuels). The fact that steam engines burning fuels are productive and a worker lacking food is not exposes the weakness of his position. By undermining Babbage and Senior's extension of Smith's distinction between physical structures that process and store materials and flows of materials and energetic powers, Mill brought the development of engineering ideas about the machines and powers in production theory to a halt.

Mill recognized the existence of increasing returns in manufacturing and argued that the causes increasing "the productiveness of [manufacturing] industry preponderate greatly over the one cause [the increasing expense of materials] which tends to diminish that productivity." Yet he treated increasing returns as a subsidiary proposition to "the general law of agricultural industry," that is, diminishing returns, which he calls "the most important proposition in political economy." Increasing returns is not given an independent development. The statement of neither proposition includes a consideration of the essential inputs required to increase output according to the conservation laws which tie output to material and energy input. Mill was aware of the complementarity between capital and natural powers and materials and output but his lack of concern with accurate scientific terminology and any systematic consideration of the physical principles of production fail to preserve or advance the physical production framework advanced by Say, Ricardo, and Senior let

alone Babbage or Whewell. Given the that Mill's work was a common point of departure for neoclassical theorists, his failure was a critical factor in the neoclassical failure to seriously grapple with energy and scientific approach to production theory.

9. How Energy Came up Missing in Neoclassical Economics

The main early neoclassical economists, including William Stanley Jevons, Leon Walras, Alfred Marshall, and Wilfredo Pareto, took their inspiration from the rational and analytical mechanics of the 18th century and 19th century. Thanks to the construction of mechanics in terms of the equilibrium between forces, mechanics provided economics with a powerful mathematical language of analysis that included a model of self-regulating activity produced by the inexorable operation of opposing forces. Neoclassical economics abandoned the production approach of the classicals to reconstruct economics as a mathematical science of behavior and markets expressing the interplay of demand and supply based on theories of marginal utility, marginal productivity, and universal scarcity, the latter being the consequence of the substitution assumption or "law of variable proportions," which assumes that any nonscarce good of interest to economics will become scarce by being substituted for other goods.

We have already noted how the standard formulation of diminishing marginal productivity violates the basic laws of thermodynamics. The adoption of the idea of marginalism necessarily pushed materials and energy and any scientific formulation of production from the scene. William S. Jevons eliminated fixed capital from his theory and provided no treatment of the role of energy in his treatment of production despite having written and entire book on the critical importance of coal to British industrialization just a few years before. Leon Walras had initially specified raw materials as an intermediate input but then proceeded to vertically aggregate production so that output was a direct function of the "original factors of production," land, labor, and capital. Capital was assumed to be like land so that adding more labor to a given amount of capital could be assumed to add more output albeit subject to diminishing returns.

But the theoretical problems of marginal productivity are not confined to a special assumption (individual inputs are productive everything else constant) that can be relaxed by adding more inputs. The assumption reflects a structural feature of the theory, which effectively prevents any realistic physical reformulation of the theory. Thanks to the adoption of the mathematics of field-theory as the structural framework informing utility and production theory, every input in the production function (which is based on the potential function of analytical mechanics) is effectively assumed to have productive capacities because of the existence of an energy gradient in the space over which the production function is defined. Each production input is effectively considered to be an axis in n-dimensional space. Thus, an increase in any input (an increase along one axis) automatically involves a movement to a higher energy gradient. Output, defined as the potential function of the field, increases. These movements, moreover, are always reversible since the field theoretical framework is nondissipative and conservative.

In the 18th and early 19th centuries, nature (economics included) had been assumed to be self-regulating as a metaphysical principle grounded in nature's design. In the late 19th century, despite the development of a new energy physics based on the laws of thermodynamics and an understanding of the inexorable processes of entropic dissipation, economics had found a new and powerful mathematical and conceptual framework in field theory for expressing and developing its vision of order, stability, harmony, and rational allocation. For economics, the virtue of a theory of electromagnetic or gravitational fields is that there is no dissipation. Unlike other energy conversions, which are dissipative, the energy principles of field theory are conservative (no dissipation takes place). This provides a framework taken from physics that isolates economics from the troubling questions about natural resources or sustainability. These issues have little meaning once the assumptions of the theory have been accepted.

Further Reading

Aspromourgos T 1986 Political economy and the social division of labour: The economics of Sir William Petty. *Scottish J. Pol. Econ.* **33**, 28–45

Ayres R U 1978 *Resources, Environment, and Economics: Applications of the Materials/Energy Balance Principle.* Wiley, New York

Christensen P 1989 Hobbes and the Physiological Origins of Economic Science. *History Pol. Econ.* **21**, 689–709

Christensen P 1994 Fire, motion, and productivity: The protoenergetics of nature and economy in François Quesnay. In: Mirowski P (ed.) *Natural Images in Economic Thought, 'Markets Read in Tooth and Claw'.* Cambridge University Press, Cambridge, UK, pp. 249–88

Christensen P 2003 Epicurean and stoic sources for Boisguilbert's physiological and Hippocratic Vision of nature and economics. In: Schabas M, De Marchi N (eds.) *Oeconomies in the Age of Newton.* Duke University Press, Durham, NC, pp. 102–29

Debeir J-C, Deleage J-P, Hemery D 1991 (John Barzman, trans.). *In the Servitude of Power: Energy and Civilization through the Ages.* Zed Books, Atlantic Highlands, NJ, London, UK

Dixon B L 1988 *Diderot, Philosopher of Energy: The Development of his Concept of Physical Energy, 1745–1769.* Voltaire Foundation, Oxford, UK

Frank R G 1980 *Harvey and the Oxford Physiologists.* University of California, Berkeley, CA

Garber D, Henry J, Joy L, Gabbey A 1998 New Doctrines of Body and its Powers. In: *The Cambridge History of Seventeenth Century Philosophy, I.* Cambridge University Press, Cambridge, UK, pp. 553–623

Golinski J 2003 Chemistry. In: Roy Porter (ed.) *The Cambridge History of Science*, vol. 4. *Science in the Eighteenth Century.* Cambridge University Press, Cambridge, UK, pp. 375–396.

Vaggi G, Groenewegen P 2003 *A Concise History of Economic thought: From Mercantilism to Monetarism.* Palgrave, New York

Vatin F 1993 *Le Travail: Economie et Physique: 1780–1830.* Presses Universitaires de France, Paris

Wicken J S 1987 *Evolution, Thermodynamics, and Information: Extending the Darwinian Program.* Oxford University Press, New York and Oxford, UK

Wise N, Smith A C 1989 Work and waste: Political economy and natural philosophy in nineteenth century Britain. *History Sci.* **27**, 263–301, 391–449.

Paul Christensen

Ecosystems and Energy: History and Overview

Glossary

biogeochemical Pertaining to the abundance and movements of chemical elements or compounds in and between various components of an ecosystem or between ecosystems.

density dependence The situation in which the number of animals or plants at a future time is dependant on the number at an earlier time.

ecosystem A unit of nature, generally from 1 hectare to some thousands of square kilometers, with boundaries defined by the investigator, that includes plants, animals, microbes, and the abiotic environment.

endogenous A factor from inside the system that influences the system.

exogenous A factor from outside the system that influences the system.

trophic Pertaining to food.

We may consider ecosystems as machines that have been selected to use solar and occasionally other energies to reorganize the raw materials of the earth into systems that support life. Ecosystems, and their component organisms, use this energy to rearrange the chemical molecules of the earth's surface and of the atmosphere into living tissue according to those patterns that have high survival potential. They are, in this context, antientropic—that is ecosystems (and life in general) work against the tendency for the molecules of the world to degrade into relatively random arrangements. As such they are self-designed, in that their organization has emerged through natural selection for structures that capture, hold on to, and invest energy into further structure in a way that appears to be self-designed) the structure is generated to continue life and enhance over time its ability to capture and use energy). More generally, life is most explicitly nonrandom arrangements of molecules, and all organisms use energy to create and maintain these specific molecular arrangements. But individual organisms cannot survive in isolation because they need the supporting context of other species, the soil or other substrate, the atmosphere or water around them, proper hydrological cycling, and so on. Thus, to our knowledge, ecosystems are the minimum units of sustainable life outside of the laboratory. The proper functioning of the earth's ecosystems is also essential for human beings—for atmospheric and climatic stability; for the production of food, fiber, and clean water; and for physical and psychological well-being.

1. Introduction

1.1 Early Studies at the Ecosystem Level

Although the English term "ecosystem" was coined in 1935 by Tansley, ecosystems have been the subject of human inquiry for millennia. The first written records that survive include the work of Aristotle, an insightful and knowledgeable natural historian, and especially his student Theophrastus, who was interested in how different species of trees grew in different locations. Of course, practical people living on (or off) the land (fishermen, ranchers, hunters, trappers, many farmers) also knew and know a great deal of a practical nature about how ecosystems operate.

Well before Tansley, the formal scientific study of ecosystems was well developed by Russian and Ukraine scientists in the early part of this century. The serious student of ecosystems should return to the seminal, often brilliant earlier writings of Stanchinskii, Vernadaskii, Morozov, and their colleagues. Vernadsky in particular was interested in how ecosystems, and even the living world as a whole (the Biosphere), functioned. Unfortunately the work, and sometimes even the lives, of these early ecologists was terminated by Stalin and his henchman Lysenko because their view of nature's limits to human endeavors was in opposition to official communist social engineering.

The practical implications of understanding energy flow in ecosystems became obvious to a number of early investigators who often attempted to determine how much food could be harvested from a unit area of land or sea. Stanchinskii, mentioned earlier, attempted to examine the quantitative importance of the flows at each trophic (food) level in an Ukranian grassland. Wells *et al.* examined qualitatively the flow of energy from the sun through the food chain to the herring fisheries of the North Sea as early as 1939. The first study in North America is usually attributed to Raymond Lindeman, who quantified explicitly the flow of energy from the sun through primary producers to higher trophic levels in a bog lake in Minnesota. Lindeman generated most of his numbers from preexisting information but was extremely creative in organizing the organisms of an ecosystem as trophic (or food) levels, rather than simply by species. Although Lindeman made some errors (for example, by subtracting predation after estimating production from turnover rate), his work remains extremely important for our conceptualization of how energy flows through nature.

An especially important group of studies on the eneregetics of ecosystems were done by Eugene and especially Howard Odum. The brothers worked together measuring photosynthesis in a coral reef at

Eniwetok. The most important of these studies was done by Howard on the ecosystem at Silver Springs, Florida. The particular advantage of this site was that the water flowing out of the main spring boil was always at a constant oxygen concentration, so that the deviation of the oxygen content of water downstream (higher during the day, lower at night) could be used to estimate photosynthesis and respiration of the entire ecosystem. Odum also measured explicitly the energy fixed or used by each trophic level by, for example, putting fish into cages and measuring their growth. Odum found that by far the largest proportion of the energy captured at a given trophic level was utilized by that trophic level for its own maintenance respiration, which was consequently unavailable to higher trophic levels. Lindeman had introduced the concept of trophic efficiency, defined as the ratio of production at one trophic level to production at the next. Trophic efficiency is commonly from 10 to 20% but occasionally may be very different. The concept is important and familiar in agriculture where beef or fish production per hectare is much less than the production of plants of that same area, due principally to the large maintenance respiration of both the plants and animals. Both Howard and Eugene Odum continued basic research on the relation of energy to ecosystem structure and function into the new millennia in astonishingly long and illustrative career in research and graduate training.

Research has emphasized that most trophic relations occur not as simple straight-line chains but as more complicated food webs, in which a given species and even different life stages of that species eat from different trophic levels. For example, a herring whose diet contained 50% algae and 50% herbivorous crustaceans would be assigned to trophic level 2.5. Many, perhaps most, organisms are omnivores rather than strictly herbivores or carnivores, and probably the majority eat dead rather than live food. The single most important attributes of food quality, other than its energy content, is its protein content, which is approximately proportional to the ratio of nitrogen to carbon in the food (Table I).

Table I
Properties of food.

Food type	Example	Kcal/gm	Percentage protein
Carbohydrate	Grains	4.25	2–14
	Beans	4.25	5–30
Fat	Butter	9.3	0
	Olive oil	9.3	0
Protein	Animal flesh	5.0	20–30

From Whittaker (1975) and USDA nutritional Web site.

1.2 Ecosystems versus Species-Oriented Ecology

During this time something of a gulf developed between ecosystem-level ecologists and many other ecologists. This was due in part to the perception by some that ecosystem ecologists were paying insufficient attention to the basic laws and mechanisms of nature, including natural selection, that had been especially well developed by biologists at the individual and population level. Some ecosystem-oriented papers appeared to these ecologists to be almost mystical in their message that the system knows best. On the other hand, ecosystem-level ecologists could point to the failure of population dynamics to do a good job in prediction and management of fisheries, in large part because the population-level ecologists tended to ignore spatial and interannual variations in the ecosystem productivity that often determined year class strength. As part of this debate throughout the literature of ecology, there has been considerable discussion and even tension about the relative importance of exogenous (meaning external to the system, such as climate variability, another phrase meaning the same thing is forcing function) controls versus endogenous (meaning internal to the system, such as density dependent survival) controls for determining the structure and function of ecosystems. To some degree the argument is specious, for obviously both are important, although the endogenous aspects have been overly emphasized. Clearly, as described here, climate (an exogenous factor) is critical in determining the basic possibilities for life, and evolution (basically an endogenous factor) is important in determining the response of living creatures in a region to the climate.

Part of the problem leading to this dichotomy was the seemingly completely disparate perceptions of the different groups on what was the important currency of ecology. Energy was perceived as the currency of the ecosystem-level ecologists, and fitness (reproductive potential over time that includes the abilities to both survive and reproduce) was seen as the currency of the population-level ecologists, and each group tended to dismiss, argue against, or ignore the perspectives of the other. These were spurious arguments, and although many previous papers hinted at the importance of energy for fitness, this has been especially well done in papers such as that by Thomas *et al.* titled "Energetic and Fitness Costs of Mismatching Resource Supply and Demand in Seasonally Breeding Birds." This is one of the best papers in ecology, and there is more good science packed into the two-and-one-half pages of this paper than in many books. Simple observations and measurements, and double labeled isotopes, were used to examine energy costs and energy gains of various populations of European tits (chickadees). Those birds that timed their migrations and reproductions appropriately to the availability of their preferred high-energy food (caterpillars, in turn dependent on the phenology of their oak-leaf food source) were much more fit in the evolutionary sense. That is, birds with optimal timing were much more successful at raising more and larger (and hence more likely to survive) offspring and were themselves more likely to survive and reproduce again the next year. Birds that timed their reproduction inappropriately relative to the caterpillar dynamics worked themselves into a frazzle trying to feed their less successful offspring. Finally Thomas *et al.* found that climate change was interfering with past successful timings because the birds were tending to arrive too late, since the phenology of the trees was responding more quickly to temperature increases than was the impetus for the birds to migrate. Presumably birds that had genetically determined behavioral patterns that were getting them on the breeding grounds too early in the past would be selected for if and as the climate continued to become warmer. This study also indicates the importance of the quality of energy. In this case, it was not simply total trophic energy available to the birds that was critical but also their quality as represented in their concentration as large food packages, showing again the importance of energy quality as well as quantity.

The paper by Thomas *et al.* shows elegantly and more explicitly than earlier studies what many suspected for a long time: probably the principal contributor to fitness is the net energy balance of an organism. The idea that energy costs and gains are related directly to fitness is not a new one, found, for example, as Cushing's match/mismatch hypothesis in fisheries, which in turn is a restatement of Hjort's 1914 paper about year class abundance in Scandinavian cod. Energetics, combined with climatic and other external forcing, can explain population dynamics much more powerfully than the density dependant population equations that still seem to dominate much of population ecology. Clearly density dependence exists. But how often is it the principal determinant of population year class strength? Perhaps the best way to consider the importance of density dependence is what Donald Strong calls density-vague relations—that is the empirical information shows that density dependence works sometimes but does not operate importantly every year. The way that we have set up many of our experiments in ecology virtually guarantees that we will find some density dependence even when it is not dominant.

1.3 General Principles of Energy for Ecosystems

Some general principles about energetics that have emerged from all of these studies are that (1) the world of nature is (of course) as beholden to the laws of thermodynamics as any other system; (2) ecosystems can be considered the products of self

design—that is, given the environmental circumstances where they are found they "create" through natural selection (operating at various levels, from genes to organisms to species to coevolved groups of organisms) the systems that we see today; (3) virtually every ecological or even biological action or interaction has an important or even dominant energy component; (4) ecological efficiency, or transfer from one place to another, is low, from one or two to a few tens of a percent, principally because of the need for each level of a food chain to support its own maintenance metabolism; (5) each organism is an energy integrator that integrates through growth and reproductive output (i.e., fitness) all energy costs and gains; and (6) each specific organism is adapted to its own biophysical environment in a way that maximizes this difference between energy costs and gains. As a consequence, the collective energy captured and used by that ecosytem is thought to be maximized; (7) species are ways of packaging morphologies, physiologies, and behaviors that allow sufficient, or even maximum, reproductive output as a consequence of having more or less optimal patterns for that given environment. The ecosystems that have survived over time are thought by many to be maximizing the power (by operating at intermediate, not maximum, efficiency) obtainable under the existing physical circumstances.

More recent research on ecosystems and energetics (i.e., since about 1980) has focused on increasingly detailed examinations of the role and activity of particular species, a greater understanding of the factors that control ecosystems, the importance of disturbance, and recovery from disturbance and the role of microbial components in decomposition and regulation. An important component of this has been the 20 or so LTER (long-term ecological research) centers where basic studies are carried out for long time periods to see how, for example, ecosystem components respond to climate variability, long-term exclusion of herbivores, and so on. This approach has allowed the comparison of the basic patterns of ecosystem.

2. Energy and the Structure of Ecosystems

Neither ecosystems nor their component species can exist without a constant supply of energy to maintain the biotic structures and their functions. The source of this energy is in almost all cases the sun. Clearly the sun runs the energy, carbon, and nitrogen fixation of green plants. Less obviously the sun does many other things for ecosystems: most important it evaporates and lifts water from the ocean and delivers it to continental ecosystems, replenishes carbon dioxide and other gases through winds, pumps water and associated minerals from the roots to the leaves via transpiration, provides nutrients through weathering, and so on.

Ecosystem studies tend to be different from much of the rest of ecology, and indeed science in general, in their emphasis on interdisciplinary and interactive studies. In most other subareas of science, including ecology, the emphasis is on understanding a specific phenomenon and on carefully controlled studies under exacting laboratory conditions. In contrast, when studying ecosystems it is generally necessary to consider aspects of physics, chemistry, geology, meteorology, hydrology, and biology, and in many cases some of the social sciences as well. Since it is impossible to replicate and control all of these factors carefully, as would be the case in the laboratory, ecosystem studies are more likely to be based on observation or correlation than experimental approaches. In these senses and in others, ecosystem science tends to be holistic—that is, it emphasizes the entirety of both a system and a problem rather than just some aspect, and it frequently uses the conceptual tools of systems analysis. For example, if we are to understand the relation between a particular ecosystem and the atmosphere, we obviously must consider the physics of the response of organisms to temperature, the chemistry of the relation of carbon exchange between plants and the atmosphere, the possible effects of changing climate on hydrology and hence the availability of the water needed for the plant to grow, the movement of gases through the atmosphere, and the social factors that are causing the carbon in the atmosphere to increase. In addition, we often need special integrative tools that are outside the traditional scientific disciplines that allow us to examine all or many of these factors in interaction. These tools include quantitative flow diagramming, computer simulation, and theories that are transcomponent.

Such a holistic, or, more accurately, systems perspective is increasingly important as human impacts are changing many aspects of the earth rapidly. Such a perspective does not obviate other approaches to science; in fact, it tries to incorporate and integrate as possible all relevant levels and techniques of analysis. The problems that the increasing human population and its growing affluence are putting on the world require a whole different approach to science—one that is difficult if not impossible to do in the laboratory. Nevertheless it must adhere to the same basic standards as the rest of science in that the results of the science must make predictions that are consistent with the real world. This can lead to certain problems when scientists concerned with the destruction of ecosystems and their species are faced with attempting to apply their results to saving nature. The science of ecology is full of generalizations about the supposed virtues of undisturbed nature that have led to theories such as diversity and stability, diversity and productivity, and the intermediate disturbance hypothesis whose validity seems to wax and wane with each new study. Separating the

robust ecological theory that will help with the genuine environmental crises we face from the good sounding but poorly extrapolateable theories has often been a difficult feat.

3. Origin and Approach of Ecosystem Studies

The word ecosystem is a combination of the words "ecology" and "system." The word ecology is derived from the Greek word *Oikos,* which means pertaining to the household, especially to the management of the household. The academic discipline of ecology refers to the study, interpretation, and management of our larger household, including that of the planet that supports our species and that provides the basis for our economy. System refers to all of the (or all of the important) components of whatever is being considered and the pathways and rules of interaction among the components. Thus, ecological system, or ecosystem, refers to the structure and function of the components that make up household earth and the interactions among those components.

Ecosystems generally are considered from the perspective of their structure, that is their composition and changes in structure (in terms of geological land forms, number, diversity, and abundances of species, biomass, height of vegetation, abundance of critical nutrients, etc.), their function, that is the pathways and rates of energy and nutrient flows, and the regulatory or control processes that govern those changes and flows. We might also want to ask many other kinds of questions about, for example, the diversity of the species that live there or how that diversity effects the function of the ecosystem. Ecosystem structure and function are related over evolutionary time. For example, tall trees functionally need to supply water and nutrients to growing leaf tips, which in the case of the rain forest might be 60 m above the ground. Consequently they have evolved elaborate structures, including a system of pipes, to pump the water to those growing tips. This in turn is possible only where the soil is generally wet and the air is dry, at least periodically.

An extremely useful basic assessment of the most important attributes of ecosystems is found in the classic paper "Relationships between Structure and Function in the Ecosystem" by E. P. Odum. Odum compared the basic attributes of an evergreen forest (such as might be found in the Southeastern United States) and a deep ocean ecosystem (such as might be found a few hundred kilometers to the west of that peninsula). Odum found that these ecosystems are very different in their physical structure—that is the physical characteristics of the ecosystem itself. The organisms themselves constitute the major physical structure in the rain forest. It is possible for the plants to be large here because the soil allows for an anchoring place for the tree, the climate is near optimal for plant growth, and there are no large grazing animals that can eat the tree. Therefore the trees can become very large and become themselves the principal determinant of the physical structure of the ecosystem.

In the ocean, by comparison, the lack of a firm substrate and the constantly changing position of a parcel of water makes it impossible for the organisms to change the nature of the physical medium (the water) significantly, and it is the water itself that forms the basic structure of the ecosystem. Since the organisms are constantly sinking or being swept away by turbulence, they must be small and be able to reproduce rapidly. The total biomass, or living weight, of the aquatic organisms is very small compared to that of the forest. Also, interestingly, the grazers (or animals that eat the plant) have nearly the same biomass as the algae in the water but are much less, as a proportion, in the forest.

One of the most interesting observations when comparing these two ecosystems is that although the biomass of the plants is thousands of times greater in the forest, the rates at which the plants capture the sun's energy (their primary productivity) is much more similar, especially in coastal areas (Table II). In fact, given the greatly varying quantity of plant biomass around the world, most ecosystems have broadly similar rates of production, at least if there is enough water. Thus, we can say that the structure appears much more variable than the function, at least with respect to these particular attributes.

From an energy perspective, we know that the structure of an ecosystem is costly to build and costly to maintain. "Good" structure—that is structure that has been selected for—will be well worth it energetically for it will draw maximum power (i.e., useful energy sequestered per time) given the environmental possibilities, and it will become more abundant than those components that draw less power. Hence, structure and function, and the controls that link them, are intimately and functionally linked through natural selection.

One of the fundamental ideas used to analyze ecosystems is the idea of food chains, which is based on the transfer of energy originally derived from the sun through the biological system. The word "trophic" comes from the Greek word meaning food, and the study of feeding relations among organisms is called trophic analysis. More specifically, the study of trophic dynamics emphasizes the rates and quantities of transfer over time. Trophic processes are the essence of ecosystem processes, for they are the ways that solar energy is passed from the abiotic world to plants and to animals and decomposers, and they allow nonphotosynthetic organisms, including ourselves, to exist. These pathways of food are also a principal way that materials such as nutrients are passed from one location to another within an ecosystem. Trophic studies are an important component of ecology because trophic processes determine energy availability and hence what is and what is not possible for a given organism, and for all organisms collectively, within an ecosystem. In addition, many management concepts and objectives important to people, such as those relating to harvesting fish, timber, or crops, are oriented toward understanding and directing the trophic relations of nature. Energy is stored for a short time in the ecosystem in the biomass of the various organisms, soil, and so on, and some of this energy is transferred from one trophic level to the next (Fig. 1). For example, some of the free energy captured by the plants is eaten by heterotrophs, although a larger part is used by the plant itself. Thus, most of the energy once captured is lost from each trophic level as waste heat as organisms use that energy to fight entropy. Eventually, all the energy originally captured by autotrophs is degraded and returned to the surrounding environment as waste heat. Ecosystems do not recycle energy; instead, they capture a small percentage of the solar energy that falls on them, concentrate it, use it to do work, and then return it to the environment in the form of low-grade heat that is no longer available to do work. The principal work that has been done is the maintenance of life, and this has required energy to sequester and rearrange chemicals, overcome

Table II

Ecosystem structure and function of some typical ecosystems as measured by approximate biomass and production levels.

Ecosystem type	Biomass (dry tons/Ha)	Net productivity (dry tons/Ha/yr)
Tropical rain forest	225 (60–800)	10–35
Tropical seasonal forest	175 (60–600)	10–25
Savanna	40 (2–150)	2–20
Temperate deciduous forest	200 (60–600)	6–25
Temperate evergreen forest	200 (60–2000)	6–25
Boreal forest	200 (60–100)	4–20
Temperate grassland	16 (2–50)	2–15
Desert	7 (1–40)	0.1–2.5
Ocean	0.03 (0–0.05)	0.02–4
Estuaries	10 (0.1–60)	2–35

Based on summary by R. L. Whittaker and G. Likens, in Woodwell and Pecan (1973), modified by Sandra Brown, personal communication.

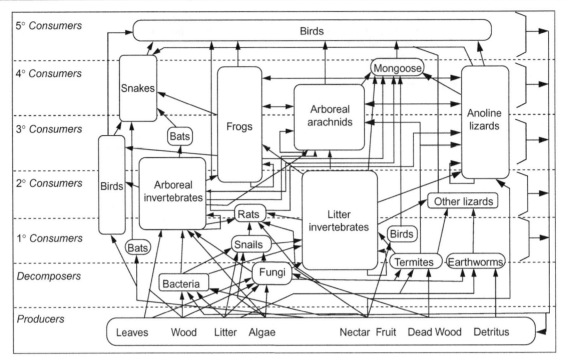

Figure 1
Energy flow diagram for the food web of the Luquillo Forest, Puerto Rica. Pictorial representation. Modified from Hall, C. A. S. (1999), Ecosystems, pp. 160–168, in D. E. Alexander and R. W. Fairbridge (Eds.), "Encyclopedia of Environmental Science." Kluwer Academic, Norwell, MA.

gravity, move either themselves (animals) or their chemical constitutes (plants and animals), generate microclimates, maintain general homeostasis, and then replicate the whole structure and functions through reproduction.

There is constant selective pressure operating on organisms to maximize the useful energy that they capture and to use it in ways that contribute to their survival and to propel their genes into the future. Many think that this process leads to the collective maximization of power at the level of the ecosystem—that is, those organisms and arrangements of nature will be selected for that most fully exploit available energy and use it most effectively compared to alternative organisms or patterns. Obviously this is a difficult idea to test at the level of entire ecosystems and on a scale of human generations, but it does offer intriguing hints that natural selection operates in ways that generate and maintain (or self-organize) entire systems. For example, although it is unlikely that many of the same species were present in the wet montane tropics 300 million years ago as are there now, it is likely that a recognizable tropical wet forest was there that had similar functions and rates of processes as modern tropical wet forests.

3.1 Trophic Processes in More Detail

The pathway of energy conversion and transfer (the eating of one organism by another) goes from the initial capture of solar energy by autotrophs, to the herbivores that eat plants, to the first-level carnivores (that eat herbivores), on to the top carnivores. The principal pathways within a food chain can be represented as energy transfer through a series of steps. The power flow of an organism per unit area, or of a trophic level, is called productivity and normally is expressed in units of kilocalorie per square meter per time.

Autotrophs. In the first step autotrophs, or green plants, use chlorophyll to capture energy from solar-derived photons and store this energy by restructuring the carbon atoms of carbon dioxide derived from the surrounding atmosphere or water into complex organic compounds. Primary production is the fixation of solar energy by green plants. Gross productivity is total energy captured, whereas net production is that minus the energy required for respiration. Net energy is then allocated into tissue increments, leaf turnover, reproductive products, and, in time and through these processes, the replacement of the first plant with a second. In the second step herbivores or primary consumers obtain energy by eating autotrophs. Secondary production is the accumulation of animal or decomposer living tissue. Heterotrophs obtain metabolically useful energy from the consumption of the organic molecules in the food they obtain from other organisms.

Grazers.   Since a substantial amount (perhaps 80 to 90%) of the energy captured by an ecosystem is used by the first trophic level for its own necessary maintenance metabolism, food chains are inherently and necessarily "inefficient." Thus, there is and must be much more production of biomass of deer than wolves or cougar in a forest, for the deer must use a large portion of the energy in the leaves it eats for its own metabolism rather than for growth; much less is left for their own growth and the potential exploitation by the predators.

Decomposers.   The world is covered with living creatures, and these creatures are continuously dying. Yet the world is not littered with carcasses, so obviously they disappear in some way. We call this process decomposition, and it is mediated principally by single-celled organisms, especially bacteria and fungi. In addition, the decomposition process gets a start initially by larger animals, from buzzards to maggots to protozoans, and in earlier days, even our own ancestors. Although humans are often disgusted by decomposition (and, in fact, have probably been selected for this quality), without decomposition the biological world would come to a stop, as all of the earth's available materials would be tied up in various carcasses!

Some food chains are complicated indeed. Think of a salmon feeding on the plankton of the mid-Pacific ocean and then swimming far up an Alaskan river, fueled by energy reserves built on this plankton, to spawn and die—or to be eaten by a bear. The offspring of the salmon that did spawn spend their first year of life in the freshwater environment, eating zooplankton that ate phytoplankton whose growth was fertilized by phosphorus leaching from the bones of the dead parents. Another way that food chains are complicated is that much of the energy flows not through live but through dead organisms. For example, in the rain forest most of the plant material is not eaten directly by live leaf munchers but instead by various insects and other small organisms that consume the dead leaf material called detritus.

3.2   Trophic Dynamics and Biomass Pyramids

Thus, of all the energy captured by green plants, progressively less and less flows to the next consumer,

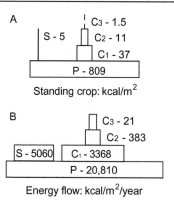

Figure 2
The trophic categories given in Fig. 1 represented as (A) an energy pyramid and (B) a biomass pyramid. Data from Silver Springs, FL. Modified from Hall, C. A. S. (1999), Ecosystems, pp. 160–168, in D. E. Alexander and R. W. Fairbridge (Eds.), "Encyclopedia of Environmental Science." Kluwer Academic, Norwell, MA.

or, as it is often called, trophic level. When the rate of flow of energy is graphed, it nearly always will look like a pyramid, with the large base representing the energy captured by the autotrophs and each successive layer representing higher trophic levels, each further removed from the base (Fig. 2). This is called the pyramid of energy. Biomass plotted by trophic level may look like this but often looks more like a square than a pyramid. How can this be? The reason is that larger organisms often found in higher trophic levels generally respire (use up) energy more slowly; in a sense they get less energy, but they hold on to what they do get for a longer time. So in the ocean there is roughly the same biomass of algae, zooplankton, small fish, and large fish. The reason is that the small algae "turn over" much faster, meaning that they have a much higher rate of metabolism and loss to other organisms that eat them.

4.   Biogeochemical Cycles

A second critical area for the study and understanding of ecosystems is the structure and function of nutrient cycles. Nutrients can mean all of the chemical elements that an organism or an ecosystem needs to grow and function, but most commonly the term is restricted to a consideration of nitrogen, phosphorus, potassium, and less commonly calcium, sulfur, iron, and micronutrients such as copper, cobalt, and molybdenum. Sometimes, but not generally, CO_2 is considered a nutrient, for it too can limit growth of plants. Finally, it also (rarely) can mean complex organic materials necessary for growth, such as vitamins. We refer to the movement of a particular nutrient through an ecosystem, often changing from one form to another, as a nutrient cycle. Nutrient cycling, like all other processes, occurs because of energy forces including evaporation, precipitation, erosion, photosynthesis, herbivory decomposition, and so on, all run by the sun.

4.1 Limiting Nutrients

The study of nutrients in ecosystems was first undertaken systematically by the German chemist Liebig, who found that the growth of plants tended to be limited by only one element at a time. For example, if a plant had insufficient phosphorus relative to its needs, its growth would stop until phosphorus was added. At some point adding more phosphorus had no further effects, but at that time some other element, perhaps potassium, might be limiting the plant's growth, as could be determined by noting the growth response, or lack thereof, of adding potassium. This led to the formulation of Liebig's law of nutrient limitation, which states that the growth of plants tends to be limited by a single nutrient at any one time, that which is least abundant or available relative to the needs of the plant. Nutrient cycles at the global level were initiated by Vernadsky and greatly encouraged by the growth of atomic energy in the 1950s, which provided radioactive tracers that helped follow nutrient pathways, funding to do that research, and also a reason as to why we might be concerned about them—that is, if there were releases of radioactive elements into the environment.

Nitrogen is especially important because plants need it in large quantities and it is often restricted in its availability and energetically expensive to get into a useful form. Plants and animals use nitrogen extensively because it is, after carbon, the most important component of proteins. Plants extract it from the soil, and animals get it by eating plants or other animals. All organisms must concentrate the normally relatively rare nitrogen to get enough to build their proteins. Waste nitrogen is released from animals as ammonia or other compounds in their urine. Since N_2, ammonia, and nitrous oxide are volatile, nitrogen is different from most other nutrients in having a gaseous phase.

The cycle of nitrogen in ecosystems is also complicated because nitrogen is a useful and versatile atom with a valence state that allows for many complex interactions with other atoms. All organisms use nitrogen to make proteins. Some use ammonia as a fuel to gain energy. Others use nitrate for an electron receptor in the same way that humans use oxygen. Still others invest energy to take it out of the atmosphere for their own use. Even though nitrogen is presently about 80% of the atmosphere, plants cannot use it in this form, only after it is fixed (i.e., incorporated as nitrate or ammonia). In nature, this can be done only by the bacteria and blue-green algae. Thus, all members of all ecosystems are dependant on these two types of life for their nitrogen, or at least they were until about 1915, when humans learned how to fix nitrogen industrially through the Haber-Bosch process.

Evolution has found phosphorus to be an especially useful element, and hence it has a particular importance in ecosystems. Phosphorus is used by all organisms for their genetic material (DNA and RNA) and for energy-storage compounds (ATP), and in vertebrates for teeth and bones. It seems to be in a rather peculiar short supply in the world's biogeochemical cycles relative to the importance given it by evolution. Phosphorus tends to be especially limiting for the growth of plants in freshwater, and even where abundant it may be held very tightly by soils so as to be virtually unobtainable by plants. Phosphorus cycles rapidly between plants and the environment and is passed along food chains as one organism eats another (Fig. 3). When organisms die, they decompose, and we say that the nutrients such as phosphorus are mineralized (returned to their nonliving state). Phosphorus is easily dissolved in water, and in many respects the cycle of phosphorus at the ecosystem level is the cycle of water.

4.2 Redfield Ratios

It is not only the abundance of each nutrient that is important, but also their concentration relative to each other. The importance of this idea was first worked out by the oceanographer Alfred Redfield, who sailed around the world's seas on the first oceanographic ship, the *Challenger*, meanwhile taking samples of the oceans and its various life forms. Redfield's chemical analysis was a rather difficult undertaking, since the surface of most of the earth's seas are low in nutrients because the constant sinking of organisms and their feces tends to deplete nutrients in the surface of the sea. Nevertheless, Redfield was rather astonished to find that both the sea and the small plants of the sea had nearly constant ratios of nitrogen to phosphorus, about 15 atoms of nitrogen for each one of phosphorus, in both the living and the nonliving environments. Since we know that aquatic plants need roughly 15 atoms of N for each one of P, then we can determine whether the environment is N or P limited by whether it has more or less than this ratio. If the water is 20 to 1 it is likely to be P limited, and if it is 10 to 1 it is likely to be N limited. Likewise agricultural extension agents can analyze different soils and determine which nutrient is limiting, saving the expense and environmental problems of pouring on unneeded nutrients.

5. Stability and Disturbance

A question that has intrigued many ecologists is the degree to which ecosystems are stable—that is

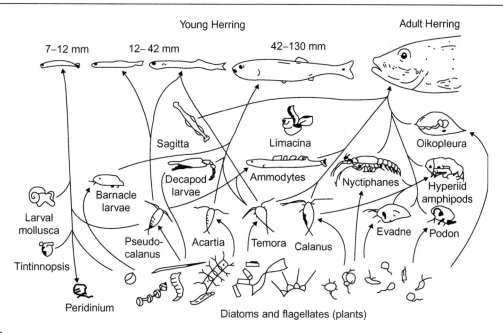

Figure 3
An early example of a food chain (from Wells *et al.*, 1939). In this case, the energy is represented as flowing from right to left, the opposite of later convention.

relatively homeostatic or self-correcting. It seems to many early investigators that natural ecosystems tended to have self-correcting properties, so if they were disturbed they would bounce back toward their original conditions.

5.1   Succession

Hence earlier ecological studies tended to emphasized the equilibrium nature of ecosystems. The idea was that there was a strong tendency for ecosystems to "evolve" toward a "climax" state determined by the climatic conditions of the site. In this view, succession was an orderly process that would bring the system back to climax, a condition of stability with no more species change. Climax was considered the normal state of ecosystems, and considerable virtue was associated with this process, a sort of "mother nature knows best." The view is evolving that ecosystems are continuously subjected to disturbance, and are continuously responding in various ways to the various disturbances they are subject to. For example, tropical ecosystems were once thought particularly unchanging. Research has shown that many tropical forested ecosystems are continuously in disequilibrium—that is, recovering in some way from some externally imposed stress or pulse, such as a hurricane, a landslide, a drought or, locally, a tree fall.

5.2   The Role of Species Diversity

An unresolved issue in ecology is to what degree all of the different species found in an ecosystem are required for it's proper functioning. Twenty-five years ago, the idea that diversity (of species) and stability (of populations and ecosystems) were positively connected was very popular. At this time, the best that we can say is that certainly an ecosystem needs many species to function properly and that through coevolution many species are tightly linked to each other. But we do no know in a general sense whether ecosystems *need* all, or even most, of the plethora of species that are found there in order to continue their basic structure and function. At the same time, we are finding an increasing number of examples where animals once thought rather superfluous with respect to ecosystem function are in fact critical for some important function such as distributing or germinating the seeds of a dominant plant. Species that have a disproportionate effect on the entire ecosystem are called Keystone species. An example is the sea otter of California-to-Alaska coastal waters, which maintains kelp beds by eating sea urchins that would other wise overgraze the kelp. Additionally we are understanding better that ecosystems themselves can be much more resilient than many of their component species, since if one species is knocked out, another (although perhaps one less desirable from a human perspective) can sometimes take over that function. This is an important question as all over the world humans are reducing the diversity of ecosystems, and they are especially impacting the large animals.

5.3 Disturbance

An important aspect of ecosystems that we have learned is that although we once thought of ecosystems as, generally, stable and focused on the tendency toward equilibrium that occurred through the process of succession, we now view ecosystems as much more transient phenomena, generally subject to, and responding to, external forces, which we normally term disturbances. For example, graduate students are often taught that tropical forests were "stable," due principally to a lack of winter and glacial-cycle impacts. In reality, many tropical forests, for example the Luquillo experimental forest in Puerto Rico, are routinely subject to three type of natural disturbances: tree falls, landslides, and hurricanes, so that the forest is almost always responding to, or recovering from, disturbances.

Some important nutrient cycling studies were done at the U.S. Forest Service experimental watersheds at Hubbard Brook, New Hampshire. Here entire watersheds were cut, and vegetation regrowth suppressed, to examine the role of the vegetation in determining nutrient retention by ecosystems (it was substantial). Even so, the scientists there also found that long-term trends in rainfall intensity were equally important in determining the nature of the ecosystem. In other words, even in a relatively undisturbed forest there was a large amount of what we might call disturbance introduced simply by the normal decade-scale cycles of drought and flood, and this greatly interfered with the evolution of the ecosystem toward what we might have once called mature. Thus, at the level of the ecosystem it is clear that both endogenous and exogenous factors are important in determining the structure of the ecosystems and the nature of nutrient movements.

5.4 The Role of Human Disturbance

We previously discussed the importance of natural disturbance in determining the structure and function of existing ecosystems. Unfortunately, often it has been frustratingly difficult to unravel the impact of natural factors on ecosystem structure and function because almost all ecosystems of the world have been heavily impacted by prior human activity. For example, it was widely believed that New England forests were in some kind of pristine condition prior to the American Revolution, or even prior to European colonization. Instead we now know that the New England landscape had been modified severely by the activity of both Native Americans and very early colonists.

Likewise Perlin has documented the intensive deforestation that has taken place in virtually all portions of the world over the past thousands of years due to human activities. The basic cycle, repeated over and over again, has been that initially human settlers had relatively little impact on the forested lands except to clear small areas for agriculture. Often this land was abandoned after fertility declined and a new patch cut while earlier patches reverted to forests. This was not too different from the patterns of natural disturbance. But then, characteristically, some metal was discovered. Huge areas of forests were cut to provide the charcoal to smelt the metal, and the net effect was extensive deforestation. Subsequently there tended to be an increase of agricultural production on these newly cleared and fertile lands, followed by large human population increases, and then substantial soil erosion, river and harbor sedimentation, and eventual agricultural and population collapse. For example, ancient Greece was deforested 4000 years ago and then recovered, only to be again deforested, recover, and be deforested again in relatively modern times. The use of chemical fertilizers has allowed agriculture to be maintained over a longer time period, so that the land is not abandoned to reforestation. Curiously, some few parts of the world, notably the Eastern United States and Western Europe, are now undergoing reforestation because the exploitation of fossil fuels has allowed the forests a respite, albeit perhaps only temporarily.

There are other ways in which people have and continue to impact natural ecosystems. Humans have changed the relative abundance of species through the introduction of new species and through the exploitation of natural populations. For example, the introduction of goats into the Canary Islands 800 years ago resulted in wholesale destruction of the natural vegetation, and the unique flora and fauna of islands such as the Galapagos and Hawaii are being changed dramatically by human-caused introductions, many of them centuries old. Curiously the flora and fauna of the island of Puerto Rico seem to be much less impacted by invasive species, and when the vegetation is allowed to recover from human disturbance, native species predominate.

One consequence of all of this is that it is difficult for ecologists to measure what is "natural" in nature, for almost all nature that we have had an opportunity to observe is already quite different from what might have existed some tens of thousands of years ago. Meanwhile, climate change, at both a local and global level, are impacting all organisms in every ecosystem, subjecting them to different temperature and rainfall regimes and to different carbon and nitrogen levels than those within which the organisms evolved. We are learning that ecosystems can adapt surprisingly well to at least some of these changes, but we do not know how long this will continue into the future.

6. The Future of Ecosystems

The natural ecosystems of the earth are tremendously and increasingly impacted by human activities. For example, Vitousek and his colleagues have estimated that fully 40% of the earth's primary production is directly or indirectly exploited by our own species, and others believe this proportion is higher. This is remarkable considering that we are but one of millions of species on this planet, and that about two-thirds of the planet is water and as such is difficult for humans to access. People are probably not doing anything conceptually different from other organisms in nature, which also tend to eat or otherwise exploit whatever resources are available as much as possible. Natural selection almost certainly has hardwired us for lust and something resembling greed. The difference is that agricultural, medical, and especially fossil fuel technology have allowed people to be enormously more effective at exploiting resources compared to other species, or indeed compared to our earlier selves. At the same time, the earth's ability to check our own population levels through predation and disease has been greatly reduced. The net result is that the human impact has grown enormously.

The consequences of this tremendous rate of exploitation are only beginning to be understood. They are not necessarily bad from the perspective of ecosystems, at least if we measure good and bad in terms of their biomass and productivity. For example, Nemani and his colleagues assessed the impact of changing climates and nutrient cycles on global vegetation and found that there had been a probable increase in productivity of ecosystems in the temperate regions and a decrease in the tropics. But in a conference on the sources and sinks of carbon, Sampson and his colleagues concluded that human activity was turning what had been natural sinks of carbon for most of the world's ecosystems into carbon sources, probably exacerbating the greenhouse effect. Although various scientists argue whether or not human activity has impacted the earth's climate to date, that is not really the issue. We have changed the concentration of carbon dioxide in the atmosphere by only about 40% so far. If civilization continues along its present path, the change could be 500 to 1000%. Thus, natural ecosystems, and the services that they provide, are likely to continue to decrease in area and naturalness, since the human population continues to grow—in fact, to increase more rapidly than ever before—and most of the earth's human citizens have increasing desires for material goods, which can come only from the earth itself. Young people reading this article will almost undoubtedly see a disappearance of most of the world's natural ecosystems during their lifetimes unless there is an extraordinary effort to reduce human impact or unless that occurs involuntarily.

7. Conclusion

Ecosystem science got a relatively slow start in America because of an early species-oriented, stamp collecting attitude toward the large, fascinating, and largely unknown diversity of nature in the new world. In addition, there was competition for funding, attention, and prestige from other levels of ecological and biological inquiry, a largely false sense of the power of formal mathematical analysis that came into vogue in ecology, and the fact that ecosystem studies often were expensive. But in the 1950s, ecosystem studies began to prosper, propelled in part by new techniques for studying the energetics of whole aquatic systems and also by curiosity and funding related to how radioactive materials might move through ecosystems if released by accident or war. Important studies of grasslands, forests, and lakes in the 1960s quantified many basic aspects of ecosystems and began the process of constructing ecosystem models. In the 1970s, an ecosystems approach to watersheds in New Hampshire (Hub-bard Brook) and North Carolina (Coweeta) showed how an experimental approach could be used to address many basic aspects of understanding ecosystems, and Howard Odum began his analyses of humans as components of ecosystems. The importance of ecosystems-level studies is becoming more and more obvious, and ecosystem studies have changed from a relatively obscure part of biology to a healthy and vibrant science all of its own. For example, the U.S. Forest Service in the early 1990s changed the entire focus of its management for our national forests from one based on timber yield to one focused on managing the forests as ecosystems. Another important trend is that ecosystem approaches are moving from the temperate latitudes increasingly to other regions, such as the tropics and tundra. One next phase, barely started, is to use ecosystem approaches to study cities and the relation of cities to natural or seminatural systems. Perhaps our understanding of how natural systems have evolved to prosper in the circumstances of relatively limited and stable energy supplies can give us some useful information as humans, presumable, will have to adapt to a more energy-restricted future.

Acknowledgments

This article is based in part on Hall, C. A. S. (1999), Ecosystems, pp. 160–168, in D. E. Alexander and R. W. Fairbridge (Eds.), "Encyclopedia of Environmental Science." Kluwer Academic, Norwell, Massachusetts.

Further Reading

Brown M T, Hall C A S (eds.). (in press). An H. T. Odum primer. *Ecological Modeling*

Craig B J 2001 Eugene Odum. *Ecosystem Ecologist and Environmentalist.* University of Georgia Press

Golley F 1993 *A History of the Ecosystem Concept in Ecology: More Than the Sum of the Parts.* Yale University Press, New Haven, CT

Hagen J P 1992 *An Entangled Bank: The Origins of Ecosystem ecology.* Rutgers University Press, New Brunswick, NJ

Hall C A S (ed.) 1995 *Maximum Power: The Ideas and Applications of H. T. Odum.* University Press of Colorado, Niwot, CO

Hall C A S 1988 An assessment of several of the historically most influential models used in ecology and the data used in their support. *Ecol. Modelling* **43**, 5–31

Likens G, Borman H 1995 *Biogeochemistry of a Forested Ecosystem.* 2nd ed. Springer Verlag, New York

Lindeman R 1942 The trophic-dynamic aspect of ecology. *Ecology* **23**, 399–418

Odum E P 1962 Relationships between structure and function in the ecosystem. *Japanese J. Ecology,* **12**(3), 108–18 Reprinted in Bobs Merrill College Division reprint series in the Life Sciences.

Odum H T 1957 Trophic structure and productivity of Silver Springs, FL. *Ecological Monographs* **27**, 55–112

Odum H T 1988 Self-organization, transformity and information. *Science* **242**, 1132–9

Odum H T (199X). Systems Ecology. Wiley Interscience. Reprinted as *General and Ecological Systems.* University Press of Colorado, Niwot, CO.

Thomas D W, Blondel J, Perret P, Lambrechts M M, Speakman J R 2001 Energetic and fitness costs of mismatching resource supply and demand in seasonally breeding birds. *Science* **291**, 2598–600

Wiener D 1988 *Models of Nature: Ecology, Conservation and Cultural Revolution in Soviet Russia.* Indiana University Press, Bloomington, IN

Woodwell G M, Pecan E V (eds.). (1973). Carbon and the biosphere. Brookhaven Symposium in Biology, 24. National Technical Information Service. Springfield, VA.

Charles A. S. Hall

Electricity Use, History of

In the 19th century, electricity was at first exotic, and then a luxury. Since then it has become a necessity, but paradoxically both the level of demand and the level of use vary considerably from one nation to another. The telegraph (1838), telephone (1876), and broadcast radio (1919) at first spread slowly; as with many later applications, consumers needed time to discover their uses. An interconnected electrical system grew rapidly from c. 1880, with lighting followed in rough sequence by commercial establishments (such as department stores), public buildings, monuments, streetcars, railways, subways, factories, homes, airports, farms, and stadiums. The precise sequence of adoption and degree of penetration varied from one nation to another, particularly in rural areas. Likewise, the intensity of use varied. In c. 1960, Norway, Canada, and the United States consumed more than twice as much per capita as England and France. Continually improved generating efficiencies did much to meet increasing demand while lowering the cost per kWh, until reaching engineering limits in the 1970s. Since then, new efficiencies have been in the appliances and light bulbs consumers use. Nevertheless, rising demand for electric power creates environmental problems that alternative energy sources such as wind mills can only partially solve.

1. Inventing and Promoting Electrification

Until the many practical applications of electricity that became available in the late 19th century, no single energy source could provide light, heat, and power. Electricity's versatility made it an enabling technology that allowed people in all walks of life—including architects, designers, homemakers, industrialists, doctors, and farmers—to do new things. These new and often unanticipated uses found for electricity mean that its history must include far more than such matters as dam construction, the extraction of fossil fuels, or improvements in dynamo efficiency. One must also consider energy history in terms of demand. Consumers have been an active force behind the expansion of electricity use. If one considers the rapid increase in electricity demand from the middle of the 19th century to the present, one is struck by how this increase parallels the growth of the middle class, the expansion of cities, the development of mass consumption, and increased demands on the environment. The uses of electricity are more than mere applications; they are inseparable from rising demand, which at times has outstripped the capacity of utilities and led to major programs in demand side management. Indeed, the same pattern repeated itself with the electric light, the electric iron, or the electric vacuum cleaner and a host of appliances and later applications. A new device first became a luxury for the rich, then a desirable item of conspicuous consumption for the middle classes, and, within little more than a generation, a necessity for all.

Practical applications were only possible in the 19th century and after. The understanding of electricity was so rudimentary in the late 18th century that Benjamin Franklin, a clever self-taught amateur in the American colonies, made an important contribution to knowledge by discovering that lightning and electricity were the same thing. The phenomenon of electricity had remained a curiosity since classical times, when static electricity was known but scarcely investigated. Few scientists focused attention on it until after the invention of the Leyden jar (1746), which facilitated a century of experiments that prepared the way for practical applications. Charles Coulomb, Michael Cavendish, Benjamin Franklin, Luigi Galvani, Alessandro Volta, Humphrey Davy, and Andre Ampere developed a rich store of observation and theory that made possible Faraday's studies of electromagnetic induction, which in turn made possible the dynamo (1831). When Faraday was asked whether his experiments would ever result in tax revenue, he boldly declared that they would.

Indeed, during the next half century a series of industries and businesses emerged based on the special properties of electricity. Its movement was so rapid as to be instantaneous, a property exploited by the telegraph, various alarm systems, and the telephone. It could produce electrolysis, making possible metal plating. Electricity could also produce light when passed through certain conductors, eventually making possible arc and incandescent lighting.

Electricity provided a way to transmit power over considerable distances, allowing falling water or burning coal to be translated into motive force by distant electric motors. Electricity could also be used to produce precisely controlled heat without fire or the consumption of oxygen, leading to applications in baking, enameling, and industrial heating and burning. Each of these innovations was eventually incorporated into devices used in the home and on the farm.

Given the many potential uses of electrification, it might seem that its adoption was virtually automatic. Such was not the case. As late as 1912, less than 3% of electrical power was residential, whereas almost 80% was used in manufacturing, with the remainder used commercially. Although it might be tempting to treat the history of electrification as the story of inventors producing a stream of practical devices, it was a far more complex developmental process that included government, manufacturing, marketing, and several different groups of consumers. Utilities quickly discovered that large, urban customers were far more profitable than smaller customers in suburbs and the countryside. The spread of electrical service therefore moved from larger to smaller customers, and from urban to rural, in a series of distinct stages. This process began in the 1870s in Europe and the United States and has since spread throughout the world. However, even in the early 21st century electricity has by no means become the dominant source of lighting or power in every location, and in most cases other sources of heat are more economical. If the world's capital cities are electrified, some of the countryside remains in darkness, as revealed by satellite photographs. Indeed, even in the nations in which the process began—England, France, Germany, and the United States—in 1900 almost all factory power still came from falling water or steam, and less than 5% of all homes had electricity. Although in retrospect electrification might appear to have been a steady and inevitable process, closer inspection reveals uneven growth, at times thwarted by competition from gas, coal, and other energy forms. In general, electricity replaced gas for lighting because the incandescent bulb was safer and cleaner than an open flame and because it was more like daylight. In contrast, gas remains a strong competitor to electricity for heating and can also be used for air-conditioning. Yet such generalizations do not apply universally because energy use was inflected by local conditions and by social class. For decades, English workers clung tenaciously to gas lighting, leaving electricity to the middle and upper classes. In contrast, wealthy Swedes were reluctant to give up traditional forms of lighting that they considered genteel, whereas Swedish workers more rapidly embraced the electric light. In 1931, less than 30% of British homes were wired, and few were all electric, at a time when 90% of Swedish homes had been wired. In the United States, most urban but few rural dwellings were electrified by that time, and wiring was less a marker of social class than of geographical location.

From a technical standpoint, electrification was not a unified process that spread like ripples in a pond from certain nodal points. Rather, there were several interrelated processes, which began at many different locations, driven not only by utilities but also by the demands and interests of consumers. Particularly in the early decades, separate systems developed to serve different needs, including small stand-alone generating plants for some hotels, department stores, theaters, and wealthy homes and large, direct current power houses owned by street railways. Until the 1930s, many factories also had their own power stations.

The degree of such fragmentation varied, being considerably greater in Britain than in the United States, for example. In the early 20th century, London had 65 electrical utilities and 49 different types of supply systems. Anyone moving within London risked entering a new grid because there were 10 different frequencies and 32 different voltage levels for transmission and 24 for distribution. Once consumers had converted all the lamp and appliance plugs and made other appropriate changes, they discovered that each utility also had its own pricing system. In the United States, where utilities worked harder to coordinate their systems and interconnect their lines in order to achieve economies of scale, consumers still faced considerable inconsistencies in plug and socket sizes. These were only ironed out in the 1920s with the help of the federal government. Even today, as every tourist knows, the electrical current and the plugs needed to access the system vary among the United States, Britain, and continental Europe, and these three different standards have been reproduced throughout much of the rest of the world.

The large electrical manufacturers, which sold generating equipment and provided technical expertise to new markets, exported these standards. With so many potential customers, the manufacturers did not attempt to enter all markets. Rather, each focused on certain regions, to a considerable degree based on political spheres of influence and colonial empires. For example, General Electric (United States) sold lighting and traction systems to Brazil and much of Latin America, whereas in most cases General Electric (United Kingdom) transferred the same technologies to the British Empire. German and Austrian electrical expertise spread throughout central Europe and the Nordic countries. To some extent, France developed its own systems but relied heavily on investment and partial ownership from the United States.

On the demand side, in all the industrial countries the middle classes developed a demand for electrification in many guises. On the supply side, the electrical industry was an expansive investment opportunity that was more important and longer lasting than the Internet bubble of the 1990s. Between 1875 and 1900, electrical equipment makers and utilities formed the most dynamic sector of the American economy, growing from a few small adjuncts to the telegraph companies into a $200 million industry. After 1900, it grew into a colossus that was inseparable from the health of the national economy. The electrical industry's potential was early recognized by J. P. Morgan and other leading investors who financed Edison's work. Once commercial development began, a flurry of mergers reduced the field from 15 competitors in 1885 to only General Electric and Westinghouse in 1892. Before this time, railroads had been America's largest corporations, demanding the greatest capital investment, but they were a mature industry compared to the emerging electric traction companies, local utilities, and equipment manufacturers.

2. *The Telegraph, Alarm Systems, and the Telephone*

The invention of the telegraph in 1838 heralded the first widespread practical application of electricity. However, potential consumers were not certain what to use the telegraph for. Conceived and patented by Samuel F. B. Morse, the telegraph astounded incredulous crowds, and the first telegraph offices often provided seating for the public, who could scarcely believe that it was possible to sever language from human presence. Twenty years after its invention, when thousands of miles of lines linked the United States, the *New York Times* declared that "the telegraph undoubtedly ranks foremost among that series of mighty discoveries that have gone to subjugate matter under the domain of mind" (August 9, 1858).

However, even the inventor of the telegraph did not understand its enormous commercial potential, and he tried to sell it to the federal government. Many European countries chose to make the telegraph part of the postal service, but the U.S. Congress refused, and only after 5 years granted Morse $30,000 to build a line from Washington to Baltimore. People seemed to have little use for it, however, and even when the first line connected New York, Philadelphia, and Washington business still developed slowly. Six months after it started operation, the line averaged less than 2000 words a day, and it was generating only approximately $600 a week. By 1848, the telegraph linked Chicago to New York, however, and grain traders with immediate access to information soon made fortunes trading in midwestern wheat. In less than a generation, Western Union became one of the largest corporations in the United States. During the same years, newspapers realized the commercial value of Mexican war news and formed an association that inaugurated the wire services. As these examples suggest, the potential of the various forms of electrification was only gradually understood.

During the second half of the 19th century a series of technologies based on the telegraph, such as burglar alarms, stable calls, and door bells, familiarized people with the idea that a current carried over wires could be instantaneously transmitted to virtually any point. However, other practical applications were few because most electricity still came from inefficient batteries. Further use of electricity was limited until large and efficient generators became available in the 1870s. Until then, "impractical" electric lighting was confined to special effects in theaters.

Like the telegraph, the invention of the telephone (1876) at first seemed a curiosity, and Alexander Graham Bell was unable to sell his invention even to Western Union. Setting up on his own, Bell concentrated on businessmen and wealthy households and did not expand service widely until his patent ran out and competition emerged in the 1890s. It took more than a generation before phone companies realized that people wanted phones not just for business and emergencies but also for gossipy conversations. Even so, the telephone took a long time to become a necessity to consumers compared to the electric light. Only after World War II did most American homes have telephones. Even then, they were all black, heavy apparatuses that the subscriber did not own but leased. They were not stylish elements of décor that varied from one home to another but utilitarian objects that were engineered to last decades. In contrast, a greater variety of models were marketed in Europe, despite the fact that universal service only came to some countries during the 1960s, and Spain and Portugal only achieved it at the end of the 1980s. On the other hand, during the 1990s Europeans adopted mobile phones far more quickly than did Americans.

Again, as late as 1919 the U.S. government failed to see the broadcasting potential of radio, understanding it only as a superior form of the telegraph, for wireless point-to-point communication. In the early 1920s, a few amateurs initiated the first public radio broadcasts. Yet even when the potential of a device is grasped, not all societies have made the same choices about how to embed electrical technologies within the social structure. Should suppliers be public or private or a mix of the two? Should pricing encourage extensive or intensive use of a system? What is the best balance between regulation and open competition? Between 1880 and 1900, such questions had to be answered first with regard to public lighting and then street traction systems.

3. *Public Lighting*

In both Europe and the United States, regular public electric lighting appeared first in the late 1870s as a spectacular form of display in city centers and at expositions and fairs. The first installations were arc lights, which shed a powerful glare over streets and the insides of some stores. Department stores and clothiers quickly adopted such lights because they were cleaner, brighter, and safer than gas and because they proved popular with customers. By 1886, 30 miles of New York's streets had arc lights, installed approximately 250 ft apart. In 1879, Thomas Edison demonstrated his enclosed incandescent bulb, which had a softer glow that was well suited to most indoor and domestic settings. Both arc and incandescent lighting competed with already existing gas systems until the first decades of the 20 th century. Edison recognized that his light, its distribution system, and its cost had to be designed to complete with gas. For as much as a generation, some new buildings had both systems. The Edison system was rapidly adopted wherever convenience and fissionability were highly valued, and it spread rapidly into the wealthiest homes, the best theaters, and the most exclusive clubs in New York, London, Vienna, and Paris. However, 20 years after its introduction, the electric light was almost entirely restricted to public places. Less than 1 home in 10 had been wired, and the average person encountered electricity at work, in the street, and in commercial establishments.

From its inception, the public understood lighting as a powerful symbolic medium. The history of lighting was far more than the triumph of the useful. Before gas and electric streetlights people had to find their way with lanterns, and the city at night seemed fraught with danger. Public lighting made the city safer, more recognizable, and easier to negotiate. However, mere functionalist explanations cannot begin to explain why electric lighting had its origins in the theater or why spectacular lighting emerged as a central cultural practice in both Europe and the United States between 1885 and 1915. As early as 1849, the Paris opera used an electric arc lamp to create a sunrise, and showmen of every description seized on the new medium to attract crowds. During the 1870s, thousands of people turned out to see the first arc lights in Cleveland, Boston, London, and Paris. The new light fascinated them not only because it was so much brighter than existing gas but also because it seemed to violate the natural order. For the first time in history, light was separated from fire. It needed no oxygen. It was unaffected by the wind. It could be turned on in many places simultaneously at the turn of a switch. In 1883, Niblo's Garden featured "novel lighting effects by the Edison Electric Light Company," which included an illuminated model of the Brooklyn Bridge and chorus girls with lights flashing from their foreheads as they flourished electrically lighted wands. Promoters and public alike demanded ever-greater public displays as use far exceeded necessity.

Intensive lighting became a form of symbolic expression at not only theaters but also world's fairs and other public events. Every exposition attempted to outdo all previous efforts by escalating both the scale and the completeness of displays. Intensive arc lighting at Chicago in 1894 gave way to thousands of incandescent lights at Omaha (1898), arrays of colored bulbs in Buffalo (1901), and artificial sunrises and electrical fireworks in San Francisco (1915). Many fairs made illuminated towers their central symbols, notably Buffalo's Electric Tower, San Francisco's Tower of Jewels, and New York's Trylon and Perisphere in 1939. Electricity became more than a theme; it provided a visible correlative for the ideology of progress, and it was placed at the apex of an evolutionary framework, as opposed to "ethnological" villages—compounds representing the life of Mexicans, Cubans, blacks in the Old South, Filipinos, or Africans in the jungle—where the visitor could see staged versions of earlier forms of social evolution. Electrification thus became embedded in a social Darwinist ideology of racial superiority, as darkness represented the "primitive," whereas artificial light exemplified Christianity, science, and progress. At the Buffalo Pan-American Exposition, "the strongest, crudest colors" were nearest the entrances, and there was "a progression from warm buff and ochre walls ... through more refined and brilliant hues to ... ivory white, delicate blue, greens, and gold" on the central tower.

During the same years, electrical advertising signs spread from a few isolated installations in approximately 1890 to nearly ubiquitous evening displays. Particularly in the United States, main streets became "great white ways" that were more alluring at night than during the day. In each nation, particular locations became famous tourist attractions because of their advertising signs, notably New York's Times Square and London's Piccadilly Circus. In these public squares, the electric sign, the spotlight, and even the streetlight became economic weapons. The competition for attention did not create the majestic serenity of an exposition, dominated by beaux-arts architecture bathed in white light. Commercial lighting produced a potpourri of styles and colors flashing against the night sky. Thousands of individual decisions produced this electric landscape, but it nevertheless had an unanticipated collective impact. Literally millions of lights seen together, particularly from a distance or from the top of a skyscraper, had a strong popular appeal. Electric lighting was also used to dramatize important monuments. One of the most effective early displays was the lighting of the Statue of Liberty in 1886, which made the statue stand out against the darkness of the New York harbor. In the following decades, search lights would be permanently trained on many of America's most important symbols, including many skyscrapers, the White House, Niagara Falls,

Old Faithful, Virginia's Natural Bridge, and Mt. Rushmore.

Between 1880 and approximately 1920, dramatic lighting transformed the night city into a scintillating landscape that was a startling contrast to the world seen by daylight. Street lighting, illuminated buildings, and electrical advertising simplified the city into a glamorous pattern that signaled more than the triumph of a new technology. By 1920, spectacular lighting had become a sophisticated medium of expression that governments and corporations used to sell products, highlight public monuments, commemorate history, encourage civic pride, and even simulate natural effects.

4. Industrial Applications

Industry found many uses for electricity, and from 1880 to 2000 it used the preponderance of power in most nations during any given year. Factory electrification began with lighting. Munitions plants and flour and cotton mills quickly adopted electric light because it was far safer than any other form of illumination. Printers, artists, and others whose work required accuracy and true color relations found gas inferior to incandescent light. Factory owners soon realized that light made possible the construction of large interior spaces without regard for windows. However, if lighting came quickly to the factory, adoption of electric drive required a generation. Existing steam and water power functioned well and cost a great deal to replace. These power sources could not easily be transmitted over more than a few hundred feet, but there were other alternatives. In 1891, to transmit water power 2 km or less, wire rope cost half the price of electricity. At that time, electricity was only cost competitive for networked power systems more than 5 km long. For example, high-pressure hydraulic transmission was a well-established alternative. Introduced by Joseph Bramah in Dublin in 1795, it was applied to cargo handling in many British ports and to centralize urban hydraulic power supply. Hydraulic power was also applied to heavy industry, such as steel production, and water motors using a city water supply could drive small appliances requiring less than 2 horsepower as well as church organs and many elevators in commercial buildings. Milwaukee alone had 600 hydraulic elevators connected to the city water system. Power from illuminating gas also briefly flourished in the 1890s. Clean, safe, compressed air, however, was the most successful alternative to electric power and was well developed in Paris, where in 1899 more than 100 miles of pipe supplied 24,000 horsepower to a wide variety of customers.

Nevertheless, electricity proved so versatile and generation costs declined so quickly that after 1900 it

was in the ascendant. An investigation by the Franklin Institute showed why it was preferred. Interestingly, savings in energy costs were considered to be of no great importance since power usually comprised only 1–3% of a workshop's budget. Furthermore, electrical equipment was somewhat more expensive than the familiar mechanical drive, but savings in maintenance and depreciation offset this expense. Electricity's advantages were primarily in creating freedom in the location of machinery, improvements in lighting and ventilation, cleanliness (there were no overhead oil drippings from shafting), reductions in shutdowns, greater precision of electrical equipment, and increases of 20–30% in output as a result of all these factors.

Adoption of electricity in industry varied considerably by nation, however, with the United States taking the lead, closely followed by Germany. Although accurate statistics are difficult to obtain, in 1928 one British expert estimated that 73% of manufacturing power was electrical in the United States compared to 67% in Germany, 48% in Britain, and less for other European countries. Britain particularly lagged behind in iron, steel, and chemicals, but this gap was partly closed during the 1930s, when its economy performed better than that of the United States.

Electricity's versatility extended into virtually all areas of manufacturing. For example, electrolysis enabled silverware manufacturers to coat materials with silver. In coal mines and rock quarries, electric drills were far less cumbersome to move around than drills driven by compressed air or steam. By 1920, many steel rolling mills had installed electric drive and controls that made it possible for one man to guide steel from the ingot furnace to the finished rail. An observer noted, "Every motor replaced a man, but the work is done better and more quickly." In food processing plants, aluminum factories, brass foundries, and potteries, electric furnaces and ovens produced no smoke, consumed no oxygen, and maintained high temperatures precisely. Such factories reduced spoilage and increased throughput.

Electric motors proved especially useful in moving materials, particularly in mines and other closed environments. Electric motors were also adapted to overhead hoists and cranes, and these began to move their loads more freely once overhead drive shafts were removed in favor of electric drive. Indeed, by 1920 electricity had become the basis for a revolution in factory design, brightening and opening up the work space and making possible new work flows, most notably on the assembly line, where electrically powered machines were arranged in combinations that were impossible using older power-delivery systems. This is not to say that electrification necessarily led to a particular form of mass production. Rather, it facilitated experiments and new designs. At the same time, electrification altered the siting requirements of factories, which no longer had to be immediately accessible to rushing streams or coal supplies. Indeed, after the successful demonstrations of alternating current in the 1890s, the short practical transmission range of direct current no longer placed a limit on where power could be delivered. Whereas early generating plants such as Edison's Pearl Street Station in New York had to be near the city or town to reach customers, during the 20th century power stations increasingly were located on the edge of the city or in the countryside. Because transmission lines could reach virtually anywhere, factories and businesses could also move away from railroads and canals to the edge of town or along main highways. Thus, electrification was used to transform both factory layouts and the location of industry.

From a worker's point of view, incandescent lighting reduced the danger of fire, reduced pollution, and made it easier to see, but it also made possible round-the-clock shifts. Furthermore, as electric motors and cranes provided more horsepower to production, they made possible radical changes in the layout of production, most strikingly in the development of Henry Ford's assembly line in 1912, an innovation partly anticipated by Edison's experiments with automating iron mining in the 1890s. The assembly line was literally impossible in any complex industry before machines were freed from the rigid organization required by overhead drive shafts and given the flexibility of electrical devices.

The connection between electrification and mass production at times has been overlooked by historians, who have emphasized scientific management, particularly the work of Frederick Winslow Taylor. It is true that Taylor systematized work and made tasks such as shoveling more efficient. In contrast, Ford's assembly line used electrified machines to eliminate shoveling altogether. Taylor retained piecework, a concept that makes no sense on the assembly line, where all share the same pace. Taylor maximized efficiency in existing production technologies; Ford transformed the means of production. Taylor saved time; Ford sped up time. In 1914, after Taylor imposed his system on the Packard Motor Company, its 4525 workers produced 2984 cars a year, less than 1 per man. This was essentially artisanal production of a luxury vehicle. Ten years earlier, well before the assembly line, Ford produced 12 cars for every worker. By 1914, 13,000 Ford employees, using conveyor belts, electric cranes, and 15,000 specialized machine tools, manufactured 260,000 cars or 20 per worker. Ford's system outperformed Taylor's by more than 2000%.

5. Electrified Transportation and the City

As industrial plants grew larger and became concentrated in cities, the urban scale expanded and electricity supplied many forms of transportation. Indeed, it made possible faster and easier movement not only at the street level but also into the sky and down into the earth. The Columbian Exposition of 1894 introduced visitors to electric boats and electrified trains and even installed some electrified moving sidewalks. Although similar sidewalks are found in few places outside airports today, a closely related technology, the escalator, has become almost ubiquitous in department stores and large public buildings.

The clean and quiet electric car initially seemed to be a promising form of urban transportation. A French coach builder, Charles Jeantaud, introduced an electric car in 1894, but it never sold well. In contrast, many American inventors enjoyed initial success, including an "Electrobat" introduced in Philadelphia in the same year. The early leaders were two Hartford-based firms, Pope Manufacturing, which made 500 vehicles in 1897, and the Electric Vehicle Company, which made 2000 electric taxis and several electric trucks the following year. For approximately a decade such vehicles were almost as numerous as fledgling gasoline cars. Even Henry Ford bought an electric car for his wife Clara in 1908. Such vehicles were easier to control than gasoline cars, and with their slower speeds (25–30 mph) and shorter range (60–80 miles) they seemed ideal for women making social calls or shopping but not for men, who demanded faster, long-distance transport. The electric car could easily strand drivers in the countryside, far from any recharging station, and even if one were found recharging took hours. If electric trucks and cars largely disappeared after the early 1920s, however, smaller electric gaffle trucks, golf carts, and indoor delivery vehicles held a niche market. Eight decades later, electric vehicles are still being designed and tested, and hybrid cars are now sold that switch automatically from gasoline motors to electric drive, depending on driving conditions.

Other forms of electric transportation were more widely adopted. Vertically, the elevator made the skyscraper practical, as it carried executives to their preferred offices high above the streets. These offices had electric lights and telephones, their plumbing was supplied with water by electric pumps, and their air could be heated or cooled electrically. The electrified office literally detached executives from the noises, smells, and temperatures of the street.

Horizontally, electric trolleys made a much larger city feasible. In the 1890s, streetcars were faster and cleaner than horsecars, and they could be stopped more precisely and climb steeper grades than cars pulled by steam. Electric cars could easily be lighted at night and heated in winter, and overall their operation cost less than any alternatives. Real estate promoters often invested in or owned traction companies, whose spreading lines defined much suburban and regional development, especially in new cities such as Minneapolis and Los Angeles but also in older metropolises, such as Boston, Copenhagen, or Amsterdam. By 1902, large cities averaged more than 200 trips per citizen per year, and Americans alone took 4.8 billion journeys by streetcar.

In the larger cities, notably Paris, New York, Budapest, and London, the technology of the electric tram was adapted to underground use. Construction began in the 1890s, and millions of commuters and shoppers were literally hidden from view. The subway showed that electric light, power, and ventilation made possible an entire new urban setting beneath the pavement. New department stores added several brightly lighted basement stories, and underground arcades connected some buildings. In Toronto and Montreal, where winter temperatures prevail, such urban designs became particularly popular.

Electrified transport transformed the city, bringing an unprecedented number of rural and urban consumers to the center, which stimulated new forms of entertainment. Vaudeville theaters and movie houses were able to fill thousands of seats several times a day, and new sports arenas (electrically illuminated) drew even larger crowds to see professional baseball, football, and basketball. Traction companies drew some of the crowd to their own amusement parks, which consumed some of their excess electrical output on evenings and weekends. The technology of electrical traction was the basis of the roller coaster, the subterranean tour, and many other rides. Trolley parks spread rapidly after 1888, and by 1901 more than half of all street railway companies operated at least one (often several). These typically generated as much as 30% of the line's overall traffic as well as provided a market for electricity. The trolley made the city the center of an easily accessible network, from the downtown shopping district to streetcar suburbs, the amusement park, and the fringe of rural hamlets on interurban lines.

Urban electrification was not merely a matter of improving existing structures. The electrified city had a fundamentally different built environment than the city before it. At the center were large office buildings, department stores, theaters, and brilliantly lighted public buildings accessible on an expanded scale reaching miles further into surrounding countryside. Practical electrical lighting, elevators, and air-conditioning likewise made it possible to create self-enclosed environments, including the later developments of indoor sports stadiums and shopping malls. The modern city is unimaginable without electricity, which enabled enormous changes in its design.

Yet these changes were by no means the same in all cultures. In the United States, where the automobile became a central form of transportation in the 1920s, the electrified city was quickly modified, sprawling out into the suburbs. In contrast, European cities retained mass transit and bicycles for a generation longer, and as a result the electrified cityscape developed and became naturalized. The urban landscape made possible primarily by electrification (i.e., without intensive use of the automobile) is therefore far more visible in a city such as Oslo than in Los Angeles.

6. Personal Uses of Electricity

The public became excited by electrical medicine well before Edison had perfected his incandescent light. Electricity seemed to be a mysterious fluid that permeated living organisms, leading Walt Whitman to write of "the body electric." Electricity seemed capable of recharging the body, helping it to overcome a host of diseases and conditions. Neurasthenia and neuralgia became fashionable complaints apparently caused by lack of nerve force, and millions bought patent medicine cures for electrical imbalances. Respected doctors hooked their patients up to generating machines and passed currents through diseased limbs or sought to (re)animate sexual organs. Patent medicines such as Dr. Cram's Fluid Lightning claimed to contain electricity in liquid form so that one could ingest it and revitalize the whole body. By the mid-1890s, magazines such as *Popular Science Monthly* asked, "Is the Body a Human Storage Battery?" So many believed the answer was yes that electrical devices were mass marketed. The *Sears and Roebuck Catalog* for 1901 sold electric belts designed to restore men's sexual powers and to help either sex restore lost energies. Some electrotherapeutic electric belts retailed for as much as $40 and promised a "triple silver plated battery of great power, silver plated nonirritating electrodes, and a very powerful electric sack suspensory." Women bought electric corsets fitted with "magnetic steel" with the expectation that they could mould their figure to any desired form. Even the U.S. Congress had a cellar room in the Capitol building fitted with an electrical medical apparatus. The *Electric Review* noted, "The members say it is splendid after they have exhausted their brain power by speechmaking or listening. A great many members take electricity, and some go to the basement of the Capitol for it every day during the season."

Although these "uses" of electricity are clearly bogus, other applications just as clearly were not. By the 1890s, it was firmly established that ultraviolet light, either from the sun or from arc lights with special filters, could kill bacteria, and in 1896 Niels Finsen discovered that lupus could be cured by repeated exposure to ultraviolet rays. He won the Nobel Prize for this work in 1903, and such successes seemed to be only the beginning. The Belgian chemist Ernest Solvay proposed that electrical stimulation might accelerate plant growth and increase the yield of crops. Similar ideas inspired experiments by the Nobel Prize-winning chemist Svante Arrhenius. He "tested the effect of electrical stimulation on the growth of children," who, he reported, grew 20 mm more than those in the control group.

When prestigious scientists made such claims, the gullibility of laymen is not surprising, and the enthusiasm for electrical cures lasted well into the 20th century. Despite the efforts of the American Medical Association to discredit them, people bought electrical cures in the 1920s and 1930s. For example, thousands of people paid $60 for an I-ON-A-CO electric collar in hopes of rejuvenation. As this popular faith in electrical medicine suggested, people regarded electricity to be a magical fluid, a nerve-tingling "juice," or an invisible tonic that could restore normal health. From the 1880s until 1920, when few people had wired their homes, electrical medicine offered direct contact with the mysterious force that was transforming work and public life. For approximately half a century, the body literally became the contact zone where men and women explored electricity's possible meanings.

By the time the electrical medicine craze was over, electricity had begun to be used to enable the body in many ways as battery technology improved. Small, lightweight batteries could deliver small but constant amounts of power for a year or more, and starting in the 1950s they first made possible electric watches and hearing aids, and later pacemakers and a host of ingenious medical devices. Slightly larger batteries combined with the new transistors made possible radios that could be carried in a shirt pocket. Consumers have not given up on using electrical devices to reshape their bodies. Some still buy electric belts, advertised on TV in the wee hours of the morning, and attach electrodes to their stomachs with the expectation that they can effortlessly tone up their muscles and look thin. Millions of people pay to use private gyms with Stairmasters and other electrical devices that tell them how long the workout has lasted, monitor heartbeat, and count calories lost.

By the 1990s, improved batteries, weighing less and lasting longer, had become an assumed part of most people's everyday life. They drove electric toothbrushes, cordless drills, and other tools. They powered tiny portable radios, disc players, mobile telephones, and palm desktop computers. They had become essential to thousands of children's toys, even if they were not usually included with those toys when purchased. In outer space, satellites and exploring machines relied on solar-powered batteries to keep them operational for years. In little more than a century, electrical devices had penetrated into the most intimate parts of the body and reached millions of miles into space and, in the process, extended human sight and hearing.

7. Domestic Electrification

In the 1880s, the few domestic consumers of electricity were almost exclusively wealthy, and they chose the new form of illumination as much for its social prestige as for convenience. Some even bought personal generating plants, especially in the countryside. Utilities did not focus on domestic customers until after a generation had passed, during which time a host of appliances appeared that made higher levels of consumption possible. Electric lights alone did not create much demand, but with the marketing of electric irons, fans, toasters, coffee makers, and other devices after 1900, the domestic consumer became more interesting. In the United States, gas continued to supply the vast majority of light and heat to private residences until after 1910, with the major shift to electricity occurring between 1918 and 1929. In Britain, the change occurred later, and as late as 1919 less than 6% of all homes were wired.

In the decade before World War I, the new profession of home economists championed the electrified house, but the domestic market only became particularly attractive to utilities once their overall load curve peaked in the day, creating a need for compensating night demand from homeowners. However, the conversion to electricity was neither automatic nor easy, and in many areas gas remained the stronger competitor, for example, in refrigerators until at least the end of the 1920s and in stoves and home heating until the present. Furthermore, it was not always easy to determine what the consumer would buy. Electric Christmas tree ornaments were a success in the United States but less so in some European countries. Even today, most Scandinavians prefer to put lighted candles on their trees. Electric cigarette lighters caught on in automobiles but not in the home. The air-conditioned bed never sold well, although there was limited success with electric milk warmers for babies.

Given the uncertainties of public demand, could the electrical utilities create this demand in practice? In Chicago, Samuel Insull showed other utilities that it was possible to build up the load through innovative pricing policies designed to attract a wide variety of customers, whose variable energy use evened out the demand curve and permitted production efficiencies. His variable pricing persuaded large customers, such as factories and traction companies, to abandon their own power systems, creating daytime demand that decreased at night. This created the generating capacity for a large domestic market, where use peaked after working hours.

In addition to Insull, Henry L. Doherty in Denver was another innovator in promotion. Their salesmen marketed electrical devices systematically, selling door to door. They also employed extensive newspaper advertising, built model electric homes that were extremely popular in approximately 1920, and paid for follow-up demonstrations by home economists in schools and at public events. Many people were afraid to use an electric iron, for example, and salesmen were trained to educate consumers and to dispel their fears of the new technologies. They emphasized cleanliness, comfort, convenience, and economy of operation—themes particularly appealing to those responsible for public buildings, especially schools. By approximately 1915, educators were convinced that the air temperature should be 68–70° and that sickness could be controlled through good ventilation. Schools also prepared a new generation of consumers. Students experienced the superior light, heat, and ventilation that electricity made possible, and in class they learned about the scientific background of the new domestic technologies. There were also gender-specific classes: newly invented home economics for girls and shop for boys. Additional domestic appliances made the home an even more attractive market as vacuum cleaners and washing machines became common during the 1920s, followed by radios, stoves, electric phonographs, and refrigerators in the 1930s.

The widespread adoption of labor-saving machines soon spread in Europe. In Britain, by the late 1920s domestic power consumption had outpaced commercial use. France lagged slightly behind and experienced the appearance of electric kitchen appliances as Americanization, although its level of consumption nevertheless was only 40% that of the United States as late as 1937. In that year, Norwegians and Canadians were by far the most avid domestic consumers. They faced long, cold winters but possessed ample hydroelectric resources and as a result used more than twice as much electricity per inhabitant as the United States and four times as much as Germany. Although electrification spread more slowly in some areas than others, once it entered the home it was seldom removed, and by the 1930s it had begun to be seen as a necessity.

U.S. consumption continued to double every decade. *Fortune* noted in a 1969 article on the dangers of brown-outs and utility overloads that "the American likes his home brilliantly lit, of course, and he has a passion for gadgets that freeze, defrost, mix, blend, toast, roast, iron, sew, wash, dry, open his garage door, trim his hedge, entertain him with sounds and pictures, heat his house in winter, and—above all—cool it in summer."

The European pattern of development was much different than the American pattern. In part because of the denser population and in part because electricity was more often considered to be a state service rather than a private business, lighting was widely dispersed because it seemed fair to do so, although the cost was generally higher than in the United States. As with gasoline, the expense of the service discouraged Europeans from adopting a wide range of appliances during the first half of the 20th century.

Since approximately 1960, however, the patterns of consumption have converged, and by 2000 a French or German home had much the same repertoire of electrical appliances as one would find in Canada or the United States. Nevertheless, differences remain. What are considered normal and acceptable levels of illumination in many European homes may seem dim to American visitors. Many a temporary resident complains that the owners must have removed most of the floor lamps. In fact, Scandinavians prefer a living room that is dim by American standards. They generally find it cozy to have only small pools of light and a few candles.

8. Rural Electrification

One of the most striking changes in the past two centuries is the shift away from agricultural life, which rural electrification abetted, although it had begun long before. In 1800, approximately 90% of all Europeans and Americans were farmers; today, farmers comprise less than 4%. Half the United States was still rural until the 1920 census, so much of the decline had little to do with electricity. However, the disappearance of millions of additional farmers was due in large part to the increased efficiency brought by electrification. Although the scale and organization of individual farms changed continually for 200 years, rural electrification exemplifies the substitution of capital equipment for labor, bringing both reductions in waste (e.g., through refrigeration) and improvements in productivity using such devices as milking machines, incubators, and electric pumps.

Not all applauded these developments. British thinkers, such as G. M. Trevelyan, Vaughan Cornish, and Patrick Abercrombie, developed environmental ideas about the relationship between man and nature. Such thinkers believed that periodic experiences of wilderness could assuage the wounds of urban civilization and therefore attacked the march of electrical pylons across the countryside. In contrast, most agrarians in the United States embraced electrification as a way to reverse flight to the city by bringing new amenities to farmers and small towns. John Crowe Ransom, for example, realized that electrification could make farming more attractive, and Herbert Agar argued that it could bring some urban amenities to the countryside and help preserve the Jeffersonian ideal of equality and independence.

Although the long view makes the process of electrification seem rational and even inevitable, in fact rural electrification was long ignored by utilities, unless prodded by government. Some of Edison's early utilities were in smaller cities, but he quickly realized that the distribution costs for scattered customers were too high. Indeed, the costs were much higher for direct current than for alternating current, which only became commercially available in the

1890s. After that time, in places where rural services were nationalized or subsidized, such as in New Zealand, the countryside was rapidly brought into the grid. In the United States, however, where power was largely privately controlled, more than 90% of all farmers still did not have electricity as late as 1935. Although equipment manufacturers and hardware stores advertised electrical wares to farmers in magazines and displayed appliances at county fairs, few people could buy them because before the 1930s private utilities used most of their resources to electrify urban homes and sponsored few experimental rural projects. The long distances between farms, with only three or four customers per mile, made transmission costs high, and farmers had little money due to the long agricultural depression that persisted throughout the 1920s.

For utilities, this rural poverty was not an incitement to action but a caution against expansion. As a result, by the 1930s electricity had become naturalized as a normal part of daily life in cities, creating a vivid contrast with a countryside still plunged in darkness. New Deal politicians, encouraged by Franklin D. Roosevelt, made rural power a national issue and created a Rural Electrification Administration (REA) that established rural cooperatives throughout the United States. Indeed, one of the selling points for this program was that the nation had fallen behind the successful efforts in Europe to provide electricity to the entire population.

Organizing farmers into REA cooperatives and getting electric lines to their homes were only first steps. A self-sustaining cooperative required higher electrical consumption, and REA coops soon were being visited by "utilization experts," Washington's bureaucratic term for salesmen. Their mission was to convince farmers to buy motor-driven equipment and larger appliances. Customers who never installed more than a few electric lights did not create enough business to enable coops to repay government loans. Washington agencies at first misjudged the farmers, however, and the utilization experts initially pushed heavy farm equipment. In practice, however, only dairies and chicken ranches discovered immediate advantages of electrical equipment. Raising wheat, tobacco, cotton, and many other crops could be done profitably much as before. However, the farmhouse was quite a different story. Farmers adopted washing machines and refrigerators more quickly than urban consumers, and they did so for sound economic and practical reasons. Cities had laundries, but in the country all washing had to be done by hand after water was pumped from a well and heated on a stove. Little wonder half of the REA's customers in the first 3 years bought washing machines, which had a much higher priority than indoor toilets. A 1939 study of the first cooperative in Indiana showed that after 2 years of service 68% had washing machines, whereas only 10% had utility motors; 45% had

vacuum machines but less than 1% had bought any machines for agricultural production. Only during World War II did farmers purchase large amounts of such equipment due to manpower shortages and high crop prices. Productivity and profits rose.

Rural electrification was not implemented to drive people off the land but to keep them there by bringing to the country the amenities enjoyed by the city. Paradoxically, however, rural electrification meant fewer farmers were needed. In the United States, people as diverse as Henry Ford, Frank Lloyd Wright, Lewis Mumford, southern Agrarians, and New Dealers all agreed on the desirability of rural electrification as a means to regenerating life in the countryside. However, in practice relocation programs and rural electrification instead helped make farmers so productive that many became redundant. In the United States, 40% of the population was on the farm in 1930, 15% remained in 1950, and there was only 8.7% by 1960. The percentage kept declining for the rest of the century and fell below 2% in 2000. This transformation cannot be attributed to electrification alone, of course, because mechanization of farm work and transport also played a role. As farmers moved out, however, others began to move beyond the suburbs into the countryside. Although some were hobby farmers, most wanted to escape the city, but they demanded the urban amenities that only electricity made possible, such as indoor running water, washing machines, vacuum cleaners, radio, and television. After 1975, migration into rural America was greater than out-migration, and the trend shows no sign of abating. Thus, rural electrification underlay not only the decline of the farm but also the deconcentration of cities and suburbs.

Most European countries did not repeat this development to nearly the same degree. During most of the 20th century, cities remained more attractive to most citizens, and until at least 1965 those who commuted were more likely to use bicycles, trains, and trolleys than automobiles. At the same time, European Union subsidies made it possible for a higher percentage of farmers to remain on the land, and national land legislation, notably in The Netherlands, often discouraged sprawl into the countryside. In Europe, where rural electrification was first realized, the electrified city remains the focus of life far more than in the United States, suggesting how much cultural values shape the consumption of electricity. Indeed, overall in the 1990s, a North American family was using twice as much energy as its Western European counterpart.

9. Air-conditioning

The skyscrapers of Dubai or Mexico City would not be habitable without air-conditioning, which has transformed possibilities of life in large areas of the world that are hot during much of the year. Stuart Cramer coined the term air-conditioning in 1904. He developed humidity controls for Southern textile factories, where excessive dryness made threads break easily, creating havoc in production. Cramer focused on ventilation and humidification more than cooling. Willis Carrier, a Cornell engineering graduate, did the most to promote the acceptance of a precisely controlled indoor environment. His Carrier Air Conditioning Company (1908) sold devices that could regulate both a room's humidity and its temperature, and these spread rapidly from textile mills to food processors and breweries.

Engineers used air-conditioning to increase their authority and control in factories. They first insisted on control of the building and later over the activities within. At times, the new technology displaced skilled workers whose tacit knowledge of materials and their response to different atmospheric conditions previously had been crucial to production of such varied things as textiles, macaroni, candy, beer, and cigarettes.

In the 1920s, air-conditioning spread into stores, skyscrapers, theaters, restaurants, and trains, familiarizing the consumer with its benefits and making people more accustomed to an enclosed, artificial environment. By the 1930s, many firms saw enormous potential for domestic sales and began a fierce competition for a market that only developed after World War II. Many early domestic machines did not regulate humidity but focused on cooling only. As a result, the public perception of air-conditioning remained confused, and in 1938 less than 1 home in 400 used it in even one room. After World War II, however, the industry overcame consumer resistance by working with builders and architects and by achieving economies of scale in producing the equipment.

Air-conditioning was not merely something added to existing structures; its use transformed housing design. Builders partially offset the cost of air-conditioning by eliminating attic fans, window screens, sashes, and high ceilings. New homes were often designed to minimize the purchase price, not the operating costs. Sealed windows combined with the heat produced by appliances, television, and the modern kitchen made air-conditioning a necessity, and the resulting electricity demand eventually helped create the national energy shortages of the 1970s.

In 1955, the federal government adopted a temperature policy that in effect required air-conditioning in most of its new buildings. The Federal Housing Association also permitted new mortgages to include it. Two forms of air-conditioning competed in the marketplace. Selling to both home and workplace, the Carrier Corporation led old engineering firms committed to central systems, whereas General Electric and other appliance manufacturers sold smaller units designed for individual rooms. Central systems began to predominate in hot cities such as Las Vegas, Dallas, and Bakersfield by the 1960s. Air-conditioning reached 24 million homes by 1970. Thereafter, it rapidly changed from being a luxury to a necessity, creating enormous summer demand for electricity. At office buildings such as the United Nations headquarters, floor-to-ceiling windows increased air-conditioning costs by 50%. By 1982, just after the end of the energy crisis, most office towers and more than 60% of all homes had some form of air-conditioning, half of which were central systems for cooling the entire structure. This was high market penetration considering that this technology was unnecessary in areas such as the Pacific Northwest, the northern tier of states, higher elevations generally, and many coastal communities.

10. Electricity Use and the Environment

Compared to burning wood or coal, electricity seems to be a clean and odorless form of energy. Indeed, for generations electricity provided a better alternative to older, more polluting technologies. Unlike gas and kerosene lamps, electric lights left no soot. Clean electric trains eliminated the cascades of soot from steam locomotives. In factories, electric motors emitted far less oil and raised far less dust than overhead drive systems. Hydroelectric plants create energy with no smoke, but unfortunately falling water provides less than 5% of U.S. energy needs and even less in some other nations. Generating electricity in coal-fired power plants causes pollution, even if it is often invisible to consumers. Such plants are usually located far from the sites of consumption, but their smoke, and the acid rain that eventually results, falls on distant consumers.

Fortunately, during the 20th century utilities developed economies of scale, transforming a larger percentage of coal into energy. Gas-fired plants are even more efficient. During the 20th century, the production of electricity became increasingly efficient, which lowered prices and reduced pollution. In New York City, for example, the average residential customer in 1930 paid $29.23 for 404 kilowatt-hours (kWh) of electricity per year, slightly more than 7¢ per kilowatt-hour. By 1956, the average New Yorker's consumption had almost quadrupled to 1536 kWh, but the cost had only doubled to $61.09. For the nation as a whole by that date, the average consumer used much more, 2969 kWh, but paid only $77.19 for it. Electricity became cheaper because engineers found ways to extract more from a ton of coal or barrel of oil. By the 1970s, however, it was difficult to achieve further production efficiencies. Thereafter, the most cost-effective savings lay not in production but rather in more efficient consumption.

Industrial research therefore focused on developing highly efficient products, notably light bulbs, motors, washing machines, stoves, and refrigerators. As a result, during the two decades after 1973 total U.S. energy use increased only 10%, an achievement based more on technological ingenuity than on changes in lifestyle. People did not use fewer appliances, but their new devices were more efficient. Home refrigerator manufacturers tripled the energy efficiency of refrigerators during these two decades, and by the 1990s washing machine electricity use had been halved. However, efforts to become more efficient resembled the mad queen in Alice in Wonderland, who had to run as fast as she could just to stay where she already was. Even greater efficiencies or lower per capita consumption will be necessary in the 21st century as electricity reaches all the world's population. Recognizing this, in 1994 the World Energy Council proposed the idea of national limits in electricity use. It noted that annual electricity use per capita higher than 6 MWh did not seem to improve life expectancy. It therefore suggested that the 30 highest consuming nations should reduce electrical consumption to this level, whereas the rest of the world should not allow consumption to rise higher than this level. Even if these targets are achieved, however, the increased use of batteries in toys and other consumer products poses another environmental problem because the discarded batteries contain highly poisonous materials. The public tends to discard them in ordinary dumps, allowing lead, cadmium, and other toxic wastes to filter into the groundwater.

Overall, demand for electricity poses environmental problems that lay at the heart of the quest for sustainable development. However, whereas coal and oil-fired power plants contribute to global warming and acid rain, new forms of wind mills, solar power, and passive solar heating promise not only to reduce pollution but also to decentralize power production, moving it closer to the consumer. In the future, as in the past, production of electricity will prove inseparable from its use.

11. The Future of Electricity Consumption

By 2010, most people in the industrial world carried electricity around with them at all times, in their watches, mobile phones, Ipods, and other miniaturized devices. This miniaturization began with portable radios in the 1950s, primarily as a fad among young people. But with each passing year more and more functions were added. By 2010 the miniature radio was just a minor feature built into new cell phones that also included clocks, email, sms messaging, digital cameras, Internet hookups, games, downloaded television programs, and much more, all packed into a device smaller than a package of cigarettes. This worldwide diffusion of cell phones, Ipods, Blackberries, and MP3 players intensified battery use, as did the spread of laptop computers. Few of these devices lasted for more than two or three years, creating a disposal problem. Discarded electronic devices became a toxic form of waste, not least because the batteries were not always removed. Thus the ubiquity of electricity was double-edged, bringing pollution along with convenience.

During the first decade of the new millennium electricity consumption once again became a political issue. Several factors were responsible. One was the Kyoto Protocol, signed by most of the world's nations beginning in 1997, which went into effect in 2005. During this eight-year interval, the signatory nations were unable to restrict consumer demand. Instead, between 1997 and 2005, the world increased its electricity consumption by an additional 500 billion kilowatts each year, rising overall by more than 30% to 15,758 billion. (All statistics are from the US Energy Information Administration.) This growth was by no means evenly distributed, ranging from Japan that nearly held even at 4% to China's rampant economy whose 223% increase in electrical consumption accounted for a larger absolute increase than Europe and the United States combined. Germany (10%), France (17%), and the United Kingdom (12%) grew more slowly than Ireland (51%), India (40%), Argentina (39%), and Brazil (27%), but everywhere the demand for more and larger televisions. A second factor was the debacle of electricity deregulation in California, which instead of creating efficiencies through competition gave ENRON and other corporations the chance to "game the system," driving up prices and triggering expensive rolling blackouts. More generally, the deregulation debate confronted the issue of whether more or cheaper electricity could be delivered without dire ecological consequences. Awareness of global warming grew. If many initially denied its reality, including the George W. Bush Administration, by 2008 it was widely understood to be a global crisis. Indeed, Barack Obama made energy a central issue in his campaign, calling for conversion to electrical and hybrid cars and for greater reliance on renewable sources of electricity.

Yet if Americans as voters responded to this message with considerable enthusiasm, Americans in the marketplace moved more slowly. In the 1990s GM became the first major modern electric car producer, with the EV1. Yet it refused to sell the vehicle, but rather leased it to 1100 consumers in California and Arizona. When GE discontinued the EV1 in 2003, it refused to sell the cars to consumers, but rather recalled and destroyed all but a few models. In contrast, Toyota developed and sold the hybrid Prius to more than 1 million consumers worldwide by 2009. Belatedly seeing its mistake, GM developed the Volt, an electric vehicle with a range-extending gasoline motor, scheduled to go into production for 2011. By that time, however, American automakers also had to compete against an ambitious Chinese government program that sought to dominate the emerging electrical vehicle market. In 2007 the diminutive Chinese Flybo Electric Car was already being sold in the US for $12,950, while the larger $22,000 F3DM made by BYD was expected in 2010. Chinese interest in electrical cars grew directly from its mushrooming domestic market, where hundreds of new coal-fired power plants supplied electricity to the more than 300 million people in its rising middle class.

In most countries, the quest for greater efficiency in energy use remains a piecemeal affair of diverse governmental and private initiatives. In the US some cities, such as Sacramento, took the lead in energy saving design and technology. Likewise, some utilities promoted the use of more efficient appliances. From 2005 the European Union took more vigorous and comprehensive action. It required that all appliances have energy efficiency labels (ranging from A to G). The EU also began phasing out incandescent light bulbs in favor of compact fluorescent lights (CFLs) that use far less energy. Substituting one million CFLs saves the equivalent of the power used by 60,000 middle-class homes during a year, and the EU decision will force consumers to install more than 1 billion CFLs. Nor is light reduction necessarily privation. By reducing electric light, the night sky becomes more visible. Most urbanites the world over scarcely see the heavens at night, because excessive artificial light reflects into the atmosphere, making it impossible to see more than the brightest stars. The Dark Sky Association has found ways to shield lights on streets and in parking lots so that they illuminate the ground, not the sky. These concerns originated during the 1960s in Tucson, where poorly designed lighting interfered with the Kitt Peak National Observatory. Retrofitting streetlights and passing zoning laws on lighting made the city more attractive at night and cut light pollution. The observatory could continue its work, the city as a whole became less garish, taxpayers saved money on electricity, and Tucson contributed a little less to global warming.

In contrast to such piecemeal efforts, only a few comprehensive projects attempt to rethink entirely how people consume electricity. For example, the Abu Dhabi government is investing $22 billion in a new city that runs on solar power and requires only one fifth as much energy as a typical industrial town of 50,000 people. In Masdar City there are to be no cars but only electric transport beneath the streets, with power coming from an electric grid. The project has been endorsed by the World Wildlife Fund and has ties to MIT. Ironically, United Arab Republic oil revenue is funding this massive experiment, which may yield new models for the future.

More immediately, voluntary restriction of electrical consumption has become popular internationally in the annual Earth Hour. Beginning in 2007 in Sydney Australia, and soon in London, in just two years "Earth Hour" spread to almost 4000 cities in 88 countries. Held on the last Saturday in March, the event asks individuals to turn off their lights and appliances for a single hour, in a voluntary global green out that rolls across the planet. For example, in 2009 more than half of all Canadians participated, and during and electricity use fell by 15% in Toronto and almost as much elsewhere. While this is largely a symbolic gesture, through temporary abstinence citizens are telling politicians that they want to change habits of consumption. Conceivably, Earth Hour may mark a turning point, putting an end to the apparently insatiable demand for ever more electricity. But if so, the event in itself is only a symbolic statement, a minimal, short-term reduction that has yet to affect surging consumer demand.

Further Reading

Beltran A, Carré P A 1991 *La fée et la servante: La société fançaise face á l'electrcité xix–xx siéele*. Éditions Belin, Paris

Hannah L 1979 *Electricity before Nationalisation: A Study of the Development of the Electricity Supply Industry in Britain to 1948*. Macmillan, London

Hughes T P 1983 *Networks of Power Electrification in Western Society, 1880–1930*. Johns Hopkins Univ. Press, Baltimore

Marvin C 1988 *When Old Technologies Were New*. Oxford Univ. Press, New York

Nye D E 1990 *Electrifying America: Social Meanings of a New Technology*. MIT Press, Cambridge, MA

Nye D E 1998 *Consuming Power: A Social History of American Energies*. MIT Press, Cambridge, MA

Platt H L 1991 *The Electric City*. Chicago Univ. Press, Chicago

Thomas de la Peña C 2003 *The Body Electric: How Strange Machines Built the Modern American*. New York Univ. Press, New York

David E. Nye

Energy in the History and Philosophy of Science

Glossary

argumentative analogy   An analogy in which the secondary term is used to bring to bear an already-organized set of equations or network of concepts on the primary term, so that the primary term becomes a mode of the secondary.

constructivism   The view that that nature is how we represent it and that the real depends on intersubjective agreement.

context of discovery   The manner in which a theory or datum was historically discovered.

context of justification   The manner in which a theory or datum is justified by reference to all the available evidence on hand.

corporeal template   The use of the first-person experiences of a sentiently felt body to forge or model new concepts.

energeia   A term in Aristotle's metaphysics constructed out of ergon, meaning act or deed, used to refer to the activity of tending toward or enacting a goal (literally, en-act-ment).

filtrative analogy   An analogy in which the secondary term calls attention to certain of its conventionally understood features in the primary term.

paradigm   A framework that provides the conditions, criteria, and examples for explaining problems and phenomena of a standard type.

realism   The view that nature has a fixed structure prior to the way human beings represent it.

vis viva   Living force, later formulated as mv^2 and thus an ancestor of the modern notion of kinetic energy, that some early modern philosophers argued existed in addition to quantity of motion or dead force (formulated as mv).

One important and controversial issue in the history and philosophy of science concerns whether the phenomena named by concepts precede the development of the concepts themselves. While realists claim that nature has a fixed structure and only concepts have histories, constructivists claim that, inasmuch is nature is how we represent it, natural phenomena are themselves histories. The history of energy sheds some light on this issue, exposing the ambiguities that lead to each position, by highlighting the significance of the fact that the environment in which scientists work is not simply a human-environment relation but one mediated by technology.

1.   The Puzzle of Energy

Few things in the history of science raise more hackles than the issue of whether the phenomena named by concepts precede the development of the concepts themselves. The natural and common-sensical realist assumption, which it is fair to say is shared by most people and nearly all scientists, is that nature has a fixed structure, and that while scientific concepts have histories—they arise, are transformed, and occasionally pass away—the phenomena to which these concepts refer do not. Conceptual evolution merely testifies to the weakness of the science of the past, and to the success of contemporary science at obtaining a better grip on the world.

The opposite constructivist extreme is the view that nature is how we represent it, and the real depends on intersubjective agreement. Sociologist Bruno Latour, for instance, has provocatively argued that things, not just words, have histories; has claimed that microbes (and not just the concept of a microbe) did not exist before Pasteur; and has remarked that Pharaoh Ramses II could not have died of tuberculosis (as now thought) because the bacillus had no real existence before Robert Koch discovered it in 1882.

The history of energy is an intriguing episode in the history of science that sheds some light on this issue. It does so because the phenomenon of energy nicely illustrates the presence of factors such as the technological transformation of the world, and even the role of more metaphysical considerations such as the changing patterns of thought that sometimes have to occur for a concept to be taken as applying to nature.

2.   The Word Energy

The word energy has a well-charted and uncontroversial history. Energy comes from the Greek energeia, or activity. For practical purposes one may say that the word was all but invented by Aristotle, who treats it as a noun form of a nonexistent verb. The word was constructed out of ergon, meaning act or deed; en-erg-eia might be literally translated as "en-act-ment." In his substance ontology, Aristotle uses it to refer to the activity of tending toward or enacting a goal, a meaning much different from today's. Aristotle saw bodies as composites of matter (hyle) and form (eidos). But for him, matter was only the possibility of a body, with no inherent structure and no perceptible or corporeal properties of its own—only a body can be soft or hard, hot or cold—and obtains a structure only via its form. Inextricably related to this distinction is another related to a body's activity. Every existing thing, he said, has an energeia, which maintains it in being and is the activity of tending toward or enacting a goal, its end or function (telos) to which it is naturally fitted. Aristotle has various names for how that activity is related to its end. A body's sheer potential or capacity for action he calls dynamis (potency), while its being at work en route to or at its telos he calls its en-ergeia. Every existing thing, then, has some activity or energeia, which maintains that thing in being. Aristotle seems only to have sensed, but not explicitly described, a difference between brute mechanical power and accumulated energy, or what we might call force and energy.

As philosophers such as Stephen Toulmin have emphasized, Aristotle's view was a product of the everyday situations with which Aristotle was specifically concerned. When he did concern himself with situations involving what we would call force and energy, they were mechanical horse-and-cart like situations in which a body (such as a horse) moves against a resistance (sources such as friction) to keep a body (cart) in motion. Aristotle compared other situations he was interested in to this type of example in formulating explanations of them. His analysis was therefore based on what Stephen Toulmin rather loosely (and prior to Thomas Kuhn's more elaborate and technical account) called a particular paradigm. It would not have occurred to Aristotle, Toulmin notes, to explain a familiar horse-and-cart situation with reference to "idealized, imaginary states-of-affairs that never in practice occur, ideals to which even the motions of the planets can only approximate." Unless we understand this context in which Aristotle was making and justifying explanations, we not only misunderstand him, but risk coming away thinking that he was guilty of elementary, unscientific blunders.

Aristotle's conception of energeia is, of course, worlds away from the modern conception of energy as the sheer capacity to do work. Energeia is not found in a specific place, but suffused everywhere and in different forms. It cannot be created and stored, converted or dissipated; it has a directionality toward forms to be realized and cannot be detached from these ends. The notion that energeia could be set up independently over and against the objects on which it does work, or stored apart from them, or collected or harnessed, would not have made sense to Aristotle.

Energy, the eventual English anglicization of energeia, came to lose the technical meaning Aristotle had given it and, for a while, virtually any value as a scientific term. In the mid-18th century, philosopher David Hume, for instance, complained that the words power, force, and energy are virtually synonymous, and "obscure and uncertain." As late as the seventh edition of the *Encyclopaedia Britannica* (1842), the word only merits a short entry: "ENERGY, a term of Greek origin, signifying the power, virtue, or efficacy of a thing. It is also used figuratively, to denote emphasis of speech."

3. The Concept of Energy

How, then, did the word come to refer to a precise technical concept? The word would become a technical term only by being linked, at the beginning of the 19th century, with a concept that itself had undergone a complex evolution, and this evolution was the product of a number of factors.

One factor was a metaphorical extension of the subjective experience individuals have of themselves as a bodily center of action, which was then projected into nonhumans as one of their properties. This illustrates what philosopher Maxine Sheets-John-stone calls "the living body serv[ing] as a semantic template," a process which, she points out, is key to the emergence of many early scientific concepts. It is a classic case of the use of analogical thinking, or the use of the familiar to understand the unfamiliar. What is remarkable in this instance is that the familiar has its basis in the tactile-kinesthetic experiences of bodily life; thus, in a corporeal template.

Yale science historian Stanley Jackson, for instance, has written about how such a subjective experience was analogously extended into the concepts of early modern physics in the works of scientists such as Johannes Kepler and others. Kepler, like many scientists of his time, analogically projected a secular version of a soul-like animistic force into his mechanics. "If we substitute for the word 'soul' the word 'force' then we get just the principle which underlies my physics of the skies," Kepler wrote, adding that though he now rejected such souls, he had come "to the conclusion that this force must be something substantial—'substantial' not in the literal sense but... in the same manner as we say that light is something substantial, meaning by this an unsubstantial entity emanating from a substantial body."

What, then, was this unsubstantial entity? In the 17th century, this question sparked a furious metaphysical and scientific controversy on the existence, nature, and measure of force. René Descartes spoke of quantity of motion (mass times velocity, or mv), while Gottfried Leibniz argued that the force was not just a quantity but a quality of matter, called living force or *vis viva*, whose proper measure was mv^2. The *vis viva* controversy, as it was known, continued through the 18th century and was the subject of young Immanuel Kant's first essay. In 1807, Thomas Young, lecturing to the Royal Institution on "Collisions," said that "the term energy may be applied, with great propriety, to the product of the mass or weight of a body, into the square of the number expressing its velocity." He thereby tied the word, apparently for the first time, to its modern concept. But Young's "energy" was not ours, only what we call kinetic energy, and not even our formulation of it ($\frac{1}{2}mv^2$).

The first half of the 18th century witnessed an intense discussion of the proper way to understand notions of activity, work, force, and *vis viva* among engineers and scientists including James Joule, Sadi Carnot, Hermann von Helmholtz, James Mayer, Emmanuel Clausius, William Thomson, and others. One of the most important motivations for this activity was to achieve a formulation adequate to the phenomena that scientists and engineers of the day were occupying themselves with: most notably, steam and heat engines, which were revolutionizing the engineering and industry of the day. In particular, scientists and engineers were concerned with the question of how to evaluate and measure the efficiency of steam and heat engines.

A second, closely allied factor involved what might be called a metaphysical shift, which opened up the possibility of seeing phenomena that look and act very differently as but different modes of one force. Thomas Kuhn emphasizes this point in an article on energy conservation, saying that such a metaphysical shift helps to explain why "many of the discoverers of energy conservation were deeply predisposed to see a single indestructible force at the root of all natural phenomena." The metaphysical shift in question itself may perhaps be specified more fully as the working out of another kind of analogical thinking, in this case, the use of an extended argumentative or scientific metaphor. Analogies and metaphors come in many different kinds, one common form being filtrative. In filtrative analogies or metaphors ("man is a wolf," "love is a rose") the secondary term (wolf, rose) calls attention to certain conventionally understood features of these things in the primary term (man, love). The focus is therefore on the primary term, and the secondary term only provides an assist. In argumentative or scientific analogies or metaphors, on the other hand, the priority of the two terms is reversed, and the secondary term is used to bring to bear an already-organized set of equations or network of concepts in order to understand the primary term—and the secondary term grows into the salient, technically correct term, often expanding its meaning in the process, with the primary term being merely one mode of the secondary. One classic illustration of an argumentative or scientific analogy occurred when Thomas Young and others began to call light (the primary term) a wave (the secondary term). Waves originally referred to a state of disturbance propagated from one set of particles in a medium to another set, as in sound and water waves. When Young and other early investigators called light a wave, they assumed that it, too, moved in some medium—but this proved erroneous, and scientists were soon saying that waves propagated in the absence of a medium. Despite this change, the same concepts and equations nevertheless governed light, which was now just another mode of an expanded conception of wave. Wave had grown into the technically correct term for characterizing light, though in the process it became transformed and expanded (and the evolution would continue when waves turned up in quantum mechanics).

This kind of argumentative or scientific metaphor was at work in the beginning of the 19th century, in allowing engineers and scientists to see the presence of energy in quite different phenomena, from falling water and expanding air to heat and electricity. Only the kind of metaphorical sensitivity allowed the early investigators to realize the significance of the conversion processes that were so critical to the development of the concept of energy. It was doubtless key in allowing William Thomson, in a famous footnote (1849) to a work addressing a dispute between Joule and Carnot over conceptual and experimental problems in the heat formation of heat engines, to intimate that something involving both work and *vis viva* was conserved, though not yet visible "in the present state of science."

The specific changes in state that then made the conserved something visible are discussed at length in other entries. If any date marks the emergence of this new conception of energy into visibility, it is 1868, with the publication of the mechanics textbook, *A Treatise on Natural Philosophy*, by William Thomson and Peter Tait. The full articulation of this insight, involving the recognition that heat was energy—and only one of many forms—revolutionized science, and by the time of the ninth edition of the *Britannica* (1899), the entry on energy was six pages long and full of technical terms and equations.

The factors just mentioned are extremely significant for the history and philosophy of science, for they illustrate an inversion of the traditional conception of the relationship between theory and practice, as well as the fact that the environment in which scientists work is not simply a static human-environment relation, but a dynamic, ever-changing one mediated by technology. The practical coping with technologies can precede and even help give birth to theoretical reflection. Technologies therefore can become a kind of philosophical springboard for scientists: familiarity with them reshapes the theoretical imagination, allowing scientists to perceive and project new kinds of concepts or extend them to larger terrains.

4. *Energy Prior to 1800?*

These developments help to shed light on the puzzle, mentioned at the outset, concerning whether the phenomena named by concepts precede the development of the concepts themselves and how the two vastly different positions mentioned can arise.

The puzzle is posed here by the question: Was there energy prior to 1800? This question would surely make most people, including and especially scientists, roll their eyes. Such people would interpret the question as asking whether the phenomenon named by the word existed. In this commonsense view, the answer is surely yes. Energy, after all, was not discovered or invented. It has powered the sun for billions of years, made organisms grow for millions, and driven industrial machines for hundreds. Our opinions, ideas, and representations of energy may have evolved, but energy apart from these—what we are trying to get at through them—does not. But if the question is interpreted as asking whether the abstract notion of a capacity to do work that comes in various forms convertible into each other existed prior to 1800, then the answer to the question is just as surely no.

Traditional philosophers of science would try to resolve the puzzle by appeal to the extremely influential distinction between the context of discovery, or the manner in which a theory or datum was historically discovered, and the context of justification, or the manner in which that theory or datum is justified by reference to all the available evidence on hand—and say that the convoluted, fortuitous, and often idiosyncratic process by which a discovery or concept enters science does not matter. There is some truth to this. But a term like energy cannot be justified in a bare environment, but only within a network or horizon of other historically given concepts (force, work, heat, etc.) and technologies (steam engines, instruments to measure temperature, etc.). These phenomena lead us to interpret other things in their light, changing our view of world and its findings, structures, and laws, leading us to project them as elements of the world—past, present, and future. This does not make scientific concepts arbitrary, an artifact of language or culture, but nor does it legitimate us thinking that these concepts refer to things apart from time and history.

The various interpretations of the question arise because the formulation of concepts such as energy is indebted both to a set of purely theoretical considerations and to a practical, technologically shaped world. This latter world is constantly changing, giving rise to new kinds of phenomena to be explained and related to phenomena that already appear. The network of theory and technologically shaped world forms a horizon in which scientific claims can be tested and be judged true or false.

If one emphasizes the values which permeate the horizon at the expense of the theoretical considerations, and the dependence on them that the invariance that phenomena such as energy show—if one emphasizes let us say the role of the development of steam and heat engines, the social and economic interest in creating more efficient types of such engines, and so forth; and if one emphasizes that a concept such as energy could not have emerged without such interests—it promotes a position such as Latour's.

On the other hand, if one emphasizes the theoretical considerations at the expense of the practical and technologically rich horizon thanks to which phenomena such as energy can show themselves as being invariant in myriad guises—if one emphasizes let us say the tightness and interdependence of the connection between concepts such as energy, power, work, and so forth—it promotes the view that science represents an ahistorical reality apart from a worldly context. But because the horizon is relatively stable, we tend to forget that it, too, can change, and that we justify things only in it.

So was there energy before 1800? If this means an abstractly defined concept of a quantity conserved in closed systems and related in precise quantitative ways to other concepts like power and work, no; if this means this is how we currently understand the world and its phenomena, yes. The history of the term energy is an episode in the history of science that helps to reveal the complexities of the history of science.

The scientific tradition as it has developed up to the present has shaped the contemporary notion of energy such that everything is a modality of energy. In cosmology, for instance, starting with radiant energy, some radiant energy becomes particles (thus mass energy), some particles become atoms, which in turn become molecules and condensed matter, and so forth. Energy is, as it were, the dynamic existential materia prima of the universe. This has both a theory side that gives to the present an illusion of wild nature that exceeds all representations and is beyond human creation, and a practical, representable, and controllable side that gives to the present an illusion of being merely socially constructed.

Further Reading

Crease R 2002 What does energy really mean? *Physics World*, July.

Jackson S 1967 Subjective experiences and the concept of energy. *Perspect. Biol. Med.* **10**, 602–26

Kuhn T 1977 Energy conservation as an example of simultaneous discovery. In: Thomas K (ed.) *The Essential Tension: Selected Studies in Scientific Tradition and Change*. University of Chicago Press, Chicago, pp. 66–104

Schönfeld M 2000 The *vis viva* debate. In: Schönfeld M (ed.) *The Philosophy of the Young Kant*. Oxford University Press, New York, pp. 17–35

Sheets-Johnstone M 1990 *The Roots of Thinking*. Temple University Press, Philadelphia

Toulmin S 1961 *Foresight and Understanding*. Indiana University Press, Bloomington, IN

Robert P. Crease

Environmental Change and Energy

Glossary

agro-ecology Crop production as a component of an ecosystem that includes not only the desired species but also, for example, the weeds, water, and nutrient relations; soil characteristics; and human inputs.

cosmogenesis The creation of the whole cosmos in a single event. In some societies it has been seen as an annual renewal of nature.

developed countries Those nation-states with an industrial economy and generally high gross demestic product and commercial energy use per capita, though not necessarily evenly distributed.

extrasomatic Outside the body. In using a bow and arrow, the energy deployed is somatic but the technology used to channel it is extrasomatic.

heterotrophic Nourished from outside—that is, of all organisms incapable of photosynthesis, which constitute most living things except green plants.

intensification Delivery of more product per unit area per unit time, for example, double-cropping of rice or shelling with heavier ordnance.

less developed countries (LDCs) Those nation states with a low level of industrial development and usually a low consumption of commercial energy. Most have high rates of population growth. Sometimes referred to as low-income economies (LIEs).

moa The large flightless herbivorous bird of New Zealand, equivalent to the ostrich, emu, and cassowary. Extinct.

noosphere Concept of Soviet biologist V. I. Vernadsky that there would be a global envelope of thought equivalent to the lithosphere and biosphere.

organochlorine A class of chemical developed in and after World War II with biocidal properties; the most famous is DDT. Without natural breakdown pathways, they have accumulated in ecosystems. Still used in LDCs.

overkill A popular but graphic term for the mass extinction of genera of animals. Originally conceived for late Pleistocene times but extended to the disappearance of the dinosaurs and held in reserve for a "nuclear winter."

pastoralism A mode of life involving herds of domesticated animals pastured over large areas, with seasonal migrations that track the availability of water and animal food.

permanent cultivation The replacement of a preexisting ecosystem with a permanently manipulated agroecology in order to produce crops as frequently as climate and soil conditions allow.

photosynthesis The chemical process of the basis of life, in which the energy of the sun is transformed into the tissues of plants.

precautionary principle A commandment of cnvironmental management in which actions are avoided if there is a probability that they will lead to environmental degradation. Like certain other Commandments, breaches are frequent.

savanna A regional-scale ecosystem of tall grasses with interspersed trees. It usually experiences fire at least once a year and in Africa may well have coevolved with the genus *Homo*.

shifting cultivation Crop production in which the land is allowed to revert to a wilder state once the soil is exhausted or weeding is too onerous. Much derided by colonial administrators but ecologically appropriate in times of low population growth.

teratogenic Of fetuses (usually but not exclusively applied to humans) that are deformed by the mother's ingestion of certain chemicals. Dioxins (of which some are more toxic than others, such as 2,3,4,7,8-TCDD) have been detected as a major cause.

1. Human Access to Energy

In the planet's ecological webs, humans are heterotrophic: they cannot fix energy from sunlight like plants. So in its 4-million year perspective, human history can be seen as a set of stages of access to energy sources. Early hominids relied on recently captured energy as edible plant and animal material, and at one stage (whose chronology is disputed but may have been as long ago as 0.5 my) gained control of fire at landscape scale. This reliance on solar energy as biomass is also characteristic of preindustrial or solar-based agriculture, but here the concentration of energy is enhanced by making it available in delineated fields or herded animals. Hence, many more people could be fed off a smaller area, especially if water control could be added to the technological repertoire. Once fossil fuels were exploitable for steam generation, solar power could be subsidized by energy from the stored photosynthesis as represented by coal, oil and natural gas. Energy concentration in these fuels was higher than more recent sources, and this trend is continued by the harnessing of the energy content of the nuclei of uranium and hydrogen.

2. Conceptual Frameworks

A number of concepts are embedded in this overview of human history. The first is that of a stage of simply garnering the usufruct in the form of plant and animal material that could be gathered, hunted, or scavenged. Most early human societies were probably mobile, so that these early species of the genus *Homo* could be said to have lived in an environmentally determined fashion. Thus, another concept is that of control and the rest of human history can be viewed as a desire to increase the degree of mastery over environmental components, of which energy has been arguably the most important. The first key step was regulation of fire away from the hearth and into the landscape, with the multiple benefits thus brought about, as amplified later in this article. Fire is also at the center of the developments of industrialization, when the combustion of coal released energy that could be controlled via a steam-using technology; its concentration was such that even mobile devices could be pioneered. Direction of the energy of the atomic nucleus has been achieved for fission but for fusion, only the awesome threat of the hydrogen bomb remains available.

Another major concept is that of technology. Apart from eating it as food, humans must access the usefulness of most energy sources via an extrasomatic device. They must learn to preserve and transport fire, for example; crop harvesting is enhanced by the use of an iron blade produced by smelting metal ores with charcoal; the steamship carried goods, ideas, and troops from one end of the earth to the other. The energy made available to growing populations in the 19th century from the potato comes not only from the carbohydrates inherent in the tubers but from its better growth when encouraged by bagged fertilizers. Electricity is a flexible carrier of energy that has facilitated all kinds of environmental change, not least in the transmission of ideas about how to control the environment. In all these devices, the hope is that the group will improve their net energy yield—that is, the ratio of usable energy harvested to energy invested as labor, draft animals' food and the making of the technology itself. The major stages in human access to energy before the 20th century and the quantitative results are summarized in Table I.

Technology, however, is not simply a set of inventions. It is the product of, and embedded within, human societies. This means that the strict rationality and "scientific" qualities associated with it are not necessarily found in every application. To begin with, much technology was prescientific, in the sense that it was the product of practical-minded men (sic?), who happened on a better way of doing things. Only after

Table I
Energy Consumption through Time.

Economy	Energy consumption (GJ cap day)	Sources
Staying alive only	0.0001	Plant and animal foods
Hunter-gatherers	2.0	Wild plant and animal foods plus fire
Preindustrial agriculture	20–40	Domesticated foods plus fire, charcoal
Eve of industrialization	70–100	Foods plus fire and plentiful fuels such as charcoal
Britain in 1800 CE	20	Foods plus wood and a little coal
Britain in 1900 CE	116	Foods plus developed coal-based economy

the 17th century did the theory precede the technology in truly scientific procedures. (It has been argued that the steam engine, for example, was pioneered by practical men with no knowledge of the theories of physics.) Certain possibilities follow: it is feasible to reject technology on religious, economic or simply prejudicial grounds, as medieval Islam is said to have regarded the wheel and sensible nations have regarded atomic weapons. Hence the distribution of energy technologies is likely to be uneven, as is shown by the access to commercial energy today, or by the way in which the adoption of fossil-fuel based technologies took off in the West after ca. 1750 AD but not elsewhere. So access to, and use of, energy in any society is mediated by technologies that are themselves subject to the organization of that society, with all its potentials for, especially, inequalities of distribution.

The relationship of all these kinds of energy use to environment has of course varied according to the intensity of available energy and the number (and density) of people, but a few historical generalizations can be made. The first is that the environment is a source of energy that may be altered as the energy is garnered. An animal population severely depleted by intensive hunting is an obvious example. A savanna fired frequently to drive out game will undergo species' shift as fire-intolerant plants are extirpated. A strip mine changes the ecology in a radius measured in tens of kilometers. Second, while many of these changes may be deliberate and desired by a society, there may well be side effects which are undesired and degradatory. The smoke from unregulated coal burning has been the source of much human ill health in cities since the 13th century, for instance; the fallout from the nuclear accident at Chernobyl in 1986 sterilized large areas of formerly productive land. The possibility of negative consequences of human actions in matters of energy access and use was pointed out in the myth of Prometheus, which retains a certain resonance today.

3. Historical Stages of Access to Energy

None of these stages, with exception of the last, can be given precise dates. Thus the chronology must be read as having wide margins for different places. Agriculture, to quote an obvious example, may have been dominant in southwest Asia by 8000 BCE but only arrived in Australia with European colonization in the 19th century. Coal was used to console Roman soldiers on Hadrian's Wall but is only a significant global influence from about 1800 AD/CE.

3.1 Hunters and Gatherers (4MY–8000 BCE)

This stage encompasses (1) human biological evolution in the transitions from early African forms of *Homo* to the anatomically modern human of *Homo sapiens* at *ca* 35 ky with a worldwide distribution and (2) cultural development through the various levels of the Palaeolithic and Mesolithic to hunter-gatherers described at and after European contact, usually described as near-recent groups. There are a few remaining groups but a way of life that comprised 100% of the human population in 10,000 BCE now accounts for something less than 0.001% and is shrinking. The term "hunting and gathering" also includes fishing and scavenging; it may also be written in the reverse to suggest the dominance of plant material in most of their life ways.

The environmental relations of hunter-gatherers are dominated by the practice of food collection rather then food production. This is not to say that there is no environmental manipulation in the cause of gaining nutrition or status, but that it is practiced in a context where there is (1) a low population density, (2) no conscious attempt to manipulate the genetics of biota with the exception of the dog, and (3) a mythological context of nonseparation from the natural world and of annual cosmogenesis. For many years, academic interpretation of these relationships postulated a "child of nature" construction in which hunter-gatherers lived in complete harmony with the natural world, taking little from it and exhibiting complete reverence. Yet the development of an ever more efficient technology was characteristic of that age, as it is of industrial and post-industrial times. Devices such as the bow, spear thrower, and blowpipe harnessed the energy of a group of human muscles to direct a missile down a narrow channel. Net energy returns were of the order of 1:5-15 and might rise to 1:40. Controlled fire seems to have been used event in wet places like the lowland tropical forests, !QJ; the Californian Coast redwoods, and the moorlands of England and Wales.

Some examples of hunter-gatherer life indeed fitted a Golden Age view. Work in the 1950s and 1960s on groups like the !Kung San Bushmen of the Kalahari seemed to show that with relatively little effort, a well-nourished population could persist even in an unforgiving environment. It is certainly true that given a large and preferably diverse territory, an abundant staple food (the case of some of the !Kung it was the mongongo nut), and a slow rate of population growth, hunter-gatherers could exert relatively little effect on their environment. Some indeed adopted conservationist attitudes to their resources. Nevertheless, many historical examples have been shown to have brought about considerable environmental changes.

The most thoroughgoing of such manipulations has been the discovery that the introduction of humans into regions hitherto lacking their presence has resulted in the extinction of large groups of animals, with an emphasis on the loss of large (>50 kg adult body weight) herbivore mammals and flightless birds. The initial scene for this revelation was the High Plains of North America in the late Pleistocene (12-11 ky) but the same phenomenon has been detected in islands in the Pacific and Indian Oceans at later dates, such as New Zealand/Aotearoa (900 CE), Hawaii (600 CE), and Madagascar (700 CE). By contrast, no such overkill phases have been noted for Africa, where humans evolved and spread gradually. Another relationship between hunters and animals sees massive kills but no long-term effects on the populations of the latter since they are so massive (or so rapid at reproduction) that humans at hunter-gatherer densities cannot bring about extinction. This seems to have been the case before European contact on the High Plains of North America. Here, as in the northern forests of Canada, such equilibria did not survive contact with the more energy-intensive economies, whether these were basically agricultural as with the Hudson's Bay Company and its fur traders or the railroads with their hungry and bored laborers. Such interrelationships have been found in recent food-collecting societies: the BaMbuti of the Zaire Basin have traded meat outside their forests to the detriment of duiker populations, for example. Some groups have internalized industrial technology, as with the Inuit, the rifle, and the snowmobile, for example, leading to controversy about their access to traditional prey such as the bowhead whale. Interestingly, no such contentions seems to have arisen over plant life, suggesting that the mythic role of dead animal flesh is deeply rooted in human societies: an aspect of control, perhaps.

At an early stage, hominid energy use must have been totally metabolic in nature, confined to that needed for growth, movement, and reproduction. As cultures became more complex, demands for additional energy grew, with any orientation toward meat consumption likely to cause more impacts than reliance on plant material. Notable technological developments seem oriented to the procurement of meat rather than plants, and in the great artistic monuments of the Upper Palaeolithic in the caves of France and Spain there are no paintings of roots and berries.

3.2 Pre-Industrial or Solar-Powered Agriculture (8000 BCE–1750 CE)

In energy terms, agriculture represents an *intensification* of access. To collect enough energy for survival, an average food collector needed some 26 km² of territory as a support system. Subsistence farmers with adequate water supplies can manage on 0.1 ha per head and achieve net energy returns of 1:20. Environmentally, this is achieved by breeding plants and animals in cultural images, by altering ecosystems so as to direct their energy flows at human societies and by coping with the unforeseen consequences of these alterations of the previous patterns. The actual biological channels through which these new patterns were directed varied in time and space. In southwest Asia in 9000 to 8000 BCE, wheat, barley, sheep, and goats were central; in southeast Asia at the same time it was rice above all; in the Americas somewhat later, the potato, tomato, maize, and llama were key elements. Yet the energetics had much in common, as had the environmental consequences.

Compared with hunter-gatherers, the energy is garnered closer to home. There is usually a permanent settlement and the food energy is produced so as to minimize expenditure on travel—hence the development of the field, where a crop is kept within bounds so that its maintenance and yield is under the eye of the farmer, be it wheat or cattle. Energy has to be invested in sowing, weeding, fencing, predator control, storage, and myriad other tasks so that the nearer at hand, the greater the probability of a positive energy balance at dinnertime. The domestication of animals meant that a seasonal round of tapping plants not edible by humans and turning them into milk, meat, blood, and skins could develop on the arid or mountainous fringes of cultivable areas.

Not surprisingly, humans enlisted help in the new diversity of tasks required by the novel economy. Draft animals are not really energy efficient in the sense that they consume food even when not working but they can deliver concentrated power at key times, such as spring ploughing or carrying metal ores to the smelter. They relieve humans of wearisome tasks and may furnish symbols of status. Equally important ameliorations of work came with the harnessing of solar energy in the forms of water and wind power. Perhaps the most important was the realization that precipitation (itself a consequence of solar energy distributions) could be stored and then released at times critical for crop growth or animal watering. Coupled with energy investment in, for example, terracing, crop production was not only intensified but made more reliable. Since storage and distribution often meant that water had to be lifted, the application of wind power to the chore was another gain. The transformation of energy in wind and falling water to rotary motion also facilitated the technologies of the mill, useful for grinding corn and metal ores as well as lifting water. So as with the bow and arrow in the Upper Palaeolithic, a whole range of extrasomatic technology is brought into action to achieve both survival and wider cultural aims.

Food production is thus the core of preindustrial agriculture and at times of Malthusian pressure, virtually any environmental manipulation becomes acceptable in order to increase the quantity of crop. There are three main types of agroecology: shifting cultivation, permanent cultivation, and pastoralism. The first of these, sometimes regarded as a primitive system, is improved by conversion to permanence, but this is probably a colonialist view of a sophisticated piece of adaptation to forest and savanna environments. Its ecology focuses on the clearance of a patch of near-natural vegetation and the take-up of the nutrient capital into crops. Cleared vegetation may be burned to add to the available stock of nutrients. As fertility falls or weeds become tiresomely prevalent, the patch is abandoned and another opening replaces it. Revegetation then builds up the nutrient stock and smothers the weeds. The crop plants in fact much less efficient at energy capture than the original vegetation, but that loss is traded for the cultural acceptability of the food, herb, spice, and medicinal plants that replace it. Moreover, a cleared patch in a matrix of near-natural vegetation does not readily leak nutrients to runoff, so energy expenditure on adding manure or other sources of nutrients is limited compared with permanently cleared areas. The system is not easily intensified, however, and has to be transformed when population growth necessitates a return to a cultivable patch before the vegetation has restored the nutrient levels. Though generally thought of as a tropical system (especially in forests and savannas), it was used in the deciduous forests of Europe in the Neolithic (7000–5500 BCE) and as late as the early 19th century CE in the Boreal forests of Finland.

The historical energetics of food production are without doubt dominated by systems of permanent cultivation, found all over the world except in polar and subpolar regions. Along with sedentary settlement, they represent one of the truly revolutionary shifts in human-environment relations and the energy relationships reflect the new conditions. The basic agroecology is that of the permanent replacement of the preexisting ecology with crop plants and animals and of progressive intensification of output as populations grow. The energy flow of the land is directed toward the crop biota and all else is labeled as weeds or pests and if possible killed or driven away. As with shifting cultivation, maintenance of nutrient levels is at the heart of the system, and both energy and materials have to be used in their upkeep. Cropping is, of course, a major factor in nutrient loss from the fields, but bare soil is vulnerable to losses from runoff and many systems have adopted strategies for minimizing such losses: cover plants are one such, but the most universal is the terrace, found in a broad belt from the equator to about 55°N and the equivalent southern latitudes.

To harvest the energy in a crop, there has to be considerable energy investment. At some time, there has to be breeding of the best strains of plants and animals for local conditions, with no necessary guarantee of success in prescientific times. More obviously, consider the average field in a preindustrial society. The soil has to be turned over for a start. A spade or digging stick may suffice, or draft animals may be harnessed to a plough with a wooden or metal-tipped ploughshare. The energy pathways can be immediately seen: human energy in the first and animal energy in the second. The animals have to be fed and may thus form part of the energy flows taken by the post-harvest crop. Or indeed they may compete for energy with the human population by needing special croplands: "horses eat people" was a proverb in early modern Hungary. If metal is used then there is an ancillary need for energy in smelting and working the iron. Sowing and weeding need investment, as does the fencing, which keeps the wrong animals out of the crop or the wolves from the sheep at night. Children are very useful for shepherding and bird scaring but have to be fed. Harvesting of plants is generally labor intensive, and only careful built storage will keep at least some of the rats out. Water management also requires energy investment, whether the problem is too much of it or too little. Lastly, many plant foods require cooking since the ability of the human gut to digest cellulose is restricted. Our species is better at processing animal tissues, but the thermodynamics of conversion of plant material to animal tissue ensure that meat in preindustrial societies is a luxury. Yet in many parts of the world, such systems were stable environmentally, nutritionally adequate and even provided surpluses.

On the fringes of settled areas, environments not conducive to permanent agriculture were often found. Characteristically, these were semiarid in nature, with long periods without rainfall and no permanent surface drainage. Mountains form

another category. These could nevertheless be tapped for human use via flocks of domesticated animals in the system known as pastoralism. The fundamental agroecology was the cropping of cellulose in plants inedible to humans by domesticates such as sheep, goats, and llamas (and many more species adapted to particular environments, such as the yak), whose sustainable products could be stored for later consumption and trade. Many dairy products have their origin in such systems and surplus animals might be sold off for meat (and skins) though eaten only rarely by the pastoral communities themselves, who, however, sometimes tapped blood from the live beasts as a source of protein. The energetics of this system are more like hunting and gathering than permanent cultivation in the sense that there has to be continual movement, in this case so as not to overcrop the vegetation. One form of the life way involves the movement up mountains in the summer. This transhumance was common in Europe until the 19th century, and a form of it persisted among people of Basque origin in Nevada until very recently. The energy flow of this system is generally completed by the selling off of most of the embedded energy in the animals and their products to sedentary societies, and there is no investment in nutrient levels. Once the pastures are visited often, then the photosynthetic efficiency declines as the species' mix shifts toward the arid tolerant and more bare ground is seen.

Fishing was often an efficient method of garnering energy since the investment came more or less free as river water, lakes, or wind in sails. Energy investment in construction of boats, weirs, and traps was generally long lasting: only nets required frequent renewal. Yet fish were usually an adjunct to diets: only in a very few places were agricultural societies dominated by their consumption, though as an addition to the palate of flavors they have long been popular.

One outcome of these systems was that they could often provide a surplus. This could be used to fuel, so to speak, population growth in the producing society. Alternatively, it could be mulcted off to feed a superior class who did not have to toil in order that they might eat. Instead they might effect environmental change in other directions. One direction might be that of developing certain nonagricultural skills full time: itinerant metalsmiths are postulated for prehistoric Europe, and presumably the astronomers of ancient Babylon did not grow their own barley. Likewise, the management of forests could become a specialized occupation, with a society recognizing that the maintenance of its chief energy source apart from food was a key element in its survival. This was especially so in the great ages of empire building and trade in wooden ships, when ship timber became a key instrument of national policy for countries such as Great Britain and France. At a less material level, philosophers and university faculties might not have to bake their own bread, though interestingly monastic houses and rural clergy were usually required to be self-sufficient to some degree, even though the latter might be entitled to a tithe of local crops. The most visible product of diverting energy into the hands of an aristocracy (who in a fascinating piece of etymology in English wielded "power") was pleasure and conquest. The latter came by feeding large numbers of soldiers, usually by local communities and on an involuntary basis, diverting the energy stored after the last harvest into the army's day-to-day maintenance. The former can be seen in landscape relics such as hunting parks and gardens. Many a European city park was once the playground of the King or the Archbishop: St. James's and Hyde Parks in London had such a royal origin. The gardens of the Alhambra Palace in Granada were the work of the governors of the moorish province of Andalucia. Lastly, energy might be invested in the afterlife: only very productive irrigated cereal-growing provided enough surplus to feed the laborers who built the Giza pyramids in Egypt.

Above all, perhaps, surpluses made possible the city. Although small by today's standards, they represented concentrations of stored energy in the form of structures. They lived heterotrophically off imported energy as food and fuel, which they transformed into all kinds of wastes. (The basic ecology of today's cities is not very different.) Though gardens might produce food and water mills some energy, their citizens were mostly dependent on their hinterlands, for which a certain nostalgic reverence turns up as early as classical Rome. Some of the energy and matter was however turned into knowledge and stored not only in people's heads but in writing. It is difficult to see how the growth of abstract thought could have taken place without the concentrations made possible by the energy surpluses allowed by well-conducted agriculture. Even before the steamship, the control of wind power allowed the improvement of food production by its facilitation of transferring new crops to new places. The role of the potato in industrializing Europe followed on from its earlier spread out of South America, for example; the penetration of all kinds of European practices into the Americas can scarcely be contemplated without the horse. So good energy trappers were taken around the world. It might be arguable that the major takeoff of human population growth usually dated to about 1650 CE was in part fuelled by improved energy capture and that better crops of all kinds were implicated in the improved nutrition. (It can be added that in some countries the use of energy to boil water to make tea and coffee has been suggested as a health-producing measure that added to survival rates. Maybe).

The theme of surplus energy produced by preindustrial agriculture has one apogee: it fed the people who conceived and built the pyramids, the Gothic cathedrals, Angkor Wat, and the Ryoan-ji. Somewhat imperfectly, it must be admitted, Mozart got fed as a result.

3.3 Industrialism (1750–1950 CE)

This is sometimes called the age of fossil fuels. Without doubting the accuracy of the label, note should be made of the exploitation of these energy sources in the years before their full flowering. Natural gas, for example, was piped in hollow bamboos from natural seepages into Chinese towns in medieval times and lit the streets. Coal exports from the port of Newcastle in northeast England fuelled breweries and laundries in London in the 14th century CE. But all this was as nothing compared with the explosion of coal use (followed by oil in the late 19th century and natural gas in the mid-20th century) once the potential for combustion of these fuels to produce steam, kinetic energy of all kinds, heat, and electricity had been realized. Further, once past the mid-18th century, the technology might follow a science in that hypothesis testing and theory building preceded the construction of the machines.

The concentrated energy content of mineralized photosynthesis is central to the idea of surpluses. The environmental relations are characterized by the impacts of getting the fuels and then of using them via, as before, the mediation of technology. The key change from the previous era is the highly positive ratio of energy investment to energy yield at, for example, the pit head and the well head, with ratios of 1:20–30. Even when postextraction processing is complex, as with oil (though not with coal, the earliest fuel of this kind to be used on a massive scale), the ratio is still highly favorable. Only with the large heat losses encountered in the generation of electricity in conventional power stations do the ratios begin to drop toward preindustrial magnitudes.

The energy-environment story of the 18th and 19th centuries is that of the adoption of fossil fuels as the basis of transport, communications, and manufacturing in a set of regions that became what are now called the developed countries (or sometimes the north or the high-income economies [HIEs] or simply the rich) and the failure of other parts of the world to undergo the same transitions. They became the less developed countries (euphemized as the developing countries, or sometimes referred to as the south, the low-income economies [LIEs], or the poor). The reasons for both are complex and outside the scope of this work. The energy available in classical Greece, in 17th century Europe, and China in 1880 CE was about the same but the adoption of fossil fuels brought about radical change. The core areas in, say 1850 CE, were western and central Europe together with eastern North America. Soon to be added were eastern Europe and western Russia, Japan, a few parts of the northern shore of the Mediterranean, and other patches of North America. On the map, the sign of the times was the railroad. Those heartlands are still significant in the sense that they acted as core areas for whole nations to adopt enough of the new technologies to ensure their future access to fossil energy supplies, so the present-day situation in which access to energy (usually measured as per capita consumption of commercial energy) replicates roughly the precedence established in the 19th century. Though wood as a fuel has declined greatly in relative terms (its energy content is about 14 MJ kg), compared with hard coal at 29 MJ kg), in absolute terms its use is still rising due to population growth in the LIEs (Fig. 1).

The myriad ways in which these fuels are extracted and utilized exert considerable environmental impacts, and many books have been written about each of them. There is a basic ecology common to all in which the hydrocarbons are lifted from the earth, processed, and stored, and then transformed into their end uses. These latter may be direct (as heat, for example, to power boilers) or as embedded energy as in fertilizers and biocides. All kinds of machines mediate at every stage and millions of different items are manufactured. The environmental impact of some is obvious and direct: the bulldozer is a good instance. Others are more subtle: stored electricity makes possible battery-operated TVs in remote Asian forests whose people are then able to see HIE consumerism at work, or indeed at play or at prayer. The bulldozer is the culmination of earth-moving equipment that started with the steam crane; microelectronics carry on the tradition of the electric telegraph. In that sense, the latter exemplifies the way in which energy as electricity above all carries ideas, which include those devoted alike to changing and protecting the environment. It has without doubt made possible the global coalescence of ideas which have on the one hand allowed corporations quickly to find states willing to accept GM crops but on the other hand pioneered agreements like the Kyoto protocols, which all civilized nations have agreed to ratify.

The historical effects of industrialization on food production are seen in many environmental changes in the 19th century. Many temperate grasslands were ploughed to grown cereals for export to the rapidly growing populations of industrializing regions. The demand for meat also allowed vegetation change in other grasslands as grazing densities increased. All such changes were facilitated by the takeover of areas previously used by preindustrial people (nowadays labeled as "indigenous"), whether officially colonized (as in South Asia where "empty" lands were converted to tea and coffee by British colonial authorities) or internally reallocated, as on the High Plains

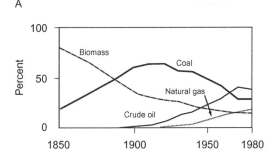

A

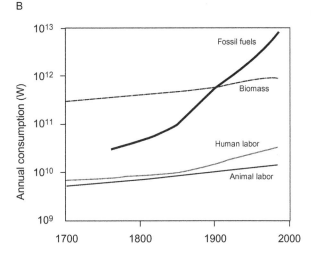

B

Figure 1
The transitions to fossil fuels. (A) shows the global proportions of primary energy for the period 1850–1980 CE and (B) the absolute amounts over a longer time period. Though biomass has declined in relative terms, the rising populations of LDCs (especially in Africa and Asia) have increased actual demand for wood fuel. From V. Smil (1991), "General Energetics." Wiley-Interscience, New York and Chichester, Fig. 10.2 (part), page 185, and Fig. 14.6, page 303. Copyright © 1991. This material is used by permission of John Wiley & Sons, Inc.

of the United States. In the case of meat, the technology of refrigeration was critical, as was the reliable transport offered by railroad and steamship. Although to be more a feature of the post-1950 period, the disappearance of the seasons in the rich world began, as asparagus, strawberries, and flowers consumed transportation energy in their journeys from the warm and sunny to the still-emerging-from-winter shops of London, Paris, and New York. The loss of the local, as energy for transport plummeted in price, has its origins in the 19th century, though clearly not confined to that period. One short-lived counter-trend came about in the 1890s in Paris, where there was enough horse-dung to fertilize and heat an export trade in winter salad crops within the city boundaries.

The theme of increased material production can be carried beyond that of food. One result of immense environmental significance was the burgeoning appetite for material possessions, leading to environmental manipulations of all kinds. This was exacerbated by the craving for the new, so that reuse and recycling (a feature of poor societies everywhere) became far less common. Again, the introduction of electricity (as late as the post-1945 in many rural parts of Europe) impelled many authorities to speed up hydropower development, with its many environmental consequences. This is just one of many environmental outcomes of the extraction of energy sources: huge open-cast coal mines in central Europe and oil wells in Mesopotamian deserts were all visible before 1950. The overall impression is of self-amplifying systems in which the extraction of energy is followed by the development of technology to use it and this in turn spawns further environmental change. The motor vehicle is a case in point, where impacts include not only the emissions to the atmosphere but also the paving over of soils, with implications for runoff rates and flood levels on lower ground. Hence, all industrial societies are replete with embedded energy: obvious in the Chicago coal yards but less visible in the suit from Jermyn Street.

As in the previous energy era, opportunities for pleasure were increased. The form of travel now called tourism was made possible by the steamship and the railroad: before the 19th century, a pleasure visit from western Europe to Egypt was too great an undertaking for almost everybody except the aristocracy. After about 1860, Britons might holiday in Switzerland, beginning a love-affair with the mountains denied them in their own small-scale islands, one which took them to the summit of Everest in 1953. (Actually it was a New Zealander and a Nepalese who first reached it, but imperial attitudes were still strong. The relevance of industrialization is in the specialized gear now used, and the fact that an expedition has had to be mounted to clear up the "litter" from such climbers including some bodies.) The ability to travel to relatively remote places led to conflicts over environmental change: the U.S. National Parks post-1945 slogan of "parks are for people" led to developments that people of later worldviews have found difficult to demolish, in all senses of the word. In a secondary frame, urban electricity made possible all kinds of entertainment, which may well help to inculcate environmental attitudes, whether these be of the exploitive or the conservative kind. Think of oil company advertisements on television or of documentaries on whales.

Only recently has the application of industrial energy to warfare been examined in an environmental context. At one level, there is the way in which calorie production must be stepped up, for soldiers need feeding well, as do the industrial workers who make their munitions. Horses, too, consume cereals and were important as late as 1945. So many grasslands are ploughed up to make way for wheat, oats, and rye crops, with downstream aftermaths for wildflower populations and soil losses. The opposite happens when relict landmines prevent land being used for agriculture: a World War II problem in Poland to this day. But the main impact of industrial warfare before 1950 was seen on the Western Front in World War I. This was in effect a huge concentration of energy use in the form of soldiers, horses, shells, and bullets. The environmental alterations were profound and almost total, as can be sensed from the paintings of John Nash, for example. Yet the land was quite quickly retransformed to agricultural use, and only a few consciously preserved remnants of trench systems are left, though the whole of them can be seen from the air at the end of summer like an archaeological crop mark. The less static warfare of 1939 through 1945 left fewer long-term environmental changes in Europe but transformed the vegetation, soils, and indeed landforms of some Pacific islands.

If some summative expression of industrialization and environment is sought, then two phenomena stand out. The first is that enough carbon was emitted into the global atmosphere to start the rise in concentration of (especially) carbon dioxide, which underpins today's concerns about climatic change. The mid-18c is usually taken as the base-level of 280 ppm, to compare with recent values of 360 ppm. The second is the explosion of the city, parasitic on the rest of the world for energy in all forms and for oxygen. The carbon loading has been mostly caused by the rich, but the city's ecology is shared by LIE and HIE alike.

3.4 Post-Industrial Economies (post–1950 CE)

After 1950, most of the indicators of environmental change start to rise in magnitude. Graphs such as those in B.L. Turner's 1990 compilation show sharp upward trends in resource use, waste production, emissions to air and water, and land use transformations. The increases are seen in Table II, where some shift toward an increased proportion being devoted simply to movement can be detected. Most of these are outworkings of increased use of oil, whose ascendancy over coal in this period is unchallenged. So post-industrial economies are in large measure the old fossil-fuel-powered systems writ large in many ways (not least in the geopolitics of oil) but in some of its energy-environment relationships there is something different about the post-1950 world.

Table II
Energy Use in Industrial and Postindustrial Societies.

Society	Total energy use (10^3 kcal cap day)	Plant and animal (%)	Home and commerce (%)	Industry and agriculture (%)	Transportation (%)
Postindustrial	230	4.3	28.6	39.5	27.3
Industrial	77	9.0	41.5	31	18

For once it is possible to date the genesis of the new world quite accurately: 2.30 p.m. on December 2, 1942, in Chicago, where Enrico Fermi conducted the first controlled fission of atomic nuclei and thus made possible both military and civilian nuclear power. (The energy content of pure ^{235}U is 5.8×10^5 MJ kg, compared with oil at 43 MJ kg.) The military applications have had some environmental effects, most notably in the fallout of radioactive material from atmospheric weapons testing before the ban in 1963. Levels of long-lived radionuclides (such as ^{137}Cs and ^{90}Sr) built up in high latitudes and have accumulated in the bodies of, for example, the Saami people. The potency of thermonuclear weapons was such that in the 1970s scientists constructed scenarios for a 5000 MT exchange that would produce at least 12 months of virtual darkness in the Northern Hemisphere ("nuclear winter"). This knowledge is sometimes credited with helping to bring about the end of the Cold War. The potency of radionuclides to affect human tissues on all time-scales from the immediate to the 240,000-year half-life of ^{240}Pu has accompanied the development of civilian nuclear power where very strict international agreements control the discharges to air and water. In spite of these, nations like Norway and Ireland object vociferously to the marine releases of the UK plant at Sellafield on the coast of Cumbria. Of greater concern, however, is the possibility of accidents resulting in the meltdown of the core of a reactor, as has happened at Sellafield in 1957, at Three Mile Island (United States) in 1979 and at Chernobyl (Ukraine) in 1986. Only the latter was truly disastrous environmentally and to human health but the memory is very vivid all over Europe. Yet of even more extensive importance is the fate of high-level radioactive wastes. Nobody yet knows really what to do with them since they must be sequestered from the environment for about one-quarter million years, a commitment rather like that of the Netherlands in keeping out the sea. One further long-term consequence of this era has been the formulation of chemicals with long-term toxicities, which can accumulate in food chains. The organochlorine pesticides and dioxin are the best known examples with the former causing high avian mortality and the latter implicated in teratogenic fetuses after its distribution in Vietnam as a contaminant of

herbicides used in the war of liberation in the 1960s and 1970s.

The adoption of nuclear power in HIEs has mostly been motivated by a fear that fossil fuels might run out, either physically, economically, or politically. The HIE lifestyle is now dependent on relatively cheap and totally constant energy supplies: to interrupt gasoline flow at the pumps for a couple of days is virtually to bring about insurrection, whereas a coal miners' strike for 3 months has virtually no effect. In a similar vein, it takes little imagination to fashion scenarios of a western world without constant electricity. Since capital formation is now a matter of colored numbers on screens, then all kinds of projects to alter the environment might come to a standstill, and the standing of the small screen as a major source of authority in western society (especially if it shows a man (*sic*) in a white coat) would be greatly diminished—no more documentaries about global warming for a while at least.

Unlike the fossil fuel era, it is difficult to designate any particular piece of technology with direct environmental effects as characteristic only of the post-1950 period. In more ways, there has been more indirect effects through capital accumulation in ways that have spun off into environmental projects like the very large dams of the late 20th century planned for China, India, and possibly Turkey. One of the diagnostic features of the post-industrial society, however, is its appellation as a knowledge economy, and there is no quibbling with the fact that the environment is a big absorber of space on hard disks the world over. Cheap and ubiquitous energy makes possible the gathering of environmental data, its storage, and even its transmission in windowless, air-conditioned classrooms.

The present is always impossible to assess in historically oriented accounts and it may well be that the post-1950 phases is not really a new era but simply an intensification of the one that started in 1750. It could equally be the start of a transition to something totally new, in which case the energy-environment linkages would be a central feature along with one of their outgrowths already in place: the creation of a truly global network of instant communication (a noosphere in fact, as envisaged by Vernadsky in 1945) such as there has never been.

4. Overview From Today's Perspectives

In the perspective of the last 10 ky, access to energy can be seen as having produced tensions in a world where access is highly differential (Table III). In a positive view, many developments have been facilitated, without which today's humans would not even be able to aspire to the cultural richness and diversity that have evolved. By contrast, access to energy has allowed impacts that if foreseen by some omniscient world power imbued with the precautionary principle would likely have been prohibited.

Looked at positively, access to energy has been one consequence of agriculture, which has brought about the possibility of adequate nutrition even for human populations of more than 6 billion: if they do not get it, then the causes are political, not to do with energy capture. In the wake of the shifts to more intensive sources of food than offered by hunting and gathering, a quite wonderful diversity of human cultures came about. (It is interesting that the more up-market the tourism, the further back along this spectrum people want to go, albeit only on vacation.) There are still elites who think that imperialism was a positive outcome of energy made available by gunpowder, the breech-loading rifle, the steamship, and by their latter-day equivalents. Travel has been one further outreach into which energy can be channeled. For humans, there is tourism and space travel (think of the rocket as an updated blowpipe), for instruments gathering data there is the satellite and the remote weather station on a buoy in the Pacific. To sum it all up, there is the city and its urbanized fringes, where the CBD stacks the decision makers in high towers (always a visible emblem of power, i.e., of personal access to energy: recollect as well the personal jet and the LDC leaders' Mercedes) and the suburbs spread out the expenditure of energy and money.

But change inevitably brings loss. Most of these have been irreversible, like the diminution of biodiversity caused by habitat change as, for example, agricultural intensification works toward monoculture or the demand for wood products devastates forests. In the period 1700–1980, forests and woodlands decreased globally by about 19%, grassland

and pasture by about 1%, but croplands increased by 466%. Industrialized warfare is better at destruction of all kinds, including life in all its forms, than a few armies raging seasonally across selected terrain. The chemical industry has produced many quality-enhancing pharmaceuticals but has also thoughtlessly allowed out persistent substances that have engendered problems for wild species, crops, and human health alike. One effect has also been to introduce uncertainty: models of climate change are in the end models and can be disregarded on those grounds by those wanting to emulate the moa rather than the ostrich. Though the multiplicity of environmental changes that are problematic for somebody or something is bewildering, the main two for humans can probably be summed up as meat and motors. The energy going into both produces all kinds of patterns that could be rearrayed to the greater good of many more inhabitants of the planet.

If, though, there is only one allowable view of the history of energy and environment, it must be that access to the energy flows and stores of the planet, especially after the 18th century, has allowed the growth of the human population—from about 720 million in 1750 to 6.1 billion of the early 21st century. Whether regarded historically, culturally, technologically, or politically, that view fills the lens.

Further Reading

Bormann F H 1976 An inseparable linkage: Conservation of natural ecosystems and the conservation of fossil energy. *BioScience* **26**, 754–60

Cottrell F 1955 *Energy and Society*. Greenwood Press, Westport, CN

Debeir J C, Deléage J-P, Hémery D 1991 *In the Servitude of Power. Energy and Civilization through the Ages*. Zed Books, London

Hall C A S, Cleveland C J, Kaufmann R K 1986 *Energy and Resource Quality. The Ecology of the Resource Process*. Wiley-Interscience, New York and Chichester

McNeill J R 1991 *Something New Under the Sun: An Environmental History of the Twentieth-Century World*. Allen Lane, New York and London

Simmons I G 1996 *Changing the Face of the Earth: Culture, Environment, History*, 2nd ed., Blackwell, Cambridge, MA, and Oxford

Smil V 1991 *General Energetics: Energy in the Biosphere and Civilization*. Wiley-Interscience, New York and Chichester

Smil V 1993 *Global Ecology: Environmental Change and Social Flexibility*. Routledge, London, and New York

Turner B L (ed.) 1990 *The Earth as Transformed by Human Action*. Cambridge University Press, Cambridge and New York

Vernadsky V I 1945 The biosphere and noosphere. *Am. Scientist* **33**, 1–12

I. G. Simmons

Table III

Energy Consumption at the End of the 20th Century, GJ Cap Year.

Region	Commercial energy	Biofuels
World	60	3.7
N & C America	243	2.0
Europe	134	1.1
Africa	12	7.2

F

Fire: A Socioecological and Historical Survey

Glossary

agrarianization The formation and spread of a socioecological regime structured around the domestication of plants and animals.

anthroposphere That part of the biosphere that has been affected by humans.

domestication The "taming" and incorporation into human society of originally "wild" nonhuman sources of energy.

industrialization The formation and spread of a socioecological regime structured around the exploitation of fossil fuels.

The ability to handle fire is, along with language and the use of tools, a universal human attainment: no human society in historical times is known to have lived without it. The ability to handle fire is also exclusively human: whereas other animals use rudimentary signals and tools only humans have learned to handle fire. From the outset, mastery of fire has strengthened the position of humans in the biosphere. At the same time, it has deeply affected the relations among humans themselves. Appreciation of the control over fire is therefore essential for an understanding of both the external expansion and the concomitant long-term internal transformation of the anthroposphere. The control over fire has been essential in the emergence of agriculture and stock raising, as well as the rise of modern industry. Its implications, including both benefits and costs, are so far-reaching as to often remain unnoticed and taken for granted. This entry intends to convey the enormous importance of the fire regime for human history and its profound impact on the biosphere.

1. The Original Domestication of Fire

1.1 Origins

Human life, like all life, consists of matter and energy structured and directed by information. All life is part of an ecosystem, and all ecosystems together constitute the biosphere—the total configuration of living things interacting with each other and with nonliving things. Every form of life continuously affects, and is affected by, its ecosystem.

As participants in the biosphere, the early hominids and their only surviving species, *Homo sapiens*, have gradually strengthened their position—at first slowly and almost imperceptibly, later at an increasingly more rapid pace with ever more striking consequences. In the process, they engrossed increasingly more terrain and incorporated increasingly more nonhuman resources into their groups: first fire, then, much later, certain selected plants and animals, and, later again, fossil fuels. As they incorporated more energy and matter into their societies, those societies grew in size, strength, and productivity, while at the same time also becoming more complex, more vulnerable, and more destructive.

During the first stage in human history and "prehistory," known in archaeology as the Paleolithic or Old Stone Age, which lasted for thousands of millennia, the overall pace of social and cultural development was slow in comparison with later stages. Yet some momentous changes took place, with great consequences for the relationships between humans and the natural environment. Humans began not only to use but also to make tools of their own, and they learned to control fire. The combination of tools and fire enabled groups of humans to leave their original habitat, the savannas of East Africa, and to migrate into other parts of the world, penetrating first into remote corners of Eurasia and then also into Australia and the Americas. The Paleolithic can thus be seen as a long run up, which was later followed by an enormously accelerating sprint of which our present age is the latest episode. It was the scene of the incipient expansion of the human realm, the anthroposphere.

A basic trend in all human history, and certainly during its earliest phases (often designated as prehistory), has been the increasing differentiation between humans and all closely related animals in terms of their behavior, power, and general orientation or attitude—their habitus. Thanks to the flexibility acquired in the course of evolution, humans were able to learn a large repertory of new forms of behavior. Especially successful were those innovations in behavior that added to human power vis-à-vis other large animals, both predators and competitors. Transmitted by learning from generation to generation, those innovations entered into the human habitus and became second nature.

The primary condition for the process of differentiation in behavior, power, and habitus has been, and continues to be, the inborn human capacity for culture as manifested in technology, organization, and civilization, each of which represents the results of social learning. Indeed social learning is the crux of culture: gathering information and passing it on to others—at first in direct interaction, at a later stage in history through written texts and today by other audio-visual means as well. The stores of cumulated information (or "cultural capital") have enabled people to tap increasingly larger and more varied flows of matter and energy and to integrate those flows into their societies.

The domestication of fire was the first great human interference with natural processes. It had numerous far-reaching implications, stretching from the first hesitant beginnings to our contemporary fuel-driven economy. It demands our attention because it reveals something of the socioecological infrastructure of the contemporary world.

In several ways, the original fire regime may be seen as a paradigm for the socioecological regimes that were developed later. It presents a paradigm in a double sense. First, in actual practice the regime by which humans learned to extend their care for and control over fire could serve as a model for subsequent forms of care for and control over other forces in nonhuman nature such as plants and animals. Second, we may regard the domestication of fire as a model case in a more theoretical fashion, since it brings out the strong link between such apparently contradictory tendencies as increases in control and dependency, in robustness and vulnerability, and in the potential for production and for destruction.

1.2   First Impact

Fire, like all natural forces, has a history. Chemically fire is a process of highly accelerated oxidation of matter (fuel) induced by heat (ignition). Three conditions are therefore necessary for it to occur: oxygen, fuel, and heat. During the first eons in the history of the earth, at least two of these, oxygen and fuel, were absent. Oxygen did not become available until, after at least a billion years, life emerged. And it was only less than half a billion years ago, during the Devonian geological age, that life assumed the form of plants, providing matter suitable for burning. From then on, most places on earth with seasonally dry vegetation were regularly visited by fire, ignited on rare occasions by falling rocks, volcanic discharges or extraterrestrial impacts, but mostly by lightning.

Its domestication by humans opened an entirely new episode in the history of fire. Humans thoroughly altered the frequency and intensity of fires. They brought fire to regions of the planet where it seldom or never burned spontaneously, and they tried to banish it from places where without human interference it would have burned repeatedly. Thus, increasingly, "natural" fire receded and made way to human or, more precisely, anthropogenic fire.

Wherever humans migrated, they took their fire along. The presence of humans with fire deeply altered the landscape, including flora and fauna. The human impact is amply documented (though still controversial) for Australia—a continent that was colonized by humans rather late. Everywhere on the planet, areas such as rain forests, deserts, and the polar regions, which were not receptive to fire proved to be hard to penetrate for humans too.

Humans are the only species that has learned to manipulate fire. Control over fire has become a species monopoly, with an enormous impact on other species, both animals and plants. It provides us with an excellent example of how new forms of behavior could change power balances—in this case, between humans and all other animals, ranging from primates to insects—and how shifts in power balances could engender changes in habitus, both among the humans who gained greater self-confidence from the presence of fire in their groups and among animals that might be bigger and stronger than humans but learned to respect and fear their agility with fire.

Control over fire, in addition to having become exclusively human, has also become universally human. We know of no human society of the past 100,000 years that lacked the skills needed to control fire.

The original domestication of fire was a dramatic transition. Palaeoanthropologists are still debating when exactly it took place. The estimates range from as many as 1.5 million to a mere 150,000 years ago. It is still an open and fascinating question whether the first steps to control over fire coincided with other changes in early human development.

In retrospect, the initial domestication of fire was an event of momentous import. A wild force of nature—blind, capricious, and hazardous—was now tended, cared for, and supplied with fuel. Our early ancestors went to all this trouble, not altruistically, but because it served them well. They put the potentially destructive and essentially purposeless force of fire to work for their own productive purposes. They managed to make fire regularly available. They no longer had to hunt for it, hoping to find somewhere the smoldering remains of a natural blaze; they made it a part (and even the center) of their own group, and they revered it as a symbol of eternal life.

The domestication of fire made humans less directly dependent on natural forces that continued to be beyond their control such as the alternation of day and night or the cycle of the seasons. It made the contrast between dark and light, between warm and cold, or between wet and dry more amenable to manipulation, and thus gave humans a greater margin of freedom from the grip of nature. It increased their power, defined as the capacity to influence the outcome of an interaction. Armed with fire, humans were able to open up impenetrable tracts of bush and to drive away animals much fiercer and stronger than they were. The gain in power made their lives more comfortable and secure. The possibilities of heating, lighting, and cooking all contributed to what we would now call a higher standard of living.

1.3 Long-Term Consequences

"Wherever primitive man had the opportunity to turn fire loose on a land, he seems to have done so from time immemorial." This statement by the American geographer Carl Sauer may sound exaggerated; but it still fails to convey the full impact that the domestication of fire has had, both on the larger biosphere and, within the biosphere, on human society itself.

The most immediate effect of the domestication of fire on the biosphere in general was an increase in the frequency with which fires occurred. Prior to its human mastery, fire was ignited mostly by lightning. From now on another source was added: even before they learned to make fire themselves, humans were able to preserve it in their hearths and to apply it wherever they saw fit. Consequently, as the number of anthropogenic fires increased, the proportion of natural fires diminished. The geologist Peter West-broek has suggested, in an admittedly speculative manner, that the earliest human fire use may have affected the planetary atmosphere and caused some of the major climate changes in the Pleistocene. Substantive evidence indicating modification of the landscape by human foragers equipped with fire has been brought forward for Australia, where most of the indigenous forests were burned down in the millennia following the arrival of the first Aborigines.

From the very beginning, humans used fire in two basic forms: the hearth and the torch. The hearth was the original site at which a fire was kept, usually at a cave entrance where it could be protected against the rain and still receive some air circulation. Since it had to be tended and since fuel had to be brought to it, it served almost naturally as a center for group life, providing heat, light, and a common focus. From the hearth developed in the course of time a variety of fire containers such as crucibles, stoves, kilns and furnaces, and, in our day, the mobile engines of cars and airplanes.

The hearthlike uses of fire have always had two kinds of environmental side effects. First of all, fuel had to be supplied. As long as human communities were small and living in areas with abundant wood, this did not cause much of a problem. When greater numbers of people started living in large urban concentrations, however, the need for fuel became a strong contributing factor to deforestation over big areas and, in our own age, to depletion of fossil resources. The second side effect of hearthlike fire consists of its waste products: ashes and smoke. Although smoke might be useful in driving away insects and other insidious creatures, it has always been considered mostly as a nuisance to be got rid of. As long as people lived in isolated caves or huts, this was relatively easy. Urbanization and industrialization have seriously aggravated the problem.

While the functions of the hearth were originally primarily turned inward, the torch was a more outward directed implement. It was used not only to provide light at night, but also by day to set fire to shrubs and grasses—an effective way to destroy obstacles for foraging and to rout animals, both predators and prey, and thus to extend the human domain. Torches undoubtedly contributed to deforestation: wood was burned wholesale, regardless of its possible value as timber or fuel. In the age of agriculture, the torch was used for slash and burn and other techniques for clearing land, and it served as the model for a whole array of fire weapons culminating in our own time in rocket-propelled missiles.

Surveying the entire trajectory of the human use of fire from its earliest beginnings, we can distinguish three stages. During the first stage, there were no groups possessing fire; there were only groups without fire. Then, there must have been a second stage, when there were both groups with fire and groups without fire. We do not know how long that stage lasted, nor do we know how often it may have recurred. All we know is that it has come to an end. It was a transitional stage, leading up to the stage in which humankind has now been living for thousands of generations: the stage when there are no longer any groups without fire. All human groups are groups with fire.

Although we lack empirical evidence for the first two stages, that very lack leaves us no choice but to accept the conclusion that societies with fire were in the long run obviously more fit to survive than societies without fire. If we then ask why it was that societies without fire disappeared, there seems to be only one plausible answer: because they had to coexist with societies with fire, and apparently in the long run such coexistence proved impossible.

This may sound like a dismal conclusion suggesting fierce contests ending in the elimination of the losers. If such contests did indeed take place, they have left no trace of empirical evidence; we only have the highly imaginative evocations of what might have happened in books and films like *The Quest for Fire*, directed by Jean-Jacques Annaud. However, we can also view the fact that possession of fire has become a universal attribute of all human societies as an important example of the general rule that changes in one human group lead to changes in related other groups. If group A had fire, and the neighboring group B did not, group B had a problem. It could either try to minimize contact with group A and perhaps move away, or do as group A had done and adopt a fire regime, which should not pose insurmountable difficulties as long as the capacity to learn from the others was sufficient. In the latter case, instead of a zero-sum elimination struggle there would have been what American freelance author and scientist Robert Wright calls a "nonzero" situation, with an outcome from which both parties benefited.

The rule that changes in one human group lead to changes in other related groups may sound like a rather tautological explanation for social change, but it is not. It is a highly generalized empirical observation, similar to an observation we can make about fire: fire generates fire and, in a similar fashion and more generally, change generates change, and social change generates social change.

Such is the nature of the dynamics of human society and culture. After the original domestication of fire, it was never humans alone who interacted with other human groups and with nonhuman nature. It was always humans with fire, equipped with fire and with the products of pyrotechniques: cooked food, pointed spears and arrows, earthenware, metal tools and weapons. Their presence put an end to humans without fire.

Another general conclusion to be drawn from these observations is the following: changes in climate and precipitation have never ceased to be important causes for humans to change their way of life. Humans are no different from other species in that they will always have to accommodate the basic conditions of earthly nature, such as the alteration of day and night or of monsoons and seasons. In the course of human history, however, in addition to these overridingly powerful extra-human conditions, conditions brought about by humans themselves have become increasingly more important to the extent that, in our contemporary world, humanity has become a major agent of ecological change.

1.4 Regimes

The domestication of fire meant that people tamed a strong and potentially destructive natural force and made it into a regularly available source of energy. In so doing they initiated changes in the natural environment, in their social arrangements, and in their personal lives. These three aspects (ecological, sociological, and psychological) are all part of the changing human relationship with fire.

In its ecological aspects, the domestication of fire affected the relations between humans and the nonhuman world so deeply that we may call it the first great ecological transformation brought about by humans, which was followed much later by the second and third of such transformations, generally known as the agricultural and industrial revolutions and better characterized by the long-term processes of agrarianization and industrialization.

Each of the three transformations spelled the formation of a new socioecological regime: the fire regime, the agrarian regime, and the industrial regime, marked by, respectively, the utilization of fire and elementary tools, the rise and spread of agriculture and animal husbandry, and the rise and spread of large-scale modern industry. The later regimes have not made the earlier regimes obsolete; rather, they have absorbed them and, in so doing, transformed them. Each new regime brought an expansion of the anthroposphere within the biosphere.

Defining the three regimes jointly in similar terms is helpful in order to better understand each of them separately as well as in their interrelations. A common conceptual model invites and facilitates comparison. The comparison allows us to explain the sequence in the emergence of the regimes and to perceive not only their similarities and differences but also their interlocking.

2. Agrarianization

2.1 Continuities

The history of the past ten thousands years may be read as a series of events accompanying the process of agrarianization of humankind, a process in the course of which humanity extended the domain of agriculture and animal husbandry all over the world and in so doing made itself increasingly more dependent on this very mode of production.

The initial transition from gathering and hunting to agriculture and animal husbandry was not necessarily abrupt. A group that started to cultivate a few crops would not have to give up its older ways altogether. There would have been only few, if any, agrarian societies from which gathering and hunting disappeared completely at once. However, the proportion of products acquired in the older way inevitably diminished as agriculture and animal husbandry advanced.

From the very beginning, the process of agrarianization was linked closely to the domestication of fire. It is hard to imagine how people could have begun to cultivate plants and to domesticate animals had the art of handling fire not already been thoroughly familiar to them. For one thing, they needed a hearth fire to cook on. The first crops cultivated on any large scale were cereal grains such as wheat, rice, and maize, which, owing to their high nutritional value and their capacity to withstand storage for long periods, formed very appropriate staple foods for a human community; to serve this purpose, however, they had to be made more easily digestible with the help of fire.

A second and very different reason why the control of fire formed a precondition for agrarianization was the human predominance over all other mammals, which was grounded partly in the use of fire. The human species' monopoly over fire was so solidly established by the time agriculture began, and is today so easily taken for granted, that it is seldom given separate attention in this context. Yet it deserves mention. Their hegemony in the animal kingdom enabled people not only to bring certain species, such as goats and sheep, under direct control, but also—at least as important—to keep most of the remaining "wild" animals at a distance from their crops and herds.

Third, experience in controlling fire may have furthered plant and animal domestication in another, even more intangible way, which our distance in time makes it difficult to assess precisely but which we are also, for that very reason, likely to underestimate. The time-honored practice of handling fire could not have failed to prepare humans for the many chores involved in agriculture and animal husbandry. It taught them that extending care upon something nonhuman could be well worth the trouble and thus made it more acceptable to them to accommodate the strains of an agrarian life, full of self-imposed renunciation for the sake of a possible future yield.

The most immediately visible link between early agriculture and ancient use of fire lay in the custom of burning off land with an eye to food production. Of old, foraging peoples were wont to apply their torches in order to keep the land open for gathering and hunting. Even in recent times, those firing practices were continued in some parts of the world, as in Australia where the Aborigines' judicious use of fire in keeping their land open for kangaroos and humans has become known as "firestick farming," a term suggesting a form of proto-agrarianization.

2.2 Sequences

The rise of agriculture was in many respects remarkably similar to the domestication of fire. Again, humans added new sources of energy to their own, this time by adopting certain plants and animals and integrating them into human societies. Plants that were formerly wild now began to be cultivated, wild animals were tamed and used for food or other purposes such as traction, and all these species were made part of the human domain, of the anthroposphere, which thus increased in size and complexity.

The transition from foraging to agriculture did not automatically make people happier and healthier. Agrarian life brought new hardships. Diets became more monotonous. Sedentary life in villages increased the susceptibility to disease. It allowed for a rise in fertility, but it also caused higher mortality. The result was large numbers of children, many of whom never reached adulthood. Not surprisingly, research on skeletons reveals that after agrarianization, human lives tended to become briefer, and bodies shorter. It is equally understandable that the texts of the great religions of Eurasia all exhibit a nostalgia for the lost paradise of a pre-agrarian era.

The very nature of agrarian life foreclosed a return to foraging, however. There are a few known cases of such a return, but those are exceptions. Almost everywhere, with so many people, and so little land, the only option for the survival of an agrarian population was cultivation. "Work" was writ large over the agrarian world.

If, at a very early stage, the domestication of fire had made human groups more productive but also more vulnerable (because from now on they had to rely on their fire), the rise of agriculture had the same twofold effect. Being able to grow more food, human groups grew more populous and became more dependent on their crops, and thus more vulnerable to failure or loss of their harvests. As the expansion of agriculture and pastoralism left increasingly less land available for foraging, the opportunities to escape from this vicious circle dwindled.

The first stage in the process of agrarianization necessarily involved clearing the land, removing any existing vegetation that would compete with the planted crops. In many cases, the most efficient procedure to accomplish this was by means of fire. As long as land remained plentiful, continued recourse was often taken to fire in a system practiced throughout the world and known under various regional names that are usually subsumed in the standard literature under the label "shifting cultivation." Shifting cultivation implies that a piece of primary forest is first cleared by the method of slash and burn and is then used for one or more harvests of crops. When, after a while, crop nutrients in the soil become exhausted and undesired plants (weeds) begin to dominate, the farmers temporarily abandon the land and turn to an adjacent lot, burn the vegetation down, and bring it under cultivation, again until they find harvesting unrewarding. Eventually they return to their first plot, which by then has become secondary forest or secondary bush, and resume their activities of burning, planting, and harvesting there. The duration of the entire cycle may vary as to time and place, but the principle remains the same.

In many parts of the world, in the course of time, burning the land and letting it lie fallow for a number of years or seasons ceased to be a regular practice and was replaced by more intensive methods of working the soil, requiring greater investments of labor but yielding a larger output per acre and thus making it possible to feed more mouths. The most common means of accomplishing higher yields by harder work were irrigation and ploughing.

2.3 Fire in Settled Agrarian Societies

In all agrarian societies with permanent settlements, new uses of fire and new attitudes toward fire developed. During the long first stage of human fire use, the main concern always was to keep the communal fire burning. From now on, however, the main concern became almost the opposite: to keep the fires that were lit for numerous purposes from spreading and from running out of control.

The uses of fire became increasingly more varied. Specialized pyrotechnic crafts emerged, such as blacksmiths and potters. Among the growing urban populations, fire was regarded with greater anxiety, for several reasons: with the proliferation of fires, the risks of conflagrations increased, and with the accumulation of property, people had more to loose.

Of course, it was fire as a natural force with its self-generating destructiveness that was feared. But more and more, this natural force manifested itself almost exclusively in the guise of anthropogenic fire. With all the fires burning in a city, one moment of carelessness might cause a conflagration. People had to rely on other people's caution. They had to oppress attempts at deliberate fire setting. And they had to reckon with the very worst danger: the organized form of murder and arson known as war.

A common problem in all advanced agrarian societies was the prevention of uncontrolled fire in cities. This problem may have been less urgent where the major construction material was stone or brick; but throughout the agrarian era we find major cities in which most houses were built from timber. Rome, Constantinople, Moscow, London, and Delhi as well as the capitals of China, Korea and Japan all suffered conflagrations; all of them faced problems of fire prevention.

Today we may tend to conceive of problems of fire prevention primarily in terms of technology: of building materials and technical equipment. However, the problems were (and still are) at least as much civilizational, or social. The crucial issue was human behavior, care, consideration. According to a well-known story, a temple in ancient Greece burned down because the servant was careless; similar events must have occurred all over the world. Everywhere it was the human factor that counted most of all.

This was also reflected in the presence or absence of fire brigades. Today, it may seem self-evident that a city should have a fire brigade. However, the power relations in earlier societies could be such that the authorities would forbid the organization of fire brigades by citizens. A dramatic example of a ban put on urban fire brigades is provided by two letters that were exchanged in the second century CE between the governor of a province in the eastern part of the Roman empire, Pliny the Younger, and his emperor, Trajan. In the first letter, Pliny informed the emperor that the capital city Nicomedia was completely ruined by a fire; people had watched the blaze passively, unable to do anything. He therefore asked permission to establish a fire brigade of citizens in order to prevent such disasters in the future. The emperor denied the request bluntly, for—as Pliny should have known—such an organization of citizens could easily turn into a club with subversive political activities.

It is still unknown to what degree Trajan's reply was typical of rulers in agrarian empires. The scattered evidence provided by local histories suggests that the situation was very different in the towns of medieval Europe, where the members of various guilds were required to take a share in fighting fires. This issue is a promising subject for comparative historical research.

3. Industrialization

3.1 The Industrial Regime

While the basic theme of human history over the past 10,000 years has been the agrarianization of the world, the history of the past 250 years has unfolded under the aegis of industrialization. In the process, the anthroposphere has become one global constellation, extending all over the planet, and its impact on the biosphere has become more and more intense.

In line with the approach to the domestication of fire and agrarianization, industrialization can be defined as the formation of a socioecological regime, structured around a new source of energy: fossil fuel—first in the form of coal, later also of oil and gas. The nature of this new energy source has made the industrial regime different in a basic way from the earlier socioecological regimes. Unlike plants and animals, and even wood, fossil fuel is not directly connected to the continuous flow of solar energy. It is a residue of solar energy from a remote past contained in geological formations. The energy stocks are not diffuse like sunlight but concentrated in particular locations from where they can be extracted through concerted human effort. They have two seemingly contradictory properties: they are abundant and finite.

The abundance is indeed great. Coal, oil, and gas represent the remains of the nonoxidized biomass—in other words, unburned fuel—of over 3 billion years. When people began exploiting those enormous reserves they entered as it were, in the words of the environmental historian Rolf Peter Sieferle, a "subterranean forest" of inconceivably large dimensions, which, moreover, in the course of 250 years of exploitation and exploration proved to contain far more riches than was originally expected.

Yet no matter how large, the hidden stocks are also finite. In contrast to plants, which are the direct autotrophic products of the photosynthetic conversion of solar energy, and to animals, which are the heterotrophic consumers of plant and animal food, geological stocks of fossil fuel do not partake in any living metabolism. They are incapable of growth or reproduction, irreplaceable, unrenewable and, as is becoming more and more evident, their use generates combustion products that enter the biosphere.

3.2 Origins and Preconditions

The genesis of industrialization raises a whole array of intriguing questions concerning its preconditions and its immediate causes as well as its functions and side effects. These are in principle the same questions that can be raised about the domestication of fire and the origins of agriculture. Because industrialization began only recently, we have the benefit of far more and far more precise empirical evidence; nevertheless, the problems remain puzzling. Any answer that comes to mind can only be tentative and subject to caveats.

One thing is certain: industrialization was not triggered by a major change in climate. At most, its beginnings more or less coincided with the end of the most recent secular dip in temperature in Europe, from 1550 to after 1700, known as the little Ice Age; but no relationship with this climatological event was evident. Industrialization was a completely anthropogenic transformation, brought about by humans in societies that were fully equipped with fire and agriculture.

The beginning of industrialization heralded a new era in human history. Yet it was clearly a continuation of the fire regime, for every industrial process rested on controlled use of fire. Industrialization also presupposed agrarianization: it could never have started without a strong agrarian basis that was able to feed numerous people who were not engaged in agriculture themselves and who could therefore be recruited for work in industry and trade. From the outset, the industrial world coexisted and coevolved with the agrarian world, in close interdependence and symbiosis, culminating in our time in forms of industrial farming or agroindustry that lean as heavily on fossil fuel as any other form of industrial production.

3.3 Further Parallels

Just as agrarianization must have begun in small farming enclaves, carved out in an environment that continued to be the domain of foragers, industrialization started with single steam-powered factories—often called mills as if they still were driven by wind or water—standing apart in the agrarian

landscape. To sketch their initially semi-isolated position, Sieferle's image of industrial archipelagos is appropriate.

From the very start, however, even as islands in an agrarian landscape, the factories had an ecological impact stretching beyond the land on which they stood. To begin with, considerable amounts of matter and energy went into building them. The usual construction material was partly timber, but to a larger extent brick, the manufacturing of which involved huge quantities of fuel. The engines were made largely of metal, also in a highly fuel-intensive fashion. Then, brick and timber were needed for the houses to accommodate the workers. So even before the factories began to operate, their tentacles were already reaching into the environment.

Once in operation, the factories had to be supplied with a continuous input of material, such as iron and cotton to be processed and of fuel to keep the engines going. The need for fuel explains the location of the early industrial plants: close to coal mines and, in the case of heavy industry, sites of iron ore deposits. Of course, a nearby sea or river port facilitated transportation; failing that, canals were built, and later railways, connecting the islands of industrial production with the areas where the raw materials were found and with the markets where the products were sold. In 19th-century Britain, an ever more extensive and intricate network of canals and railways was formed through which the industrial regime spread its reach over the entire country.

3.4 Industrialization as Intensified Land Use

Our earliest human ancestors treated the land they lived in as essentially all of the same nature—a territory for both collecting and hunting, for foraging food as well as fuel. In agrarian societies, most of the land was parceled up into three distinct parts with distinct functions: fields or arable planted with crops; pastures or meadows where cattle and horses grazed; and heath and wood land providing fodder for pigs and sheep and, even more important, fuel and timber. All three areas were controlled by humans, who were able to make the land thus divided up more productive for their own specific purposes, but who also found themselves faced with a continuously decreasing wild acreage.

Industrialization offered an escape from this predicament. The exploitation of coal was a highly intensified form of land use, which relieved the pressure to use large areas extensively for growing wood for fuel. The virtual expansion of land attained through the use of coal helps to explain the unstoppable advance of industrialization, once it was underway. Like the agrarian regime, the industrial regime made offers that were disagreeable in many ways to a great many people, but in the long run for all of them impossible to ignore or to refuse. The huge gains in energy could be transformed into economic, political, and military advantages that proved to be irresistible. All over the world, societies without fossil energy industry made way for societies with fossil energy industry, a process which could not fail to have profound consequences for the biosphere.

In our own age, each year more than 100,000 square kilometers of remaining wild land in Africa, Asia, and South America are subjected to human exploitation. One of the causes is the need for fuel in the Third World; in many densely populated areas, fuel for the fire to cook on has become as proverbially scarce as the food to be cooked. In addition, there is a rising demand for timber and pulp from the most highly industrialized countries. The greatest threat to the continued existence of natural forests does not lie in felling the trees, however, but in the indiscriminate burning down of entire tracts in order to clear the ground for raising crops and cattle for commerce. The world economy generates a rising demand for products of tropical agriculture. To meet this demand, garden plots are converted into plantations, and the small farmers have to leave. At the same time, population continues to grow, so that pressure on land becomes even greater. As a result, the contradictory situation arises that in a world suffering from severe shortage of wood, each year many tens of thousands of hectares of forest are set on fire.

In Australia, exceptionally, anthropogenic fire has diminished since the establishment of a European colonial regime. From the time of their first arrival, the British had a sensation of the Aborigines going about burning the land incessantly. The British took every measure they could to repress these burning practices, with such success that rainforest has been regaining ground again in Australia over the past 200 years. An unintended side effect has been the increasing incidence of large wildfires during dry seasons, for the many fires of limited scope lit by the Aborigines prevented the vast fires that now, when they are raging, constitute a serious threat to the metropolitan suburbs.

3.5 Current Uses and Manifestations of Fire

The 20th and 21st centuries have seen a common shift in preference for energy sources all over the world. Wherever electricity is now available, it is the favorite means of generating light, and combustion engines are, along with electricity, the favorite means of generating motion.

The shift from steam engines to electricity and internal combustion is reflected in the changed industrial landscape, which in the Western world is no longer dominated by endless rows of factory chimney stacks. Yet the methods of production in modern industry, and in agriculture as well, continue to be highly fuel-intensive. Most of the energy consumed, including most electricity, is derived from the fossil fuels coal, oil, and gas. Combustion processes thus still play a central role, but they are relegated to special containers so that most people are not directly confronted with them. Soot, smoke, and fire risks are reduced to a minimum. The furnaces and combustion chambers in which enormous heat is concentrated remain cool on the outside.

Typical products of modern fuel-intensive industry are cars, with engines designed to be propelled by finely tuned and minutely controllable combustion processes. Indeed, the car may almost serve as a symbol of the highly complex and differentiated ways in which, in our day, thermal energy is being used. Cars are set in motion by burning fossil fuel. They are made of steel, plastic, and glass—materials that are produced and processed at high temperatures. Yet no one who gets into his or her vehicle and turns on the electrical ignition to start the engine needs be consciously aware of using fire and products of fire. When driving, people do not perceive the processes of combustion that keep their car going: they do not see the petrol gas burning under the bonnet, nor have most of them even remotely sensed the fire in the factories and power plants without which their cars would never have been produced at all.

A very different example to the same effect is farming. In the early 19th century, when Britain was already beginning to industrialize, practically all the energy consumed on the farm was still produced within the confines of the farm and its fields in the form of human and animal labor; the open fire that was burning in the hearth was fuelled with wood from the immediate surroundings. By the turn of the 21st century, the situation has become very different, with practically all the energy used now brought in from outside the farm, in the form of fertilizer, oil and petrol, and electricity.

A major advantage of the new sources of energy is their flexibility. The fuels are easier to transport and to distribute than wood or coal, and the combustion can be regulated more precisely. Given the technical facilities, gas, oil, and electricity provide for very even and accurately controllable flows of energy. Electricity has the additional advantage of being totally "clean" at the place of destination. Domestically, a few simple actions and a negligible risk suffice to provide people with an immense array of services: some substituting for old chores such as cleaning the floor and washing dishes, others based on entirely new appliances such as television sets and computers. Industrially, the same advantages apply at a much larger scale, permitting a far more diversified use of power than was possible with steam power.

The impact of electricity and internal combustion engines makes itself felt in every sector of social life: in agriculture, industry, traffic and transportation, domestic work and leisure. Everywhere it is possible to mobilize large quantities of energy with very little physical effort. The result is to make life more comfortable in many respects, enhancing the sense that physical processes can be mastered and also, at times, fostering the illusion of independence.

An illusion it clearly is. Regardless of whether people can avail themselves of energy in the form of a petrol engine, a battery, or a connection to an electric circuit or a gas main, in each case they are able to do so only because they are part of a complex and far reaching network of social interdependencies, connecting them eventually to the energy stored in fossil fuels. As long as the supply lines are functioning, and as long as people are able to meet their financial obligations, they do not need to bother much about the entire constellation. They are immediately confronted with it, however, the moment something goes wrong with any of the conditions.

In this way the exploitation of the new sources of energy clearly continues a trend that has begun with the first domestication of fire. Dependence on the forces of nature has become more indirect (which is not to say less important!), and by the same token dependence on cultural and social resources has increased. A complicated technical and organizational apparatus is needed to continuously maintain the supply of energy. Most of this apparatus is located behind the scenes of industrial society, invisible to the ordinary consumer. Energy is made available in such a convenient fashion that it is easy to forget the social effort required to produce it.

That social effort is spent, first of all, at the drilling wells and in the mines where the energy is won and, next, during the operations of processing it into consumable gas, oil, or electricity, and of transporting and distributing it. While the many provisions needed for the undisturbed flow of energy are often taken for granted, they cannot fail to exert permanent pressures on those who benefit from it as customers. The bills have to be paid, financially and otherwise.

The permanent availability of electricity, at every hour, in many parts of the world, has led to a diminution of the contrast between day and night. By the middle of the 19th century, the large investments made in their factories impelled many owners, under the pressures of competition, to let the engines run day and night. Gaslight illuminated the workplace. In the 20th century, night life has steadily extended, especially in the cities. Water mains, sewage, gas, electricity, telephone, fax, internet, radio, police, fire brigade, hospitals are all generally expected to operate day and night. International interdependencies never come to a halt. This is one of the reasons why many people turn on the news as soon as they wake up in the morning; before resuming their daily activities they wish to learn what has happened while they were asleep—in their own country, where it was night, and elsewhere, where it was day time.

Once in a while there is a hitch. Sometimes the local supply of electricity breaks down, as happened for a number of hours in the blackout in New York on July 13, 1977, and for longer periods at a regional scale in California in 2001 and in the northeastern United States and southern Canada in 2003. In New York the failure was due to technical causes; in California its primary causes lay in disturbances in the financial sector. A combination of economic and political complications brought about the international oil crisis of 1973, when the majority of oil-producing countries jointly managed to enforce a drastic increase in the world price of crude oil.

Disturbances are remarkably rare, in view of the prodigious growth of energy consumption since 1950. The industrial economy is a fuel economy, revolving around the regular supply of fuel that can be easily converted into energy. Light, warmth, motion, and even coolness are produced with fuel. The rising supply of all these fuel-intensive amenities in turn constantly stimulates demand from customers, who are eager to enhance both their physical comfort and their social status.

In a large part of the world, access to the benefits of modern technology is still restricted to a small upper crust of society. Moreover, while in the western world and parts of Asia effective measures have been taken to reduce the polluting effects of industrial and domestic fuel use in densely populated areas, the quality of the air has only deteriorated in the rapidly growing megacities of Africa, Asia, and Latin America. As a consequence, according to an estimate of the World Health Organization in 1997, air pollution killed about 400,000 people worldwide annually.

Still, economic growth has continued in the poor parts of the world, and this is bound to affect the rich countries as well. Wealth attracts poverty; history abounds with examples of this general rule. When the opportunity offers itself, many people from poorer regions will try to migrate to regions with a higher standard of material comfort.

Meanwhile, global fuel consumption continues to rise. In rich countries, advanced technologies permit customers to enjoy the use of their appliances and vehicles without any physical inconvenience caused by the combustion processes. Poorer people generally are prepared to put up with noise and smell to profit from fossil fuel energy.

3.6 Fire and War

With growing interdependence among people, dependence on natural forces has become more indirect: longer and more ramified social chains are formed between the production of things and their use. Even the threat of violent destruction of lives and property comes far less often from natural forces than from forces unleashed by one human group against another. The most powerful groups are those that command the organizational and technical means to mobilize huge amounts of energy and matter against their enemies.

In our day, the search for the most effective means of violence has led to the exploitation of a new source of energy: nuclear fission. For the first time in human history, a new source of energy was first applied at a large scale in war, with the sole intent of massive destruction. It was the result of an enormously concentrated accumulation of technical, scientific, and economic capital, invested in a single one-purpose enterprise: the production of an atomic bomb.

But no matter how single-focused the effort, the invention of the atomic bomb followed a general rule in the history of technology and produced unintended side effects. One of these was that the development of the bomb by the United States became the first step in a process of escalation and proliferation, which has become the greatly enlarged version of the paradigmatic case *par excellence* of an arms race. Another unintended consequence has been the rising level of risk of fatal consequences in case of a breakdown in any of the nuclear plants that were built in the second half of the 20th century.

War has always been an important factor in the relations between humans and their physical environment. In most cases, it was directed at destruction of a part of the anthroposphere: especially that part in which the organizational basis of the enemy was supposed to be most vulnerable. The more investments a group made in controlling its environment, the more susceptible it became to losses through violence. Advanced agrarian communities could suffer a severe setback in case their rice fields, vineyards, or terraces were destroyed. Many cities in agrarian societies underwent drastic reduction or were even totally annihilated after military surrender: the greater the concentration of physical wealth, the more irreparable the destruction.

In industrial society, enormous means of destruction were developed even before the invention of the atomic bomb. During World War II, air raids brought devastation to a great many cities. Global industrial society proved to be sufficiently resilient and affluent, however, so that after the war every bombed city was rebuilt at its original site. Such were the powers of recuperation of the industrialized parts of the anthroposphere.

4. Symbolic and Religious Aspects

4.1 The Blessings of Fire

We may safely assume that fire, having played such a crucial role in the development of human group life,

has always kindled the imagination and given rise to rites and beliefs. Unfortunately, for the earliest stages we have only material remains to go by, and it is very difficult to reconstruct the symbolic and religious meanings attached to fire solely on that basis. Since the invention of writing, however, many facts relating to these intangible aspects have been recorded. More recently, anthropological fieldwork has yielded a profusion of data and insights.

Thus, from all over the world myths have been collected relating how man first came to master fire. Typically these myths represent the initial conquest of fire not as a gradual long-term process but as a single adventurous event, often marked by cunning and deceit. According to most myths, fire was at first possessed by the gods until, at a given moment, they gave it to humans or, alternatively, it was stolen from them, either by some animal who then passed it on to humans or by a semidivine Promethean culture hero.

All these myths convey a sense of the preciousness of fire. Fire is regarded as a unique possession, which separates humans from the animal world. As a corollary of this conception, many peoples used to tell that other peoples did not have fire and, consequently, could not cook their food so that they had to live like animals. Needless to say, fire has been a universal element of culture ever since the emergence of *Homo sapiens*, and all these stories are spurious and slanderous.

Directly related to such beliefs is the idea that fire steered by human hands has purifying effects. The domain of human culture is seen as coinciding with the range of domesticated fire. Thus, cooked meat is regarded as edible for humans, raw meat as only fit for beasts. Similarly, land cleared by burning is trusted as belonging to the human domain, whereas uncleared bush and forest are feared as wild and dangerous, a hideout for predators, serpents, insects, and evil spirits.

The sense of preciousness of fire also found expression in rites. The communal fire would be the center of group life, and people would treat it as sacred, surrounding it with ceremony and taking great care that it would never cease burning. The Roman cults of the hearth of Vesta, the fire goddess, show how such ancient rites could persist in an urban setting. In many societies the perpetual burning of the communal fire would be interrupted once a year; all fires were then ritually extinguished and replaced by new, pure fire that had to be solemnly ignited by the chief priest.

4.2 The Curses of Fire

When, after the rise of agriculture, people increasingly came to live in villages and cities, the number and variety of fires also increased, and so did the risk that any one of these fires would cause a conflagration. Fire continued to be a cherished necessity, but since it was so abundantly available, the sense of its preciousness was often superseded by anxiety and fear of the dangers it posed.

The transition was reflected in some of the great religions that emerged in Eurasia during the past 3000 years. The veneration of fire was most strongly preserved in Hinduism, with its cult of the fire god Agni, and in Zoroasterism, with its cult of Atar. Both religions kept alive the age-old worship of fire as a primeval force of vital importance to humans; but this was combined with awe and anxiety for its destructive powers.

Christianity and Islam stand much further away from the ancient worship of fire. Some old traditions persist, as in the burning of incense and candles, including the renewal at Easter in Roman Catholic churches. Fire no longer plays any part in beliefs about the origins of human society, however. Instead, it is strongly associated with evil. Sinners and unbelievers are threatened with a life after death in which they will be tormented by eternal burning. It seems as if the agonies of city conflagrations as described by several pagan authors from ancient Rome have found a religious symbolization in the vision of hell.

Fire festivals, such as those held over Europe at Saint John's day when things considered impure would be committed to the flames, were occasions of combined joy and horror. Modern literature and art abound with evocations of the terrors of fire. When novelists describe the nightmare of a conflagration, they attribute the disaster, quite realistically, not to natural forces such as lightning but to human action—either to individual negligence, to madness or malice, or to collective hostilities in war.

4.3 Fire as an Element

A more neutral way of regarding fire, not in terms of its good or bad effects on human beings but as a natural phenomenon in its own right, was first proposed by philosophers of ancient Greece. They developed, out of various antecedents, a generally accepted doctrine according to which the universe consisted of four elements: earth, water, air, and fire. In China and India, similar systems of thought were worked out. With only minor modifications the worldview in which fire constituted one of the major elements remained the dominant cosmology for natural scientists and physicians well into the modern era.

Alchimists and chemists, as the direct successors of the metallurgists and fire masters of a previous age, made the study of fire into their central concern. Just as modern mechanical industry in its initial stages owed a great deal to the steam engine, the rise of modern physical science would have been inconceivable without the numerous efforts of investigators experimenting with and theorizing about fire.

5. Conclusion: The Eclipse of Fire?

The controlled use of fire has been a human species monopoly since time immemorial. It constitutes an integral part of the apparatus with which humans have established and continue to maintain their dominance over other species.

Ever since the earliest stages of its domestication, the same phenomenon has recurred again and again in the human relations with fire: deliberately sought advances in control lead to unintended increases in dependency, and as the dependency sinks deeper into the social and economic infrastructure, it tends to become increasingly less clearly perceptible. Like agrarianization, industrialization began with conspicuously huge applications of fire. As industrial production became more specialized and highly organized, so did the use of fire. In everyday life in highly industrialized contemporary societies, flames are visibly present only in such highly domesticated guises as cigarette lighters, candles for ceremonial use, or wood fires intended to create an atmosphere of comfort and relaxation. When, on the other hand, a fire is shown on television or in the press, it almost always spells war and disaster. The regularly controlled combustion processes on which industrial production largely rests are now mostly relegated away from public view.

In accordance with this apparent eclipse, natural scientists and philosophers no longer regard fire as one of the four elements out of which the world is composed. The very concept of fire has disappeared from scientific literature, to be replaced by the more abstract concept of energy—something that cannot be directly seen, heard, smelled, or felt. Similarly, many a day may pass in the life of the average citizen of a modern industrial state during which he or she does not perceive any physically observable fire. Yet it continues to be an integral part of the anthroposphere.

Further Reading

De Vries B, Goudsblom J (eds.) 2002 *Mappae Mundi. Humans and Their Habitats in a Long-Term Socio-Ecological Perspective.* Amsterdam University Press, Amsterdam

Elias N 2000 *The Civilizing Process. Sociogenetic and Psychogenetic Investigations,* Rev ed., Blackwell, Oxford

Goudsblom J 1992 *Fire and Civilization.* Allen Lane, London

Goudsblom J, Jones E, Mennell S 1996 *The Course of Human History. Economic Growth, Social Process, and Civilization.* M. E. Sharpe, Armonk, NY

Grübler A 1998 *Technology and Global Change.* Cambridge University Press, Cambridge, UK

McClellan J E III, Dorn H 1999 *Science and Technology in World History*. Johns Hopkins University Press, Baltimore

Pyne S J 1991 *Burning Bush. A Fire History of Australia*. Henry Holt and Co., New York

Pyne S J 2001 *Fire. A Brief History*. University of Washington Press, Seattle, WA

Rossotti H 1993 *Fire*. Oxford University Press, Oxford, UK

Sieferle R P 2001 *The Subterranean Forest. Energy Systems and the Industrial Revolution*. The White Horse Press, Cambridge, UK

Simmons I G 1996 *Changing the Face of the Earth. Culture, Environment, History*. 2nd ed. Blackwell, Oxford

Turner B L II, Clark W C, Kates R W, Richards J F, Marhews J T, Meyer W B (eds.) 1990 *The Earth as Transformed by Human Action Global and Regional Changes in the Biosphere over the Past 300 Years*. Cambridge University Press, Cambridge, UK

Wright R 2000 *Nonzero. The Logic of Human Destiny*. Pantheon, New York

Yergin D 1991 *The Prize: The Epic Quest for Oil, Money and Power*. Simon and Schuster, New York

Johan Goudsblom

G

Geographic Thought, History of Energy in

Glossary

econometric analysis   A test of economic theory with linear or nonlinear regression and other multivariate statistical methods using socioeconomic and technical data.

geography of energy   The study of energy development, transportation, markets, or use patterns and their determinants from a spatial, regional, or resource management perspective.

location theory   Spatial theory developed to solve optimization problems for selecting a preferred zone or location for a facility, which involves trading off transportation costs with production and processing costs, production balances, and market-delivery levels.

mathematical programming model   An operations research technique designed to determine an optimal solution to a well-defined problem by maximizing or minimizing an objective function subject to technical, economic, and often environmental constraints.

regional input-output analysis   An analysis of the economy at the urban, state (provincial), or multistate (provincial) levels based on a fundamental identity that equates supply and demand using fixed proportions of intermediate inputs.

Three Mile Island (TMI)   The name of the nuclear power plant near Harrisburg, Pennsylvania, that was the site of the most serious nuclear accident in the United States on March 28, 1979.

Since the field of geography encompasses broad study of interactions between humans and the environment, the concept of energy has been operationalized and studied in several ways. These approaches have included energy balance studies in geographical climatology, energy resource availability and conservation in applied climatic research, cultural geography, and economic geography of energy resources. In the past few decades, new geographic perspectives on energy have been introduced from spatial analysis and modeling to applied, political, transportation, hazards, geographic information systems, global environmental change, and ecological economics. Some of these areas are even considered separate fields.

Understanding and modeling solar radiation and the energy budget of the climate at a variety of spatial scales have long been central concerns of several types of climatological studies: modeling, synoptic and dynamic climatology, climate change, and applied climatology. Analytical approaches have become increasingly mathematical over time, and this work has been somewhat separate from other branches of geography. It is consequently more convenient to trace the evolution of the energy concept in geographic thought by focusing on the contributions of human geography, which initially considered energy from economic and cultural perspectives. The article begins with a discussion of pre-World War II studies of the coal industry. Subsequent sections consider the rise of petroleum as a global political force, commercial nuclear power, and additional work on the location and spatial distribution of fossil fuels. A new period in geographic thought followed the Three Mile Island nuclear power plant accident in 1979, when geographers gave attention to technological hazards and risks of energy systems, the back end of the nuclear fuel cycle, the diffusion of energy conservation technologies, and the increasing scarcity of fossil fuels. The article concludes with a discussion of the growing need to develop sustainable energy resources in the context of concerns about global climatic change and rising energy demands in developing nations.

1. Early Research on the Coal Industry, 1934–1955

Coal, as the fossil fuel of greatest abundance and widest distribution, has long attracted attention from geographers. In the early years, this attention focused on the location of deposits and their transport and much of it mirrored the major coal fields of Britain, Germany, Poland, and the United States. The accounts in geography journals were largely descriptive, economic, and cartographic. Little or no interest was evident regarding the environmental costs. Later, when the public mood shifted and impacts were widespread, more attention was paid to environmental impacts, especially air pollution and its dispersal, impacts of acid rain, landscapes altered by strip mining and waste heaps, and the forced relocation of residents. This emphasis came with the increasing knowledge not only of the damage that was created by our demand and use of coal but also that we should and could do something to reduce or reverse the impact it was having on air, land, and water. From this point forward, research articles were much more widely dispersed, within a range of environmental journals and books, and the type of writing was more analytical.

2. Fossil Fuels and Electric Power Generation, 1955–1971

Parallel to the growth and development of the oil and natural gas industries in the United States and other advanced economies, geographers began to give more attention to the other fossil fuels, their development, and use patterns in the mid-1950s. Another exciting new energy source gained attention with the commercialization of atomic power plants in Britain, Russia, and the United States during this period. Nuclear power development attracted considerable attention among geographers in North America and Europe. This topic is considered separately since the economics, spatial diffusion, risks, social acceptance, waste problems, and other dimensions of nuclear power were highly uncertain in the 1950s and thus it was treated differently.

Although a quantitative revolution in geography first appeared in the mid-1950 s there was little evidence of a change in methods in energy research until the early 1960s. Indeed, most geographical publications on energy issues in the 1950s were studies on the regional geography of energy (i.e., "areal differentiation"). Although important, this work was overly descriptive and often handicapped by scant understanding of the complex interplay between resources and technical, economic, and social factors in energy markets. Several detailed analyses were completed on the oil, gas, and electric power industries of the United States, Britain, Europe, and the former Soviet Union during this time. A few researchers went beyond the single-region scale and assessed the development of power interconnections in Europe, North America, and the Soviet Union. A 1960 paper addressed the conflict between hydro-electric power generation and salmon fisheries in the Pacific Northwest, which is still relevant today. This study suggested the use of (but did not apply) optimization criteria to help resolve the conflict. The same year marked the inaugural publication of *Soviet Geography: Review and Translation*, which was to become a major scholarly outlet for energy studies of the former Soviet Union and its various regions.

Neoclassical studies of the regional geography of energy continued throughout the 1960s at a variety of spatial scales. Several books and articles on the economic geography of the oil and gas industries analyzed national and international patterns of location, economic growth, development, and trade but did not suggest general market patterns. The first truly international energy study, a book on the economic geography of the world oil industry by the British geographer Peter Odell, was published in 1963, just a few years after the formation of the Organization of Petroleum Exporting Countries (OPEC). Articles on coal geography also assessed economic growth and development but considered declining or less prosperous coal fields and their socioeconomic effects as well. Political factors (i.e., national policies, government regulation, political stability, etc.) began to be mentioned more frequently in this literature but were still secondary to economic considerations.

Of special significance was the work by Gerald Manners in the 1960s to incorporate location theory into energy geography. General locational principles were identified and applied to the oil, gas, and electric power sectors. Transportation costs were shown to be of fundamental importance because they determine the energy facility location between a resource market(s) and final demand center(s). Also important are factor costs of production, resource availability, weight and quality, scale economies, technical efficiency, modal and interfuel competition, and the load factor(s). Similar principles were used to explain the changing pattern of pit location within a coal field. The first such analytical, general approach for studying the industrial geography of energy was published in book form by Manners in 1964. (A 1950 book on the geography of energy by Pierre George received little notice outside of France.) Owing to data availability and the scale of the economies, his research focused on Western Europe and North America. National studies and atlases for the United States and United Kingdom were to come later.

As the 1960s closed, a few prominent authors completed seminal works on the international patterns of energy geography. A landmark book by Odell addressed oil and world power; this book was first published in 1970 and updated seven times. The monograph considered this leading industry's production, economic growth, and trade in the context of its prominent role in national and international political affairs. By analyzing the global structure of the oil industry along with regional perspectives and company–nation state relationships, Odell provided the most complete story of this vital resource. Three more general energy papers by geographers in the September 1971 issue of *Scientific American* showed the close connection between energy resource type and technology at the various life-cycle stages and the pattern of energy flows in primitive and advanced industrial economies.

Two final (although brief) energy books from the perspective of world economic geography closed out this period, both of which were written by Nathaniel

Guyol. One of these monographs provided, for the first time, a system of energy accounting that made it possible to compare energy statistics in multiple countries on a consistent basis, a subject he had worked on since the late 1940s. An accounting worksheet, suggestive of a first cut of a simple input-output table, was developed for The Netherlands as an example. The other volume was narrower but more detailed: a systematic survey of the electric power industry in 162 nations. Important observations were made by Guyol on the regional variations in electricity supply and use.

3. The Birth of Commercial Nuclear Energy, 1951–1973

The use of commercial nuclear power grew up as the stepchild of the military program. After spending an enormous amount of money to develop and construct the three bombs that were detonated in 1945, there was a concerted effort to turn this research to peaceful use. There followed almost a decade of research and development that led to the construction of electrical generating stations at Calder Hall, England, Shippingport, Pennsylvania, Eureka, California, and eventually more than 400 other commercial units.

Geographers showed little interest in the dispersed siting of the nuclear research program of the war years, but afterwards they addressed the possibility of using nuclear bombs to reshape the earth's surface, the location of nuclear fuels, and the siting of nuclear power plants. It was electricity generation that was to achieve the greatest momentum as commercial enterprises began appearing in several countries: Britain, the United States, the Soviet Union, and later France, Germany, and Japan. Researchers during this period published a steady, if thin, series of geographical studies, only rarely turning their attention to the spatial pattern of the fuel cycle or dispersal of the technology.

For more than 15 years, there seemed little uncertainty or public misgivings about the wisdom of a nuclear power program, and the geographic literature did not contradict this trend. Industry, government, and the public seemed to all be moving in the same direction, one that looked to nuclear power to generate a growing, perhaps dominant, proportion of the great amounts of new generating capacity that many countries were demanding. With U.S. growth rates for electricity exceeding 7% per year in the 1960s, nuclear power not only seemed like the most likely supplement to coal-fired power plants but also seemed to be the eventual replacement technology as concerns grew over the pollution that such fossil power plants produced. Moreover, it was recognized that by themselves coal stations would not be able to keep ahead of the tidal wave of demand for electricity. The public, bolstered by the reassurances of government and contractors, regarded the use of nuclear power as the only way to meet the steep rise in need.

During the early civilian period, power plant siting was influenced mostly by seismic stability and the availability of cooling water. Little consideration was given to isolating power plants from urban areas; indeed, many were located close to load centers, the better and more conveniently to service their needs. Thus, the pattern that was emerging favored the densely settled northeastern United States and Chicago. Gradually, additional power plants appeared in the southeast. In Britain, all but one (in Wales) were sited along coasts. France used both coastal and riverine sites, starting mostly with the latter locations, and in contrast to the United States and the United Kingdom, the distribution of nuclear power plants had no obvious locational concentration. However, the benign public reaction to nuclear power was to be short-lived.

4. Modeling the Spatial Development and Transportation of Fossil Fuels and Power Plant Siting

By the early 1970s, the quantitative revolution was firmly entrenched in energy geography. Researchers began to apply several mathematical and statistical techniques to problems of spatial energy resource development and transportation, power plant siting, and socioeconomic and environmental impact analysis. These important research developments, spanning the 1970s through the mid-1990s, are discussed in turn.

4.1 Resource Development and Transportation

Beginning in the 1970s, geographers developed and applied a variety of simulation, mathematical programming, and econometric models to oil and gas issues. Among the early work was a rather extensive analysis of North Sea oil development through 2080. A less ambitious study assessed the optimal oil pipeline network in the British sector of the North Sea. North American oil industry studies also used mathematical programming models, with the site selection and storage levels at the Strategic Petroleum Reserve in Louisiana as an example. Since the mid-1980s, a series of related papers have used econometric methods to update the oil industry models of geologist M. King Hubbert. This work often differentiated resource quality and used more than 50 years of historical data to assess the bleak future prospects of the U.S. oil industry. Somewhat less attention has been given to oil development in other regions. A few econometric studies have analyzed foreign investment, locational analysis, and resource-based industrialization in the Middle East, other developing countries, and the former Soviet Union. Additional studies have highlighted the main result of these developments for the United States: growing dependence on foreign sources of oil.

Mathematical programming methods also began to dominate research on the development and marketing of natural gas in North America and the Soviet Union. Since the production of this fossil fuel generally lagged that of petroleum in the United States and was decontrolled starting in the late 1970s, new research opportunities were pursued for modeling greater development, especially in Alaska, and trade with Canada and Mexico. Similarly, natural gas emerged as the fuel of choice in the Soviet Union in the 1980s. In order to advance this research area, it was necessary to improve upon previous models of the pipeline network and capacities. Following the decline in oil and gas prices and plentiful gas supplies in the United States since 1986 this research area has been largely abandoned. Only one major analysis of U.S. natural gas development has been completed since then, an econometric analysis of the declining yield per effort and resource scarcity.

Not surprisingly, the quantitative research methods that were first used in the 1970s for oil and gas studies also were applied to coal problems, along with the use of cluster analysis and spatial interaction models. Given the wide distribution of coal reserves in the United States and their often long distances from markets, transportation bottlenecks have occasionally arisen. Such problems are ideal cases for the use of operations research methods, such as linear, nonlinear, mixed integer, and multiobjective programming. An early 1970s study of the Great Lakes region was a landmark in the modeling of the whole coal resource delivery system. This was followed by a series of coal studies: regional industry analyses, such as for Britain, western U.S. states, and Pennsylvania; transshipment problems involving railroads and ports; coal exports; competition for water resources; other environmental constraints; and structural and technological change.

Geographers, along with scholars in other fields, have also studied the development potential of synthetic fuels from fossil fuels. This work has analyzed the spatial distribution patterns and development potential of coal gasification and liquefaction plants, oil shale, tar sands, and alcohol fuels. The huge water and other resource requirements and environmental effects were given prominent attention. These studies tracked the government policy interest and short-lived massive subsidies to potential synthetic fuels industries in North America in the late 1970s and early 1980s, which were briefly considered a viable alternative to the growing dependence on foreign oil supplies. Among these resources, only alcohol fuels (from ethanol and methanol) and the Canadian tar sands industry have established commercial viability.

4.2 Power Plant Siting

The siting patterns of electric power generating stations have been a major research area among energy geographers. Although early papers on this subject were published in 1960, the vast majority of these studies were done in the mid-1970s through the mid-1980s. Perhaps surprisingly, a retrospective discriminant analysis of U.S. power plant siting patterns from 1912 to 1978 showed that fuel choice was more important than regional differences in the siting process. Development of site suitability models for fossil-fueled and nuclear power plant location analysis was pioneered at the Oak Ridge National Laboratory in the late 1970s. These and later siting models were applicable to efforts to avoid or mitigate inevitable locational conflicts surrounding proposals for many new energy facilities. Geographers developed land use screening procedures to generate local site suitability scores by estimating a land use compatibility index and using importance weights offered by decision makers. A large range of engineering, ecological, and socioeconomic criteria were analyzed. These models were applied to the Mid-Atlantic states. Despite the value of these methods for initial site screening, it has been noted that the approach largely failed to be implemented because of a lack of need for new power plant sites at that time. On the other hand, the screening models amount to a stepwise heuristic and can select inferior sites. Improvements to the method were made through the development and application of optimal transshipment and multi-objective programming models, which were also applied to the Mid-Atlantic region. Another study considered the challenges in siting coal-fired power plants in western states, where there are stricter air quality standards.

The siting and safety of nuclear power plants received the most attention by geographers of energy during the early to mid-1980s. Ironically, following the Three Mile Island (TMI) accident in 1979, North American energy geographers shifted their research from power plant siting to reactor decommissioning and waste disposal, whereas British colleagues focused on nuclear plant siting. Computer programs were written in England to analyze alternative power plant siting criteria while emphasizing remote areas and the potential human dose from radioactive releases during accidents. Government policy was shown to downplay the siting criterion of proximity to human population centers, and thus arguably safety, despite the existence of ample remote sites in Britain. Cluster analysis was used to show that local population density received greater attention with regard to siting in the United States than in Britain. Broadening this discussion from a Canadian perspective, it was shown that existing nuclear power expansion plans in the mid-1980s were not as robust as construction of small-scale hydroelectric stations.

Expansion of cheap hydroelectric power in eastern Canada, whether small or large, fits in well with plans for increased electricity trading with the United States. Indeed, further study has highlighted the parallel socioeconomic and environmental challenges accompanying the recent "outsiting" of nonnuclear power plants in Mexico and Canada that have been dedicated to U.S. markets.

Since U.S. electric utilities had a greatly diminished need for power plant siting in the late 1980s as independent power producers and demand-side management picked up the slack, energy geographers abandoned regional siting analysis and developed broader, more detailed, and sophisticated industry resource planning models. A few major, large linear programming (LP) models were built with a full range of technical options for meeting future electricity demand under environmental constraints. One of these used multicriteria decision making and applied it to demand-side management planning in British Columbia, Canada. Most of the LP models were applied to sulfur dioxide (SO_2) emissions control. There was some limited regional detail to them, which overlapped with the work of regional scientists and regional economists.

4.3 Socioeconomic and Environmental Effects

Research by geographers on the mixed effects of coal mining dates back to at least the early 1950s. Initially, this work focused on the eastern United States, but by the late 1970s and 1980s it had shifted to the arid West (along with an increasing share of the coal production). This research analyzed the potential socioeconomic impacts of "boom-town" development of coal or synthetic fuels. The declining economic fortunes of some of the eastern coal mining areas such as anthracite towns of Pennsylvania continued to receive attention in the 1980s. Regional input–output (IO) analysis was demonstrated to be an especially useful tool for analyzing economic effects of spatial shifts in the coal industry. Similarly, most coal research in Britain during this period focused on the uneven economic development or shift between declining and developing regions and a short-lived effort to revive the industry. When U.S. western energy resource development plans were scaled back, research emphasis shifted to the economic and ecological effects and control of acid rain emissions of SO_2 and NO_x. This subject continued to receive the attention of a few geographers through the 1990s as the Acid Rain Program of the Clean Air Act Amendments was implemented, including (interstate) SO_2 allowance trading.

Many energy geographers have examined the socioeconomic and environmental effects and planning issues surrounding petroleum development in disparate regions such as the Middle East, Scotland,

Norway, and the U.S. Gulf Coast. The U.S. studies have also considered impact mitigation in the context of the National Environmental Policy Act and the macroeconomic effects of a future large supply disruption by OPEC. Econometric analyses by geographers for the United States have shown since 1989 that policies to expand domestic oil production would have adverse economic and ecological effects. Although less sweeping in its findings, a regional economic analysis in Canada of potential oil resource development in the nearby Beauford Sea found that little or none of the net economic benefits would accrue to Canada's Northwest Territories. This research used a multiregional IO model.

5. Hazards, Risks, and Technological Diffusion: Behavioral Research by Geographers Since the TMI Accident

Public disinterest in nuclear power was shaken by the successful campaign to raise awareness about its purported safety risks. It was a movement that slowly gained momentum, but by 1976 an anti-nuclear referendum in California received ample public notice and approximately 40% of the votes. Although the referendum failed, it was a watershed for the nuclear industry because it raised many more questions than the industry could successfully answer about the safety and the financial feasibility of the technology. In addition, it prompted more restrictive and demanding siting criteria, and more than anything else it put nuclear power on the map of regular public debate. Like colleagues from other disciplines, geographers became more involved, and it was a natural topic of interest because most of the issues were spatial in nature.

The increased attention by geographers first manifested itself in questions about the adequacy of the siting criteria that had been unquestioned during the previous decade. Since much of the impetus for the nuclear debate had come from California, it was not surprising that the strongest opposition to power plant siting appeared there as well. Geographers played a quiet role at first, mostly related to siting, but their participation grew as the public debate expanded to other spatial issues, such as radioactive waste transport and disposal, dispersal of radioactivity by air currents, and identification of evacuation routes that might be needed in the event of an accident. These developments were all part of a simmering debate about nuclear power in the United States and elsewhere, but they would come to a boil with the TMI Unit 2 accident near Harrisburg, Pennsylvania, on March 28, 1979. Ironically, Pennsylvania had enthusiastically commissioned the earliest U.S. nuclear power plant 22 years earlier.

It was the TMI accident that firmly placed geographers within the nuclear power debate. For several years, many geographers turned their attention to the myriad questions that the accident raised, including siting criteria, the dispersal pattern of possible contamination, evacuation behavior, and the adequacy of emergency planning and preparedness. The effects that the accident had on nuclear power were fundamentally geographical. Most noticeably, no new operating licenses were granted afterward, largely because of the elevated public concerns about the appropriateness of virtually every suggested site. Emergency requirements that were promulgated became so demanding that power plants already completed (e.g., the Shoreham plant on Long Island) were unable to develop an approvable evacuation plan.

Geographers participated in these studies and were most heavily involved in evaluating and measuring public risk perceptions. They asked questions about relative risk, the level of risk required to trigger evacuation, and the use of distance as a buffer to risk. These issues came into play because of the obvious inadequacy of emergency preparations at TMI, but the interests of geographers were not limited to them. Issues that had long been beneath the surface came into full view, and they included all forms of "nuclear geography," including a striking increase in the attention paid to radioactive waste transport and disposal.

Although attention to these matters was initially concentrated in the United States, it did not remain within its borders. Geographers in other countries, particularly Britain, heard the call as well, and geographers there had been making similar intellectual preparation to participate. There was a different attitude in the United Kingdom, however, in large part because their system of government had not been as receptive to an open debate on nuclear power as in the United States. As it soon became clear, its citizens had more practical reason for concern. Great Britain is a country smaller than Arizona, without the vast distances and isolation possible in the United States. Moreover, when the inquiry into the 1957 Windscale accident in northwest England confirmed public suspicions about nuclear hazards and ill-considered official secrecy, a phalanx of geographers was to lead the charge for a much wider nuclear debate.

All the nuclear plants that currently exist in the United States had already been sited by 1979, but this was not the case everywhere. In the United Kingdom, for instance, great attention was paid to siting and safety issues of the planned addition to the Sizewell nuclear station on the Suffolk coast. There was already a Sizewell A, but Sizewell B was to be a pressurized-water reaction of American design, ostensibly safer than the Magnox and Advanced Gas-Cooled reactors that comprised all the early British reactors. In response to the work of many groups, and especially geographers in the United

States and Britain, the government conducted the longest and most detailed public inquiry in the history of the country.

Like the TMI accident, the Sizewell inquiry galvanized geographical attention. Soon, geographers in England were writing about waste transport through London, waste disposal under the Irish Sea, and elevated leukemia frequencies in Cumbria. Perhaps due to the smaller size of the country, much of the work of geographers in Britain attracted more attention than similar work did in the United States. Geographers regularly appeared on TV and in various newspapers and magazines, which gave their comments more visibility. Eventually, geographers from Britain and the United States began collaborating on articles and books. Even France, which operates nuclear power plants that are closer to London than those in Wales, was not immune from criticism, and although the debate was more subdued there, geographers tracked and reported on various acts of public opposition.

Attention of geographers to nuclear issues peaked in the early 1980s and then began to subside, only to be shaken into a higher degree of agitation 7 years later by the accident at the Chernobyl nuclear power plant in Ukraine. This was a far more serious accident, and it had global implications. Climatologists tracked the movement of radioactive materials, but access to data was difficult owing to longstanding policies of the Soviet Union. Nevertheless, several geographers focused their attention on the disaster, with some examining issues of the power plant's location adjacent to the vulnerable Pripyat Marshes and 90 miles upwind from the heavy concentration of humans around Kiev, others observing how land had to be reshaped and forests leveled for safety reasons, and still others focused on the forced evacuation of more than 135,000 people who lived near the plant. Others were intent on observing the various political ramifications, including how the accident helped hasten the collapse of communism.

The Chernobyl accident also pounded several more nails in the nuclear coffin. In 1988, Swedish voters reaffirmed that nation's decision to phase out nuclear power, and Germans voted to take much the same direction. The RBMK design used at Chernobyl was largely abandoned, and many of those that had been built in the former Soviet Union and satellite countries have been closed. Moreover, attention to the various elements of nuclear power plant decommissioning gained increasing momentum.

Although it was obvious that nuclear plants would eventually be "turned off," the questions were when, by what process, and to what effect. There had been little public attention to these questions prior to 1985, but geographers helped bring attention to them during the next decade, emphasizing postoperational land use, transport and storage of the resulting low-level radioactive waste, the spatial and temporal sequence of targeted plants, and the socioeconomic consequences of the decommissioning process.

At approximately the time that interest in nuclear power was on the decline among geographers, attention was increasing in the area of energy conservation and efficiency. It was not causative, but when problems began appearing and proliferating with nuclear power, several electric utility demand-side management strategies were waiting in the wings. The logic was that if the electricity that nuclear power had been expected to supply was not actually going to be available, shortfalls would occur unless demand could be cut. It was cut, across the board and across the country, in every end-use sector, with the most noticeable changes occurring in the commercial and industrial parts of the economy and gradual progress in the residential sectors as well. Geographers worked with other professions to identify where these changes could be made and evaluated program effectiveness. There were several obvious uses to emphasize, including increased efficiency in lighting, insulation, appliances, and transportation.

6. Development of Sustainable Energy Resources

Although physical geographers have long included solar energy in its various manifestations in their work, only since the late 1970s have geographers actively researched renewable forms of energy. Even so, this research interest tapered off in the late 1980s, seemingly coincident with the crash in world oil prices in 1986. This hiatus was short-lived because by the early 1990s, there was renewed interest in fossil fuel scarcity, namely that of oil and gas. Geographers also began to explore the energy policy implications of the growing concern regarding global climate change, with an emphasis on energy efficiency and renewable energy development.

6.1 Oil and Gas Scarcity

Although acknowledgment of the ultimate scarcity of petroleum resources dates back to the 1956 work of M. King Hubbert, who predicted with startling accuracy the ultimate size and production peak of oil resources for the United States and the world, there is much greater concern with its "net energy" from production rather than ultimate depletion. Cutler Cleveland and Robert Kaufmann, among others, in a series of papers since 1991 have convincingly forecast using econometric analyses that the amount of oil added to proven reserves per effort from additional well drilling (yield) in the United States will cease to be a net energy source early in this century. This line of research builds on the work of Hubbert, who was the first to recognize that the historical production pattern would follow a classic bell-shaped curve, reflecting the interplay of resource depletion, real oil prices, technical innovation, and political decisions on long-term production costs. Consequently, domestic oil discovery, production, and proven oil reserves are on an inexorable downward path. These findings call into question energy policy decisions based on assumptions that these geophysical realities of the U.S. resource base can be altered.

With the conventional oil resource base increasingly scarce and dependence on it unsustainable, the cleaner option of natural gas has received more attention, especially since estimates of its proved reserves were greatly increased in the early 1990s. Here too, analysis by Cleveland and Kaufmann is instructive. Using a double-log econometric model of yield per effort of gas drilling in the United States, they found an exponential decline for gas discoveries not associated with oil fields in the lower 48 states from 1943 to 1991. Their model accounts for cumulative drilling, real gas price, changes in drilling effort, and shifts between onshore and offshore gas. Thus, natural gas development lags oil but is subject to similar factors controlling its scarcity.

6.2 Greenhouse Gas Reduction and Renewable Energy Development and Use

Coal, the third fossil fuel, is well-known to have the highest emission rate of carbon dioxide (CO_2), the principal greenhouse gas. By the early 1990s, a growing cadre of geographers began to actively work on energy policy responses to the threat of planetary warming and climate change. Indeed, they recognized that many preferred policy options also reduce conventional air pollutants and have a variety of other ancillary benefits. These energy options include switching to gas, energy efficiency and energy conservation, and rapid development of renewable and possible nuclear energy sources (although no major energy geographers have promoted the latter option through their research).

Three major types of greenhouse-related energy studies have dominated the work of energy geographers in the past few years: detailed technology assessments at the U.S. national level, sectoral-specific studies, and foreign case studies. One leading study provided an integrated assessment of the potential benefits of advanced energy technologies for greenhouse gas reductions in the United States. The authors found that large CO_2 reductions are possible at incremental costs below the value of energy saved, especially efficiency measures. Many sectoral-specific studies by geographers have highlighted the technological improvements and potential roles of energy efficiency, wind, solar, and solar–hydrogen technologies in buildings, transportation, manufacturing, and electric power. Much of this research has been conducted at the Oak Ridge National Laboratory.

123

Since global warming and climatic changes do not respect political boundaries, case studies of other nations can be valuable. For instance, several geographers have studied the broad range of policies that would be required to significantly reduce CO_2 emissions in Germany, Canada, and China. These studies not only highlight the great difficulty of significant emission reductions but also emphasize the necessity to implement policy reforms outside of the energy sectors (e.g., addressing urban design and transportation policies). To accelerate the process of emissions reduction in developing countries and economies in transition, the 1992 Framework Convention on Climate Change approved a pilot program in joint implementation of the treaty, whereby two or more nations cooperatively reduce or sequester emissions in their countries. Implementation of this pilot program was slowed, however, by concerns among developing countries that national sovereignty would be violated and that Western nations or Japan would gain a competitive economic advantage. Geographers have generally been sympathetic to these concerns, questioning the policy's fairness and cost-effectiveness, while identifying constructive solutions that involve compromises by the applicable countries.

Renewable energy technologies have different spatial characteristics than those of fossil or nuclear fuels (e.g., with respect to availability, transportability, reliability, and storage), which have yet to be fully explored. Among conventional energy resources, the two renewable sources that are well established are hydroelectricity and biomass. Hydroelectric installations, however, are not always sustainable because gradual siltation of their reservoirs may limit facility lifetime to 50–200 years. Since the 1990s, hydroelectric power has attracted increasingly negative attention in the United States and elsewhere, where the large part of its potential has already been tapped. Identified shortcomings include forced migration of people, displacement and destruction of salmon stocks, and downstream ecological changes.

Geographers have written a modicum of papers and monographs on hydropower facility construction in the past decade, which have focused on trends in North America and a few mammoth dam projects in developing countries such as Brazil, China, and India. Research emphasis in the United States has shifted from dam construction to dam removal and downstream restoration. Several dams have been suggested for demolition, especially smaller structures. Of the large dams, the most attention has been given to the proposed removal or "neutralization" of Glen Canyon Dam, located upstream from the Grand Canyon.

Biomass has long been the major (potentially) renewable energy source in developing nations, and it has received the most attention among the renew-ables from geographers. It has also received research attention in Scandinavia, where it is commonly used but somewhat controversial in Sweden. Along with its relative abundance, biomass also has challenging limitations owing to its wide variety of forms, ranging from wood to crops and garbage.

Wind energy has been the fastest growing energy source in the world in the past decade, and in the late 1990s Germany overtook the United States as the global leader. An important book printed in 2002 includes 10 chapters by geographers that compare and critique the varying prospects and policy approaches to renewable energy development, especially wind, biomass, and solar energy, in the United States and northern Europe. Wind energy use and public perception in California, the location of its most extensive U.S. development, have been examined by geographers in articles and a 2002 book that examines the issue of aesthetic intrusion on the landscape. Extensive study of wind energy has also been conducted by climatologists. For example, a 1997 paper analyzed the influence of climatic variability on wind resources in the U.S. Great Plains, the area of greatest wind power potential.

Like wind and hydropower, immobility presents geographic challenges for the development of geothermal energy, particularly the pressure it puts on natural resources. Geographers have given the least research attention among the renewable energy sources to geothermal (because this research area is dominated by geologists and engineers), after significant contributions were made by Pasqualetti in the late 1970s and 1980s.

6.3 Energy in the Developing World

Faster than other professions, geography is predisposed to grasp differences between the developed and the developing world—differences that are particularly apparent in considerations of energy. The developed world relies on fossil fuels, whereas the developing world relies on renewables. The developed world searches for energy to maintain a high standard of living, whereas the developing world must sell its resources to stave off poverty. The developed world worries about the environmental impacts of energy supply, whereas the developing world worries about having any energy supply at all. The developed world has achieved a high standard of living by consuming prodigious amounts of energy, whereas the developing world resents any attempts to curtail their upward mobility by doing the same, even though the environmental impacts are now so much more obvious, troubling, and dangerous. The developed world is in the minority, whereas the developing world is in the majority. Such differences form the basis for much of the work geographers have pursued in developing countries.

Although many developing countries share a need to trade environmental quality for energy supply, they are in other ways profoundly different. For example, China and Mexico are industrial countries, whereas Haiti and Ghana have primarily agrarian economies. In the former, government concentrates on how to increase gross domestic product; in the latter, government tends to worry more about matters of day-to-day survival. The work of geographers has focused on issues of environmental quality, economic growth, equity, and sustainable development and in a more speculative mode the consequences of bringing per capita energy consumption in these countries closer to that of the United States or the European Union.

The work of energy geographers in developing countries is often closely tied to environmental considerations. China, for example, is intent on using its great reserves of coal to help fuel its climb toward greater economic prosperity. As a consequence of pursuing this goal, air pollution has increased in many locations within the country, as has concern about the possible impact of such additional coal burning on global warming. Geographers, among many scientists, have monitored such trends and have expressed opinions ranging from how one might minimize such impacts to suspecting that the impacts on climate change are insignificant.

Other topics of interest to geographers include energy trade and transportation; the use and scarcity of biomass; hydropower development; and the development of geothermal, wind, and solar power. Of these topics, energy transport and trade have had perhaps the most natural and longest interest. The focus of such attention by geographers evolved from matters of shipping routes and port facilities to the location of lines for the movement of energy by rail, pipe, and wire. Shipping routes tend to change infrequently, and then usually as the result of military conflicts and new discoveries. Modern examples include the closing of the Suez Canal and the development of oil fields in the Gulf of Campeche, both in the 1970s.

Geographers continue to track the discovery and development of new sources of energy, although such activities trend toward the expansion of existing areas of extraction rather than new developments. Exceptions to this rule include oil developments in African countries such as Chad and South American countries such as Ecuador, Peru, and Bolivia. In the latter country, interest has been drawn especially to natural gas fields on the eastern side of the Andes, with the prospect of piping through Chile to the Pacific Coast for liquefaction and transport to markets. One market is the United States, and Bolivia is considering shipping to a planned regasification facility in Baja, California, and then piping the gas to California. In so doing, they will be avoiding the environmental encumbrances of trying to locate regasification plants within the United States while still having a U.S. market. In all cases, these and similar projects hold the hope for improving the economic prosperity of citizens of the producing countries, although the usual questions exist as to whether a significant proportion of the extracted wealth will remain local. Geographers have noted this ethical issue and in a few cases have proffered recommendations.

Although these examples rest on matters of resource sales and environmental impacts, geographers have sustained an interest in the transportation routes, including the economic costs and benefits of the developing countries that are physically touched along the way. Chad needs to move its oil through Nigeria, for example, and the vast resources of the landlocked Caspian Sea must cross through several countries to gain access to the sea, which is a diplomatic challenge. Locally, Mexico and the United States have been working to establish closer ties through increased land-based trade of electricity, coal, oil, and natural gas, in both directions, across their common border. This has caught the attention of geographers in both countries, who have pointed to the shared advantages. They have also identified some of the shared environmental costs when the impacts of energy development cross international borders, such as when air pollution wafts northward from power plants in the Salinas district south of Eagle Pass, Texas.

Hydropower projects always hold a particular appeal for developing countries. Not only do they provide electricity, flood protection, and, often, irrigation water but also they create jobs and national pride. In developing countries, however, they also produce some problems that have not been as publicized elsewhere. Often, these problems have been associated with the displacement of large numbers of people as reservoirs fill. In the past, the most notable examples of this pattern have been in Africa and Brazil, but currently world attention has focused on the Yangtze River in China, the Naranda River in India, and the Euphrates River in Turkey. Even in places where few people live, hydroelectric projects are often controversial, either for reasons of environmental disruption, such as on the border of Uruguay and Brazil, or for reasons of native land claims (as with the huge projects on James Bay in Canada). Although all these hydro projects have inherently geographic applicability, only a few geographers are involved in studying them.

The oldest and most traditional energy resource is biomass, usually in the form of wood from shrubs and trees, crops, and animal dung. Biomass can be used for heating, cooking, and transportation. Wood fuel use and sustainable development have been extensively researched in case studies of India, Pakistan, Tanzania, Kenya, Zimbabwe, Ghana, and Haiti. Brazil, in contrast, has supplemented wood and charcoal use with sugarcane and bagasse for ethanol fuel and cogeneration. The heat content of biomass is low by weight compared to coal and other fossil fuels, so for areas of quickly growing population many geographers have reported how such dependency quickly transforms landscapes, which often do not recover their original appearance or productivity. The relentless and growing dependency on such resources has secondary impacts on soil erosion, microclimates, and other aspects of local and regional ecological balance, which geographers have long noted and studied, especially as it pertains to the creation and implementation of government policies.

Developing countries are among the most logical candidates for the development of renewable energy resources, especially direct solar and wind power. From a geographical perspective, direct solar power, either for hot water or for electricity, holds the greatest advantage because it is not site specific; photovoltaic panels can be installed and used with little need for local technical competence or development of a delivery infrastructure. Geographers have written about the relative advantages of such sources of power for isolated people in India and several other developing countries in Latin American and Africa, and the experiences in these locations can be transferred to just about any place in the tropics. Regarding wind power, some developing countries, such as Mongolia, have great potential, but wind (like geothermal) is spatially concentrated and immobile. The number of studies should increase in the coming decades as the quest for sustainable development continues.

Further Reading

Cleveland C J, Kaufmann R K 1991 Forecasting ultimate oil recovery and its rate of production: Incorporating economic forces into the model of M. King Hubbert. *Energy J.* **12**(2), 17–46

Cook B J, Emel J L, Kasperson R E 1990 Organizing and managing radioactive-waste disposal as an experiment. *J. Policy Analysis Management* **9**(3), 339–66

Cook E F 1976 *Man, Energy, Society.* Freeman, San Francisco

Cuff D J, Young W J 1986 *The United States Energy Atlas*, 2nd ed., Macmillan, New York

Elmes G A, Harris T M 1996 Industrial restructuring and the changing geography of the United States coal-energy system, 1972–1990. *Ann. Assoc. Am. Geogr.* **86**(3), 507–29

Guyol N B 1971 *Energy in the Perspective of Geography.* Prentice-Hall, Englewood Cliffs, NJ

Hosier R H (Ed.). 1993. Urban energy and environment in Africa [Special issue]. *Energy Policy* **21**(5), 434–558

Jackson E L, Kuhn R G, Macey S M, Brown M A 1988 Behavioral energy research in geography. *Can. Geogr.* **32**(2), 162–72

Majumdar S K, Miller E W, Panah A I (eds.) 2002 *Renewable Energy: Trends and Prospects.* Pennsylvania Academy of Science, Easton, PA

Manners G 1971 *The Geography of Energy*, 2nd ed., Hutchinson, London

Odell P R 1986 Oil and World Power. 8th ed., Penguin, Harmondsworth, UK

Pasqualetti M J (ed.) 1990 *Nuclear Decommissioning and Society: Public Links to a New Technology.* Routledge, London

Pasqualetti M J 2000 Morality, space, and the power of wind-energy landscapes. *Geogr. Rev.* **90**, 381–94

Sagers M J, Green M B 1986 *The Transportation of Soviet Energy Resources.* Rowman & Littlefield, Totowa, NJ

Smil V 1988 *Energy in China's Modernization.* Sharpe, Armonk, NY

Smil V 1998 *Energies: An Illustrated Guide to the Biosphere and Civilization.* MIT Press, Cambridge, MA

Solomon B D, Ahuja D R 1991 International reductions of greenhouse-gas emissions: An equitable and efficient approach. *Global Environ. Change,* **1**, 343–50

Wilbanks T J 1994 Sustainable development in geographic perspective. *Ann. Assoc. Am. Geogr.* **84**(4), 541–56

Zeigler D J, Johnson J H Jr. 1984 Evacuation behavior in response to nuclear power plant accidents. *Professional Geogr.* **36**(2), 207–15

Barry D. Solomon and Martin J. pasqualetti

H

Hydrogen, History of

Glossary

electrolysis The process of splitting water into its components, hydrogen and oxygen, by means of an electrical current.

energy carrier A form of matter that carries energy from one point to another, as distinct from an energy source, for which hydrogen is often mistaken.

fuel cell An electrochemical device that combines hydrogen and oxygen, producing an electrical current and water as a by-product.

hydrogen economy A term first introduced in the 1970s to describe the concept of an energy system based primarily on the use of hydrogen as an energy carrier.

proton-exchange membrane One type of fuel cell, featuring a membrane through which protons but not electrons can pass, forcing the latter to move along an electrode and generate a current.

The history of hydrogen as an energy carrier begins with its discovery and the comprehension that it is one of the basic building blocks of the universe. When the early Greek philosophers named water, air, fire, and earth to their list of elements, they did not realize that water in fact consisted of two elements. One of these elements is oxygen, the respiratory prerequisite for human life. The other is hydrogen, which modern science has revealed as the lightest and most abundant element in the universe.

1. Discovery of Hydrogen

By the most recent estimates, hydrogen accounts for approximately 75% of the mass of the entire universe and accounts for more than 90% of all molecules. According to the Harvard astrophysicist Steven Weinberg, between 70 and 80% of the observable universe is composed of hydrogen, with the rest attributable to helium, the universe's second lightest element. This is why hydrogen is called *Wasserstoff*—the stuff of water—in German.

The first recorded production of hydrogen took place in the 15th century, when the Middle Age physician Theophrastus Paracelsus dissolved iron in the acid vitriol, producing hydrogen gas. Paracelsus did not note that hydrogen was flammable; this was left to Turquet de Mayeme and Nicolas Lemery, French scientists who, in the 17th century, mixed sulfuric acid with iron, yet did not think that the resulting gas could be an element unto itself, only a burnable form of sulfur.

The actual classification and formal description of hydrogen would have to wait until the 18th century. As the early Greeks would have no doubt appreciated, the modern discovery of hydrogen was tied to advances in identifying its fellow component in water, oxygen. Such advances would, of course, require first that scientists reconsider the conventional notion of air as a basic element. The first to question this longstanding characterization was Herman Boerhaave, a Dutch doctor and naturalist who believed that air contained an ingredient that made breathing and combustion possible. Wrote Boerhaave in 1732, "The chemist will find out what it actually is, how it functions, and what it does; it is still in the dark. Happy he who will discover it." Boerhaave had a modicum of intellectual support in the 17th-century writings of British scientists Robert Boyle and John Mayo, who suspected that some substances in the air were responsible for the process of combustion. But this ran counter to the phlogiston theory, another 17th-century idea. The phlogiston theory, one of the first attempts to explain the process of combustion, was first published in 1697. Its origins lay with a German scientist named Georg Ernst Stahl, who contended that a substance known as phlogiston gave all matter the ability to burn. Because phlogiston disappeared from material during the combustion process, and because it was considered impossible to reduce the substance to a pure state, this was not a theory easily disproved. Modern chemistry has since shown that Stahl had it precisely backwards: the process of burning arises from adding a substance—oxygen. The idea that combustion consisted of the release of phlogiston served as an impediment to the recognition of oxygen and hydrogen as gases.

Fortunately for scientific progress (but rather unfortunately for Stahl), the late 18th century would witness a scientific race of sorts to successfully isolate oxygen. Among those who discovered, but did not name, the element were Joseph Priestley, a British minister; Carl Wilhelm Scheele, a doctor of Swedish–German descent; and Antoine Laurent Lavoisier, by then France's leading chemist. Although Scheele was the first to produce pure oxygen, between 1771 and 1772, the other two beat him to the press, publishing their findings in 1774. Because the scientists studying air were also investigating water, the isolation and identification of hydrogen proceeded in parallel fashion. But most chemists, such as Boyle, still saw the hydrogen they produced as another type of air. That would change with the English nobleman Henry Cavendish, the first to discover and describe many of the important characteristics of hydrogen. Ironically, Cavendish did not name hydrogen because he was not fully free from the phlogiston theory—he thought he had discovered the substance in its pure state. But his key breakthrough was to identify two kinds of "factitious air": "fixed air," or carbon dioxide, and "inflammable air," or hydrogen. In a 1766 paper to the Royal Society of London, Cavendish provided exact measurements of the weights and densities of both gases, revealing the inherent lightness of hydrogen. (Recognition of this led the French physicist Jacques Alexandre Cesar Charles to fly a hydrogen-filled balloon to a height of close to 2 miles in 1783.) In his *Experiments on Air* treatise of the 1780s, Cavendish showed it was possible to mix hydrogen with air, ignite the mixture with a spark, and produce water.

Cavendish's findings attracted Lavoisier, who tried his experiments in reverse. Splitting water molecules into hydrogen and oxygen in an experiment considered to be a breakthrough, Lavoisier combined hydrogen and oxygen to produce 45 g of water (which are still preserved in the French Academy of Science). In February 1785, Lavoisier conducted, before a large group of scientists, definitive experiments proving that hydrogen and oxygen are the fundamental constituents of water. His greatest work, *The Method of Chemical Nomenclature*, labels the "life-sustaining air" oxygen and the "inflammable air" hydrogen.

The first large-scale production of hydrogen was helped along by France's historical circumstances: the storming of the Bastille in 1789 and the subsequent warfare of the French Revolution. Guyton de Norveau, chemist and member of the Comité de Salut Public (Committee for Public Salvation), proposed that the army use hydrogen-filled captive balloons as observation platforms, a proposal that was approved after he and Lavoisier repeated the latter's 1783 experiment. The scientist Jean Pierre Coutelle built the first hydrogen generator; this was a furnace with a cast iron tube, containing iron fillings, with steam piped in at one end and hydrogen emerging at the other. A standard generator was later developed and included temperature-control systems, precursors to the coal gas generators that would later be used for lighting and heating in the early

19th century; coal gas would often be misnamed "hydrogen gas."

2. Origins of Interest in Hydrogen Energy

In 1800, six years after Lavoisier went to the guillotine in the melee of the French Revolution, another important scientific discovery was made. This was electrolysis, the splitting of water into hydrogen and oxygen by the passage of an electric current through it. William Nicholson and Sir Anthony Carlisle, British scientists, had discovered electrolysis, only a few weeks after Alessandro Volta built the first electric cell. As the 19th century progressed, other proposals to put hydrogen to practical use emerged, and from intriguing sources—including the clergy. In 1820, Reverend W. Cecil, a Fellow of Magdalen College and of the Cambridge Philosophical Society, read "On the Application of Hydrogen Gas to Produce Moving Power in Machinery." His treatise described the space and time limitations of water-driven and steam engines, and the advantages of a hydrogen engine in providing motive force in any location and with little preparation needed. The transactions of the society provide considerable detail of the engine, but no evidence as to whether it was ever built by Cecil.

Hydrogen, and its unusual properties, became a topic of growing interest among not only scientists but also science fiction writers. One of the most well-known fictional discussions of hydrogen is found in Jules Verne's *The Mysterious Island*, one of the writer's final works, published in 1874, almost precisely one century before modern hydrogen research took off. Set in the American Civil War, the book at one point depicts five characters speculating on the future of the Union, and the impact on commerce and industry were coal supplies to dwindle. Whereas four of the characters express fear about the loss of coal-powered machinery, railways, ships, manufacturing, and other forms of modern civilization, the other character, the engineer Cyrus Harding, asserts that they will burn "water decomposed into its primitive elements":

Yes, my friends, I believe that water will one day be employed as fuel, that hydrogen and oxygen which constitute it, either singly or together, will furnish an inexhaustible source of heat and light, of an intensity of which coal is not capable. Some day the coalrooms of steamers and the tenders of locomotives will, instead of coal, be stored with these two condensed gases, which will burn in the furnaces with enormous calorific power. There is, therefore, nothing to fear. As long as the earth is inhabited it will supply the wants of its inhabitants, and there will be no want of either light or heat as long as the productions of the vegetable, mineral, or animal kingdoms do not fail us. I believe, then, that when the deposits of coal are exhausted we shall heat and warm ourselves with water. Water will be the coal of the future.

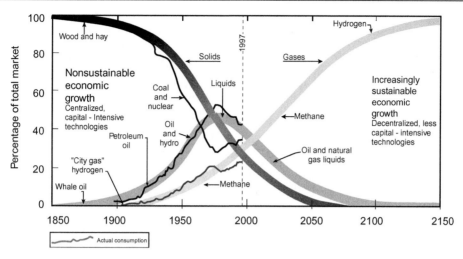

Figure 1
In the "age of energy gases," the transition in global energy systems. Reprinted from Dunn (2001), with permission.

Verne may have skipped two energy transitions along the path from coal to hydrogen—oil and natural gas—and therefore he missed what energy expert Robert Hefner would in 2000 depict as the "age of energy gases" (Fig. 1). Nor did Verne mention what the primary source of the hydrogen might be. But it is a prescient passage, given the state of late-19th-century science.

Another adventure story, *The Iron Pirate*, authored by Max Pemberton and published in England in 1900, featured speedy battleships using hydrogen engines. The book influenced the British scientist W. Hastings Campbell, as he related to colleagues at Britain's Institute of Fuel in 1933. His reference was noted in the *Journal of the Institute of Fuel* as "another instance of the very annoying persistence with which art always seemed to anticipate discoveries."

Interest in hydrogen as a fuel had begun to blossom in the wake of the First World War, prompted in part by concern about access to energy supplies. At first, interest was strong in Germany and England, as well as in Canada. The early 20th century had seen the emergence of several new companies with a stake in hydrogen production, such as the Canadian Electrolyser Corporation Limited (now Stuart Energy Systems, one of the world's leading electrolytic hydrogen plant manufacturers). The firm arose from the interest of its founder, Alexander Stuart, in utilizing the nation's considerable excess hydroelectric capacity. The first commercial uses of its electrolyzers were not, however, to make hydrogen fuel, but rather to make hydrogen and oxygen for steel-cutting procedures. Other uses included fertilizer production, and in 1934 Ontario Hydro constructed a 400-kW electrolysis plant. But World War II and growing use of natural gas in Canada halted the project.

Back in Britain, a young Scottish scientist, J. B. S. Haldane, prophesied, in a lecture at Cambridge University in 1923, hydrogen as the fuel of the future—derived from wind power, liquefied, and stored. Future British energy needs would be met by metallic windmills that decomposed water and stored the gases in vast underground reservoirs for later use. Britain's other early researchers and proponents of hydrogen as a fuel included Harry Ricardo, one of the developers of the internal combustion engine, and R. O. King, a member of the British Air Ministry Laboratory.

Germany counted numerous hydrogen supporters among its engineers and scientists, some of them influenced by Verne. Franz Lawaczeck, a turbine designer, sketched concepts for car, trains, and engines and is considered the first to propose the transport of hydrogen energy by pipeline. More whimsical was Hermann Honnef, who envisioned huge steel towers with windmills producing hydrogen. Use of liquid hydrogen in aircraft attracted attention in Italy, the United States, and Germany in the late 1930s. Lecturing before the American Institution of Electrical Engineers, Igor Sikorski cited the fuel's potential for improving the speed and performance of long-distance aircraft. But the most intriguing application was for the German zeppelins, in wide use in the 1920s and 1930s. As zeppelins lost weight through fuel consumption, they typically blew off hydrogen; engineers proposed burning the hydrogen as extra fuel, increasing output and saving energy. Although tests in the Mediterranean were promising, there is no evidence that the principles were applied to typical flights.

A prominent figure in early hydrogen advocacy was the German Rudolf Erren. Erren spent much of the 1930s in Britain, where he developed advanced combustion processes that would allow the use of hydrogen as a fuel or "booster" additive. In the mid-1930s, he and the engineer Kurt Weil proposed to the Nazi government that it convert from the internal combustion engines to these multifuel systems. At the time, Germany was interested in economic self-sufficiency and reducing dependence on imported liquid fuel. Eventually, between 1000 and 4000 cars and trucks were converted to the new system. Erren's efforts extended elsewhere. In Dresden, a hydrogen rail car ran for several years. In England, delivery vans and buses were tested; a prototype plane was readied and hydrogen-propelled submarines and torpedoes drew public interest in the early 1940s. But Erren's possessions were confiscated during World War II, after which he was repatriated to Germany. None of his engines survived the war years.

The war spurred direct interest in hydrogen in areas of the world where fuel supplies were in danger of being cut off. As Australia's wartime fuel needs grew and supplies from Borneo were lost to Japan, the government of the state of Queensland authorized construction of a hydrogen plant in Brisbane, using off-peak electricity. But the Allied victory in 1945 and the reversion to cheap oil and gasoline halted progress—as may be said about postwar hydrogen progress more generally.

3. Roots of the Hydrogen Economy Concept

The early 1950s saw a resumption of interest in hydrogen. The British scientist Francis Bacon developed the first practical fuel cell. Though invented in 1836, the fuel cell had yet to see significant applications outside the laboratory, a situation that would soon change with the U.S. space program.

In the 1960s, the idea of hydrogen as a medium for energy storage spread. German physicist Eduard Justi proposed in 1962 the recombination of hydrogen and oxygen in fuel cells, and in 1965 the production of solar hydrogen along the Mediterranean, for later piping to Germany and other nations. John Bockris, an American electrochemist from Australia, proposed in 1962 the energizing of U.S. cities with solar energy from hydrogen, an idea he expanded on in his 1975 book, *Energy: The Solar-Hydrogen Alternative*. Bockris traces the coining of the term hydrogen economy to a 1970 discussion at the General Motors (GM) Technical Center in Warren, Michigan. As a consultant to GM, Bockris was discussing possible alternatives to gasoline in the wake of growing environmental awareness, and the group agreed that "hydrogen would be the fuel for all types of transports." GM began experimental work on hydrogen, but would later be eclipsed by overseas carmakers.

Various visions of the hydrogen economy began to emerge. The Italian scientist Cesare Marchetti, an influential advocate in Europe, promoted the large-scale production of hydrogen from the water and heat of nuclear reactors. Derek Gregory and Henry Linden of the Institute of Gas Technology, leaders in early U.S. hydrogen research and development, were motivated by the prospect of hydrogen as a substitute for natural gas. Space scientists and engineers continued to pursue liquid hydrogen potential. Despite the limited efforts of GM, Ford, and Chrysler, interest among automotive engineers grew. The Army, Navy, and Air Force pursued liquid hydrogen and oxygen applications, their experiments spilling over into the space program. During the 1970s, aided by the two oil shocks, both environmental concerns and energy security accelerated interest in hydrogen. Popular journals such as *Business Week, Fortune*, and *Time* ran stories on hydrogen development. Researchers formed groups such as the International Association for Hydrogen Energy (which currently publishes the *International Journal of Hydrogen Energy*).

In the 1980s and 1990s, governments and international organizations began to pay additional attention to the idea of a hydrogen system and its components of production, delivery, storage, and use (Fig. 2). The United States, West Germany, European Community, and Japan began to lay out funding for hydrogen research and development in the mid-1970s, but would cut back as the effects of the oil shock waned. Nevertheless, the institutionalization of hydrogen into energy policy and strategy was underway. The 1980s were a paradoxical period for the hydrogen prospect. Official government support by and large waned, particularly within the U.S. Administration, which significantly cut funding for alternative fuel budgets to make room for its expanding Cold War weapons buildup. One unintended consequence of this benign neglect was the decision of a creative but frustrated engineer named Geoffrey Ballard, then under the employ of the Department of Energy, to leave Washington, D.C. for the west coast of Canada. In Vancouver, British Columbia, Ballard and several like-minded scientists decided to focus on developing electric batteries—a focus that soon switched to the proton-exchange membrane (PEM) fuel cell (Fig. 3).

Although the fuel cell had, by the 1980s, enjoyed several decades of use in the U.S. space program, efforts to bring the technology down to earth had been hampered by issues of cost and efficiency. Most fuel cells relied on significant amounts of the expensive metal platinum and were large and bulky, minor considerations for space programs but major problems for terrestrial commercial applications. Through a combination of ingenuity, government seed money (from the Canadian defense agency), venture capital, and persistence, Ballard's company achieved a 20-fold reduction in platinum requirements for its PEM fuel cell, with commensurate cost

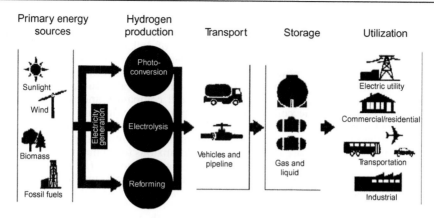

Figure 2
The components of a hydrogen energy system. Reprinted from Dunn (2001), with permission.

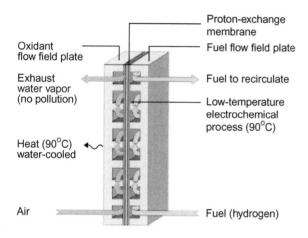

Figure 3
A proton-exchange membrane fuel cell. Reprinted from Dunn (2001), with permission.

reductions. In 1993, Ballard rolled out the first PEM fuel cell bus at its headquarters near Vancouver. That same year, it entered into a cooperative agreement with Daimler Benz (later to become Daimler-Chrysler) to jointly develop fuel cells for cars and buses, a partnership that was joined by Ford in 1997, with an overall estimated investment of $300 million. Daimler announced a $1 billion, 10-year commitment to fuel cell development; Daimler's founders were the first to bring the internal combustion engine car to market, and the company appears intent on doing the same for the fuel cell vehicle.

The late 1990s were characterized by an auto race of sorts, as companies paraded demonstration fuel cell vehicles at auto shows and the media buzzed over whether and when carmakers would unveil their first commercial vehicles. Executives also seemed to be staging another competition based on who could sound most bullish, even a bit visionary, about the hydrogen prospect. Echoing the GM engineers of three decades ago, Executive Director Robert Purcell announced in 2000, at the annual meeting of the National Petrochemical and Refiners Association, that "our long-term vision is of a hydrogen economy." As with the introduction of fuel-efficient vehicles in the 1980s, and hybrid electric vehicles in 2000–2002, the Japanese automakers quietly poured money into efforts that threatened to put them ahead of their U.S. and European competitors. In late 2002, Honda and Toyota announced they would be the first to introduce fuel cell cars to the U.S. market.

Energy suppliers were also giving more attention to hydrogen as the century came to a close. In 1999, Royal Dutch/Shell established a core business, Shell Hydrogen, to pursue business opportunities related to the hydrogen economy—an announcement that caused British Petroleum, ChevronTexaco, and ExxonMobil to follow with similar but smaller investments—and advocacy. Testifying before the U.S. Congress, Frank Ingriselli of Texaco (soon to become ChevronTexaco) asserted that "greenery, innovation, and market forces are propelling us inexorably towards hydrogen. Those who don't pursue it, will rue it." Shell Hydrogen would become involved in one of the most interesting hydrogen developments of the late 20th century: the announcement by Iceland that it planned to become the world's first hydrogen economy. The idea of using hydrogen in Iceland was first proposed in the 1970s by a chemistry graduate student named Bragi Arnason as a way to utilize the island's abundant geothermal and hydro resources to reduce oil import reliance. As oil price volatility resurfaced in the 1990s, it coincided with the nation's struggle to address greenhouse gas emissions and with growing global interest in fuel cells and hydrogen. Acting on the recommendations of a panel chaired by Arnason (who had by then earned the nickname "Professor Hydrogen"), the Icelandic government formed in 1999 a consortium, Icelandic New Energy, with Shell, DaimlerChrysler, and the Norwegian power company Norsk Hydro, to investigate the potential for transforming the nation's energy economy to hydrogen between 2030 and 2040. The consortium planned to convert gradually, from petroleum to hydrogen, the buses of the capital, Reykjavik, followed by the island's entire fleets of buses, cars, and fishing boats.

Other islands, likewise vulnerable to oil price volatility, soon followed Iceland's lead. Hawaii approved in early 2001 a jump-start grant to support a public/private partnership in hydrogen research and development, intended eventually to make Hawaii a major player in hydrogen exports. The South Pacific island of Vanuatu launched a feasibility study for transitioning to hydrogen, and plans to be completely renewable-energy-based by 2020. Cuba is also contemplating a move to hydrogen.

These developments shed useful light on the converging forces that are renewing interest in hydrogen. Technological advances and the advent of greater competition in the energy industry are part of the equation. But equally important motivations for exploring hydrogen are the energy-related problems of energy insecurity, air pollution, and climate change—problems that are collectively calling into question the fundamental sustainability of the current energy system. In the 21st century, it is likely that there will be five drivers for hydrogen—technology, competition, energy security, air pollution, and climate change. They are far more powerful drivers than those that existed during the 20th century, and they help to explain why geographically remote islands, stationed on the front lines of vulnerability to high oil prices and climate change, are in the vanguard of the hydrogen transition.

4. Globalization of the Hydrogen Movement

Iceland and other nations represent just the tip of the iceberg in terms of the changes that lie ahead in the energy world. The commercial implications of a transition to hydrogen as the world's major energy currency will be staggering, putting a $2 trillion energy industry through its greatest tumult since the early days of Edison, Ford, and Rockefeller. During 2002, over 100 companies were aiming to commercialize fuel cells for a broad range of applications, from cell phones, laptop computers, and soda machines to homes, offices, and factories, to vehicles of all kinds. Hydrogen was also being researched for direct use in cars and planes. Particularly active were fuel and auto companies, who were spending between $500 million and $1 billion annually on hydrogen.

Leading energy suppliers had created hydrogen divisions, and major carmakers continued to pour billions of dollars into a race to put the first fuel cell vehicles on the market between 2003 and 2005. Buses were being test-driven in North America, Europe, Asia, and Australia. In California, 23 auto, fuel, and fuel cell companies and seven government agencies had partnered to fuel and test drive 70 cars and buses. Hydrogen and fuel cell companies had captured the attention of venture capital firms and investment banks anxious to get into the hot new space known as or energy technology (ET).

The geopolitical implications of hydrogen are enormous as well. Coal fueled the 18th- and 19th-century rise of Great Britain and modern Germany; in the 20th century, oil laid the foundation for the unprecedented economic and military power of the United States, of which the fin-de-siècle superpower status, in turn, may be eventually eclipsed by countries that harness hydrogen as aggressively as the United States tapped oil a century ago. Countries that focus their efforts on producing oil until the resource is gone will be left behind in the rush for tomorrow's prize. As Don Huberts, Chief Executive Officer of Shell Hydrogen, has noted, "The Stone Age did not end because we ran out of stones, and the oil age will not end because we run out of oil." Access to geographically concentrated petroleum has also shaped world wars, the 1991 Gulf War, and relations between and among Western economies, the Middle East, and the developing world. Shifting to the plentiful, more dispersed hydrogen could alter the power balances among energy-producing and energy-consuming nations, possibly turning today's importers into tomorrow's exporters.

The most important consequence of a hydrogen economy may be the replacement of 20th-century "hydrocarbon society" with something far better. The 20th century saw a 16-fold increase in energy use—10 times more than the energy used by humans in the 1000 years preceding 1900. This increase was enabled primarily by fossil fuels, which account for 90% of energy worldwide. Global energy consumption is projected to rise by close to 60% over the next 20 years. Use of coal and oil is projected to increase by approximately 30 and 40%, respectively.

Most of the future growth in energy is expected to take place in transportation, where motorization continues to increase and where petroleum is the dominant fuel, accounting for 95% of the total. Failure to develop alternatives to oil would heighten growing reliance on oil imports, raising the risk of political and military conflict and economic disruption. In industrial nations, the share of imports in overall oil demand will rise from roughly 56% today to 72% by 2010. Coal, meanwhile, is projected to maintain its grip on more than half the world's power supply. Continued increases in coal and oil use will exacerbate urban air problems in megacities such as

Delhi, Beijing, and Mexico City, which experience thousands of pollution-related deaths each year. And prolonging petroleum and coal reliance in transportation and electricity would increase global carbon emissions from 6.1 to 9.8 billion tons of carbon by 2020, accelerating climate change and the associated impacts of sea level rise, coastal flooding, and loss of small islands; extreme weather events; reduced agricultural productivity and water availability; and the loss of biodiversity.

Hydrogen cannot, on its own, entirely solve each of these complex problems, which are affected not only by fuel supply but also by factors such as population, over- and underconsumption, sprawl, congestion, and vehicle dependence. But hydrogen could make a major dent in addressing these issues. By enabling the spread of appliances, more decentralized "micropower" plants, and vehicles based on efficient fuel cells (of which the only by-product is water), hydrogen use would dramatically cut emissions of particulates, carbon monoxide, sulfur and nitrogen oxides, and other local air pollutants. By providing a secure and abundant domestic supply of fuel, hydrogen would significantly reduce oil import requirements, providing the independence and energy security that many nations crave.

Hydrogen has also been increasingly recognized by the scientific community as facilitating the transition from limited nonrenewable stocks of fossil fuels to unlimited flows of renewable sources, playing an essential role in the "decarbonization" of the global energy system needed to avoid the most severe effects of climate change (Fig. 4). The United Nations 2001 *World Energy Assessment* emphasized "the strategic importance of hydrogen as an energy carrier"; the accelerated replacement of oil and other fossil fuels with hydrogen could help achieve "deep reductions" in carbon emissions and avoid a doubling of pre-industrial CO_2 concentrations in the atmosphere, a level at which scientists expect major, and potentially irreversible, ecological and economic disruptions. Hydrogen fuel cells could also help address global energy inequities, providing fuel and power and spurring employment and exports in the rural regions of the developing world, where nearly 2 billion people lack access to modern energy services.

Despite these potential benefits, and despite early movements toward a hydrogen economy, its full realization has faced an array of technical and economic obstacles. The feasibility of production, delivery, and use all awaits further improvement. Hydrogen is only beginning to be piped into the mainstream of the energy policies and strategies of governments and businesses, which still tend to aim at preserving the hydrocarbon-based status quo and expanding fossil fuel production. Market structures have been typically tilted toward expanding fossil fuel production. Subsidies to these energy sources, in the form of direct supports and the "external" costs of

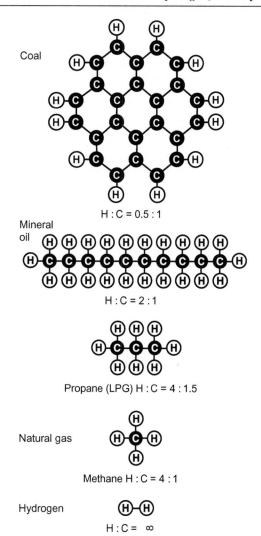

Figure 4
The atomic hydrogen:carbon ratios of selected energy sources. LPG, Liquid propane gas. Reprinted from Dunn (2001), with permission.

pollution, were estimated at $300 billion annually in 2001. The perverse signals in the energy market that lead to artificially low fossil fuel prices and encourage their production and use make it difficult for hydrogen and fuel cells (for which the production, delivery, and storage costs are improving but look high under such circumstances) to compete with the entrenched gasoline-run internal combustion engines and coal-fired power plants. This could push the broad availability of fuel cell vehicles and power plants a decade or more into the future. Unless the antiquated rules of the energy economy, aimed at keeping hydrocarbon production cheap by shifting the cost to consumers and the environment, are reformed, hydrogen will be slow to make major inroads.

One of the most significant obstacles to realizing the full promise of hydrogen is the prevailing perception that a full-fledged hydrogen infrastructure——the system for producing, storing, and delivering the gas—would immediately cost hundreds of billions of dollars to build, far more than a system based on liquids such as gasoline or methanol. This is tied to the "chicken-and-egg" dilemma that confronts any new infrastructure. Automakers have been hesitant to mass-produce fuel cell vehicles without assurance of a sufficient fueling network. Energy suppliers, meanwhile, have been resistant to ramping up hydrogen production without assurance of a sufficient number of vehicles to use the fuel. Because of this dilemma, auto and energy companies have been investing millions of dollars into the development of reformer and vehicle technologies that would derive and use hydrogen from gasoline and methanol, keeping the current petroleum-based infrastructure intact. To some analysts, this incremental path—continuing to rely on the dirtiest, least secure fossil fuels as a bridge to the new energy system—represents a costly wrong turn, both financially and environmentally. If manufacturers "lock in" to mass-producing inferior fuel cell vehicles just as a hydrogen infrastructure approaches viability, trillions of dollars of assets could be wasted. Furthermore, by perpetuating petroleum consumption and import dependence and the excess emission of air pollutants and greenhouse gases, this route would deprive society of numerous benefits. By the late 1990s, fossil fuels were the source of some 95% of the hydrogen being produced worldwide—approximately 400 billion cubic meters in 1999, primarily not for energy but for the refining of petroleum and the manufacture of resins, plastics, solvents, and other industrial commodities. Over the long run, critics argued, this proportion would need to be shifted toward renewable sources, not maintained, for hydrogen production to be sustainable.

The "fuel choice" question heated up in the 1990s, complicated by competing life-cycle and "well-to-wheels" studies by government, industry, and nongovernmental organizations. A growing number of scientists in government and industry openly challenged the conventional wisdom of the incremental path. Their research suggested that the direct use of hydrogen would be the quickest and least costly route—to the consumer and the environment—toward a hydrogen infrastructure. Their studies pointed to an alternative pathway that would initially use the existing infrastructure for natural gas (the cleanest fossil fuel and the fastest growing in terms of use) and employ fuel cells in niche applications to buy down their costs to competitive levels, spurring added hydrogen infrastructure investment. As the costs of

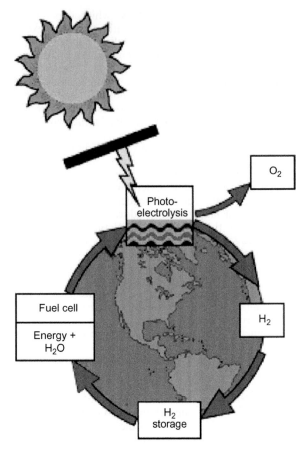

Figure 5
A renewable-hydrogen cycle. Reprinted from Dunn (2001), with permission.

producing hydrogen from renewable energy fell, meanwhile, hydrogen would evolve into the major source of storage for the limitless but intermittent flows of the sun, wind, tides, and Earth's heat. The end result would be a clean, natural hydrogen cycle, with renewable energy used to split water into hydrogen, which would be used in fuel cells to produce electricity and water—which then would be available to repeat the process (Fig. 5). Some of the experts argued that there were no major technical obstacles to the alternative path to hydrogen. As one researcher put it, "If we really wanted to, we could have a hydrogen economy by 2010." But the political and institutional barriers appeared formidable. Both government and industry had devoted more resources toward the gasoline- and methanol-based route than to the direct hydrogen path, though they were open to the latter: remarked one industry executive, "everyone is placing bets on several horses."

Meanwhile, hydrogen still received a tiny fraction of the research funding that was allocated to coal, oil, nuclear, and other mature, commercial energy sources. Within energy companies, the hydrocarbon side of the business continued to argue that oil will be dominant for decades to come, even as other divisions prepare for its successor. Hydrogen also posed a management challenge, that of looking beyond continuous, incremental improvement of existing products and processes, and toward more radical, "disruptive" technological change. And little had been done to educate people about the properties and safety of hydrogen, even though public acceptance, or lack thereof, would in the end make or break the hydrogen future. The societal and environmental advantages of the cleaner, more secure path to hydrogen illustrated the essential—and under-recognized—role for government in the hydrogen transition. Indeed, without aggressive energy and environmental policies, the hydrogen economy is likely to emerge along the more incremental path, and at a pace that is inadequate for dealing with the range of challenges posed by the incumbent energy system. Neither market forces nor government fiat could, in isolation, move society down the more direct, more difficult route. The challenge appears to be for government to guide the transition, setting the rules of the game and working with industry and society toward the preferable hydrogen future.

This catalytic leadership role, some advocates have argued, would be analogous to that played by government in launching another infrastructure in the early years of the Cold War. Recognizing the strategic importance of having its networks of information more decentralized and less vulnerable to attack, the U.S. government undertook critical research, incentives, and public/private collaboration into the development of what we now call the Internet. A similar case could be made for strategically laying the groundwork for a hydrogen energy infrastructure that best limits vulnerability to air pollution, energy insecurity, and climate change. Investments made at present would heavily influence how, and how fast, the hydrogen economy emerges in coming decades. As with the Internet, putting a man on the moon, and other great human endeavors, the dawning of the hydrogen age was unlikely to result from business as usual.

5. Future of Hydrogen History

While its effects continue to ripple outward, the events of September 11, 2001 will become an important reference point in the history—written and unwritten—of hydrogen's development as an energy source. Interest in hydrogen surged in the aftermath of the 2001 terrorist attacks on the United States. Subsequent discussions of the appropriate military response revealed the extent to which dependence on Middle East oil had complicated U.S. foreign policy

choices in that region. By chance, Shell had planned to unveil its latest long-term energy scenarios in New York City the following month. Their exposition of uncertainties, discontinuities, the "unthinkable," and the "unknowable" struck a strong chord at a moment when an unthinkable event had just happened—one that, to more business and political decision makers, made the vision of a hydrogen economy even more desirable than before.

Presenting the scenarios, Shell chairman Philip Watts presented two scenarios, "Dynamics as Usual" and "Spirit of the Coming Age," and three factors that would drive changes in the energy system: resource scarcity, new technology, and shifting social and personal priorities. Under the first scenario, there is an evolutionary progression or "carbon shift" from coal to gas to renewable energy. The second scenario involves "something more revolutionary: the potential for a truly hydrogen economy, growing out of new and exciting developments in fuel cells, advanced hydrocarbon technologies and carbon dioxide sequestration." The prominence of hydrogen in this scenario—not a prediction, but a tool for helping managers plan for different possible futures and make better strategic decisions—is one indication of the growing attention being given to hydrogen by the world's leading multinationals.

Hinting at hydrogen's ascendancy as well is *Tomorrow's Energy* by Peter Hoffmann, a journalist who had followed hydrogen and fuel cell developments since the 1970s and who edited the well-respected industry periodical *Hydrogen and Fuel Cell Letter*. His work originally began as an update of his 1981 book *The Forever Fuel*. But it soon became a new book, due to the progress in hydrogen developments that had since been made. Among the late-20th-century advances reported by Hoffmann was the latest discovery related to one of hydrogen's biggest public relations problems: the Hindenburg. The prospect of hydrogen as a fuel often raises safety questions, in part because of its association with the German airship that exploded in 1937, taking 36 lives. For years, it was widely believed that the cause of the explosion was the ignition of the hydrogen gas used for lifting the ship. Bain Addison, a retired National Aeronautics and Space Administration (NASA) scientist who doubted this explanation, carefully studied film footage and documents relating to the incident. In 1997, he publicized his surprising finding that the zeppelin's cover had been painted with iron oxides and aluminum, compounds used in rocket fuel. Addison also uncovered evidence that the Nazi government was aware of the design flaw but suppressed this information.

Beyond the Hindenburg, the safety of hydrogen can be compared with that of other fuels in use today. Hydrogen does have a wide range of limits for flammability and detonability, which means that a broad range of mixtures of hydrogen in air can lead to a flame or explosion. However, the lower limits are most relevant to transportation uses, and in this regard, hydrogen is comparable to, or better than, gasoline and natural gas. At this lower flammability limit, the ignition energy, i.e., the energy in a spark to ignite a fuel mixed in air, is about the same for hydrogen and methane. Hydrogen is also nontoxic, unlike methanol or gasoline in higher concentrations. And hydrogen is very buoyant, escaping quickly from leaks, whereas gasoline puddles, causing its fumes to build up. The prevention, detection, and management of hydrogen leaks comprise an important safety issue, requiring that areas where hydrogen is stored and dispensed be well ventilated.

Although the chemical industry routinely handles large quantities of hydrogen safely, the question is whether this safety record will be transferred to hydrogen vehicle and refueling systems. Several studies in the 1990s explored this question. A 1994 report by researchers at Sandia National Laboratories stated that "hydrogen can be handled safely, if its unique properties—sometimes better, sometimes worse, and sometimes just different from other fuels—are respected." A 1997 study by Ford Motor Company concluded that, with proper engineering, the safety of a hydrogen fuel cell vehicle would potentially be better than that of a gasoline or propane vehicle. To ensure safe and standardized practices for using hydrogen, several national and international organizations have begun to develop codes and standards for hydrogen and fuel cells.

Another indicator of the increased attention given to hydrogen is the publication of *The Hydrogen Economy* by social critic Jeremy Rifkin. Rifkin, who has written extensively on the social and economic implications of scientific and technological trends, writes that hydrogen and the communications revolution will create a relationship that "could fundamentally reconfigure human relationships in the 21st and 22nd centuries. Since hydrogen is everywhere and is inexhaustible if properly harnessed, every human being on Earth could be 'empowered,' making hydrogen energy the first truly democratic energy regime in history." The resulting hydrogen energy web "will be the next great technological, commercial, and social revolution in history." Rifkin's utopian vision has engendered some justifiable skepticism from the corporate world. A review in *Business Week* opined that "the hydrogen hype... may be too much, too soon. While hydrogen's abundance and cleanliness make it the favorite as the fuel of the future, fossil fuels still have a cost advantage. And it's not obvious that even when hydrogen does come into widespread use, democracy will follow." The article argues that sharp cost declines in the cost of renewable energy, not a shortage of oil, will be "the factor that forces the transition to a hydrogen economy.... The hydrogen economy will dawn when systems based on renewables and

hydrogen beat hydrocarbons in straight competition for consumers' dollars."

Indeed, hydrogen still faces considerable cost hurdles at all stages of the energy system, from production to transportation to storage to use. In addition to major advances in renewable energy, improvements in storage technologies such as carbon nanotubes and metal hydrides, development of hydrogen-compatible pipelines, and further strides with fuel cells are among the prerequisites to making hydrogen competitive with hydrocarbons. The most economic method of producing hydrogen at present, via the reformation of natural gas, has been mostly limited to feedstock production; extracting hydrogen from electrolysis from solar or wind power costs three to five times more.

At the same time, and as recognized to varying degrees by business and government, renewables and hydrogen do not face a level playing field today, making concerted public action essential to the hydrogen transition. Japan had set the pace in the global race to hydrogen with its 1993 announcement of a 10-year, $2 billion commitment to promote hydrogen internationally through its World Energy Network (WE-NET) Project. But the stakes had been raised by 2002. The European Commission (EC) unveiled a $2 billion, 4-year investment in hydrogen technology research. In announcing the initiative, EC President Romano Prodi argued that the program would be as important to Europe as the space program was for the United States in the 1960s. "It's like going to the moon in a series of steps.... We expect an [even] better technological fallout."

The U.S. government, often criticized for supporting hydrogen less aggressively compared to its European and Japanese counterparts, also moved forward in 2002, announcing a "Freedom Car" partnership with the "big three" automakers to develop fuel cell vehicles. The National Academy of Sciences commissioned a panel study to examine the technical potential of transitioning to a hydrogen economy. Soon after, the Department of Energy unveiled an ambitious hydrogen road map, a blueprint crafted by public and private experts to coordinate the long-term development of a hydrogen economy. In an introduction to the report, Secretary Spencer Abraham sounded almost reminiscent of Jules Verne: "To talk about the 'hydrogen economy' is to talk about a world that is fundamentally different from the one we now know. A hydrogen economy will mean a world where our pollution problems are solved and where our need for abundant and affordable energy is secure... and where concerns about dwindling resources are a thing of the past."

It has taken more than a century, but the idea of hydrogen as an energy source has moved off the pages of science fiction and into the speeches of business and political leaders. In other words, hydrogen energy has evolved from a matter of scientific and technological curiosity to a major strategic issue for political and corporate decision makers. Whether it continues to sustain such interest will determine whether a hydrogen economy truly emerges or whether hydrogen remains, as enthusiasts have called it for several decades, "tomorrow's energy" or "the fuel of the future."

The history of hydrogen as an energy source is in many respects only beginning, with its future path and direction full of many possibilities. However it unfolds, the story of hydrogen in the 21st century promises to be as exciting as *The Prize*, Daniel Yergin's epic about oil in the 20th century. Whether, how, and how fast the hydrogen economy is realized hold enormous consequences for humanity's social, economic, political, and environmental future. That, in turn, will hinge on whether hydrogen becomes not only an interesting possibility, but also a compelling imperative. The hydrogen endeavor is a truly an historic one, bringing to mind President John F. Kennedy's impassioned appeal for the Apollo space program: "There are risks and costs to a program of action, but they are far less than the long-range risks and costs of comfortable inaction."

Further Reading

Dunn S 2000 The hydrogen experiment. *World Watch* **13**(6), 14–25

Dunn S 2001 *Hydrogen Futures: Toward a Sustainable Energy System.* Worldwatch Pap. 157, Worldwatch Institute, Washington, D.C.

Hoffmann P 2001 *Tomorrow's Energy: Hydrogen, Fuel Cells, and the Prospects for a Cleaner Planet.* MIT Press, Cambridge, Massachusetts

Jensen M W, Ross M 1999 The ultimate challenge: Building an infrastructure for fuel cell vehicles. *Environment* **42**(7), 10–22

Koppel T 1999 *Powering the Future: The Ballard Fuel Cell and the Race to Change the World.* John Wiley & Sons, New York

McNeill J R 2000 *Something New Under the Sun: An Environmental History of the Twentieth-Century World.* W W Norton & Co, New York

Ogden J M 1999 Prospects for building a hydrogen infrastructure. *Annu. Rev. Energy Environ.* **24**, 227–79

Rifkin J 2002 *The Hydrogen Economy.* Penguin Putnam, New York

Royal Dutch/Shell Group 2001 *Energy Needs, Choices and Possibilities: Scenarios to 2050.* Royal Dutch/Shell, Amsterdam

U.N. Development Programme, U.N. Department of Economic and Social Affairs, and World Energy Council 2000 *World Energy Assessment Report.* United Nations, New York

U.S. Department of Energy 2002 *National Hydrogen Energy Roadmap.* Department of Energy, Washington, D.C.

Veziroglu T N 2000 Quarter century of hydrogen movement, 1974–2000. *Int. J. Hydrogen Energy* **25**, 1143–50

Yergin D 1991 *The Prize: The Epic Quest for Oil, Money, and Power.* Simon & Schuster, New York

Seth Dunn

Hydropower, History and Technology of

Glossary

axial-flow turbine   A collective term for turbines with axial flow through the runner blades axially to the turbine shaft; both propeller turbines and Kaplan turbines are axial-flow turbines.

base load   Typically, the minimum load over a given period of time.

capacity   The greatest load that a piece of equipment can safely serve.

dam   A massive wall or structure built across a valley or river for storing water.

design head   The head at which the turbine is designed to operate at maximum efficiency.

draft tube   The diffuser that regains the residual velocity energy of the water leaving the turbine runner.

energy   The power of doing work for a given period, usually measured in kilowatt-hours.

forebay   The upstream part of the bay-like extension of the river for the location of a powerhouse.

Francis turbine   A radial-inflow reaction turbine where the flow through the runner is radial to the shaft.

generator   A machine powered by a turbine that produces electric current.

gross head   The difference between headwater level and tailwater level at the powerhouse.

headrace   The portion of the power canal that extends from the intake works to the powerhouse.

headwater   The water upstream from the powerhouse, or generally the water upstream from any hydraulic structure creating a head.

headwater elevation (or headwater level)   The height of the headwater in the reservoir.

hydraulic efficiency   An efficiency component of the turbine, expressing exclusively the power decrement due to hydraulic losses (e.g., friction, separation, impact), including the losses in the scroll case and the draft tube.

hydroelectric power   The electric current produced from water power.

hydroelectric power plant   A building in which turbines are operated, to drive generators, by the energy of natural or artificial waterfalls.

hydropower plant (or hydropower development)   The comprehensive term for all structures (one powerhouse and pertaining installations) necessary for using a selected power site.

hydropower station   A term equivalent to the powerhouse and sometimes including the structures situated nearby.

hydropower system   Two or more power plants (and therefore two or more powerhouses) that are cooperating electrically through a common network.

intake (or intake works or headworks)   A hydraulic structure built at the upstream end of the diversion canal (or tunnel) for controlling the discharge and preventing silt, debris, and ice from entering the diversion.

Kaplan turbine   An axial-flow reaction turbine with adjustable runner blades and adjustable guide vanes.

kinetic energy   Energy that a moving body has because of its motion, dependent on its mass and the rate at which it is moving.

load   The amount of electric energy delivered at a given point.

load demand   A sudden electrical load on the generating units, inducing the rapid opening of the turbines.

load factor   The ratio of the annually produced kilowatt-hours and of the energy theoretically producible at installed capacity during the whole year.

load rejection   A sudden cessation of electrical load on the generating units, inducing the rapid closure of the turbines.

main shaft   The rotating element that transmits torque developed by the turbine runner to the generator rotor or that transmits torque developed by the motor to the pump impeller.

mechanical efficiency   An efficiency component of the turbine, expressing exclusively the power losses of the revolving parts due to mechanical friction.

needle valve   A streamlined regulating body moving like a piston in the enlarged housing of the valve.

net head   The part of the gross head that is directly available for the turbines.

nozzle (or jet nozzle)   A curved steel pipe supplied with a discharge-regulating device to direct the jet onto the buckets in impulse runners.

peak load   The greatest amount of power given out or taken in by a machine or power distribution system during a given time period.

Pelton turbine   The main type of turbine used under high heads.

penstock (or pressure pipe)   A pressurized pipeline conveying the water in high-head developments from the headpond or the surge tank to the powerhouse.

plant discharge (or plant discharge capacity)   The maximum discharge that can be used by all turbines of the power plant with full gateage (i.e., the entire discharging capacity of the turbines).

pondage   That rate of storage in run-of-river developments that can cover daily peaks only.

power   The rate at which work is done by an electric current or mechanical force, generally measured in watts or horsepower.

powerhouse   The main structure of a water power plant, housing the generating units and the pertaining installations.

propeller-type turbine   The collective term for axial-flow reaction turbines; in this terminology, it denotes two types: fixed-blade propeller turbines and adjustable-blade propeller turbines (i.e., Kaplan turbines).

pumped-storage development   A combined pumping and generating plant; it is not a primary producer of electrical power, but by means of a dual conversion, it stores the superfluous power of the network and returns it during peak load periods, as would a battery.

reaction turbine   A collective term for turbines in which the water jet enters the runner under a pressure exceeding the atmospheric value. The water flowing to the runner still has potential energy, in the form of pressure, that is converted into mechanical power along the runner blades.

reservoir   An artificial lake into which water flows and is stored for future use.

run-of-river plant   A development with little or no pondage regulation such that the power output varies with the fluctuations in the stream flow.

runner   The rotating element of the turbine that converts hydraulic energy into mechanical energy; for reversible pump-turbines, the element is called an impeller and converts mechanical energy into hydraulic energy for the pump mode.

scroll case (or spiral case)   A spiral-shaped steel intake guiding the flow into the wicket gates of the reaction turbine.

semi-scroll case (or spiral case)   A concrete intake directing flow to the upstream portion of the turbine with a spiral case surrounding the downstream portion of the turbine to provide uniform water distribution.

setting   The vertical distance between the tailwater level and the center of a turbine runner.

specific speed   A universal number that indicates the machine design: impulse, Francis, or axial.

tailrace   The portion of the power canal that extends from the powerhouse to the recipient watercourse.

tailwater   The water downstream from the powerhouse, generally the water downstream from any hydraulic structure creating a head.

tidal power plant (or tidal power station)   A power station that uses the potential hydraulic power originating from the tidal cycles of the sea.

Turbine   A device that produces power by diverting water through blades of a rotating wheel that turns a shaft to drive generators.

turbine discharge capacity   The maximum flow that can be discharged by a single turbine at full gateage.

turbine efficiency   The entire efficiency of the turbine (i.e., the product of hydraulic mechanical and volumetric efficiencies).

water power   A general term used for characterizing both power (kilowatts) and energy (kilowatt-hours) of watercourses, lakes, reservoirs, and seas.

Hydropower has a long and rich history, being important to the development of modern industrial society and essential to the electrification of the world. This article reviews some of this history and the hydropower potential that remains in the world today. Next, the water cycle and how hydropower is extracted from this cycle will be discussed. Finally, some of the ecological considerations and the advantages and disadvantages of hydropower development will be addressed.

1. Introduction

Falling or flowing water has been used to perform work for thousands of years, with the particular uses varying with the social and political conditions of the times. Although the Greeks and Romans knew of waterwheels since the 3rd century BC, these labor-saving devices were not used extensively until the 14th century. Early tasks included grinding grain, sawing wood, powering textile mills, and (later) operating manufacturing plants. Mills or factories were located at the hydropower sites to directly use the available energy. By the end of the 18th century, there were approximately 10,000 waterwheels in New England alone. The power output of these early plants, usually limited to about 100 kW, is compared with other power sources in Fig. 1.

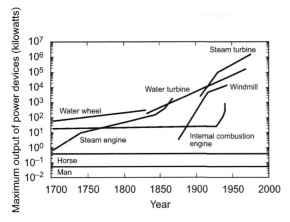

Figure 1
The maximum power output of selected power devices over the period from 1700 to 1970. From Chappel, 1984.

During the 19th century, hydropower became a source of electrical energy, although some form of hydroelectric turbine development can be traced back as far as 1750. Benoit Fourneyron is credited with developing the first modern hydroelectric turbine in 1833. The first hydroelectric plant is documented as coming online September 30, 1882, in Appleton, Wisconsin, and is still functioning. However, there is some dispute over this; Merritt cited the Minneapolis Brush Electric Company as beginning operation of a hydroelectric plant some 25 days earlier. The generation of electricity from falling water expanded the need for larger hydroelectric plants because the energy did not need to be used on-site. The transmission of power over long distances became economical in 1901 when George Westing-house installed alternating current equipment at Niagara Falls in New York, further expanding the potential uses of hydropower.

As Fig. 1 indicates, the power capabilities of water turbines became larger as the need grew. During the 1930s, large dams and ever-increasing turbine capacities became the norm. The power capacity of steam turbines was also increasing rapidly, and the relative cost of electricity continued to fall. Finally, during the period between 1940 and 1970, the cost of operating and maintaining older, smaller hydroelectric plants became greater than the income they could produce, and many were retired. This is seen in Fig. 2, where small hydropower capacity decreased as overall hydropower capacity climbed rapidly in the United States. A similar trend occurred in European countries. Hydropower development in other parts of the world was insignificant before 1930, as indicated by world hydropower production in Fig. 3.

The largest hydropower facility in the world is currently the Itaipu Dam on the Parana River, located between Brazil and Paraguay. This hydro-plant's

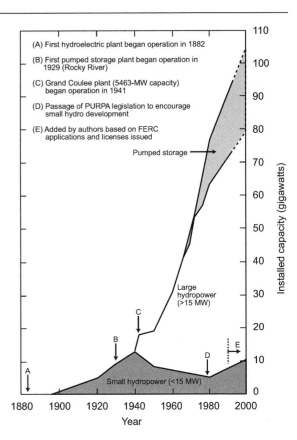

Figure 2
Installed hydroelectric capacity in the United States for the period from 1882 to 2000. FERC, Federal Energy Regulatory Commission. Assembled from Federal Energy Regulatory Commission, 1992, and International Waterpower and Dam Construction annual yearbooks.

12,600-MW capacity is the equivalent of 12 large thermal powerplants. It is upstream from the incomparable Iguasu Falls. China is building an even larger project at Three Gorges. The world's largest capacity hydroelectric plants in 2000 are listed in Table I.

There are two other basic sources of hydropower besides that which is extracted from the world's rivers: tidal power and wave power. Small tidal mills to provide mechanical power existed hundreds of years ago. However, the first hydroelectric tidal plant was developed during the 1960s on the La Rance estuary in northern France. Producing up to 240 MW of power, this tidal plant uses a dam across a cove mouth to form a pond. Sluice gates open to let water flow in during the rising tide and then close with the returning tide as water is directed through a standard hydroturbine.

The feasibility of tidal power depends on the range of tide experience and on finding a location where an inordinately long dam does not need to be built. Thus, only a handful of tidal plants have been developed since the La Rance plant: a 10-MW plant in 1986 and a number of smaller plants in China; an 18-MW plant in 1984 at an existing flood structure at Annapolis Royal in Nova Scotia, Canada; and a 400-kW plant in the Soviet Union. The turbines at Annapolis Royal are of interest because they can turn in both directions, capturing energy from both the incoming and outgoing tides.

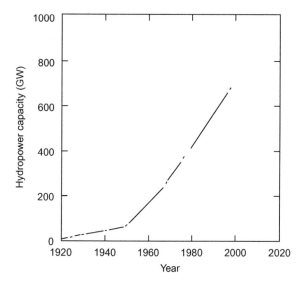

Figure 3
World hydropower production. Assembled from International Water Power and Dam Construction annual yearbooks.

The United Kingdom has studied the feasibility of building a dam across the Severn estuary, generating up to 7000 MW of power with 192 hydroturbines. Canada has conducted a similar study on the Bay of Fundy.

All tidal power projects are low head, up to 36 feet (11 m). Gulliver and Dotan found that the cost per kilowatt produced is approximately proportional to $H^{-0.53}$, where H is the hydraulic head across the dam. Obviously, any low-head hydropower project must be planned carefully.

2. Hydropower Potential

On a worldwide basis, hydropower represents roughly one-quarter of the total electrical energy generated. There is approximately 3800 GW of technically feasible hydropower potential existing in the world that could produce 14,370,000 GWh per year. Of this amount, 2100 GW (8,082,000 GWh/year) is feasible under current economic conditions and 700 GW (2,645,000 GWh/year) was installed in 2001. The technically feasible potential is that considered developable based on physical considerations without considering current economics or other environmental issues. It represents 39% of the total energy in the world's rivers. The available and developed power in 2001 is given by continent in Fig. 4. The tremendous potential in Asia, Africa, and South America is apparent, amounting to more than half of the world's total hydropower potential. On the other hand, hydropower developed or under construction in Europe and North American is at 73 and 70% of the economically feasible potential, respectively. The future hydropower development in these two

Table I
World's Largest Capacity Hydroelectric Plants in 2000.

| Name of dam | Location | Rated capacity (MW) | | Year of initial operation |
		Present	Ultimate	
Itaipu	Brazil/Paraguay	12,600	14,000	1983
Guri	Venezuela	10,000	10,000	1986
Grand Coulee	Washington State (United States)	6494	6494	1942
Sayano-Shushensk	Russia	6400	6400	1989
Krasnoyarsk	Russia	6000	6000	1968
Churchill Falls	Canada	5428	6528	1971
La Grande 2	Canada	5328	5328	1979
Bratsk	Russia	4500	4500	1961
Moxoto	Brazil	4328	4328	n.a.
Ust-Ilim	Russia	4320	4320	1977
Tucurui	Brazil	4245	8370	1984

Source: Mermel, T. W. (2000). The world's major dams and hydro plants. "International Water Power and Dam Construction," 2000 Handbook. Wilmington, Kent, UK.

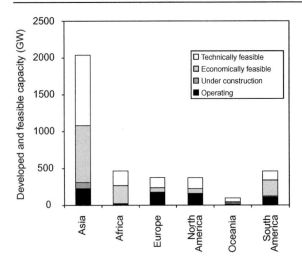

Figure 4
World hydropower resources by continent. Assembled from "International Hydropower and Dams," 2001 Handbook.

continents is likely to be in the form of upgrades in power and pumped storage.

3. What Nature Gives Us

Hydroelectric power, similar to many of humanity's activities, is produced by insertion into the hydrologic cycle, where water evaporates from the oceans, seas, lakes, and the like into the atmosphere, forms clouds of condensed water drops, falls as precipitation (e.g., rain, sleet, snow), and flows into rivers and eventually into the ocean. This cycle has significant variability given that floods and droughts are common. A dam placed in a river forms a reservoir with a given hydraulic head or a difference in elevation between the water behind the reservoir and the water below it. This hydraulic head is used to create power according to the relation

$$P = \eta\gamma QH, \tag{1}$$

where P is the power produced (in watts), H is in meters, Q is the discharge (flow) routed through the hydroelectric turbine (in cubic meters/second), γ is the specific weight of water (in Newtons/cubic meter), and η is the overall efficiency of the hydroelectric facility (fraction between 0 and 1).

The parameter that is highly variable in Eq. (1) is the discharge, Q. At a typical dam and reservoir on a river, the discharge over the dam during floods can be 1000 times greater than that during droughts. Predicting this discharge for a typical year represents much of the risk in developing a hydroelectric facility because it typically requires 30 years of

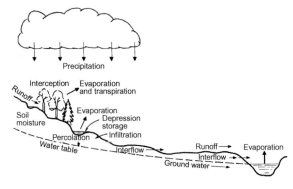

Figure 5
Processes affecting the relationship between precipitation and runoff.

records to reduce the uncertainty in mean discharge to $\pm 15\%$.

A watershed or drainage basin is the region contributing flow to a given location such as a prospective hydropower site on a stream. Any watershed is composed of a wide variety of foliage, soils, geological formations, streams, and the like and, therefore, is unique. Because no two watersheds are the same, comparisons of runoff from two separate watersheds and generalized relationships for watershed runoff are not accurate. The data obtained from stream gauges in the watersheds of interest should be used whenever possible.

The ratio of stream runoff to precipitation within a given watershed depends primarily on seven factors that are illustrated in Fig. 5 and described as follows:

• *Interception.* Interception is the precipitation stored on vegetative cover and then evaporated. Interception accounts for between 10 and 20% of annual precipitation in a well-developed forest. The interception of crops varies greatly. However, some approximate values for a 25-mm storm are as follows: cotton, 33%; small grains, 16%; tobacco, 7%; corn, 3%; alfalfa, 3%; meadow grass, 3%.

• *Depression storage.* Depression storage is the rainwater retained in puddles, ditches, and other depressions in the surface. It occurs when the rainfall intensity exceeds infiltration capacity. At the end of the storm, water held in depression storage is either evaporated or infiltrated into the soil. The depression storage capacity of most drainage basins is between 13 and 50 mm of precipitation. A specific type of depression storage that is handled separately is a blind or self-enclosed drainage basin, that is, a portion of the drainage basin that does not drain into the stream network but rather is self-enclosed, usually with a lake, marsh, or bog at the center. Blind drainage basins are normally excluded from the hydrologic analysis of a hydropower site.

• *Surface runoff.* Surface runoff is the precipitation that moves downslope along the soil surface until it reaches a stream or lake. It is primarily associated with flood events, although in larger watershed basins the effects of surface runoff can be felt for up to a month. A small portion of any precipitation event will fall directly into the channel or stream. This immediately becomes runoff. It is grouped together with runoff from impervious areas, such as parking lots, streets, and buildings, because this runoff occurs rather quickly after precipitation.

• *Infiltration.* Infiltration is the passage of water through the soil surface. It usually implies percolation, which is the movement of water through unsaturated soil. Infiltration capacity depends on soil porosity, moisture content, and vegetative cover. Sandy or highly organic soils will have a greater infiltration due largely to increased porosity. A wet soil will have a lower infiltration capacity. Vegetation cover increases infiltration by retarding surface flow, increasing soil porosity with the root system, and reducing rain packing of the soil surface.

• *Soil moisture.* The precipitation that infiltrates into the soil will first be used to replenish soil moisture. Over time, the soil moisture is taken by plant root systems and eventually is transpired from plant foliage as part of the photosynthetic process. Water used to replenish soil moisture will not appear as stream flow. Soils with a high percentage of decayed plant material have a large capacity to retain moisture.

• *Interflow.* Interflow is water that infiltrates through the soil and moves laterally in the upper soil layers until it reemerges as surface runoff. A thin soil surface covering rock, hardpan, or plow bed will usually have large quantities of interflow. Interflow often emerges as a spring in riverbanks. It will not usually affect flood peaks but will increase stream flow at a steady rate for some time after the peak.

• *Groundwater flow.* If the infiltrated water percolates downward until it reaches the water table (i.e., zone saturated with water), it will eventually reach stream as groundwater flow, which is the primary source of base flow for streams. Ground-water flow influences stream flow on a seasonal, rather than a weekly, time scale.

A schematic diagram of the segmentation of rainfall for an extensive storm in a relatively dry basin is given in Fig. 6. The shaded area indicates the quantity of rainfall that will eventually become stream flow. The general order in which the various types of flow reach the stream is as follows: channel precipitation, surface runoff, interflow, ground water.

The seven parameters that influence the runoff process, which depends on foliage, soil type, geological formation, and watershed geomorphology, can vary greatly within a given watershed and will certainly vary between two distinct watersheds.

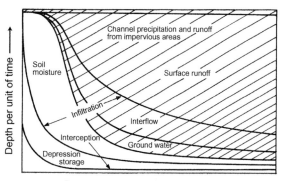

Figure 6
Schematic diagram of the segmentation of rainfall for an extensive storm in a relatively dry basin. The shaded area indicates precipitation that eventually results in a runoff.

Therefore, it should be no surprise to discover that two adjacent watersheds of similar drainage areas have entirely different runoff characteristics.

4. How Hydropower is Captured

The broad variety of natural conditions at hydropower sites has resulted in development of many different types of hydropower schemes. This is a disadvantage so far as engineering costs are concerned, but it provides a great deal of stimulus to those involved in hydropower development. A longitudinal section of a typical scheme is provided in Fig. 7.

In the run-of-river development, a short penstock or dam directs the water through the turbines. The powerhouse is often an integral part of the dam, as shown in Fig. 8. The natural flow of the river remains relatively unaltered. A more complex development occurs at diversion or canal projects, where the water is diverted from the natural channel into a canal or long penstock, as shown in Fig. 9. This results in a significant change in the flow of water in a given reach of the river, sometimes for a considerable distance.

Storage regulation developments are defined as those in which an extensive impoundment at the power plant, or at the reservoir upstream of the power plant, allows for regulation of the flow downstream through storage. Water is stored during high-flow periods and is used to augment the flow during low-flow periods. This allows for a relatively constant supply of energy over the course of the year.

Pump storage facilities are normally large developments in which water is pumped from a lower reservoir during off-peak hours when the cost of energy is low. The pumps are run in reverse as

143

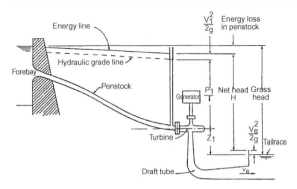

Figure 7
Schematic of a hydropower installation. $V_1^2/2g$ is head lost before the turbine, P_1/γ is the positive head extracted by the turbine, Z_1 is the suction head on the turbine, and $V_e^2/2g$ is the velocity head remaining at the draft tube outlet.

Figure 8
The 8.4-MW St. Cloud hydropower facility is a run-of-river development with the powerhouse composing a portion of the dam. Net head is 5.2 m. (A) Upstream view of the spillway and intakes into gates and powerhouse. (B) Downstream view of the powerhouse and spillway. Courtesy of M. A. Mortenson Company.

Figure 9
The Mayfield project on the Cowlitz River in Washington. It is composed of an arch dam and spillway (background), a power tunnel leading to surge tanks (left), and penstocks leading to a 122-MW powerhouse (center front). Notice how the layout maximizes efficient use of the terrain. Courtesy of Harza Engineering.

turbines during the hours of peak demand to augment the power supplied from other sources. Thus, the pumped storage facility serves the same function as a large battery. It is used to allow large base load facilities to operate at continuous power output in cases where it is uneconomical to allow the power output of a large plant to fluctuate.

Tidal power plants have also been developed or considered. These are located in areas where there are large tidal fluctuations. Low-head turbines are used to harness the energy in the tidal cycle, and an entire bay or estuary is enclosed by a low dam.

Hydropower schemes can either be single purpose, having as their only purpose the production of electricity, or multipurpose, having hydropower production as just one aspect of the total use of the facility. Multipurpose facilities include those in which hydropower is developed in conjunction with irrigation, flood control, navigation, and water supply. Hydropower plants are also categorized by the type of use. For example, a base load plant is one in which the power is used to meet all or part of a sustained and constant portion of the electrical load. Energy from these plants that is available at all times is referred to as "firm power." The need for power varies during the course of the day, and power requirements over and above the base load requirement are met by peak load facilities. These are plants in which the electrical production capacity is relatively high and the volume of water discharged through the units can be changed readily to meet peak demands. Storage or pondage of the water supply is necessary for these load demands because such plants can be started and stopped more rapidly and economically than can fossil fuel and nuclear power plants.

A typical hydropower facility consists of the following components:

- The dam or diversion structure with the associated reservoir
- The powerhouse structure and its foundation
- Hydraulic conveyance facilities, including the head race, headworks, penstock, gates and valves, and tailrace
- The turbine-generator unit, including the guide vanes or wicket gates, turbine, draft tube, speed increaser, generator, and speed-regulating governor
- Station electrical equipment, including the transformer, switch gear, automatic controls, conduit, and grounding and lightning systems
- Ventilation, fire protection, communication, and bearing cooling water equipment
- Transmission lines.

The selection of the proper turbine for a given site depends on the head and flow available as well as the use that is to be made of the facility. Because there is such a broad variation in available conditions in the field, many different turbine configurations have been developed and are available. It is essential that a hydropower engineer understand in detail the rationale for selection of hydroturbines. The other items listed, although not unique to hydropower production, are also crucial to a properly operating facility.

5. Hydroturbines

The water turbine has a rich and varied history. It has been developed as a result of a natural evolutionary process from the waterwheel. Originally used for direct drive of machinery, the use of the water turbine for the generation of electricity is a comparatively recent activity. Much of its development occurred in France, which did not have the cheap and plentiful sources of coal that England had and that sparked the industrial revolution during the 18th century. During the 19th century, France found itself with its most abundant energy resource being water. To this day, *Houille Blanche* (literally "white coal") is the French term for water power.

In 1926, the Société d'Encouragement pour l'Industrie Nationale offered a prize of 6000 francs to anyone who would "succeed in applying at large scale, in a satisfactory manner, in mills and factories, the hydraulic turbines or wheels with curved blades of Bélidor." Bélidor was an 18th-century hydraulic

and military engineer who, between 1737 and 1753, authored a monumental four-volume work, *Architecture Hydraulique*, a descriptive compilation of hydraulic engineering information of every sort. The waterwheels described by Bélidor departed from convention by having a vertical axis of rotation and being enclosed in a long cylindrical chamber approximately 1 m in diameter. Large quantities of water were supplied from a tapered sluice at a tangent to the chamber. The water entered with considerable rotational velocity. This preswirl, combined with the weight of water above the wheel, was the driving force. The original tub wheel had an efficiency of only 15 to 20%.

Water turbine development proceeded on several fronts from 1750 to 1850. The classical horizontal axis waterwheel was improved by engineers such as John Smeaton (1724–1792) of England and J. V. Poncelet of France (1788–1867). This resulted in waterwheels having efficiencies in the range of 60 to 70%. At the same time, reaction turbines (somewhat akin to modern lawn sprinklers) were being considered by several workers. The great Swiss mathematician Leohard Euler (1707–1783) investigated the theory of operation of these devices. A practical application of the concept was introduced in France in 1807 by Monnoury de Ectot (1777–1822). His machines were, in effect, radially outward-flow machines. The theoretical analyses of Burdin (1790–1893), a French professor of mining engineering who introduced the word "turbine" into engineering terminology, contributed much to our understanding of the principles of turbine operation and underscored the principal requirements of shock-free entry and exit with minimum velocity as the basic requirements for high efficiency. A student of Burdin, Benoit Fourneyron (1802–1867), was responsible for putting his teacher's theory to practical use. Four-neyron's work led to the development of high-speed outward-flow turbines with efficiencies on the order of 80%. The early work of Fourneyron resulted in several practical applications and the winning of the coveted 6000-franc prize in 1833. After nearly a century of development, Bélidor's tub wheel had been officially improved. Fourneyron spent the remaining years of his life developing some 100 turbines in France and Europe. Some turbines even found their way to the United States, the first around 1843.

As successful as the Fourneyron turbines were, they lacked flexibility and were efficient over only a narrow range of operating conditions. This problem was addressed by S. Howd and U. A. Boyden (1804–1879). Their work evolved into the concept of an inward-flow motor as a result of the work of James B. Francis (1815–1892). The modern Francis turbine, shown in Fig. 10, is the result of this line of development. At the same time, European engineers addressed the idea of axial-flow machines, which

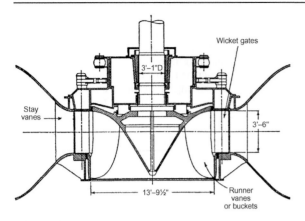

Figure 10
Francis turbine. From Daily (1950).

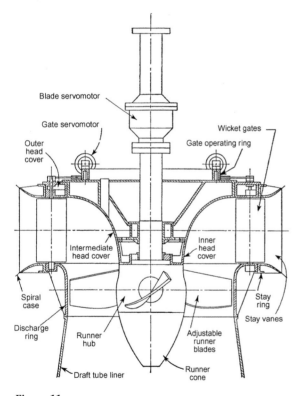

Figure 11
Smith–Kaplan axial-flow turbine with adjustable-pitch runner blades. From Daily (1950).

today are represented by "propeller" turbines of both fixed-pitch and Kaplan types, as shown in Fig. 11.

Just as the vertical-axis tub wheels of Bélidor evolved into modern reaction turbines of the Francis and Kaplan types, development of the classical

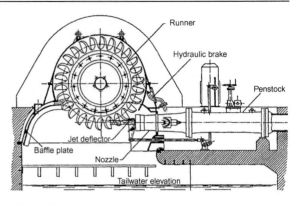

Figure 12
Cross section of a single-wheel, single-jet Pelton turbine. This is the third highest head Pelton turbine in the world ($H = 1447\,\text{m}$, $n = 500\,\text{rpm}$, $P = 35.2\,\text{MW}$, $N_s \sim 0.038$). Courtesy of Vevey Charmilles Engineering Works. Adapted from Raabe, J. (1985), "Hydro Power: The Design, Use, and Function of Hydromechanical, Hydraulic, and Electrical Equipment." VDI-Verlag, Dusseldorf, Germany.

horizontal-axis waterwheel reached its peak with the introduction of the impulse turbine. The seeds of development were sown by Poncelet in 1826 with his description of the criteria for an efficient waterwheel. These ideas were cultivated by a group of California engineers during the late 19th century, one of whom was Lester A. Pelton (1829–1908), whose name is given to the Pelton wheel (shown in Fig. 12), which consists of a jet or jets of water impinging on an array of specially shaped buckets closely around the periphery of a wheel.

Modern impulse units are generally of the Pelton type and are restricted to relatively high-head applications (Fig. 3). One or more jets of water impinge on a wheel containing many curved buckets. The jet stream is directed inward, sideways, and outward, thereby producing a force on the bucket, which in turn results in a torque on the shaft. All kinetic energy leaving the runner is "lost." A draft tube is generally not used because the runner operates under approximately atmospheric pressure and the head represented by the elevation of the unit above tailwater cannot be used. Because this is a high-head device, the loss in available head is relatively unimportant. The Pelton wheel is a low-specific speed device. Specific speed can be increased by the addition of extra nozzles, with the specific speed increasing by the square root of the number of nozzles. Specific speed can also be increased by a change in the manner of inflow and outflow. Special designs such as the turbo and crossflow turbines are examples of relatively high-specific speed impulse units.

Over more than 250 years of development, many ideas were tried. Some were rejected, whereas others were retained and incorporated into the design of the hydraulic turbine as we know it today. This development has resulted in highly efficient devices, with efficiencies as high as 96% in the larger sizes. In terms of design concept, these fall into roughly three categories: reaction turbines of the Francis design, reaction turbines of the propeller design, and impulse wheels of the Pelton type.

Most Pelton wheels are mounted on a horizontal axis, although newer vertical-axis units have been developed. Because of physical constraints on orderly outflow from the unit, the maximum number of nozzles is generally limited to six or less. Although the power of a reaction turbine is controlled by the wicket gates, the power of the Pelton wheel is controlled by varying the nozzle discharge by means of an automatically adjusted needle, as illustrated in Fig. 12. Jet deflectors, or auxiliary nozzles, are provided for emergency unloading of the wheel. Additional power can be obtained by connecting two wheels to a single generator or by using multiple nozzles. Because the needle valve can throttle the flow while maintaining essentially constant jet velocity, the relative velocities at entrance and exit remain unchanged, producing nearly constant efficiency over a wide range of power output.

5.1 Performance Comparison

The physical characteristics of various runner configurations are summarized in Fig. 13. It is obvious that the configurations change with speed and head. Impulse turbines are efficient over a relatively narrow range of specific speed, whereas Francis and propeller turbines have a wider useful range. An important consideration is whether or not a turbine is required to operate over a wide range of load. Pelton wheels tend to operate efficiently over a wide range of power loading due to their nozzle design. In the case of reaction machines that have fixed geometry, such as Francis and propeller turbines, efficiency can vary widely with load. However, Kaplan turbines can maintain high efficiency over a wide range of operation conditions. The decision of whether to select a simple configuration with a relatively "peaky" efficiency curve or take on the added expense of installing a more complex machine with a broad efficiency curve will depend on the expected operation of the plant and other economic factors.

Note in Fig. 13 that there is an overlap in the range of application of various types of equipment. This means that either type of unit can be designed for good efficiency in this range but that other factors, such as generator speed cavitation and cost, may dictate the final selection.

5.2 Speed Regulation

The speed regulation of a turbine is an important and complicated problem. The magnitude of the problem varies with size, type of machine and installation, type of electrical load, and whether or not the plant is

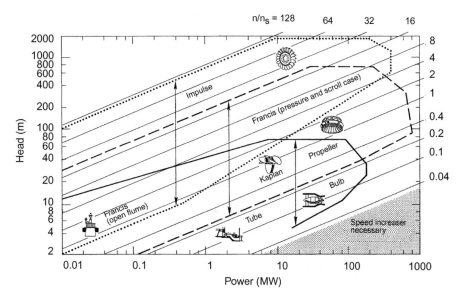

Figure 13

Application chart for various turbine types. n/n_s ratio of turbine speed in revolutions per minute n to specific speed defined in the system; $n_s = nP^{1/2}/H^{3/4}$ (with P in kilowatts). From Arndt, R. E. A. [1991]. Hydraulic turbines. *In* "Hydropower Engineering Handbook" [J. S. Gulliver and R. E. A. Arndt, Eds.]. McGraw-Hill, New York.

tied into an electrical grid. It should also be kept in mind that runaway or "no load" speed can be higher than the design speed by a factor as high as 2.6. This is an important design consideration for all rotating parts, including the generator.

The speed of a turbine has to be controlled to a value that matches the generator characteristics and the grid frequency,

$$n = \frac{120f}{N_P}, \qquad (2)$$

where n is the turbine speed in revolutions per minute (rpm), f is the required grid frequency in hertz, and N_p is the number of poles in the generator. Typically, N_p is in multiples of 4. There is a tendency to select higher speed generators to minimize weight and cost. However, consideration has to be given to speed regulation.

It is beyond the scope of this subsection to discuss the question of speed regulation in detail. Regulation of speed is normally accomplished through flow control. Adequate control requires sufficient rotational inertia of the rotating parts. When load is rejected, power is absorbed, and this accelerates the flywheel; when load is applied, some additional power is available from deceleration of the flywheel. Response time of the governor must be carefully selected because rapid closing time can lead to excessive pressures in the penstock.

A Francis turbine is controlled by the opening and closing of the wicket gates, which vary the flow of water according to the load. The actuator components of a governor are required to overcome the hydraulic and frictional forces and to maintain the wicket gates in a fixed position under steady load. For this reason, most governors have hydraulic actuators. On the other hand, impulse turbines are more easily controlled. This is due to the fact that the jet can be deflected or an auxiliary jet can bypass flow from the power-producing jet without changing the flow rate in the penstock. This permits long delay times for adjusting the flow rate to the new power conditions. The spear or needle valve controlling the flow rate can close quite slowly, say in 30 to 60s, thereby minimizing any pressure rise in the penstock.

Several types of governors are available that vary with the work capacity desired and/or the degree of sophistication of control. These vary from pure mechanical, to mechanical–hydraulic, to electrohydraulic. Electrohydraulic units are sophisticated pieces of equipment and would not be suitable for remote regions. The precision of governing required will depend on whether the electrical generator is synchronous or asynchronous (i.e., induction type). There are advantages to the induction type of generator. It is less complex and, therefore, less expensive; however, it typically has slightly lower efficiency. Its frequency is controlled by the frequency of the

grid it is feeding into, thereby eliminating the need for an expensive conventional governor. It cannot operate independently and can only feed into a network, and it does so with a lagging power factor that may or may not be a disadvantage, depending on the nature of the load. For example, long transmission lines have a high capacitance, so in this case the lagging power factor may be an advantage.

Speed regulation is a function of the flywheel effect of the rotating components and the inertia of the water column of the system. The start-up of the rotating system, t_s, is given by

$$t_s = \frac{I\omega^2}{P} \qquad (3)$$

where I is the moment of inertia of the generator and turbine (in kilogram-meters), ω is the rotational speed of the unit (in radians/second), and P is the power generated (in watts).

The starting-up time of the water column, t_p, is given by

$$t_p = \frac{\Sigma L V}{g H} \qquad (4)$$

where L is the length of water column segment, V is the velocity in each segment of the water column, and g is the acceleration of gravity.

For good speed regulation, it is desirable to keep $t_s/t_p > 4$. Lower values can also be used, although special precautions are necessary in the control equipment. It can readily be seen that higher ratios of t_s/t_p can be obtained by increasing I or decreasing t_p. Increasing I implies a larger generator, which also results in high costs. The start-up time of the water column can be reduced by reducing the length of the flow system, by using lower velocities, or by adding surge tanks, which essentially reduce the effective length of the conduit. A detailed analysis should be made for each installation because, for a given length, head, and discharge, the flow area must be increased to reduce t_p, leading to associated higher construction costs.

5.3 Cavitation and Turbine Setting

Another factor that must be considered prior to equipment selection is the evaluation of the turbine with respect to tailwater elevations. Hydraulic turbines are subject to pitting due to cavitation. For a given head, a smaller, lower cost, high-speed turbine runner must be set lower (i.e., closer to tailwater or even below tailwater) than does a larger, higher cost, low-speed turbine runner. Also, atmospheric pressure or plant elevation above sea level is a factor, as are tailwater elevation variations and operating requirements. This is a complex subject that can be accurately resolved only through model tests. Every runner design will have different cavitation characteristics. Therefore, the anticipated turbine location or setting with respect to tailwater elevations is an important consideration in turbine selection.

Cavitation is not normally a problem with impulse wheels. However, by the very nature of their operation, cavitation is an important factor in reaction turbine installations. The susceptibility to cavitation is a function of the installation and the turbine design. This can be expressed conveniently in terms of Thoma's sigma, defined as

$$\sigma_T = \frac{H_a - H_v - z}{H}, \qquad (5)$$

where H_a is the atmospheric pressure head, H_v is the vapor pressure head (generally negligible), and z is the elevation of a turbine reference plane above the tailwater (Fig. 7). Draft tube losses and the exit velocity head have been neglected.

The parameter σ_T must be above a certain value to avoid cavitation problems. The critical value of σ_T is a function of specific speed. The Bureau of Reclamation suggested that cavitation problems can be avoided when

$$\sigma_T > 0.26 N_S^{1.64} \qquad (6)$$

where N_S is the specific speed of the turbine or $N_S = n(\text{rpm}) \, P(\text{kW})^{1/2}/H^{5/4}$ (feet). Equation (6) does not guarantee total elimination of cavitation; it guarantees only that cavitation is within acceptable limits. Cavitation can be totally avoided only if the value σ_T at an installation is much greater than the limiting value given in Eq. (6). The value of σ_T for a given installation is known as the plant sigma σ_P. Equation (6) should only be considered as a guide in selecting σ_P, which is normally determined by a model test in the manufacturer's laboratory. For a turbine operating under a given head, the only variable controlling σ_P is the turbine setting z. The required value of σ_P then controls the allowable setting above tailwater:

$$z_{allow} = H_a - H_v - \sigma_p H \qquad (7)$$

It must be borne in mind that H_a varies with elevation. As a rule of thumb, from the sea level value of 10.3 m, H_a decreases by 1.1 m for every 1000 m above sea level.

6. Ecological Considerations

The ecological impacts of hydroelectric power production that are perceived depend on the scale at which attention is focused. At the global environmental scale, hydropower is still seen as a clean form of energy. It is a replacement of fossil fuel power production, which is an inherently dirty source of energy. Terrestrial ecology is disturbed by mining activities, transportation of the fuel, and plants that burn the fuel. Water from rivers and lakes that is used to cool power generation systems returns hotter than normal, altering the ecology of these water bodies. Finally, atmospheric impacts of fossil fuel power production are substantial, including the production of nitrous oxides, sulfur compounds, heavy metals, and carbon dioxide. The future state of the earth and its atmosphere is affected by our power production through burning fossil fuels.

Although hydroelectric power production on the global scale is a relatively clean alternative to power produced through fossil fuels, on the local scale (i.e., the river system) the ecological impacts of hydropower can be significant. The dam and reservoir, which most hydroelectric power facilities require, significantly alter the aquatic habitat and the species present. A dam will typically hinder migrating fish species in their spawning activities and cause water quality problems such as dissolved gas bubble disease. The reservoir used for storage can alter flows, such that floods and droughts are less frequent, and will catch most of the sediment. This will alter the river morphology in the downstream reach because the flood flows are important to forming river morphology. In tropical climates, parasite transmission can increase below a dam (e.g., below the Aswan High Dam). With extensive development of dams, a river will become a series of reservoirs and the river habitat will have disappeared entirely, such that many clams or mussel species will no longer exist and the dominant fish species will have been altered.

There are only selected locations on a river that are appropriate for dams and reservoirs. These locations are also the best for recreational activities such as canoeing, rafting, and fishing. Therefore, there is significant public interest in any proposed dam/hydropower facility that is proposed. As the prospective locations for dams and hydropower facilities are developed, there are fewer remaining locations for the recreational activities that need to compete for these sites. From Fig. 4, it can be seen that between 70 and 73% of the economic hydropower potential of Europe and North American has been developed. The recreational and environmental interests of the two continents need to be considered, and it is unlikely that much of the remaining hydropower potential of these two continents will be developed. However, in Asia, Africa, and South America, there is substantial hydropower remaining, and the power needs of the 21st century will be significant. The building of new dams and hydropower facilities in these regions is more likely.

7. Advantages and Disadvantages of Hydropower Development

It is obvious that hydropower development is a renewable energy source, and this is a significant advantage. When hydropower development occurs at existing dam sites, the environmental impact is often minimal. However, significant changes can occur when an existing dam site is developed for power generation. For example, water that would normally flow over a spillway, where a large amount of aeration would take place, is now funneled through turbines, where little or no aeration occurs. The resulting substantial difference in the rate of aeration at a given point in the river can have a notable influence on the dissolved oxygen content of a considerable reach of river. In certain areas, especially those near large municipalities, the environmental impact can be significant.

Although the fuel costs of hydropower plants are negligible, their construction and capital equipment costs are usually substantially greater per unit of installed capacity than those of thermal power plants. This can be seen as either a benefit or a detriment. The economic feasibility of hydropower development is very sensitive to the difference between a discount rate, used to bring future income and costs to present value, and an assumed escalation rate, used to predict the future cost of electricity. The discount rate is taken as the load interest rate, which is usually fixed. The escalation rate is an estimate, which can vary greatly. For example, during a period of escalating inflation, a hydroelectric project that is already in place looks like a very good investment.

When considering hydropower development within a given region, it is also important to look at the overall economics of that region. For example, it should be noted that hydropower development means that a substantially larger percentage of the investment capital can stay within a given region because much of the developmental work can be done by local engineers and contractors. The more sophisticated coal-fired and nuclear power plants are designed and built by specialized contractors, and this often means that large amounts of capital leave the local economy. In many instances, the same is true for the amount of capital necessary for fuel for thermal power plants. This substantial drain on the economy can be very significant. In addition, hydropower facilities require minimal maintenance and do not have the same requirements for skilled personnel as do the more sophisticated thermal power plants.

There are other advantages of hydropower, and especially of small hydropower suitable as appropriate technology in less developed economic regions. Many future possibilities in small hydropower development will depend on the economic climate. There is a definite market for small turbine technology. If this market is developed, one could expect that significant improvements, both in operational characteristics of turbines and in reduced cost, can lead to small hydropower facilities being more cost-effective. There are many other aspects of small hydropower that have not yet been explored. In many cases in developing nations, small hydropower stations becomes the catalysts for the development of small manufacturing facilities. One case in point is electrolytic manufacturing of fertilizer. It should also be noted that the current miniprocessor technology has developed to the point where it is feasible to operate a system of small hydropower plants on a completely automated basis, with only a traveling crew of workers needed for maintenance. In some cases, the number of small sites developed could supply the same amount of power as could one large nuclear power plant, without the safety or security hazards normally involved with a large-scale development. If the total amount of power is distributed over several small plants, overall reliability can increase because it is very unlikely that all plants would suffer outages at the same time.

However, there are also several disadvantages to small hydropower development. The most obvious is the fact that economy of scale does not prevail. This results in high initial cost for a relatively low installed capacity. In many cases, these plants are run-of-river; that is, their capability for generating power fluctuates wildly with the seasons, and this prevents a system of small power plants from acting as an equivalent base load plant. In many areas of the world, peak power is available during the late spring, whereas peak demand occurs during the midsummer or midwinter. This mismatch of power need and availability can be quite serious.

In addition to the lost economy of scale, there are other disadvantages that relate to the head available. For example, low-head facilities are those in which the available head is less than approximately 20 m. Because available power is proportional to the product of flow and head, larger amounts of flow must be handled to generate a given power level at lower head. Thus, the size of the machine increases, producing a disproportionate increase in cost for the amount of power developed.

Further Reading

Armstrong E L 1985 The global outlook for additional hydropower use. In: *Waterpower '85*. American Society of Civil Engineers, New York

Arndt R E A 1981 Cavitation in fluid machinery and hydraulic structures. *Annu. Rev. Fluid Mech.* **13**, 273–328

Chapman D W 1986 Salmon and steelhead abundance in the Columbia River in the nineteenth century. *Trans. Am. Fish. Soc.* **155**, 662–70

Chappel J R 1984 *The Future of Hydropower*. EG&G Idaho, Idaho Falls, ID

Daily J W 1950 Hydraulic machinery. In: Rouse H (ed.) *Engineering Hydraulics*. John Wiley, New York

Deudney D 1981 *Rivers of Energy: The Hydropower Potential, Worldwatch Paper 44*. Worldwatch Institute, Washington, DC

Federal Energy Regulatory Commission 1992 *Hydroelectric Power Resources of the United States, Developed and Undeveloped*. FERC-0070. FERC, Washington, DC

Gulliver J S, Arndt R E A 1991 *Hydropower Engineering Handbook*. McGraw-Hill, New York

Gulliver J S, Dotan A 1984 Cost estimates for hydropower at existing dams. *J. Energy Eng.* **110**, 204–14

International Electrotechnical Commission. 1963. *International Code for the Field Acceptance Tests of Hydraulic Turbines*, Publication 41

Isom B G 1969 The mussel resource of the Tennessee River. *Malacologia* **7**, 397–425

Merritt R H 1979 *Creativity, Conflict, and Controversy: A history of the St. Paul District*. U.S. Army Corps of Engineers, 008-022-00138-7. U.S. Government Printing Office, Washington, DC

Rouse H, Ince S 1963 *History of Hydraulics*. Dover, New York

Shea C P 1988 *Renewable Energy: Today's Contribution, Tomorrow's Promise*. Worldwatch Paper 81. Worldwatch Institute, Washington, DC

Smith N 1980 January The origins of the water turbine. *Sci. Amer.* **242**(1).

Van der Schalie H 1972 World Health Organization Project Egypt 10: A case history of a schistosomiasis control project. In: Farver J T, Milton J P (eds.) *The Careless Technology: Ecology and International Development*. National History Press, Garden City, NY

Wirth B D 1997 Reviewing the success of international flooding of the Grand Canyon. *Hydro Rev.* **16**(2), 10–6

John S. Gulliver and Roger E. A. Arndt

M

Manufactured Gas, History of

Glossary

carbureted water gas   Gas produced by spraying blue gas (water gas) with liquid hydrocarbons and thermally cracking it to form carbureted water gas and tars.

coal gas   A method of heating bituminous coal in a retort to destructively distill volatiles from the coal in order to produce fuel gas for illuminating and heating purposes.

water gas   A fuel gas composed primarily of hydrogen and carbon monoxide produced through the interaction of steam with incandescent carbon, usually from anthracite coal or coke.

Welsbach mantle   A mantle made of cotton fabric impregnated with rare earth that becomes incandescent when exposed to a gas flame.

This article examines the development of the manufactured gas industry in the United States from its origins early in the 19th century to its demise in the post-World War II period. It discusses the production technologies of coal gas, water gas, and oil gas and their methods of distribution. It also discusses industry innovations, such as the development of carbureted water gas and of the Welsbach mantle, and their effect on regulatory standards. It explores the shift from the use of manufactured gas to natural gas in the period from the 1920s through the 1950s, with the consequent demise of the industry. Finally, the article examines the environmental effects of the industry and the issues of the cleanup of former MGP sites.

1. Origins and Characteristics of The Industry

Manufactured gas was one of the most critical energy sources and fuels that provided American cities, as well as other cities throughout the world, with light and energy during much of the century from 1850 to 1950. Manufactured gas is produced by heating coal or other organic substances such as wood, pitch, or petroleum, driving off a flammable gas made primarily of carbon monoxide and hydrogen as well as other gases. The discovery that flammable gas could be produced in this manner was first made in Great Britain by John Clayton in the late seventeenth century followed by the work of Frenchmen Jean Pierre Minckelers and Philippe Lebon in the late 18th and early 19th centuries. At the turn of the century, the British engineer William Murdock developed methods of gas lighting that were used in cotton mills.

Demonstrations and applications of gas lighting began in the United States in 1802, based on European developments. Gasworks were installed to provide better illumination for cotton mills and lighthouses during the first decade of the century. In 1817, the Baltimore municipality granted the Gas Light Company a franchise to lay gas pipe in the city streets and to light them. Others cities quickly followed: New York, 1825; Boston, 1829; New Orleans, 1835; Philadelphia, the first municipally owned plant, 1836; and San Francisco, the first western plant, 1854. The U.S. Census reported 30 manufactured gas plants in 1850, 221 in 1859, 390 in 1869, 742 in 1889, 877 in 1904, and 1296 in 1909, but the actual number was probably much larger.

The manufactured gas utilities that emerged ranged in size from very small plants in small towns (cf. 5000 population) that served a limited number of consumers (200–300) to very large plants with extensive distribution systems that served thousands of customers with gas for home lighting, cooking, and heating, served municipalities by illuminating city streets, and provided gas for a number of industries. Urban gas utilities were systems that generated fuel for their customers from a central plant (or plants in the case of the very large utilities) and distributed it through a piped system. By far, it was most widely used for lighting purposes. Individual gas production units were also manufactured and were used for isolated factories and homes (spirit gas lamps). From the late 19th century onward, various industries in need of low-Btu (British thermal unit) gas frequently installed coal-using producer gas units in their factories.

Municipal and state governments regarded gas manufacturing plants as public utilities and required that they acquire franchises to use the city streets for their distribution lines. In order to profitably survive, gas manufacturing utilities required long rather than short franchises—the exact term, however, produced many heated political battles. Franchises were originally local but frequently became the concern of state authorities because the natural monopoly characteristics of the utilities were resistant to market forces. Battles over issues such as price, luminosity and energy ratings, and technology, as well as price and duration, frequently took place between utility firms and municipal and state authorities. Although some municipally owned gas works existed, the industry remained largely private.

2. The Period of Coal Gas Domination, 1816–1875

In the period from 1816 to 1875, the number of centralized gas works grew to between 300 and 400, primarily using technology developed in Great Britain. The most important innovations during these years concerned fuel, distillation technology, distribution, and metering technology.

Although most firms eventually used coal as their feedstock for gas production, other substances were also initially used. The Baltimore Gas-Light Company, for instance, used pine tar as the feedstock for its initial gas production, the New York Gas Works used rosin, the Brooklyn Gas Works consumed British coals, and the Wilmington, North Carolina Gas Works used wood from the Carolina Pine Barrens. Several small eastern cities used distillate of turpentine to manufacture rosin gas and Philadelphia experimented with wood as late as the 1850s. Utility managers originally avoided coal because of its impurities, but in the period from 1815 to 1820, British gas manufacturers developed methods to remove such impurities. In 1822, Baltimore became the first American city to consume coal to produce gas, using bituminous coal from Virginia mines for its feedstock.

Since bituminous coal mines were not easily accessible for most cities, the cost of the fuel kept prices relatively high or forced the use of organic materials such as wood. Expensive fuel, therefore, provided a reverse salient (a concept originated by historian of technology Thomas Hughes) in terms of extending the size of the market for improved lighting. Whereas bituminous coal, considered "the best in the world for gas light," existed in western Pennsylvania and was used in Pittsburgh for gas production and for home heating, transportation costs over the Allegheny Mountains were extremely high, limiting its use in eastern markets.

The major transportation breakthroughs came in the late 1840s and 1850s with the slack watering of the Pittsburgh area rivers and the completion of railroad connections from Philadelphia to Pittsburgh in 1852. These transport improvements facilitated the shipping of high-quality bituminous gas coal from western Pennsylvania to gas works in eastern cities at greatly reduced transportation costs. By the late 1850s, for instance, the Westmoreland Coal Company, a large western Pennsylvania producer of gas coal, was supplying 58 gas works.

Responding to the lower fuel costs and the anticipation of a larger market for cheaper light, a number of gas works were founded in the 1850s and 1860s. Between 1850 and 1860, the U.S. Census reported that the number of manufactured gas establishments jumped from 30 to 221, a rise of over 600%—an increase much greater than that of urban population growth. By 1870, the total number of manufactured gas plants reported was 390, another large jump.

The manufacture of coal gas consisted of three operations: distillation, condensation, and purification. Although there were many different processes patented for the production of coal gas, by the 1850s those most widely used operated by heating bituminous coal in a retort to destructively distill volatiles from the coal. The leading types of coal-carbonization retorts were the horizontal, inclined, and vertical. Horizontal retorts were most commonly used; until approximately 1850, they were made of cast iron, when superior clay refractories were substituted. A typical horizontal bench consisted of six retorts and a producer gas furnace that used coke or coal for heating the retorts. Coal tar condensed from the gas was also used as a retort fuel. Wet and dry lime processes were used in purification methods, as was iron oxide, beginning in 1870.

Proper modes of metering, gas storage, and distribution were critical to the successful operation of a gas firm. Gas holders, or gasometers, as they were called, were essentially inverted iron bells in a tub of water. The early gas holders were usually made of masonry and often set in the ground, with iron plates for the bell. A number of firms initially used wooden mains for distribution, but in ~1820, cast iron pipes manufactured in British foundries were introduced into the United States; however, some wooden mains of tamarack logs lasted into the late 19th century. By the 1830s, American foundries were producing their own cast iron pipe, reducing dependence on the British. From ~1820 through 1850, pipe was made in horizontally cast sections (4–5 ft in length). These were subject to considerable core sagging and slag accumulations but in 1850 foundries shifted to casting in vertical molds, producing a much stronger and longer (12 ft) pipe product. Nineteenth century cast iron gas mains often lasted 50 to 75 years.

The distribution system was another major cost for the gas companies. Firms occasionally used smaller diameter mains than was ideal, causing problems and poor service as the system expanded. In Philadelphia, for instance, the city-owned utility used pumps to force gas through mains too small to accommodate the demand, often resulting in customers receiving gas under excess pressure. Though the gas utilities seldom extended conduits to undeveloped streets, they quickly installed pipe on streets in growing areas in order to benefit from rises in real estate prices.

Cities initially used gas primarily for street illumination but cost reductions plus innovations such as improved burners and fixtures and metering led to a broader market. Gas companies originally charged flat rates per burner, leading to expansion limits for street lights and domestic sales. Meters were essential for price reduction for both municipalities and domestic consumers but were fought in many city councils. The original meters were "wet" meters, using water, but in the 1850s firms began replacing them with more accurate dry meters.

The manufacture of coal gas provided the potential for firms to recover and sell by-products such as coke, tar, light oil, and ammonia. However, unlike Great Britain, there was no market for such products in the United States (aside from the coke) before the late 19th century. The United States lacked a coal-based chemical industry and was almost completely dependent on imports for dyestuffs, drugs, and explosives and other products of coal tar and light oil origin. Ammonia was regarded as a nuisance until after the Civil War, when ammonium sulfate was used for fertilizer. Gasworks themselves burned by-product coke under their retorts and sold some on the open market, but it had limited appeal.

By the 1870s, over 400 manufactured gas utilities existed and the industry had spread throughout the country. Over half of the firms, with over 50% of the labor and capital, were located in just four states: New York (71), Massachusetts (52), Pennsylvania (43), and Ohio (32). Price, however, still limited home gas service to a relatively affluent segment of the market. Gas companies had originally hoped to make large profits from street lights, but municipal franchises applied stringent regulations as to the numbers of public street lights and rates paid, and many firms defaulted on this part of the contract or were continually petitioning city councils for rate increases. Because costs were high and margins low, gas companies were reluctant to expand to lower income markets, putting the industry at risk from lower priced fuels. In the 1850s, coal oil lamps provided much of the lighting needs of those who could not afford gas, the fuel supplied from refineries concentrated in Pittsburgh and Ohio. These lamps furnished up to 14 cp and were quite flexible to use. In the post-Civil War period, however, two new competitors for manufactured gas emerged to challenge the industry for the lighting market—kerosene and electricity.

3. Innovations and Challenges from Competing Fuels, 1875–1927

Substantial competition for urban lighting markets developed in the half century from 1875 to 1927.

Municipalities and states also subjected the industry to increased regulation. The first competitor for the lighting market was the kerosene lamp, an outcome of the discovery of oil in Pennsylvania in 1859, which largely captured the low end of the illumination market. Improvements in petroleum refining and transportation techniques and the discovery of new sources of supply resulted in reduced oil prices. By the 1870s and 1880s, most of the urban working class, as well as a substantial part of the middle class, used kerosene lamps for domestic illumination.

Electric arc lamps began competing with manufactured gas for the street lighting market in the 1880s. Since gas companies found these lighting contracts unprofitable, it was a limited loss. By the 1880s, domestic lighting accounted for 90% of gas revenues and arc lamps posed no threat in this market because their intensity made them unusable in the home. In 1881, however, Thomas Edison, the inventor of the incandescent lamp, developed a central power station in New York City that supplied power for home lighting that directly competed with interior gas light. Edison had wisely based his distribution system on that of the gas industry, constructing an "electrical analogue to the gas system."

The manufactured gas industry responded to this challenge with the development of carbureted water gas. The concept behind water gas, which was first observed in 1780 and patented in France in 1834, involved the action of steam on incandescent carbon, usually anthracite coal or coke. When the steam, composed of oxygen and hydrogen, came into contact with the incandescent carbon, it produced a fuel gas composed of hydrogen and carbon monoxide, as well as other products. The gas had a heating value of approximately 300 Btu/ft^3, burned with a blue flame, and had little illuminating power. The last characteristic diminished the utility of water gas for commercial use, because the market for manufactured gas in most of the 19th century was almost entirely in lighting.

In 1875, however, Thaddeous S. C. Lowe, a Pennsylvania inventor, patented a process to enrich the blue gas by spraying it with liquid hydrocarbons and thermally cracking the gas to form carbureted water gas and tars. The Lowe process was acquired by the United Gas Improvement Company (UGI) of Philadelphia, formed in 1882. Before this time, utilities had been slow to adopt the water gas process and the states of Massachusetts and New Jersey actually outlawed water gas because of gas poisoning incidents. UGI, however, aggressively pushed the technology and the rapid growth of the petroleum industry made oil available at a relatively cheap price. The gas industry proceeded to shift a substantial amount of its manufacturing processes from the production of coal gas to carbureted water gas. Originally carbureted water gas manufacturers used naphtha as the process oil because it was relatively inexpensive, but after 1895, as rising demand caused the price of naphtha to rise, gas manufacturers switched to the petroleum fraction known as gas oil. The invention and adoption of the automobile increased the demand for the higher petroleum fractions even further and in ∼1930 the carbureted water gas industry switched to a yet cruder grade of petroleum.

The use of oil rather than coal as the feedstock for gas production was another major step by the manufactured gas industry. Oil gas was made by the thermal decomposition or cracking of petroleum. Whale oil had been used in several gas plants early in the 19th century, but it was not until 1889 that a modern refractory process to use petroleum was invented. L. P. Lowe, the son of the discoverer of carbureted water gas, was the inventor. The first oil gas plant was constructed in California and in 1902 an oil gas plant was erected in Oakland, California for lighting purposes. The so-called Pacific Coast oil gas process was used mostly in Pacific coast cities—an area that was poor in coal but rich in petroleum.

The number of carbureted water gas plants in the United States grew steadily. In 1890, the census reported returns from 742 gas manufacturing firms, probably approximately 75% of the total. In this census, coal gas represented 46.7% of the total product, whereas water gas made up 38% of the total. Between 1890 and 1904, coal gas production dropped precipitously to 11.2% of total gas produced, straight water gas rose to 48.6%, and a new category, mixed coal and water gas, absorbed 36.4% of the market. Between 1904 and 1914, coal gas declined even further, from 11.2 to 5.2% of total gas production; carbureted water gas also declined from 48.6 to 44.2% and mixed gas advanced from 36.4 to 42.4%. Because Great Britain (the first water gas plant was built in 1890) and other European nations did not have access to inexpensive petroleum, they did not adopt carbureted water gas to the same extent as did the United States. The U.S. gas industry, therefore, became increasingly dependent on oil, whereas the European industry remained largely based on coal.

Carbureted water gas had a number of advantages over coal gas. Its production required less space, less capital, and less labor, as well as a smaller storage capacity, because plants could be started up in a relatively short time. Many coal gas plants installed auxiliary carbureted water gas machines to increase their flexibility in regard to load, mixing the two gases. Without the auxiliary water gas equipment, firms would have had to add additional storage or construct additional retorts, which would have been idle for many hours in the day. Carbureted water gas also had a higher illuminating value and a flexible heating value; a plant could produce gas varying in heating value from 300 to 800 Btu/ft^3. Utility operators could produce carbureted water gas with widely varying Btu contents by changing the amount of oil cracked into the blue gas. Higher Btu carbureted water gas could be mixed with gas with a lower Btu grade from blue gas or coal gas retorts, thereby increasing the available gas volume without the necessity of increasing the number of water gas sets.

The addition of petroleum hydrocarbons to water gas provided a gas that had illumination values equal to those of coal gas but at a lower cost. Most municipal franchises to gas companies required illuminating power of between 15 and 17 cp. The manufactured gas burners most commonly used before 1886 were the bat-wing, the fish-tail, and the Argand burners. These burners depended on the luminosity of the flame in order to fulfill legal candlepower requirements. Therefore, when carbureted water gas was burned as a fuel, the burners produced an equal or better flame at a lower cost. The lamps, however, still depended on the incandescence of the carbon particles in the flame itself for their illuminating power.

In 1885, Carl Auer von Welsbach of Germany, working in the laboratories of Robert Wilhelm von Bunsen, produced another major innovation in the consumption of manufactured gas. Welsbach patented a gas mantle made of cotton fabric impregnated with rare earth. When the mantle was placed near a gas flame from a Bunsen burner, it "became incandescent," producing a steady white light of 30–50 cp. This increased the light that could be obtained from a given quantity of gas approximately six times, utilizing the principle of incandescence rather than the direct illuminating properties of the gas. Thus, it mirrored the innovation of the gas industry's chief competitor, the incandescent electrical lamp. The Welsbach mantle provided an inexpensive and superior incandescent lamp to meet the competition of the electric light and helped modernize an industry whose existence was under severe competitive threat.

Although carbureted water gas was necessary for the older burners in order to meet existing legal candlepower and calorific requirements, it did not have to be used with the Welsbach mantle, which produced an equally brilliant light with cheaper blue water gas. Yet, legal candlepower requirements for manufactured gas required that higher priced carbureted water gas rather than blue gas had to be used with the Welsbach mantle. At the time of the mantle's introduction, municipal franchises usually stipulated a candlepower standard for gas ranging from approximately 14 to 16 for coal gas and 18 to 36 for water gas. The illuminating power of the Welsbach Mantle, however, which was not dependent on a higher candlepower gas rating, made these standards obsolete.

Statutory change to accommodate the new technology occurred slowly and some utilities actually opposed shifting to a different standard because they feared a possible profit decline. Beginning with the Wisconsin Public Utilities Commission in 1908, however, states and municipalities began requiring that gas be measured in Btu or heat units per cubic foot of gas (as was done in Great Britain) rather than lighting units. Of the 1284 gas firms reporting in the 1914 Manufacturing Census, 1055 reported candlepower standards and 811 reported standards expressed in Btu heating values. By 1919, there had been a substantial change. Of 1022 establishments reporting, only 389 reported candlepower standards, whereas 737 reported Btu heating value standards.

The use of gas solely for lighting resulted in poor load factors for gasworks. Although they needed generating, storage, and distribution equipment capable of handling demand during peak lighting periods, their capital equipment was often utilized for only 10–12 h per day. The use of larger gasholders (gasometers) was one way utilities could reduce investment in gas-producing equipment. Smaller plants frequently ran their equipment during part of the day and relied on their gasholders to supply gas at night. Larger companies could not risk supply cutoffs and had to run separate units of gas producers, even though they too used gas holders for reserve. From the 1870s through the 1890s, several important improvements were made in gasholders, including the use of iron and steel tanks and the installation of the first three-lift holder tank. Because of limited capacity, however, gas holders were not a satisfactory solution to the load factor.

Competition from electricity, despite the success of the Welsbach mantle, consistently ate into the gas industry's market share. Utilities could attempt to spread their loads as a way to meet competition and to use their technology more efficiently. The gas industry moved increasingly into markets outside of lighting, such as cooking stoves, water heating, and house heating, although firm aggressiveness varied from place to place. In both Baltimore and Philadelphia, for instance, both large markets, the gas companies only gradually moved away from lighting as their only product. In other cities, however, such as Denver and Los Angeles, utilities introduced gas-consuming appliances more rapidly. Two technical innovations aided in the expansion of these markets: the Bunsen burner (1885), which premixed gas with air, and thermostatic controls, initially applied to water heating in 1899. This movement away from light and into household appliances reflected the recognition that manufactured gas was a fuel or energy source. It was also a "technical mimicry" of electricity and an attempt by an older technology to survive by copying a more recent technology, in this case electricity.

Even though the gas industry developed a substantial market share in the home utility domain, it still faced excess productive capacity. In ~1900, gas firms began to try to capture industrial customers to absorb this capacity. Efforts were made to market gas engines as well as to sell gas for heating purposes. Potential industrial customers, however, often found it more economical to use low-Btu gas to meet their energy requirements rather than high-Btu carbureted water gas. Frequently they used producer gas, a low-grade but inexpensive gas made on the firm's site from coal. Gas companies also found it difficult to enter industrial markets because of state requirements for a fixed number of Btu per cubic foot (usually 450–550 Btu) in their gas, which raised gas prices. This Btu rating was higher than many industries desired or would pay for, limiting the number of industrial customers. The gas industry, therefore, began to push for the sale of gas with Btu as a standard and sold by the therm (100,000 Btu), as was done in Great Britain, rather than by the cubic foot. Industries could then obtain less costly low-Btu gas to supply the heat they needed in their processes. Changes in this regard were reported in the 1920 Census of Manufacturers. The census noted that the prevailing Btu standard was lower in 1919 than it had been in 1914, as the number of gas firms providing high-Btu gas dropped sharply. Between 1919 and 1927, the industrial use of manufactured gas increased from 70.4 billion to 136.4 billion ft^3.

Gas utilities could also compete with electricity by investing in electrical generating equipment or by acquiring electrical companies. In 1887, approximately 40 gas companies supplied electric lighting, usually arc lights; by 1889, the number was 266, or approximately 25% of U.S. gas companies, and in 1899, 362 or nearly 40% of gas companies provided electric lighting. A counter trend was for electrical utilities to acquire gas utilities, and large electric firms or holding companies made acquisitions of small gas companies increasingly in the 20th century.

Water gas possessed another disadvantage compared to coal gas in terms of potential profits—the character of its by-products. Markets had developed in the late 19th century for coal gas by-products such as tar for creosote in railway tie preservation, pitch for roofing and waterproofing, and refined tar for road surfacing. The carbureting process produced gaseous, liquid, and solid by-products, but they were not as salable as those from coal gas. The liquid was "water-gas tar," which was lighter and contained less valuable constituents than coal tar. The solid was coke, the sale of which could substantially affect the profit position of plants. Water gas coke, however, was lower in thermal efficiency than coal gas coke. In addition, water gas firms, rather than entering the coke markets, consumed the coke they produced as water gas fuel. Finally, unlike coal gas, the water gas process did not produce ammonia, cyanides, or phenolic compounds as by-products. The ammonia was especially valuable and could constitute 6–10% of the cost of the coal used. Before the invention of the Haber process for the synthetic production of ammonia, coal carbonization was the main source of fixed nitrogen, and coal carbonization processes found increasing favor in the 1920s among some gas producers.

Between 1914 and 1929, the amount of manufactured gas distributed expanded from approximately 204 billion to over 450 billion ft^3. A little more than 50% of the gas produced in 1929 was water gas, approximately 40% was classified as coal gas, although the sum also included gas from by-product ovens, and oil gas was 6.2%. The most significant new addition to the census categories was gas from by-product coke operations.

Between approximately 1916 and 1929, a revolution occurred in the technology used by the iron and steel industry to make coke. From 1850 through 1910, most of the coke manufactured for use in blast furnaces was carbonized in beehive coke ovens. These ovens produced a high-quality coke from bituminous coal, but wasted all of the coal by-products. The byproduct coke oven had originally been developed in Europe, and beginning in the 1890s several byproduct coke plants were constructed in the United States. The real expansion of the United States industry, however, occurred just before World War I and in the following decade. The by-product coke oven captured the valuable coal chemicals for use and also produced a gas surplus. That is, by-product ovens could be heated by 35 to 40% of the coal gas produced in the oven, leaving approximately 60% to be marketed. Coke oven gas had a heating value of approximately 560 Btu and could be easily sold to other customers. By 1932, by-product coke firms produced 25% of the manufactured gas distributed by utilities to domestic, commercial, and industrial customers in the United States.

During the 1920s, the manufactured gas utilities themselves moved into the by-product industry. In order to meet peak period requirements, utilities with carbureted water gas technology had to manufacture large amounts of coke. By utilizing by-product ovens rather than retorts, utilities could combine gas manufacture with coke manufacture to meet peak load requirements and to produce a higher quality of coke for sale than that produced in retorts. During the 1920s, construction firms such as Koppers Company and Semet-Solvay built coke oven batteries for a number of large utilities, including Brooklyn Union Gas, Consolidated Gas Company of New York, New Jersey Public Service Gas & Electric, Rochester Gas & Electric Company, and People's Gas & Coke Company of Chicago. In addition, entrepreneurs built a number of merchant plants that had as their chief products foundry coke, domestic coke, and coke for making water gas, as well as producing gas to sell under contract to public utilities. These plants had the advantage of flexibility in meeting demand peaks for either gas or for coke as

well as lowering the cost of gas generation. By the end of the decade, a clear trend existed for manufactured gas utilities to adopt carbonization equipment that produced both high-quality gas and high-quality coke, and by 1932, 18.7% of the manufactured gas produced by gas utilities was coke oven gas. Thus, as a 1926 text on fuels noted, "the gas and the coke industry gradually are merging as far as equipment requirements are concerned, and… the tendency is toward the same type of plant for both branches."

4. The Transition to Natural Gas, 1927–1954

Natural gas is a substance that occurs in nature under three conditions: in porous domes that form a gas field (nonassociated gas), in association with petroleum deposits (casinghead or associated gas), and mixed with the oil so that it must be trapped out of it. Chemically, there is no difference between natural gas and manufactured gas—each is composed of hydrocarbons—but they differ in heating value. Manufactured gas usually averages between 550 and 600 Btu, whereas natural gas is approximately 1000 Btu. Natural gas is not distributed evenly throughout the environment. In the late 19th and early 20th centuries, there were four regions where natural gas was piped in for use in cities and industries: western Pennsylvania and West Virginia, northern and central Indiana, locations around Los Angeles including the San Joaquin Valley, and eastern Kansas. Towns and cities close to these fields utilized the natural gas until it was exhausted.

In the first decades of the 20th century, additional natural gas discoveries were made in the Southwest, but gas from these fields was initially wasted because of the absence of nearby markets and because the technology to produce pipelines long enough to reach markets was nonexistent. In the 1920s, however, significant breakthroughs occurred in the making and laying of welded pipe technology. In addition, the mathematical understanding of the role of tensile steel strength, pipe thickness and diameter, pipeline capacity, gas pressure, and compressor station spacing to pipeline capacity greatly increased. Pipeline construction passed beyond the craft stage into the stage where decisions were made by mathematical analysis.

From 1925 to 1935, pipeline companies and utilities constructed a number of pipelines ranging from 200 to over 1000 miles in length, connecting southwestern and California gas fields to urban markets. Cities such as Houston and Beaumont (1925), Wichita (1927), Denver (1928), St. Louis (1928), Atlanta (1929), San Francisco (1929), Minneapolis (1930), Chicago (1931) and Washington, DC (1931) were supplied with natural gas, driving out manufactured gas. The depression and the war slowed pipeline construction, but in 1944 the Tennessee Gas Transmission Company completed a 1265-mile pipeline from the Texas Gulf Coast to West Virginia as part of a wartime push to replace diminished Appalachian supplies.

The last major regions to be linked by pipeline with natural gas fields were the Middle Atlantic States and New England. New York City and Philadelphia were considered especially prized markets, since New York City alone consumed 40% of the nation's manufactured gas. These cities were supplied with natural gas in 1947 through the conversion of Big Inch and Little Inch, wartime government-constructed petroleum pipelines, to natural gas and the sale of these pipelines to private enterprise. New England was the last major region to receive natural gas, with Boston being the most distant urban market to draw from the southwestern gas fields. The postwar period also saw a boom in the construction of pipelines serving California and Florida, as well as new major pipelines to the midwest.

In these new markets for natural gas, the logistics of conversion of manufactured gas distribution systems into natural gas distribution systems was a major undertaking. Manufactured gas utilities did not always view conversion with favor, even though there seemed to be many advantages to the change. Some utility managers worried that natural gas supplies would prove insufficient to maintain their service commitments. Corporations were also reluctant to abandon their substantial investments in gas manufacturing plants and equipment.

In the conversion to natural gas as a fuel, the gas could be circulated in the same distribution system and consumed by the same appliances used by manufactured gas. However, because of the different Btu ratings of the gases, conversion required gas main and appliance adjustments. In addition, engineers and managers had to decide whether to use natural gas or natural gas mixed with manufactured gas after conversion. Occasionally regulatory commissions overturned the decisions made by utility managers.

The 1946 Federal Natural Gas Investigating Committee recommended that natural gas be used primarily for enrichment purposes. This strategy would expand the production capacity of existing plants, provide backup in case of supply interruptions, and furnish peak load shaving abilities. A number of firms followed the recommendation for mixed gas. In 1945, for instance, 301.4 million therms of manufactured gas and 988.5 millions therms of natural gas were used to produce mixed gas; by 1954, this total had reached 936 million therms of manufactured gas and 2012.7 million therms of natural gas. Other utilities, especially after World War II, went directly to natural gas, usually to avoid the costly necessity of adjusting gas appliances twice. In many cases, utilities kept their gas-making

equipment for emergencies and for peak shaving purposes.

Once the decision concerning the type of gas to be used had been made, utilities faced major organizational tasks of conversion and adjustment, regardless of whether mixed or natural gas had been chosen. In 1931, an American Gas Association Subcommittee issued a report on "The Study of Factors Involved in Change-over from Manufactured to Natural Gas." The committee made four major recommendations: (1) that cities be divided into districts for conversion purposes and that each district be converted to natural gas separately; (2) that those handling the conversion task be thoroughly trained and that the work force be large enough so that the largest district could be covered in 1 week; (3) that customers be kept fully informed at all times of the conversion process; and (4) that a universal adapter be used to adjust orifices. The committee also warned of the drying-out effects of the natural gas on joints and valves in a system that had formerly only carried wet manufactured gas, with resultant leaks. The committee concluded, however, that compared to manufactured gas, natural gas presented fewer hazards. In approaching the conversion process, utilities communicated extensively with one another regarding the pitfalls to be avoided.

The largest single task facing a utility converting to natural gas was adjusting domestic and commercial appliances to the new gas. All gas-using appliances had to be adjusted at the orifice because of natural gas' higher Btu rating and its slower rate of flame propagation compared to 550–600 Btu for manufactured gas. Management of conversion could involve thousands of specially trained workers. In order to carry out the conversion tasks, the larger utilities hired engineering management firms. Some utilities, however, managed their own transitions rather than hiring outside contractors, forming schools and training men specifically for the job and relying on experience from other utilities.

With the transformation in the gas industry, the fuel flow now originated from a distant source rather than traveling a relatively short distance from gasworks to customer. From a regulatory point of view, this shifted responsibility to the federal government because pipelines often crossed state boundaries. Pipeline companies became subject, after the Natural Gas Act of 1938, to federal control, which had a constraining and shaping effect on the industry.

5. Environmental Impacts of the Manufactured Gas Industry

The various by-products produced by the manufactured gas industry, if not captured and disposed of in a safe manner or sold, could become environmental externalities, damaging surface and groundwater quality, poisoning soils with toxic substances, and producing odors and smoke. British gas engineers were aware during the industry's early years of the potential for environmental pollution. In London, for instance, in 1821, effluents from gasworks killed fish and eels in the Thames. In 1841, in his *Practical Treatise on the Manufacture of Gas*, published in London, engineer Thomas Peckston warned of the possible escape of "tar and ammonical liquor" from the "tar cistern" because it could "percolate through the ground" and render water in springs and wells "unfit for use."

Americans also became aware of the injurious environmental effects of manufactured gas effluents on the environment as the industry expanded. New York State, for instance, passed a statute in 1845 restricting the disposal of gas house wastes in streams in the New York City area; Pittsburgh, Pennsylvania passed a similar statute in 1869 and New Jersey approved legislation limiting the disposal of gas house wastes in streams in 1903 and 1911. Plaintiffs filed numerous nuisance suits against manufactured gas companies in the late 19th and early 20th centuries for offenses such as odors, pollution of wells, damage to trees, and the killing of finfish and shellfish. Various city councils and state bodies undertook investigations of the environmental effects of manufactured gas wastes as well as their pricing policies beginning in the middle of the 19th century. Warnings about the pollution potential of gas house wastes appeared with increasing frequency in the professional literature during the first decades of the 20th century.

Because of the potential costs of nuisance suits as well as the damage to the industry's reputation, in 1919 the American Gas Association established a committee on the "Disposal of Waste from Gas Plants." The 1920 report of the committee noted that gas plant wastes could drive "away fish and damage oyster beds," create "damage to paint on pleasure boats," cause "objectionable odors," result in "pollution of wells [by] contaminating ground water," create "deposits in sewerage systems," and cause "pollution of drinking water supplies where water is chlorinated." The committee provided numerous recommendations to help manufactured gas plant operators deal with pollution disposal questions and even volunteered to visit sites with problems. The committee continued to meet throughout the 1920s and to explore these issues.

The gradual phasing out of the manufactured gas industry in the 1950s and 1960s because of the conversion to natural gas resulted in the elimination of many plants and a gradual loss of institutional memory concerning their operations. The renewed interest in land and groundwater pollution that developed after Love Canal in 1978 and the subsequent passage of the Superfund Acts, however, focused renewed attention on the sites of former manufactured gas plants (MGPs), many of which had been stripped of gas-making technology and structures. Investigation of these sites by state and federal environmental officials identified a number of toxic by-products remaining in the soil and in groundwater from past MGP operations. Some of the MGPs had operated for a century or so and most sites contained accumulated wastes from the gas-making processes. Thus, an industry largely of the past that had operated approximately 2000 plants in towns and cities throughout the nation and provided it with fuel for light and heat for over a century was discovered to have left a heavy environmental burden for the future.

6. Conclusions

This article has focused on the evolution of the manufactured gas industry in the United States and the transition to natural gas as well as its environmental legacy. Manufactured gas systems were urban public utilities that, because they utilized city streets for their distribution systems, required first municipal and later state approval for their operations and rates.

Manufactured gas systems shared a number of characteristics with other urban networks, such as water supply, sewerage, and district heating. These were "grid-based systems" that required a special physical network devoted to supplying users. They were also centralized rather than decentralized and had natural monopoly characteristics, a factor that affected their rate-making and consumer relations and, at times, resulted in their becoming deeply involved in politics.

The substitution of natural gas for manufactured gas resulted in the gradual death of the manufactured gas industry as an entity that produced a product rather than merely distributing it. However, much of the original industry's grid remained in the form of inner-city distribution systems. The substitution of natural gas for manufactured gas did not require network reconfiguration but only adjustments because the delivery system and appliances from the first system were retained for use with the natural fuel.

The industry had large environmental effects, both during its period of operation and after, because of wastes discarded during its processing of coal and oil for purposes of producing a fuel for illumination and energy. As an industry it furnished both illumination and energy to American cities for many decades, but its significance today is measured primarily in terms of the environmental burdens it has left to contemporary society.

Further Reading

American Gas Association 1956 *Historical Statistics of the Gas Industry*. American Gas Association, New York

Anderson N Jr., DeLawyer M W 1995 *Chemicals, Metals and Men: Gas, Chemicals and Coke: A Bird's-Eye View of the Materials That Make the World Go Around.* Vantage Press, New York

Binder F M 1974 *Coal Age Empire: Pennsylvania Coal and Its Utilization to 1860.* Pennsylvania Historical and Museum Commission, Harrisburg, PA

Blake A E 1922 Water gas. In: Bacon R F, Hamor W A (eds.) *American Fuels.* McGraw-Hill, New York, Vol. 2, pp. 995–1015

Block E B 1966 *Above the Civil War: The Story of Thaddeus Lowe, Balloonist, Inventor, Railway Builder.* Howell-Nort Books, Berkeley, CA

Castaneda C J 1993 *Regulated Enterprise—Natural Gas Pipelines and Northeastern Markets, 1938–1954.* Ohio State University Press, Columbus, OH

Castaneda C J 1999 *Invisible Fuel: Manufactured and Natural Gas in America, 1800–2000.* Twayne Publishers, New York

Castaneda C J, Pratt J A 1993 *From Texas to the East: A Strategic History of Texas Eastern Corporation.* Texas A&M University Press, College Station, TX

Collins F L 1934 *Consolidated Gas Company of New York.* The Consolidated Gas Company of New York, New York

Davis R E 1935 Natural gas pipe line development during past ten years. *Natural Gas* **16**, 3–4

Eavenson H N 1942 *The First Century and a Quarter of American Coal Industry.* Privately Printed, Pittsburgh, PA

Elton A 1958 Gas for light and heat. In: Singer C, *et al.* (eds.) *A History of Technology.* Oxford University Press, New York, Vol. IV, pp. 258–76

Harkins S M, *et al.* 1984 *U.S. Production of Manufactured Gases: Assessment of Past Disposal Practices.* Environmental Protection Agency, Research Triangle Park, NC

Hasiam R T, Russell R P 1926 *Fuels and Their Combustion.* McGraw-Hill, New York

Hughes T P 1983 *Networks of Power: Electrification in Western Society, 1880–1930.* Johns Hopkins University Press, Baltimore, MD

Hunt A L 1900. Gas, manufactured. In: *U.S. Bureau of the Census. Twelfth Census of Manufacturers (1900), Part IV: Reports on Selected Industries.* p. 19. U.S. Government Printing Office, Washington, DC.

Hyldtoft O 1995 Making gas: The establishment of the Nordic gas systems, 1800–1870. In: Kaijser A, Hedin M (eds.) *Nordic Energy Systems, Historical Perspectives and Current Issues.* Science History Publications, Canton, OH, pp. 75–100

Jacobson C D, Tarr J A 1998 No single path: Ownership and financing of infrastructure in the 19th and 20th centuries. In: Mody A (ed.) *Infrastructure Delivery: Private Initiative and the Public Good.* The World Bank, Washington, DC, pp. 1–36

Kaijser A 1990 City lights: The establishment of the first Swedish gasworks. *FLUX* **1**, 77–84

Kaijser A 1993 Fighting for lighting and cooking: Competing energy systems in Sweden, 1880–1960. In: Aspray W (ed.) *Technological Competitiveness: Contemporary and Historical Perspectives on the Electrical, Electronics, and Computer Industries.* IEEE Press, New York, pp. 195–207

Kaijser A 1999 Striking bonanza: The establishment of a natural gas regime in the Netherlands. In: Coutard O (ed.) *The Governance of Large Technical Systems.* Routledge, London, UK, pp. 38–57

King T 1950 *Consolidated of Baltimore 1816–1950.* The Company, Baltimore, MD

Leinroth J P 1928 Industrial gas in the United States—Growth and future trends. In: *Transactions of the Fuel Conference, Vol. II, The Carbonisation Industry Utilization of Fuels.* Percy Lund, Humphries & Co, London, UK, pp. 1211–7

Matthews D 1987 The technical transformation of the late nineteenth-century gas industry. *J. Econ. History* **47**, 970–80

Passer H C 1953 *The Electrical Manufacturers 1875–1900.* Harvard University Press, Cambridge, MA

Passer H C 1967 The electric light and the gas light: Innovation and continuity in economic history. In: Aitken H G J (ed.) *Explorations in Enterprise.* Harvard University Press, Cambridge, MA

Platt H L 1991 *The Electric City: Energy and the Growth of the Chicago Area, 1880–1930.* University of Chicago Press, Chicago, IL

Rose M H 1995 *Cities of Light and Heat: Domesticating Gas and Electricity in Urban America.* Pennsylvania State University Press, State College, PA

Sanders M E 1981 *The Regulation of Natural Gas Policy and Politics, 1938–1978.* Temple University Press, Philadelphia, PA

Schivelbusch W 1988 *Disenchanted Night: The Industrialization of Light in the Nineteenth Century.* University of California Press, Berkeley, CA

Stotz L, Jamieson A 1938 *History of the Gas Industry.* Stettiner Bros, New York

Tarr J A 1998 Transforming an energy system: The evolution of the manufactured gas industry and the transition to natural gas in the United States (1807–1954). In: Coutard O (ed.) *The Governance of Large Technical Systems.* Routledge, London, UK, pp. 19–37

Tarr J A and Dupuy G (eds.) 1988 *Technology and the Rise of the Networked City in Europe and America.* Temple University Press, Philadelphia, PA

Thoenen E D 1964 *History of the Oil and Gas Industry in West Virginia.* Education Foundation, Charleston, WV

Thorsheim P 2002 The paradox of smokeless fuels: Gas, coke and the environment in Britain, 1813–1949. *Environ History* **8**, 381–401

Troesken W 1996 *Why Regulate Utilities: The New Institutional Economics and the Chicago Gas Industry, 1849–1924.* University of Michigan Press, Ann Arbor, MI

Tussing A R, Barlow C C 1984 *The Natural Gas Industry: Evolution, Structure, and Economics.* Ballinger, Cambridge, UK

Weber F C 1922 The future of the artificial gas industry. In: Bacon R F, Hamor W A (eds.) *American Fuels.* McGraw-Hill, New York, Vol. 1, pp. 1095–106

Williams T I 1981 *A History of the British Gas Industry.* Oxford University Press, Oxford, UK

Williamson H F, Daum A R 1959 *The American Petroleum Industry: The Age of Illumination 1859–1899.* Northwestern University Press, Evanston, IL

Joel A. Tarr

N

Natural Gas, History of

Glossary

Appalachia   The mountainous region stretching from northern Mississippi to southern New York and commonly characterized by rural communities and poverty.

British thermal unit (Btu)   The amount of heat required to change the temperature of 1 lb of water 1°F at sea level.

Federal Energy Regulatory Commission (FERC)   The successor to the Federal Power Commission; created in 1977.

Federal Power Commission (FPC)   The federal regulatory agency responsible for regulating the interstate natural gas industry; created in 1920 and abolished in 1977.

hydrocarbons   Organic compounds that are composed entirely of carbon and hydrogen. Petroleum products are composed of hydrocarbons, and methane, or natural gas (CH_4).

Insull, Samuel (1859–1938)   A leader in the American public utility industry during the early 20th century. He was born in London and served as Thomas Edison's secretary as a youth.

manufactured coal gas   A fuel gas for illuminating and heating purposes produced by heating coal in a retort and capturing the resulting vapors for distribution.

waste gas   Natural gas, considered to be a nuisance in the production of oil.

Natural gas is a vital fuel for modern society. During the last 50 years of the 20th century, natural gas satisfied approximately 25% of U.S. energy demand. It has been used for electric power generation, industrial heat processes, domestic heating and cooking, and transportation fuel. Natural gas is composed primarily of methane, a hydrocarbon consisting of one carbon atom and four hydrogen atoms (CH_4). As a "fossil fuel," natural gas is rarely pure. It is commonly associated with petroleum and often contains other hydrocarbons, including butane, ethane, and propane. In the United States, substantial natural gas utilization did not begin until after the discovery of large quantities of both crude oil and natural gas in western Pennsylvania during 1859.

1. Origins

Natural gas was observed and utilized in limited quantities during ancient times. References in literature to burning springs, burning bushes, or perpetual lights suggest that natural gas was used, albeit rarely, for heating. In ancient China, burning gas springs heated brine water in order to extract salt, and there were flaming gas springs in Greece and Rome. Recorded observations of burning springs in France, Italy, and Russia also exist. The philosopher Plutarch and theologian St. Augustine described lights that may have been produced by burning natural gas.

In colonial America, both George Washington and Thomas Jefferson observed natural gas springs. During the autumn of 1770, Washington participated in an expedition along the Ohio and Kanawha rivers in West Virginia and Ohio. Near the present-day town of Pomeroy, Ohio, Washington described a location "wch. the Indians say is always a fire." About perhaps the same site, Thomas Jefferson recorded his observations of "a hole in the earth... from which issues constantly a gaseous stream." Other visitors to these springs reported that hunters used them to cook food.

Through the early 19th century, these "burning springs" had little practical or widespread use. Most importantly, there was no practical method of either capturing the gas emanating from the springs or storing and redirecting its flow through piping.

2. Natural Gas in Fredonia, New York

Residents of Fredonia, New York, were perhaps the first Americans to use natural gas for lighting on a regular basis. Gas springs in the vicinity of Fredonia had been observed in the early 1800s, but it was not until the mid-1820s that a local gunsmith named William Aaron Hart organized an apparently successful effort to utilize gas from the local gas spring to provide light for local homes and establishments.

Some accounts written much later state that citizens of Fredonia used natural gas to illuminate their town when the French military leader Marquis de Lafayette visited. Lafayette toured America during the years 1824–1825, and he traveled to New York in the summer of 1825. Of his visit to Fredonia, Lafayette's private secretary, A. Levasseur, recorded that they had observed a great many lights in the town. The local newspaper featured a story on the same events and noted lamps and chandeliers that provided illumination in the town during Lafayette's visit. Contemporary reports of Lafayette's visit do not mention gaslights at Fredonia; only the accounts of this event written much later mention gas lighting. Lafayette's secretary did note gas lighting at other locations. While in a Boston theater, Levasseur recorded observations of "gas blazing abundantly from numerous pipes, and throwing floods of dazzling light over the hall." These lights were fueled by manufactured coal gas, however, and not natural gas.

After Lafayette's visit, William Hart continued to develop his interest in natural gas in Fredonia. During 1827, he began work on a plan to supply natural gas to a lighthouse at nearby Barcelona Harbor. After the U.S. government granted him a contract for this service, he installed a primitive gas works. It consisted of a fish barrel placed over the gas spring located at Westfield along Lake Erie. The barrel served as a "gasometer," or gasholder. Hart sealed the gasometer and transported the gas for one-half mile through hollowed out pine logs to the lighthouse. Gas from the spring provided enough fuel to illuminate 144 burners and create a bright light.

3. Early Commercial Utilization

It was not until Colonel Edwin Drake discovered oil in Titusville, Pennsylvania in 1859 that natural gas became a significant source of energy in the United States. Although Drake had been searching for oil, he found natural gas as well; oil and natural gas are often found in the same geologic structures. Natural gas discovered in eastern Pennsylvania was marketed to regional customers. Therefore, the Drake discovery heralded the beginning of both the modern U.S. oil and natural gas industries.

Prior to Drake's discoveries, there were few successful long-term attempts to utilize natural gas for either industrial or commercial purposes. By the mid-19th century, only those factories or towns located very near a natural gas well could utilize the fuel. The difficulty of containing a natural gas spring, storing the gas, and transporting it over long distances limited its utility. For example, significant natural gas discoveries such as the high-volume well discovered by William Tomkins in 1841, near Washington's burning spring on the Canadaway Creek, attracted some attention but little commercial interest. Alternatively, manufactured coal gas plants could be built and operated anywhere as long as coal, or the feedstock, was readily available. Thus, the manufactured

Table I
Introduction of Manufactured Gas to Major Cities[a].

Year	City
1816	Baltimore
1825	New York City
1829	Boston
1832	Louisville
1835	New Orleans
1836	Philadelphia
1843	Cincinnati
1846	St. Louis
1849	Chicago
1854	San Francisco
1867	Kansas City
1867	Los Angeles
1871	Minneapolis
1873	Seattle

[a]The dates primarily reflect the year manufactured gas was first produced in the city for commercial use. In some cases, however, the date reflects when a city charter was granted, and charters were sometimes granted before and even just after gas service began.

coal gas industry developed much more quickly than that of natural gas in the 19th century. By the mid-19th century, many towns and cities had a manufactured gas plant and a local distribution system that provided some coal gas for residential and business lighting. (see Table I).

The earliest recorded use of gas for industrial purposes in the United States occurred in 1840, near Centerville, Pennsylvania. The gas was used to distill salt from brine water. Gradually, in the 1860s and 1870s, local deposits of natural gas were utilized for a variety of industrial heating applications. Even in Fredonia, New York, where some residents and shop owners utilized natural gas for lighting beginning in the mid-1820s, a formal natural gas company was not organized for many years. In 1858, businessmen established the Fredonia Gas Light and Water Works Company to operate the local gas wells and discover new ones.

Natural gas was not used on a large scale until the 1880s, and gas wells were most likely to be abandoned when oil was not concurrently discovered. An example of early abandonment occurred in 1865 when a 480-foot drilling effort struck a natural gas reservoir near West Bloomfield, New York. The operators estimated the gas flow to be about 2000 cubic feet (mcf) per day; they directed the gas into a large balloon and attempted to measure the flow by calculating the time required to fill it. Because the investors were disappointed that oil was not discovered, they abandoned the project. Not everyone was disappointed that this well contained only natural gas. Several businessmen formed the Rochester Natural Gas Light Company and purchased the same gas well in 1870. The nearest town desiring natural gas was Rochester, about 25 miles away. The company constructed a pipeline system to connect the well with the town. They built a pipeline out of Canadian white pine. The 2- to 8-foot log segments were planed to a uniform 12.5-inch exterior diameter, and they were bored for an 8-inch interior diameter. Construction and maintenance of the wood pipeline system was particularly problematic, but the company began transporting natural gas during the winter of 1872. Consumers in Rochester discovered quickly that hotter burning natural gas was not easily interchangeable in their burners with manufactured coal gas. This situation resulted in lower than expected natural gas demand. Continuing problems with gas transportation facilities caused significant problems for the company; rotting and leaking wood pipelines simply prevented the adequate transportation of natural gas from well to consumer. Soon, the company stopped operations.

A more successful attempt to transport natural gas took place in 1872. New natural gas discoveries created a demand for specialized gas pipelines. In this case, a 2-inch wrought-iron line was constructed and used to connect a gas well 51 miles away. The line transported "waste gas" from nearby oil fields to Titusville. This pipe transported 4 million cubic feet (mmcf) per day to 250 customers, both residential and industrial.

The primary obstacle to the expansion of the natural gas industry in the mid-19th century was inadequate pipeline facilities and technology, not lack of supply. Hollow log pipelines leaked and disintegrated, but cast and wrought iron lines also suffered from significant intrinsic defects. Wrought iron lines in the period 1872–1890 were typically less than 8-inches in diameter, and the pipe segments were attached with couplings tightened with screws. Gas leaks were common problems. Most of the gas transported in pipelines during this period flowed under the natural pressure of the well without the aid of additional compression.

4. Natural Gas in Pittsburgh

Pittsburgh became the first major U.S. city in which industry utilized large volumes of natural gas for industrial heat processes. Abundant Pittsburgh area coal deposits and the importation of iron ore from the Juanita region in central Pennsylvania (and later from the Mesabi range) facilitated development of a substantial iron industry originally fueled by coal. Extensive coal burning for industrial heat created significant air pollution, and Pittsburgh became known as the "Smoky City." Contemporary newspaper articles noted the black smoke produced by burning coal. In 1884, *The New York Times* reported that natural gas would be used in Pittsburgh's industries to reduce coal smoke pollution.

The earliest recorded use of natural gas in a Pittsburgh iron works occurred in 1870–1871, but widespread natural gas utilization did not commence until the early 1880s, after the development of nearby gas wells. Entrepreneurs then organized new regionally based gas firms. One group of Pittsburgh area manufacturers established the Chartiers Valley Gas Company in 1883 to transport natural gas from local gas fields to their glass and steel plants. This company's first line extended from the Hickory gas field to Pittsburgh. The wrought-iron line was the first "telescoping" pipeline, meaning that a smaller diameter pipe installed at the well's origin led to a larger diameter pipe in the city. The telescoping line system was useful for lowering the gas line pressure as gas flowed into the city. For pipe less than 12 inches in diameter, the typical connection was a screw coupling. Pipe segments were threaded on the outer edge of each length end that turned into a screw coupling.

As of 1886, the Chartiers firm also laid claim to operating the largest continuous pipe in the world. The company installed a 16-inch pipe extending from the Murrysville gas field to Pittsburgh. After 8 miles, the 16-inch line was fed into a 6-mile-long 20-inch pipe, and it in turn fed into a 5-mile section of 24-inch cast-iron tubing, tested at 300 pounds per square inch (psi). The National Tube Works constructed this line; J. P. Morgan controlled National Tube.

By the late 1880s, Pittsburgh had become the locus of the American steel and coal industry, and it was also the center of the natural gas industry. In 1886, there were 10 iron and steel mills using natural gas in their puddling furnaces, with many more planning to convert to gas. Six glass-making factories and reportedly every brewery in Pittsburgh used natural gas instead of coal. The Sampson Natural Gas Crematory also used the invisible fuel. Pittsburgh received its natural gas from the lines of six companies tied into 107 regional gas wells. Five hundred miles of pipeline transported natural gas from wells to the city, including 232 miles of line within the Pittsburgh city limits.

As natural gas utilization increased, local engineers addressed technological problems associated with its transportation. Solomon R. Dresser focused attention on drilling and oil-field-related technology. In 1880, he formed S. R. Dresser & Company and conducted pioneering work in pipe coupling. In 1887, Dresser received a patent for using a rubber ring in pipe joints to create a leakproof coupling. Although this method proved not entirely satisfactory, less than a year later Dresser designed a second coupling that was more effective. He developed a two-part mechanical device that pulled the pipe segments together. Between the tightened sections, an internal rubber ring created a seal. Dresser proved the leakproof qualities of this coupling method when he developed his own gas field near Malta, Ohio and used the couplings in a gas line that extended into town.

The Malta Natural Gas Line established Dresser as a leader in the natural gas business, and his couplings attracted widespread favor; gas companies located throughout the country ordered them. As much as 90% of the gas pipeline industry used these couplings into the 1920s.

Improved couplings not only reduced leakage, they also lessened the possibility of explosions. In Pittsburgh, city ordinances prohibited gas lines from operating at pressures higher than 13 psi. This pressure limitation was intended to reduce the leaks prevalent in more highly pressurized lines. Leaks often resulted in accumulations of gas in cellars, leading to explosions and fires. Within the city, regulating valves further reduced the gas pressure. To prevent leaking gas from ending up in residential cellars, the Chartiers Valley Company used its patented "broken stone escape system." This system involved laying a pipe in a trench filled with dirt to the center of the pipeline. Workers then placed about 9 inches of broken stone on top of the line. A layer of tarpaper was then placed over the stone; dirt covered the tarpaper. The stone barrier was placed adjacent to every city lamppost so that escaping gas could vent through the stone. In addition, gas firms used "escape pipes," very small diameter lines leading from each pipe joint to a lamppost. Inspectors checked each escape pipe for possible leaks and identified the joint to which each escape line was connected. A system of 4-inch pipes distributed gas to individual residences. In these pipes, gas pressure was limited to about 5 psi. As the gas entered homes, an additional regulator/shutoff valve lowered gas pressure again to about 5 ounces per square inch, so that gas could be burned satisfactorily in gaslight fixtures.

George Westinghouse, inventor of the railroad air brake and a resident of Pittsburgh, also became involved in the expanding natural gas industry. He explored for natural gas close to home. He drilled for gas in his own backyard located in a fashionable Pittsburgh neighborhood. In late February 1884, a small volume of gas began flowing from the well. The workers continued drilling to a depth of about 1560 feet. At 3 a.m. one morning, Westinghouse awoke to the sound of a tremendous explosion and the loud sound of hissing gas from the well.

Westinghouse needed a company organization to proceed with his new plan of selling his natural gas to Pittsburgh-area customers. He purchased a moribund company, the Philadelphia Company, to produce the fuel. As President and Director of the company, Westinghouse watched the firm become one of the largest gas businesses in the Pittsburgh area. For additional supply, the company leased substantial gas production acreage in western Pennsylvania. By 1887, the Philadelphia Company supplied approximately 5000 residential and 470 industrial customers with gas from about 100 natural gas wells located on 54,000 acres.

Westinghouse's financial participation in the natural gas business brought his inventive mind in touch with some of the major problems of this new industry. Between the years 1884 and 1885, he applied for 28 gas-related patents, and during his lifetime he applied for a total of 38 gas equipment patents. Some of Westinghouse's most important inventions for natural gas included a system for enclosing a main gas line in residential areas with a conducting pipe to contain gas leaks. Westinghouse also developed a method for "stepping down" the highly pressurized gas in main trunk lines to lower pressure in residential areas. To prevent accumulations of gas in homes and shops after gas service was shut down and then restarted, Westinghouse patented a pressure regulator and cutoff valve that automatically restricted gas flow when the pressure dropped below a particular point. Tragedies nonetheless occurred. On the morning of January 31, 1885, two major explosions at Thirty-fifth and Butler streets in Pittsburgh nearly leveled an entire city block, killing two and injuring 25 others, some severely. The first explosion occurred at George Hermansdorfer's butcher shop after an accumulation of gas in his cellar; two or three people were badly burned. People rushed to investigate the explosion when a second, larger explosion occurred nearby. Subsequent explosions caused substantial injury to life and property, damaging as many as 15 buildings. Local residents threatened a riot against the gas company, and a representative of the Fuel Gas Company made a stunning admission: the pipes had not been tested before the gas was turned on. Efforts to develop gas regulators and emergency shutoff valves were absolutely required to ensure that this fuel could be utilized safely.

Andrew Carnegie, Pittsburgh's foremost entrepreneur, understood that natural gas had superior heating characteristics. He wrote: "In the manufacture of iron, and especially in that of steel, the quality is also improved by the pure new fuel. In our steel rail mills we have not used a pound of coal for more than a year, nor in our iron mills for nearly the same period." The iron and steel maker also noted that natural gas burned much more cleanly compared to coal.

By 1885, 150 companies had charters to sell gas in Pennsylvania, but the future of natural gas was not certain. Gas fields tended to exhaust themselves within several years after discovery. Selwynn Taylor, a Pennsylvania mining engineer, believed that most regional natural gas fields would soon be exhausted, and the price of coal would rise to the levels existing prior to the discovery of regional gas fields. His beliefs were typical of the time: existing natural gas fields and current production, transportation, and distribution systems simply could not supply enough gas to satisfy the demand and natural gas was ultimately unreliable. Fears of short-lived wells aside, gas discoveries in other Appalachian states, first in

Ohio and then West Virginia, made this fuel economically significant to the entire region. Industries located in cities such as Buffalo, Cleveland, Toledo, and Cincinnati all began using natural gas from nearby wells. Waste and poor planning, however, led to many failed ventures. In one episode, the Indiana Natural Gas & Oil Company had built the longest pipeline to date in 1891. The transmission system consisted of two parallel 8-inch lines extending from northern Indiana gas fields to Chicago, a distance of approximately 120 miles. These lines transported natural gas at 525 psi. The supply quickly declined and the lines were soon removed from service. Episodes such as this but on a smaller scale were repeated throughout the region.

Similar supply problems in Indiana continued. During the late 19th century, an area covering 7000 square miles included a large number of producing natural gas fields. Despite attempts to regulate the production and flow of natural gas, unrestrained gas demand soared in the state. By 1907, many of Indiana's once productive natural gas fields had expended their valuable fuel, and many natural gas customers had to return to manufactured gas utilization. Gas discoveries in Oklahoma and in the eastern and southern Kansas gas fields suffered similar stories of rapid development followed by depletion. Episodes such as these characterized the natural gas industry, as opposed to manufactured gas, as fairly undependable.

By the turn of the century, the natural gas industry was most developed in the Appalachian region. Productive gas wells in West Virginia, Pennsylvania, New York, Kentucky, Tennessee, and Ohio led to the establishment of regional gas firms that built pipelines to serve local markets in the entire region. Natural gas was used primarily for industrial purposes, but, where available, its higher heating content meant that it was a superior cooking and heating fuel, although appliances for these purposes were still not widely available until later in the 19th and early 20th centuries. Natural gas was a promising fuel, but its limited availability and dependability forced entrepreneurs to proceed cautiously with plans to develop fields and build pipelines.

5. Natural Gas in the Southwest

The discovery of massive southwestern natural gas fields and technological advancements in long-distance pipeline construction dramatically altered the early 20th-century gas industry market structure. In 1918, drillers discovered a huge natural gas field that became known as the Panhandle Field, situated primarily in North Texas. In 1922, drillers discovered the Hugoton Field, located in the common Kansas, Oklahoma, and Texas border area (generally referred to as the midcontinent area).

The combined Panhandle/Hugoton fields became the nation's largest gas-producing area, comprising more than 1.6 million acres. It contained as much as 117 trillion cubic feet (tcf) of natural gas and accounted for approximately 16% of total U.S. reserves in the 20th century.

As oil drillers had done earlier in Appalachia, they initially exploited the Panhandle Field for petroleum only while allowing an estimated 1 billion cubic feet/day (bcf/d) of natural gas to escape into the atmosphere. As new gas markets appeared, the commercial value of southwestern natural gas attracted entrepreneurial interest and bolstered the fortunes of existing firms. These discoveries led to the establishment of many new southern firms, including the Lone Star Gas Company, Arkansas Louisiana Gas Company, Kansas Natural Gas Company, United Gas Company, and others, some of which evolved into large natural gas companies.

The sheer volume of these southwestern gas fields and their distance from distant urban markets emphasized the need for advancements in pipeline transport technology. In particular, new welding technologies allowed pipeline builders in the 1920s to construct longer lines. In the early years of the decade, oxyacetylene torches were used for welding, and in 1923, electric arc welding was successfully used on thin-walled, high-tensile-strength, large-diameter pipelines necessary for long-distance compressed gas transmission. Improved welding techniques made pipe joints stronger than the pipe, and seamless pipe became available for gas pipelines beginning in 1925. Along with enhancements in pipeline construction materials and techniques, gas compressor and ditching machine technology improved as well. Long-distance pipelines became a significant segment of the gas industry beginning in the 1920s.

These new technologies made possible for the first time the transportation of southwestern natural gas to midwestern markets. Soon, the southwest supplanted Appalachia's position as the primary region for marketable gas production. Until the late 1920s, virtually all interstate natural gas transportation took place in the northeast, and it was based on Appalachian natural gas production. In 1921, natural gas produced in West Virginia accounted for approximately 65% of interstate gas transportation whereas only 2% of interstate gas originated in Texas. Most interstate gas flowed into western Pennsylvania and Ohio. Appalachian fields experienced serious depletion in the 1920s, however, and various state legislators attempted to prohibit out-of-state gas exportation. These attempts to corral natural gas for intrastate utilization were largely unsuccessful.

Between the mid-1920s and the mid-1930s, the combination of abundant and inexpensive southwestern natural gas production, improved pipeline technology, and increasing nationwide natural gas demand led to the creation of the new interstate gas pipeline industry. Metropolitan manufactured gas distribution companies, typically part of large holding companies, financed most of the pipelines built during this era. Despite the high cost of the long-distance lines, access to natural gas even for mixing with existing manufactured gas could be a profitable venture. Natural gas was so abundant it was often substantially less costly than coal gas.

In 1927, Cities Service built the first long-distance line originating in the Panhandle field. This 250-mile, 20-inch pipeline connected the Panhandle field with a Cities Service gas distributor in Wichita, Kansas. Standard Oil (New Jersey) also participated in several significant pipeline ventures during these years. The first of these was Colorado Interstate Gas Company. Organized in 1927 by Standard, Cities Service, and Prairie Oil & Gas, this firm built a 350-mile, 22-inch line originating at the Texas–New Mexico border and extending to Denver. In California, natural gas from the Buena Vista field in the San Joaquin Valley fueled industry and commercial establishments in Los Angeles, and in 1929, Pacific Gas & Electric (PG&E) constructed a 300-mile pipeline from the Kettleman field north of Los Angeles to bring natural gas to San Francisco. San Francisco was one of the first major urban areas to switch from manufactured gas to natural gas. Because the same volume of natural gas had nearly twice the heating content as coal gas, burners and airflow valves in stoves and water heaters had to be adjusted to accept the natural fuel. With near military precision, PG&E divided San Francisco into 11 districts that were successively converted to natural gas. Six hundred trained men divided into 35-member crews converted PG&E's service area within 5 months. The conversion of 1.75 million appliances cost $2 million, but natural gas was less costly for the utility to market compared to coal gas.

New long-distance gas lines and expensive conversion programs were necessary if gas utilities were going to meet consumer demand. The new holding companies marshaled tremendous amounts of capital to build pipelines, extend service, and promote gas utilization. They also became adept at advertising and marketing. Trained salesmen, company servicemen, and even co-opted plumbers touted gas. During the 1920s, utility companies offered for sale a wide variety of gas-powered appliances, including space heating units, water heaters, stoves, and even gas-powered refrigerators. By 1926, about 50,000 automatic water heaters had been installed in homes, but gas appliances were not inexpensive.

Another use for natural gas beginning in the late 19th century was carbon black production. Produced by burning natural gas, carbon black was used for coloring in paint and inks. It was also used as a reinforcing agent in rubber and automobile tires. Natural gas produced in fields not connected by pipelines to urban markets was a prime candidate for carbon black production. Even by the late 1930s,

Table II

Estimated Waste of Natural Gas in the United States in Billions of Cubic Feet[a].

Year	Natural gas waste[b]		Total U.S. natural gas consumption
	Total U.S.	Texas panhandle	
1919	213	n/a	256
1920	238	n/a	286
1921	193	n/a	248
1922	233	n/a	254
1923	416	n/a	277
1924	343	n/a	285
1925	324	n/a	272
1926	417	220	289
1927	444	405	296
1928	412	351	321
1929	589	294	360
1930	553	252	376

[a]*Source.* Federal Trade Commission, "Report to the Senate on Public Utility Corporations," Senate Document No. 92, 70th Congress, 1st Session, Part 84-A, 1935, pp. 93 and 95.
[b]Waste means gas production that was flared or vented and otherwise not utilized. n/a, data not available.

about two-thirds of the amount of marketable gas produced was either flared, vented, or used to make carbon black. (see Table II). But greater profits awaited entrepreneurs willing to finance pipelines connecting gas fields to urban and industrial gas markets. Urban natural gas utilization also brought forth efforts to develop a standardized odorant. Unlike coal gas, which typically has a distinct smell, natural gas is odorless. Thus, a leak or inadvertently opened valve might allow odorless gas to accumulate in an enclosed space and asphyxiate people, or explode. Experiments with odorants date to at least 1885, and in 1930, the Bureau of Mines conducted experiments with mercaptan, which later became the standardized gas odorizer.

6. Long-distance Pipelines

By the late 1920s, four public utility holding companies dominated the U.S. gas industry and sought to control interstate gas transportation as well. Two of the largest holding companies, Columbia Gas and Standard Oil (New Jersey), distributed more than half of the gas sold in the entire Appalachian region. Henry Doherty's Cities Service dominated the lower midwest. The largest public utility conglomerates included Middle West Utilities, Inc. and Insull Utility Investments, Inc., both controlled by Samuel Insull and headquartered in Chicago. By the late 1920s, Insull's empire included 248 gas, coal, and electric power firms serving 4741 communities in 30 states.

Planning for the first 1000-mile pipeline began in 1926 when Samuel Insull and associates discussed the possibility of building a natural gas pipeline connecting southern gas fields with Chicago area gas utilities. They sponsored engineering studies, considered a pipeline route, and examined potential gas acreage. In April, 1930 they first incorporated as the Continental Construction Corporation; a year later the company changed its name to the Natural Gas Pipeline Company of America (NGPL). NGPL's proposed 24-inch line would extend 980 miles from north Texas to Chicago. Commonly referred to as the "Chicago pipeline," this line would allow Insull to convert Peoples Gas Light & Coke Company's service area from dependence on manufactured coal gas to cleaner, hotter burning, and less expensive natural gas.

The NGPL venture was jointly planned, financed, and controlled by three utility holding companies and three other oil firms. The three holding companies were Samuel Insull's Insull & Sons, Henry Doherty's Cities Service, and Standard Oil of New Jersey. NGPL purchased its gas supply from gas fields controlled by the pipeline's owners. Standard Oil (NJ) agreed to furnish 25% of NGPL's requirements indirectly through the Canadian River Gas Company. Canadian River was a partnership of Cities Service and Prairie Oil & Gas, Standard's partners in the Colorado Interstate line. The Texoma Natural Gas Company supplied the remaining 75% of NGPL's gas requirements.

Henry L. Doherty & Company contracted to build the NGPL line. Construction began in August, 1930, and the main line was completed 12 months later. A total of 418 million pounds of steel pipe buried 6 feet transported gas at 600 psi. Construction costs for the main line, nine compressor stations, and telephone lines totaled $35 million. Although NGPL's major market was Insull's Chicago area utilities, some gas was also sold to gas distributors in Kansas and other states. The first gas deliveries in Chicago commenced on October 16, 1931, and by January 1, 1932, the line was delivering 55 mmcf/d with an originally designed total capacity of 175 mmcf/d. With access to abundant volumes of natural gas, Chicago became the largest U.S. city to convert its utility distribution system to "mixed gas," and later to straight natural gas. Peoples Gas Light and Coke Company first began producing a mixed gas with a 800-Btu content. Mixed gas, a mixture of lower Btu coal gas and higher Btu natural gas provided a hotter burning flame than did coal gas alone, for both cooking and heating. Peoples Gas Light and Coke Company began charging for gas based on a price per "therm" (1 therm = 100,000 Btu) basis, rather than by volume; natural gas had nearly twice the Btu rating compared to an equal volume of manufactured gas.

Peoples Gas Light and Coke Company organized a massive campaign to merchandise gas house-heating equipment. The company placed full-page and three-quarter-page advertisements in newspapers serving Chicago and in 50 outlying communities; advertisements appeared on billboards, streetcars, and shop windows. In addition, the utility hired 270 company-trained salesmen, 60 heating engineers, and 14 sales directors to promote gas consumption. Within the first 10 weeks of the promotion, Peoples Gas Light and Coke Company installed about 10,000 conversion burners, and the company made 30,000 gas installations during the gas sales promotion. Servicemen adjusted existing residential furnaces to accept the higher Btu mixed gas. In order to convert appliances, gas mains required cleaning to remove oil residue and other impurities from the manufactured gas.

Also during this time, a consortium led by North American Light & Power Company, which owned gas and electric properties throughout the midwest, purchased from Odie R. Seagraves and William L. Moody III (Moody–Seagraves Interests) the beginnings of the pipeline these two men had planned to build from Seagraves's Hugoton gas field properties to Omaha, Nebraska. The North American Light & Power Company joined the Lone Star Gas Company and United Light & Power Company in a partnership to purchase the Moody–Seagraves project and rename it the Northern Natural Gas Company. North American financed the construction of Northern Natural, which was completed 1931. The 1110-mile, 24- and 26-inch line transported gas to various cities along its path to Minneapolis via Omaha.

During the 1930s, a third group of entrepreneurs formed a third pipeline to connect southwestern gas fields with midwestern customers. They incorporated the Panhandle Eastern Pipe Line Company. By 1936, it was transporting gas from the Texas Panhandle through an affiliated firm to Detroit, Michigan.

7. Natural Gas in the Great Depression

In the late 1920s and early 1930s, the most well-known public utility figure was Samuel Insull, a former personal secretary of Thomas Edison. Insull's public utility empire headquartered in Chicago did not fair well in the economic climate that followed the 1929 Wall Street stock market crash. His gas and electric power empire crumbled, and he fled the country. The collapse of the Insull empire symbolized the end of a long period of unrestrained and rapid growth in the U.S. public utility industry. In the meantime, the Federal Trade Commission (FTC) had launched a massive investigation of the nation's public utilities, and its work culminated in New Deal legislation that imposed federal regulation on the gas and electric industries. The Public Utility Holding Company Act (1935) broke apart the multitiered gas and electric power companies and the Federal Power

Act (1935) and the Natural Gas Act (1938), respectively, authorized the Federal Power Commission (FPC) to regulate the interstate transmission and sale of electric power and natural gas.

During the Depression, the gas industry also suffered its worst tragedy in the 20th century. In 1937, at New London, Texas, an undetected natural gas leak at the Consolidated High School resulted in a tremendous explosion that virtually destroyed the Consolidated High School, 15 minutes before the end of the school day. Initial estimates of 500 dead were later revised to 294. Texas Governor Allred appointed a military court of inquiry that determined an accumulation of odorless gas in the school's basement, possibly ignited by the spark of an electric light switch, created the explosion. This terrible tragedy was marked in irony. On top of the wreckage, a broken blackboard contained these words, apparently written before the explosion: "Oil and natural gas are East Texas' greatest mineral blessings. Without them this school would not be here, and none of us would be here learning our lessons." Although many gas firms already used odorants, the New London explosion resulted in the implementation of new natural gas odorization regulations in Texas.

8. Appalachian Gas and Federal War Planning

During World War II, the Pittsburgh, Youngstown, and Wheeling areas contained hundreds of steel mills and metallurgical factories, as well as rubber and chemical plants that required large volumes of natural gas. Natural gas was vital to these factories because it burned at a constant specific temperature, providing high-quality product manufacture. Approximately 660 Appalachian area factories used a substantial amount of natural gas, and wartime energy demands put further pressure on Appalachian gas reserves. Appalachian natural gas production had peaked in 1917 at 552 bcf of natural gas, or about 63% of total U.S. gas production; this percentage declined to approximately 15% by the late 1930s. The decline resulted from diminishing Appalachian gas reserves as well as a proportionate increase in southwestern-produced gas. By 1943, Appalachian production alone was insufficient for meeting regional industrial, commercial, and residential demand. (see Table III).

The intense drain on Appalachian reserves stimulated private entrepreneurial efforts to increase production and build new pipelines. At the same time, some industry executives were already looking forward to a burgeoning gas industry after the war. During one meeting held during 1942, J. French Robinson, a prominent gas utility executive, stated that "in the postwar sunshine of abundant materials for our use, we will be able to realize the potential values of natural gas to all this nation as never

Table III

Natural Gas Production by Region, 1912–1970[a].

| Year | Region (%)[b] | | | Total marketed production (tcf)[c] |
	Appalachia	Southwest	Other	
1912	74	22	2	0.56
1920	55	34	11	0.80
1922	46	37	17	0.76
1924	31	45	24	1.14
1926	26	50	24	1.31
1928	21	57	22	1.57
1930	17	61	22	1.94
1935	16	65	19	1.92
1940	15	68	17	2.66
1945	10	73	17	3.91
1950	6	80	14	6.28
1960	3	87	10	12.80
1970	2	90	8	21.90

[a] *Source.* U.S. Bureau of Mines, "Natural Gas Annuals and Minerals Yearbook (Government Printing Office, Washington, D.C.), various years; and Energy Information Administration, "Natural Gas Production and Consumption," Energy Data Reports, DOE/EIA-0131 (Government Printing Office, Washington, D.C., 1978). Also see, David Gilmer, "The History of Natural Gas Pipelines in the Southwest," *Texas Business Review* (May–June, 1981), p. 133.
[b] Appalachia includes Pennsylvania, Ohio, West Virginia, and Kentucky (and New York for 1920 only). Southwest includes Texas, Louisiana, Oklahoma, and Kansas.
[c] tcf, trillion cubic feet.

before." Patriotic fervor aside, the business of war stimulated both industrial production and entrepreneurial ambition.

To direct the federal government's wartime energy policy, Roosevelt chose Harold I. Ickes, who was then Secretary of the Interior. On May 28, 1941, Ickes assumed his new position as the first Petroleum Coordinator for National Defense; this agency was later renamed the Petroleum Administration for War (PAW). In this role, the new "oil czar" exercised special emergency powers over much of both the oil and gas industries. Despite initial industry fears, Ickes implemented a cooperative relationship with the energy industry during wartime. The PAW created a Natural Gas and Natural Gasoline Division to be responsible for the gas industry. E. Holley Poe, a former executive of the American Gas Association, headed the division. His charge was maintaining natural gas production and deliverability, particularly in the Appalachian region. Poe also attempted to marshal support for joint-industry cooperation while administering the wartime industry. The PAW's authority over natural gas was relatively modest compared to that of the Supply Priorities and Allocation Board (SPAB). The SPAB, which later merged into the War Production Board (WPB), had much broader powers over industry. Regarding natural gas, the agency dictated specific gas sales allocation orders to gas pipelines.

During late 1941, representatives of the natural gas industry, military, PAW, WPB, and the American Gas Association met several times in different cities to discuss recommendations for restricting some classes of natural gas consumption and maintaining production levels during the war. J. A. Krug, Chief of the WPB Power Branch, was particularly concerned about potential shortages in Appalachia, southern California, and the midcontinent areas. He proposed a special "Limitation Order" for conserving natural gas. The order had two major goals: (1) to increase production and (2) to curtail nonessential consumption. Major General H. K. Rutherford wrote a letter of support and noted the critical situation faced by war industries dependent on natural gas.

In early February, 1942, the WPB issued Order L-31. This action called for voluntary compliance with pooling arrangements "to achieve practicable maximum output in the area or areas in which a shortage exists or is imminent." The order authorized the WPB to integrate natural gas systems, curtail gas sales when necessary, and reallocate existing gas sales. The WPB actively encouraged pipelines to transport gas at 100% load factor, to use gas storage fields whenever possible to free up pipeline capacity for gas transmission, and to develop curtailment schedules. Six months later, the WPB issued Order L-174, which imposed the same restrictions on the manufactured coal gas industry.

The PAW and WPB also addressed the Appalachian gas production problem. First, the PAW set guidelines for a new drilling program, M-68, for developing a nationwide oil and gas drilling program "consistent with the availability of material and equipment." This program limited drilling of gas wells to not more than 1 every 640 acres. Industry leaders objected to M-68, believing that it would stymie efforts to maintain current production levels. In response, the PAW issued new spacing provisions that permitted drilling one well on each 160 acres for specified deep horizons and one to each 40 acres for shallow wells.

The importance of Appalachian natural gas supply to the war effort was reflected in the disproportionate number of wells drilled there. Between 1942 and 1945, approximately 70% of all gas wells drilled in the country were drilled in Appalachia, even though overall production levels did not rise significantly. Wartime demand simply sped up the depletion of Appalachian gas fields. Government drilling and consumption regulations could not reverse this situation.

9. Gas in the Postwar Era

In the period following World War II, the natural gas industry expanded rapidly. A new round of

Table IV
Natural Gas Prices and Demand, 1945–1970[a].

| Year | Marketed production | |
	Trillions of cubic feet	Average wellhead price (cents/million cubic feet)
1945	4	4.9
1950	6	6.5
1955	9	10.4
1960	13	14.0
1965	16	15.6
1970	22	17.1

[a]*Source.* American Gas Association, Gas Facts. (Various years.)

long-distance pipeline construction made natural gas available throughout the nation. Natural gas fueled factories, electric power-generating facilities, and provided heat for homes and cooking. Demand for gas fuel rose dramatically as it became available. (see Table IV). In this postwar era, entrepreneurs organized several long-distance gas pipeline firms to connect southwestern gas supply with northeastern markets. New pipeline firms organized to sell natural gas to northeastern markets. One group of entrepreneurs purchased the so-called Big Inch and Little Big Inch pipelines from the United States government and converted them for natural gas transportation. The government had financed these lines during World War II to transport oil from the Texas Gulf Coast to the New York refinery area. Under new private ownership, the newly named Texas Eastern Transmission Corporation and affiliated lines delivered natural gas for the first time to Philadelphia, New York, and Boston. Two other new pipelines built either during or immediately after the war, the Tennessee Gas and Transmission Company and Transcontinental Gas Pipe Line Company, also began delivering southwestern-produced natural gas to northeastern customers in major urban areas.

Other pipelines extended from southwestern gas fields to growing urban markets on the West Coast and in the Southeast. California is a large producer of natural gas, but rapid population and infrastructure growth fueled the demand for more of it. El Paso Natural Gas became the first interstate pipeline to deliver natural gas to California, followed by Transwestern Pipeline Company in the early 1960s. The Northwest Pipeline Company began transporting natural gas produced in the San Juan Basin in Colorado and New Mexico to customers in Seattle after 1956. In 1959, Florida Gas Transmission Company delivered the fuel to Floridians. By the mid-1950s, therefore, the beginnings of a national market for natural gas emerged. During the last half of the 20th century, natural gas consumption in the U.S. ranged from about 20 to 30% of total national energy utilization. However, the era of unrestricted natural gas abundance ended in the late 1960s.

The first overt sign of serious industry trouble emerged when natural gas shortages appeared in 1968–1969. Economists almost uniformly blamed the shortages on gas pricing regulations instituted by the so-called Phillips Decision of 1954. This law had extended the FPC's price-setting authority under the Natural Gas Act to the natural gas producers that sold gas to interstate pipelines for resale. The FPC's consumerist orientation meant, according to many economists, that it kept gas prices artificially low through federal regulation. Gas producers consequently lost their financial incentive to develop new gas supply for the interstate market, and shortage conditions developed.

10. Deregulation

The 1973 OPEC oil embargo exacerbated the growing shortage problem as factories switched boiler fuels from petroleum to natural gas. Cold winters further strained the nation's gas industry. The resulting energy crisis compelled consumer groups and politicians to call for changes in the regulatory system that had constricted gas production. In 1978, a new comprehensive federal gas policy dictated by the Natural Gas Policy Act (NGPA) created a new federal agency, the Federal Energy Regulatory Commission, to assume regulatory authority for the interstate gas industry.

The NGPA also included a complex system of natural gas price decontrols that sought to stimulate domestic natural gas production. These measures appeared to work almost too well and contributed to the creation of a nationwide gas supply "bubble" and lower prices. The lower prices wreaked additional havoc on the gas pipeline industry because most interstate lines were then purchasing gas from producers at high prices under long-term contracts. Some pipeline companies had also invested tremendous amounts of money in expensive supplemental gas projects such as coal gasification and liquid natural gas (LNG) importation. The long-term gas purchase contracts and heavy investments in supplemental projects contributed to the poor financial condition of many gas pipeline firms. Large gas purchasers, particularly utilities, also sought to circumvent their high-priced gas contracts with pipelines and purchase natural gas on the emerging spot market.

Once again, government was forced to act in order to bring market balance to the gas industry. Beginning in the mid-1980s, a number of FERC orders, culminating in Order 636 (and amendments), transformed interstate pipelines into virtual common carriers. This industry structural change allowed gas utilities and end-users to contract directly with

producers for gas purchases. FERC continued to regulate the gas pipelines' transportation function, but pipelines ceased operating as gas merchants as they had for the previous 100 years. Restructuring of the natural gas industry continued into the early 21st century as once-independent gas pipeline firms merged into larger energy corporations.

Natural gas is a limited resource. Although it is the most clean burning of all fossil fuels, it exists in limited supply. Estimates of natural gas availability vary widely, from hundreds to thousands of years. Such estimates are dependent on technology that must be developed in order to drill for gas in more difficult geographical conditions, and actually finding the gas where it is expected to be located. Methane can also be extracted from coal, peat, and oil shale, and if these sources can be successfully utilized for methane production, the world's methane supply will be extended another 500 or more years.

Since 1970, natural gas production and consumption levels in the United States have remained reasonably stable. During the 1980s, both consumption and production levels dropped about 10% from the 1970 levels, but by the later 1990s, production and consumption were both on the rise. (see Table V). In the absence of aggressive conservation programs, unexpected shortages, or superior alternative energy sources, natural gas consumption will continue to increase.

For the foreseeable future, natural gas will continue to be used primarily for residential and

Table V

Natural Gas Production and Consumption in the United States[a].

Year	Total dry production	Total consumption
1970	21,014,292	21,139,386
1972	21,623,705	22,101,452
1974	20,713,032	21,223,133
1976	19,098,352	19,946,496
1978	19,121,903	19,627,478
1980	19,557,709	19,877,293
1982	17,964,874	18,001,055
1984	17,576,449	17,950,524
1986	16,172,219	16,221,296
1988	17,203,755	18,029,588
1990	17,932,480	18,715,090
1992	17,957,822	19,544,364
1994	18,931,851	20,707,717
1996	18,963,518	21,966,616
1998	19,125,739	21,277,205
2000	19,072,518	22,546,944

[a]*Source.* Energy Information Agency, "Supply and Disposition of Natural Gas in the United States, 1930–2000," Historical Natural Gas Annual, Government Printing Office, Washington, D.C. In millions of cubic feet.

commercial heating, electric power generation, and industrial heat processes. The market for methane as a transportation fuel will undoubtedly grow, but improvements in electric vehicles may well dampen any dramatic increase in demand for engines powered by natural gas. The environmental characteristics of natural gas, however, should retain this fuel's position at the forefront of desirability of all fossil fuels, while supplies last.

11. Current and Future Implications

In the twenty-first century, natural gas will continue to be a major source of energy in the United States and globally as natural gas demand increases throughout the world. When the price of oil temporarily reached more than $147 per barrel during July 2008, natural gas prices rose proportionally. While these prices also later declined with those of oil, such developments highlighted the potential for extreme volatility in the energy market and the urgent need to develop more efficient means for producing, transporting and utilizing natural gas. In addition, efforts to develop commercial grade methane from other sources including coal continue.

Globally, natural gas consumption continues to increase significantly. It has risen from 53 trillion cubic feet (Tcf) in 1980 to about 105 Tcf in 2006. The U.S. Energy Information Agency (EIA) projects that global natural gas consumption following current trends will reach about 134 Tcf in 2030. Much of the increase will be due to demand for natural gas in electricity generation and industrial applications.

In the United States, natural gas continues to account for approximately one-fourth of nation's total energy supply, and U.S. natural gas consumption accounts for about 21 percent of worldwide natural gas utilization, followed by Russia which accounts for about 16% of worldwide use. From 1985 to 2008, domestic U.S. natural gas consumption has increased from approximately 17 to 22 Tcf, or thirty percent. This consumption rate has not matched the rate of domestic production. This disparity has been due to depleting wells and fewer producing wells.

U.S. natural gas production since 1985 has been relatively constant between about 16 Tcf to 20 Tcf per year, but as a percentage of worldwide production U.S. natural gas production has declined dramatically. In 1985, the U.S. accounted for about 36% of worldwide natural gas production, but by 2006 the U.S. was producing only 17% of global natural gas produced.

To augment domestic natural gas supply, the U.S. is importing more natural gas. From 1973 to 2008, imports have increased from 1 billion cubic feet (Bcf) to 4.6 Tcf in 2007. Natural gas transported through pipelines from Canada and Mexico has increased from 926 Bcf in 1985 to 3.6 Tcf in 2008, and imports from Canada have accounted for about 90 percent of

all U.S. gas imports. Liquefied Natural Gas (LNG) is another growing source of natural gas imports although still significantly less than imports through pipeline. LNG imports in 1985 amounted to 24 Bcf and have increased to 352 Bcf in 2008. During the first decade of the 21st century, the U.S. was importing about 60–75% of its LNG annually from Trinidad and Tobago. The EIA forecasts that U.S. LNG imports may reach approximately 3 Tcf by 2030. Five U.S. LNG import facilities were operating in January 2008, and four new facilities were then planned for the Gulf Coast and two on the Atlantic Coast.

In 2007, the EIA reported that the U.S. had proved dry natural gas reserves of 237 Tcf, with approximately four times that amount being technically recoverable. Alaska has about 12 Tcf in proved reserves with as much as 32 Tcf identified as technically recoverable. World natural gas reserves for the same year are roughly estimated at 6,200 Tcf. Since the early 1970s, there have been plans to build a gas pipeline from Prudhoe Bay in Alaska to tap Alaskan natural gas and transport it through Canada to U.S. markets. As of 2008, the plan is for TransCanada Alaska to develop and build the proposed 1,715-mile natural gas pipeline that would extend from Prudhoe Bay on the North Slope to Alberta, Canada. If the pipeline project progresses as schedule, TransCanada estimates the first gas deliveries by November, 2017.

In Europe, long-distance natural gas pipelines are required to transport natural gas from producing to consuming regions, and natural gas pipelines there have become embroiled in political conflict; several significant episodes have occurred. In March 2007, Azerbaijan began exporting natural gas through the newly built South Caucasus pipeline to Georgia, and exports through this line also reached Turkey in July 2007. Turkey, in turn, exported this gas to Greece after completion of a new pipeline between Turkey and Greece began operating in November 2007. However, when Turkmenistan cut natural gas exports to northern Iran in January 2008, initially citing technical issues, Iran then cut its natural gas exports to Turkey presumably in retaliation for the Turkmenistan cuts.

Another serious international natural gas supply dispute developed in March 2005 between Russia and the Ukraine. In January 2006, Russia temporarily halted gas deliveries to the Ukraine. While Russia resumed shipments within several days, this incident evidenced the political natural of natural gas markets. Tensions increased once again during 2007 and 2008 between the two parties. Then in early 2009, Russia cut back substantially on gas deliveries to 18 European nations in another dispute involving Russia's gas deliveries to, and through, the Ukraine.

Natural gas continues to be heralded as an environmentally friendly fuel, and there are substantial reserves, particularly in the Middle East. The United

States will continue importing additional volumes of natural gas to augment domestic production. Internationally, natural gas consumption will grow as will the likelihood that political as well as economic events will create disruptions in the natural gas energy market.

Further Reading

Bragdon E D 1962. *The Federal Power Commission and the Regulation of Natural Gas: A Study in Administrative and Judicial History*. Ph.D. Dissertation. Indiana University.

Castaneda C J 1999 *Invisible Fuel: Manufactured and Natural Gas in America, 1800–2000*. Twayne Publishers, New York

Castaneda C J, Smith C M 1996 *Gas Pipelines and the Emergence of America's Regulatory State: A History of Panhandle Eastern Corporation, 1928–1993*. Cambridge University Press, New York

DeVane D A 1945 Highlights of the legislative history of the Federal Power Act of 1935 and the Natural Gas Act of 1938. *George Washington Law Rev.* **XIV**, (Dec. 1945)

De Vany A S, Walls W D 1995 *The Emerging New Order in Natural Gas: Markets vs. Regulation*. Quorum Books, Westport, Connecticut

Frey J W, Ide H C 1946 *A History of the Petroleum Administration for War, 1941–1945*. Government Printing Office, Washington, D.C.

Herbert J H 1992 *Clean Cheap Heat: The Development of Residential Markets for Natural Gas in the United States*. Praeger, New York

MacAvoy P W 2001 *The Natural Gas Market: Sixty Years of Regulation and Deregulation*. Yale University Press, New Haven

Peebles M W H 1980 *Evolution of the Gas Industry*. New York University Press, New York

Rose M H 1995 *Cities of Light and Heat: Domesticating Gas and Electricity in Urban America*. University of Pennsylvania Press, University Park

Sanders E 1981 *The Regulation of Natural Gas: policy and politics, 1938–1978*. Temple University Press, Philadelphia

Stotz L, Jamison A 1938 *History of the Gas Industry*. Stettiner Brothers, New York

Tarr J A 1998 Transforming an energy system: The evolution of the manufactured gas industry and the transition to natural gas in United States 1807–1954. In: Coutard O (ed.) *The Governance of Large Technical Systems*. Routledge, London, pp. 19–37

Tussing A R, Barlow C C 1984 *The Natural Gas Industry: Evolution, Structure, and Economics*. Ballinger Publ., Cambridge

Christopher J. Castaneda

Nuclear Power, History of

Glossary

Atomic Energy Commission   The powerful executive branch agency created in 1946 to oversee the U.S. atomic energy program; abolished in 1975.

atoms for peace   The U.S. program launched in 1954 to share atomic information with European allies and to develop peaceful applications of atomic energy.

Joint Committee on Atomic Energy   An unusually powerful congressional committee that dominated nuclear energy policy in the United States from 1946 to 1977.

Nuclear Regulatory Commission   The agency created in 1975 to assume the regulatory duties of the Atomic Energy Commission.

Price–Anderson Act   Legislation enacted in the United States in 1957 to encourage private sector interest in nuclear power by providing indemnification to the nuclear power industry in the event of a reactor accident. The law has been renewed several times over the program's history.

subgovernment   Also known as policy monopolies or subsystems, these are small, stable groups of actors, both public and private, that dominate policy in specific issue areas.

The 50-year history of commercial nuclear power has been punctuated by dramatic policy changes. The first 20 years, marked by limited public participation, tight government control, and promises of clean, abundant energy, were followed by a period of intense social and political conflict over the technology's environmental and safety implications. Nuclear policy in the United States and most European nations shifted from all-out support to a more ambivalent posture, which led to a dramatic slowdown in the construction of new plants. Although nuclear plants continue to be built, public opposition and high costs are obstacles to a large-scale comeback.

1. Dawning of the Nuclear Age in the United States, 1946–1954

1.1 The Military Roots of Commercial Nuclear Power

From its birth in the highly secretive Manhattan Project during World War II, atomic energy was defined and perceived in military terms, which meant that information regarding the atom was subject to elaborate security precautions. This concern with maintaining the "secret" of the atom set the tone for the atomic program far into the future. From the beginning, both public involvement in the development of atomic energy programs and access to information about atomic energy would be severely restricted.

When the war ended, the sense of national emergency that had driven the atomic program dissipated, and many of the top scientists who had worked on the bomb project returned to their positions in universities and the private sector. With virtually no stockpile of atomic weapons at the war's end, government officials worried that if the atomic program ground to a halt, the nation's security would be at risk. A vigorous weapons program, it was believed, was essential to the nation's defense and security, especially because it was widely recognized that the American monopoly on atomic weapons would not last forever.

1.2 The Atomic Energy Act of 1946

In pursuit of these general goals, Congress in 1946 passed the Atomic Energy Act, which established the institutional framework within which atomic energy decisions would be made for approximately the next 30 years. The legislation created two bodies, the Atomic Energy Commission (AEC), which inherited ownership of all atomic materials, facilities, and information from the Manhattan Project, and the Joint Committee on Atomic Energy, which was to oversee AEC operations.

The AEC was to be the primary actor in formulating the nation's atomic energy policies. Its principal duties, as defined by Congress, were the production of atomic weapons and the fissionable materials required for their manufacture. In addition to establishing the AEC, the Atomic Energy Act mandated a government monopoly in atomic energy. Congress stipulated that responsibility for all atomic programs and information be transferred from the army to the AEC. Furthermore, the AEC was granted ownership of all fissionable materials and all facilities that used or produced them. The size and scope of the AEC's operations, together with the sweeping powers granted by Congress, clearly indicated that it was intended to be a powerful agency.

Composed of nine members from each chamber, the Joint Committee on Atomic Energy (JCAE) was clearly intended to be a powerful committee. It was, for example, the only joint committee created by statute rather than by the rules of both houses. The JCAE was also the only permanent joint committee granted the full legislative and investigative powers of a regular standing committee. Its formal status alone suggests that Congress intended to exercise control over atomic energy in general and over the AEC in particular. The Atomic Energy Act required the AEC to keep the JCAE "fully and currently informed with respect to the Commission's actions," mandated that all bills concerning the atom were to be referred to the JCAE, and it set no limits on the number of staff members or consultants the committee could hire. These sweeping and unusual grants of power, along with the JCAE's status as a special joint committee with exclusive jurisdiction over a glamorous and highly technical issue, made the committee one of the most powerful congressional bodies in the nation's history.

The normal tendency of Congress to defer to the advice of its specialized committees and subcommittees was reinforced in the case of atomic energy by several additional factors. First, atomic energy was largely perceived and justified as a defense program, with primary emphasis on its military applications. As late as June 1961, approximately two-thirds of all research and development and construction activities of the AEC's reactor program continued to serve weapons and military applications. The definition and perception of the atomic program as primarily military in character lent credibility to both the program and the Joint Committee and legitimized its exclusive policymaking role.

1.3 Secrecy and Security

Because the United States was the only nation with the atomic "secret," security was considered to be an essential part of the effort to retain the atomic monopoly. The JCAE's inclination, like that of other security and defense committees, was to keep most information classified. With its exclusive control over atomic energy matters in Congress, the committee was able to dominate the legislative aspect of policymaking in these early years. The security restrictions meant that the atomic program was obscured by a blanket of secrecy that effectively limited the number of participants in the atomic program. Nuclear power would be relatively immune to challenge so long as access to information concerning the program was so closely held.

1.4 The Creation of an Atomic Subgovernment

In many respects, the atomic energy program in the 20 years following World War II was the archetypal subgovernment. A small, cohesive, and stable group of actors consisting of the Atomic Energy Commission, the Joint Committee on Atomic Energy, the nuclear power industry, and the scientists and engineers working on the program in universities and national laboratories exercised considerable autonomy in policymaking. Each of the participants involved in nuclear policymaking during this period had either an economic, political, or organizational stake in the development and use of atomic energy. There is, however, one highly unusual aspect of this policy community. Unlike the familiar example of subsystems in which government actors had been "captured" by the very business interests they were supposed to monitor, with nuclear power there was no industry "client" group until the government created one. It was only after policymakers in the mid-1950s took steps to overcome the serious technological, economic, and political obstacles confronting the generation of electricity from fission that a commercial nuclear power industry emerged. This unusual government-industry partnership is crucial to understanding the history of nuclear power in the United States, as well as in most other nations.

The atomic energy subgovernment was endowed with additional prestige and power because of the program's identification with national security issues. The actors in this tightly knit monopoly were united by the conviction that the development of atomic energy, first as a weapon but later as a means of generating electricity, was both necessary and desirable for the nation's welfare. Indeed, this commitment was enthusiastically embraced by those involved in formulating and implementing policy and drove the program for over 20 years. For the most part, this consensus meant that only supporters of nuclear power would be mobilized for political action. Presidents rarely became involved in the program, Congress readily deferred to the advice of the experts on the Joint Committee on Atomic Energy, and, with the exception of the nuclear power industry, interest group activity was virtually nonexistent. This onesided mobilization ensured that generous subsidies and the absence of political conflict marked government policy during this long period.

1.5 Nuclear Power and the National Interest

In the years following the World War II, nuclear power was equated with the national interest, and it was often said that politics and atomic energy did not mix. The cold war, the prevailing faith in science, and the recognition that the atomic monopoly could not last forever all contributed to a consensus that atomic energy was instrumental in securing the nation's future. To some, the term "national interest" could be used to promote almost anything pertaining to nuclear power. Furthermore, it was often claimed that nuclear energy would revolutionize the nature of industrial society by providing a relatively cheap and abundant source of power. Nuclear energy could lead to the discovery of new production techniques, it was said, and provide cheap electricity to areas where supplies were scarce. Others claimed that nuclear power could drastically reduce the costs of production, leading to increased economic growth and an improved standard of living. In short, a vigorous atomic energy program was thought to be necessary for a nation seeking to exert military, economic, and technological leadership in the postwar era. The national interest seemed to demand that the United States explore the atom's many possibilities.

2. Creating a Commercial Atom, 1954–1965

The dawn of the 1950s brought a subtle but significant shift in the nation's atomic energy plans. The principal focus remained on the atom's military aspects, but government officials recognized that the so-called peaceful uses of the atom could also serve a variety of important objectives in both domestic and foreign policy. Although the cold war dictated that the program's primary purpose would continue to be weapons production, the Joint Committee began to press the AEC to put greater emphasis on developing and promoting the peaceful uses of atomic energy. At the time, the most likely prospect for peaceful applications was the production of electricity from atomic fission.

2.1 The Peaceful Atom and Postwar Politics

Despite this subtle shift in the program's avowed goals, the program was still largely justified by national security arguments. Aside from enhancing the nation's scientific reputation and providing an abundant energy source, finding a constructive use for the atom would also further the nation's postwar geopolitical aims. The explosion of the Soviet atomic bomb in 1949 had punctuated the end of the American nuclear monopoly. American officials worried that the Soviets would try to develop the peaceful atom before the United States in an attempt to reap propaganda and political benefits. This prospect was not a pleasant one for those responsible for managing the atomic program.

Although American officials were preoccupied with the Soviets, they were also concerned that other nations would seek to develop their own atomic capabilities. The events of 1949 had shown that the United States no longer had a monopoly on the atom, and that the knowledge and ability to build atomic weapons would inevitably spread to other nations. It was commonly accepted that it was only a matter of time before Great Britain and France would join the atomic club. Other nations would surely seek to do the same and launch their own atomic programs, whether the United States wanted them to or not. Faced with this unpleasant reality, the United States wanted to be in a position to control nuclear proliferation to ensure that it was compatible with American perceptions of world order.

Government decision makers recognized that in order to avoid losing influence around the globe, the United States would have to assert its leadership and share some atomic information, and secrets, with other nations. With these objectives in mind, in a speech to the United Nations in December 1953, President Eisenhower launched an initiative that came to be known as "Atoms for Peace." The program was clearly an effort to distinguish between the peaceful and military applications of the atom and to identify the United States with the peaceful applications. The Atoms for Peace plan also served a number of American foreign policy goals. The Atoms for Peace proposal, according to two researchers, was "in fact nothing less than a coherent global strategy for protecting Western Europe from Soviet domination." If the United States could develop the atom as a source of electric power, it was argued, it would be a boon to Western European industries and economies while helping to contain the Soviets.

The Atoms for Peace proposal had the added attraction of assisting American industry by providing access to potentially lucrative markets. According to the terms of the bilateral agreements signed pursuant to the program, European nations contracted with American firms for all atomic technologies and materials. The attraction of these agreements, as one official from the American contractor Babcock and Wilcox wrote to the joint committee, was that they were a "means of securing a good and early foothold" in the potentially lucrative European market for American reactor vendors. One 1955 study estimated that Europe represented a $30 billion market.

2.2 The Atomic Energy Act of 1954

In order to make more progress in developing the peaceful atom, Congress revised the Atomic Energy Act to permit greater access to nuclear information and materials. Although American firms were expressing an interest in atomic power, the calls for revising the 1946 law came primarily from the U.S. government and not from American industry, which was unprepared to assume the full responsibility for nuclear development because of the financial risks involved. In contrast to coal plants, the initial costs for nuclear plants would undoubtedly be very high, profits uncertain, and the technology new and potentially hazardous. In addition, projected demand for electricity in the early 1950s provided utilities with little incentive to build additional generating capacity. Furthermore, there were no reliable estimates of the possible damages stemming from a serious reactor accident. Nevertheless, development of the peaceful atom was assumed to be vitally important to addressing a number of problems confronting the United States at the time. If the private sector would not take the necessary steps to solve these problems, the U.S. government would. The solution to the government's problems was the Atomic Energy Act of 1954, which effectively created both commercial nuclear power and a commercial nuclear power industry. The development of commercial nuclear power in the United States, in short, emerged from an effort initiated by government rather than by private industry.

To open the door to greater industrial and international participation, Congress in 1954 loosened the restrictions and access to previously classified information, permitted the private ownership of nuclear facilities and allowed the private use of fissionable materials through a system of AEC licensing, liberalized patent laws in the atomic energy field, and offered a number of economic incentives to private industry.

2.3 The AEC's Conflict of Interest

The law also made the AEC responsible for developing, promoting, and regulating nuclear power. From the outset, this statutory conflict of interest would be problematic; it became clear that the AEC would consistently emphasize its promotional and developmental responsibilities at the expense of its regulatory duties. Congress, through the Atomic Energy Act of 1954, offered little guidance on how to resolve this dilemma. The law, for example, directed the AEC to "protect the health and safety of the public," but the statute offered little insight into how to achieve that broadly defined goal. Although such an expansive grant of authority was not unique, it would later become a problem, as the atomic program moved into its commercial phase.

2.4 Subsidies

Although the Atomic Energy Act of 1954 provided the AEC with little guidance on establishing regulations or defining safety goals, it did make clear that Congress did not want the AEC to impose burdensome regulations during the early stages of the development process. The section of the law under which all of the reactors in the AEC's Power Reactor Demonstration Program were licensed instructed the AEC to "impose the minimum amount of such regulations and terms of license as will permit the Commission to fulfill its obligations under this Act to promote the common defense and security and to protect the health and safety of the public." In addition, members of the Joint Committee consistently pressed the AEC to move ahead with reactor development. Because reactors were not economically competitive with fossil fuel plants, however, the nation's electric utilities did not have the normal incentives to build them. As a result, the AEC had to offer a wide array of incentives, including ample

financial assistance, to the nuclear industry. The AEC thus embarked on several programs designed to encourage industry participation in the commercial power program. The AEC's Power Reactor Demonstration Program, for example, assisted firms by waiving for 7 years the charges for the loan of source and special materials; by performing in its laboratories, free of charge, certain research and development work; and by agreeing to purchase technical and economic data from the participants under fixed-sum contracts.

Industry's initial response to these inducements was tepid, so the government resorted to a mix of threats and incentives to induce interest. For example, even though the AEC retained ownership of all fissionable materials, through a system of AEC licensing the Atomic Energy Act of 1954 made these materials available to firms, including both the fuel for nuclear reactors and the fissionable materials produced by nuclear reactors. Because there was still a shortage of fissionable materials for atomic weapons, the AEC agreed to purchase all of the plutonium produced in private reactors at a guaranteed price that provided firms with handsome returns.

All told, by 1962, the federal government had spent over \$1.2 billion to develop reactor technology; this figure was double the amount spent by private business. One Department of Energy study conducted in the 1980s concluded that without federal subsidies nuclear electricity would have been 50% more expensive. These calculations did not include federal tax write-offs for utilities nor did they include money spent on programs with possible military implications.

2.5 The Price–Anderson Act

Even with these subsidies, the private sector was reluctant to assume responsibility for the risks involved with nuclear power. The underlying reason for this reluctance was the fear that they would be held liable for any and all damages resulting from a reactor accident. Although the AEC had an impressive safety record with its reactors, the technology was new and largely untested, and because the potential damages from an accident were unknown, the reactor vendors and electric utilities were unwilling to make a full-scale commitment to nuclear power. The industry sought, and soon received, government protection from liability claims before proceeding with any plans to build reactors. The problem was that the potential liability from a reactor accident was so large that it was unlikely that either nuclear vendors or utilities would be able to obtain the necessary insurance from private insurers. The fear of financial disaster led the industry to seek legislation limiting their liability in the event of an accident. Because the potential damages from an accident could be so high, the industry argued that indemnification, in which the government would give general protection to the atomic industry against uninsured liability to the public, was needed to allow development to proceed. Private insurers agreed to provide $60 million of protection for each nuclear facility, which was far more than had been made available to any other industry. Because it was entirely possible that damages from a reactor accident would far exceed that amount, Congress agreed to provide an additional $500 million of protection for the industry.

2.6 The First Generation of Nuclear Reactors

After adoption of the Price–Anderson Act, private sector interest in nuclear power slowly picked up. Although the liability limit eased industry fears, industry participation remained at low levels for the next few years because nuclear electricity was still much more expensive to generate than electricity produced by conventional fuels like oil and coal. However, the words and actions of government officials convinced the nuclear industry that in the future there would be nuclear power in America, with or without private sector involvement. A number of utilities decided to buy reactors in order to gain an advantage for the future; others purchased reactors because they wanted to keep the government out of the power business. The big fear of many utilities was that the federal government would build reactors on its own and give the power to public corporations. Consequently, the fear of public power led utilities to invest in reactors before they were economically competitive.

A critical moment came in December 1963, when Jersey Central Power and Light signed a contract with General Electric to purchase a 515-MW boiling water reactor for a plant in Oyster Creek, New Jersey. This agreement seemed to herald both the arrival of nuclear power as a viable alternative for generating electricity and the beginning of widespread commercial acceptance of nuclear power in the United States. The Oyster Creek reactor was to be the first built without any direct government subsidy, and the utility claimed the plant would be more economical to operate within 5 years, compared to conventional power sources. Although utilities began to order reactors in increasing numbers, in reality nuclear fission was not yet economically competitive with fossil fuels, primarily because of the higher capital costs for nuclear plants. In fact, the Oyster Creek plant was a "loss leader" designed to persuade utilities of the viability of reactor technology. Although it has been estimated that reactor vendors lost considerable sums of money on the earliest plants, the strategy worked—utilities began purchasing increasing amounts of nuclear generating capacity.

The sudden interest of the utilities in reactors can be attributed to several factors. First, demand for electricity in the 1960s was growing at an annual rate of 7%; at that rate of growth, demand would double every 10 years. Utilities thus needed to order new plants to meet projected demand. At the same time, many public utility commissions were lowering prices, which further increased demand for electricity. Next, because profits were determined by the size of the utility rate base, utilities had an incentive to build additional generating capacity to expand their rate base. Nuclear was an especially attractive option because it was by far the most capital-intensive source of electricity. Utilities thus had a number of financial incentives to build nuclear plants. Consequently, there was a rapid increase in both the number and the size of plants ordered after Oyster Creek. For example, American utilities ordered 70 reactors between 1963 and 1967, with approximately 80% of the orders being placed in 1966 and 1967. During that same period, the average capacity per reactor jumped from 550 to 850 MW. In 1967, nuclear power accounted for slightly more than half the total generating capacity ordered that year. The market for nuclear reactors was so good that some observers refer to this period as the "Great Bandwagon Market."

But the Bandwagon Market was based on exceedingly optimistic projections and expectations of construction costs and operating performance. Because commercial generation of electricity from fission was in its infancy and still largely untested, there was little in the way of hard data concerning actual costs and performance. In the early years of the commercial program, the utilities, the Commission, and the Joint Committee lacked the technical resources to assess the projections of reactor costs and efficiency offered by the reactor vendors. Not only did they lack the resources, but the AEC and JCAE also lacked the desire to evaluate the claims carefully. In their rush to commercialize reactor technology, government officials uncritically accepted the cost and efficiency estimates.

Electric utilities in the United States and abroad, caught up in the flood of good news about nuclear power, purchased more reactors and fewer fossil plants. The rush to nuclear power on the basis of the reactor vendors' early and exceedingly optimistic price estimated eventually came back to haunt utilities that had purchased reactors. Many utilities possessed neither the technical staff nor the financial resources to build and operate nuclear reactors. It would also become clear that nuclear plants were not as efficient or reliable as vendors had claimed. The optimistic predictions for nuclear power thus succeeded in establishing a market for the technology, but they also set the stage for the future economic problems that would cripple the industry.

3. *Nuclear Power Comes Under Attack*

In 1965, the politics of commercial nuclear power were of little concern to anyone not having a direct economic or political stake in the programs' success. By the middle of the next decade, however, the policymaking arena was crowded and complex, and few issues in American politics were more visible or controversial. What changed is that, as nuclear reactors began to be built, more people came to believe that the issue affected them. In the process, nuclear power came to be understood in new, less positive ways. Long perceived as a national security issue, in the late 1960s, nuclear power came to be seen in the context of debates over environmental protection, public health and safety, and energy supplies. Eventually, it became part of even larger debates about government regulation of business, citizen participation, and democratic governance. As a result of these changing perceptions, subgovernment members lost the power to define the issue and control the parameters of debate. The subgovernment and nuclear policy were soon radically transformed.

As understandings and perceptions shifted and became increasingly negative, new actors were drawn to the nuclear issue. The new participants included other federal agencies, officials from state and local governments, and a number of prominent citizen groups. Many of the new participants would be critical of nuclear power. The influx of new participants, who brought their own opinions, shattered the consensus that had existed within the small atomic subgovernment, disrupted established patterns of policymaking, and created a decidedly less supportive political environment for the nuclear program.

3.1 *The Emergence of an Antinuclear Movement*

The boom in the market for nuclear plants coincided with the rise of the environment as an issue of national importance. Subsystem members initially perceived the growing concern with environmental values as a boon to the prospects of nuclear power. The AEC and the nuclear industry actively promoted the notion that nuclear power was cleaner than coal and other sources of electricity. This makes it all the more ironic that the first consistent opposition to nuclear power emerged in response to the potential environmental consequences of reactors. For the most part, the early opposition was essentially local in nature, being directed at particular reactors and not nuclear power. More specifically, concern during this period tended to focus on the thermal pollution caused by nuclear plants' discharge of heated water into nearby lakes and rivers. The AEC initially refused to consider the environmental effects of thermal pollution, arguing that it lacked statutory authority over nonradiological environmental matters. After a 3-year battle, and much bad publicity,

Congress in 1970 adopted a law requiring federal agencies, including the AEC, to consider thermal pollution issue in the course of their licensing reviews. This controversy fundamentally altered the course of nuclear politics. It was the first critical step in the redefinition of the nuclear power issue; as concerns emerged about possible radioactive emissions from reactors and the issue of nuclear waste disposal, nuclear power was increasingly understood as an environmental issue, not a national security matter.

The most significant factor in the expansion of the debate and in the demise of the atomic subgovernment was the emergence of reactor safety as a prominent issue. The controversy over safety energized the burgeoning opposition to nuclear power and thrust the issue into the media spotlight, making it impossible for public officials to ignore. The safety issue emerged and developed in much the same way the environmental issue had—as a response by local citizens to particular nuclear reactors—before rapidly escalating to include questions about the safety of nuclear plants in general. But concerns over the safety of nuclear reactors attracted more attention and generated more controversy compared to other environmental issues, largely because nuclear technology was still relatively new and unfamiliar to the American public. When questions of reactor safety emerged, the dramatic nature of the potential consequences of reactor accidents lent the nuclear issue a sense of drama and urgency that earlier concerns about thermal pollution could not. Furthermore, because the issue was so dramatic, it captured the attention of the national media. Shortly thereafter, the conflict over the safety of nuclear reactors occupied a prominent position in the public agenda. The AEC's credibility suffered a fatal blow when experts, including its own, began to disagree in public. Once some insiders began raising questions about reactor safety and the AEC's commitment to protecting it, the conflict rapidly expanded.

Antinuclear activists in the late 1960s and early 1970s were also extremely critical of the AEC's licensing procedures and rules. Critics charged, for example, that AEC procedures precluded meaningful public participation in agency proceedings while granting the nuclear industry privileged access to AEC personnel, especially in the critical early stages of the licensing process. Critics also claimed that AEC regulations were primarily responsive to the needs and desires of the industry, reflecting the commission's enthusiasm for building and licensing reactors. Perhaps most important, critics charged that the decision to approve reactors had usually been made before citizens were given an opportunity to register their opinions. They also charged that the AEC's public hearings, the principal forum in which citizens were allowed to participate, were nothing more than stacked-deck proceedings designed to foster the illusion of citizen input.

By the middle of the 1970s, the subgovernment had lost its ability to define the nuclear issue; the debate over nuclear power thus widened to include a variety of new issues. Some opponents of nuclear power were suggesting that rather than being merely a technical matter, the nuclear issue involved important social, political, and moral questions that scientists and engineers had no special ability to resolve. Antinuclear forces argued that citizens could learn enough about nuclear power to participate in a responsible manner if given the opportunity. In this way, antinuclear forces were able to link their concerns about nuclear safety to more general concerns about accountability and responsiveness in government.

By the end of the 1970s, citizen groups were challenging, in court and in the streets, the licensing of almost every reactor. Indeed, some the largest post-Vietnam protests and demonstrations concerned nuclear power. Several rallies drew crowds estimated to be in the hundreds of thousands. Nuclear power was plagued by similar developments in Europe, especially Germany, Great Britain, France, and Sweden. In most of these nations, as in the United States, the government became less supportive of nuclear power. France, which had fewer energy options, was a notable exception to this trend, and moved ahead with an aggressive nuclear program despite widespread public opposition.

3.2 Nuclear Power and the Courts

For most of the late 1960s and early 1970s, both the AEC and the Joint Committee consistently rebuffed persons or groups seeking to participate in the formulation of nuclear policy. In a textbook case of venue shopping, critics of nuclear power turned to the courts for assistance. One of the most significant consequences of increased oversight by the courts was that the commission, and its staff, devoted greater attention to procedural rights in an attempt to ensure that its procedures were seen as fair and capable of generating a record that could withstand judicial scrutiny. Rather than run the risk of being overturned by a reviewing court, the commission would try to show that it had solicited and considered many points of view.

The consequence of increased judicial oversight was a longer and more detailed review process that led, in turn, to more stringent environmental and safety standards. In an effort to satisfy its overseers, who were demanding more comprehensive and rigorous reviews, the commission upgraded the size and quality of its technical review staff. The larger, better, and more experienced regulatory staff began demanding more detailed information from utility applicants. Moreover, the commission was cognizant of the fact that the courts were more likely to overturn standards if the commission's decision-making

process was procedurally deficient. As a result, the commission standardized its licensing review process in 1972. Specifically, the commission developed standard review plans for both safety and environmental analyses. The standard review plans identified the information needed by the staff in performing their technical review of applications and suggested a format for its presentation.

Another factor contributing to the commission's increasingly tough regulatory standards was a dramatic increase in the size of nuclear reactors. At the end of 1966, the largest reactor in operation was under 300 MW, but by the end of the decade, utilities were placing orders for plants two to three times larger. The problem was that larger, more powerful reactors posed more troublesome questions from a safety perspective. The fuel in larger reactors would overheat more quickly in the event of a failure in the plant's cooling system, and larger reactors contained higher levels of radioactivity. In short, larger reactors seemed to pose a greater risk to the public health and safety, especially because utilities were pressuring the commission to approve reactors closer to population centers.

3.3 The Rise of Energy Issues

The Arab oil embargo in October 1973 and the ensuing energy crisis profoundly altered the perception of the nation's energy policies in the eyes of the public and government officials and catapulted the energy issue to the top of the nation's political agenda. Although this turn of events might have been expected to bolster the prospects of nuclear power, it actually undermined them by subjecting the various energy subgovernments to intense scrutiny. With this increased scrutiny, energy interests were placed in direct competition for federal research and development funds; nuclear power, which had received the bulk of such monies over the years, would now face stiff competition. At the same time, government officials, seeking to establish a comprehensive and coordinated national energy plan, responded by reorganizing both the administrative and legislative branches. These organizational changes, in turn, would prove to be incompatible with continued domination of energy politics by subsystems.

3.4 The Fall of the AEC and JCAE

Largely as a result of the controversies over thermal pollution and radiation, by the early 1970s the AEC was on the defensive, suffering from a deteriorating public image and an almost complete lack of credibility. Recognizing the inevitability of the separation of the AEC's regulatory and developmental functions, members of the AEC, the JCAE, and the nuclear industry supported the Energy Reorganization

Act of 1974 because they viewed it as an opportunity to emphasize the contribution nuclear power could make to solving the nation's energy problems. Hoping to enhance the prospects of nuclear power, subsystem members favored the proposal even though it meant sacrificing the AEC, whose responsibilities would be transferred to two new agencies—the Energy Research and Development Administration (ERDA) and the Nuclear Regulatory Commission (NRC).

The abolition of the AEC deprived the nuclear industry of a powerful patron and created a more complicated institutional framework for nuclear matters. Where there was once a small group of insiders restricting participation to program supporters, by the middle of the decade the range of governmental participants would be decidedly broader and less exclusive. In addition to the Nuclear Regulatory Commission and ERDA, the courts, the Environmental Protection Agency (EPA), the Fish and Wildlife Service (FWS), and state and local governments claimed jurisdiction. The AEC, for example, had not been very concerned with the environmental impacts of reactors, but the EPA and FWS were. More and more, licensing decisions were challenged in court, and state and local governments became involved through their siting and rate-making powers. For opponents of nuclear power, these multiple venues translated into new opportunities to shape nuclear decision making and to appeal unfavorable decisions to other, more receptive governmental actors. For these reasons, institutional shifts were a critical factor in the expansion of the conflict over nuclear power and in eventual shifts in government policy. Like the AEC, the Joint Committee on Atomic Energy came under attack in the 1970s for being overly protective of the nuclear power industry. The growing wave of criticism over nuclear power, coupled with the demise of the AEC, prompted increased scrutiny of the Joint Committee and fueled calls for its abolition as well. Over the next few years, the once-invincible Joint Committee struggled to ward off challenges to its authority and to the nuclear program it had supported for so long. The joint committee would not win this battle, however, and it was eliminated by Congress in 1977.

With the demise of the JCAE, longtime supporters lost control of the strategic junctures of policy-making. Responsibility for nuclear policy was transferred to some two dozen subcommittees, some of which were sympathetic to the concerns voiced by antinuclear groups. For opponents of nuclear power, this decentralization reinforced previous venue changes and created additional access points to nuclear decision making. The new structural arrangements in Congress rendered the institution less exclusive and more permeable to outsiders, but also made it more complicated. Although nuclear policy once had been shaped exclusively the Joint Committee, it now seemed that everyone had some claim to nuclear oversight. With responsibility for nuclear policy dispersed so widely, it became increasingly difficult for Congress to resolve major policy issues.

Congress in the 1970s and 1980s was inhospitable to decisive initiatives from either side, preferring to approach nuclear policy in an ad hoc manner. Still fearful of energy shortages, members of Congress were unwilling to close the door on the nuclear option, deeming it an essential short-term energy source, but at the same time they were unwilling to proceed with a full-scale nuclear program because of concerns about reactor safety. For the next few years, Congress would increase funding for both reactor safety and licensing programs and also would insist on greater oversight of the NRC and the nuclear power industry. In this way, legislators could claim that they were doing everything possible to provide the nation with safe and secure energy supplies.

4. Regulatory Failure and the Changing Economics of Nuclear Power

In the late 1960s, the costs of producing electricity from coal and nuclear plants were comparable; nuclear plants were more expensive to build but had lower fuel and operating costs. By the early 1970s, utilities were ordering roughly equal amounts of coal and nuclear capacity. Within a few years, though, the bottom fell out of the market, sparing no sector of the industry. Significantly, nuclear power began its demise well before the accident at Three Mile Island in March 1979. Orders for new nuclear plants actually began to decline in 1974, and no orders would be placed after 1978. Furthermore, between 1974 and 1984, utilities canceled plans for over 100 reactors, many of which were already under construction. Some of the abandoned reactors were more than 50% complete.

The industry's economic woes were attributable to internal and external forces that caused construction and maintenance costs to skyrocket and utilities to scale back their plans to build additional nuclear generating capacity. Among the exogenous forces, none was more important than the Organization of Petroleum Exporting Countries (OPEC) oil embargo, which sparked sudden price hikes in fossil fuels and fundamentally transformed energy consumption patterns. Higher oil prices and the desire for secure energy sources might have been expected to make nuclear-generated electricity more attractive. Instead, fears of energy shortages and rising apprehension about American dependence on imported oil prompted Americans to use less energy, thus undercutting the need for new sources of electricity.

The nuclear industry was also hurt by rapidly rising construction costs. Construction costs for all types of power plants increased between 1968 and 1977, but those for nuclear plants rose at twice the rate of coal plants and continued to increase into the next decade. By almost any measure, nuclear plants became prohibitively expensive to build. Reactors coming on-line during the 1980s, for example, cost an average of $3500/kilowatt-hour in 1983 dollars, compared with an average of $600/kWh for the 57 reactors completed before 1981. A significant portion of the increased construction costs was attributable to new regulatory requirements that increased the amount of materials, labor, and time needed to build nuclear plants. Although the nuclear industry argued that most of their problems, especially the rising construction costs, were the result of overly zealous NRC regulators, in reality, the claim of excessive regulation is simply inaccurate and ignores other more compelling explanations.

It seems clear that earlier regulatory failures and industry mismanagement were responsible for some of the cost increases. The AEC neglected its regulatory duties for most of its history and devoted the bulk of its attention and resources to working with the private sector to develop and promote the new technology. Although this process did facilitate reactor licensing in the short term, the case-by-case approach ultimately undermined the industry's long-term prospects. During this period of rapid scale-up, no two reactor designs were alike, so every application was unique, and every application review was unique. In addition, the larger plants were more complex and posed greater potential risks. This increase in the number and complexity of applications, along with the failure to standardize reactor designs, posed tremendous difficulties for the AEC staff, which lacked the resources and experience to determine if the designs were adequate and if existing regulatory standards would assure public health and safety. The staff could not keep up, and licensing review times subsequently increased. The AEC responded to the growing licensing workload by nearly doubling its staff and by developing standard review plans to formalize the review process. As the licensing and regulatory staffs became larger and more sophisticated, they adopted new environmental and safety reviews and demanded ever more detailed information from applicants. Furthermore, as the quality and quantity of reviews increased, the staff discovered that existing standards were often inadequate and began to impose more stringent safety and environmental requirements on both new and existing plants.

Other regulatory changes were a product of increased operating experience. Once a sufficient number of reactors actually came on-line in the 1970s, the commission and the industry had a chance to evaluate the sufficiency of existing standards, designs, and operating procedures. Over time, previously unanticipated equipment and design problems

were discovered, and remedial steps were taken to correct them.

What these examples show is that many of the industry's economic wounds were either self-inflicted or were the result of regulatory failures by the AEC. If the industry and the AEC had not rushed to commercialize nuclear power before the technology was mature, much of the political controversy, licensing delays, and economic ills might have been avoided. A strong regulatory program might have detected potential problems early on, before reactors were built, and even more important, before they started producing electricity. If such a program had been in place, costly design changes and construction stoppages would not have been unnecessary. But that is not what happened. Instead, the discovery of unanticipated problems and reluctant admissions that some existing standards were insufficient undermined confidence in the agency and the industry and fueled the growing safety controversy.

5. Nuclear Power in the 1980s

5.1 The NRC after Three Mile Island

At approximately 4:00 a.m. on 28 March 1979, there was a serious accident at the Three Mile Island (TMI) nuclear facility near Harrisburg, Pennsylvania. Although it would take months to reconstruct the events and circumstances that caused the accident, it was much easier to calculate the political damage to the nuclear power industry and the NRC. Put simply, the TMI accident renewed and deepened concerns about the safety of reactors. A blue-ribbon panel appointed by President Carter to investigate the accident at Three Mile Island was highly critical of the NRC and the nuclear industry and recommended major changes in the practices and attitudes of both. After concluding that there was "no well-thought-out, integrated system for the assurance of nuclear safety within the NRC," the Kemeny Commission recommended a "total restructuring" of the agency. That did not happen, and the NRC never became an effective regulator.

In part, this was because the NRC was not really a new agency—it was merely a spin-off of the regulatory branch of the AEC. Although the Energy Reorganization Act succeeded in abolishing the Atomic Energy Commission, it never entirely displaced the commission's deeply entrenched belief in the value of nuclear power. Part of the explanation stems from the fact that the NRC was essentially a carryover from the AEC in terms of personnel, regulations, and attitudes. Indeed, the NRC's first official action was to adopt all of the AEC's rules, regulations, and standards. Throughout the 1980s, the NRC behaved much like its predecessor. The agency consistently tried to reduce licensing delays, blocked the imposition of new safety requirements, and worked to curtail public involvement in agency proceedings. To a large extent, the NRC was responding to signals from Presidents Reagan and Bush, both of whom supported nuclear power and pushed a deregulatory agenda. Reagan, in particular, tried to control the NRC by appointing commissioners who shared his aversion to regulation.

When necessary, the NRC waived and changed rules that threatened to close existing plants or to delay the licensing of new ones. In several cases when plants did not meet the regulations, the NRC simply changed the regulations to facilitate licensing. The most notorious rule change involved the Shoreham, Long Island and Seabrook, New Hampshire reactors. After the accident at Three Mile Island revealed serious flaws in emergency and evacuation planning, the NRC issued new rules that required an emergency-planning zone within a 10-mile radius of nuclear plants. The rules also mandated state and local government approval of evacuation plans for all people living within the zone. When state and local officials adjacent to the two plants refused to prepare emergency plans, the NRC issued a new rule waiving the requirement that state and local officials participate in evacuation planning.

5.2 The Problem of Nuclear Waste Disposal

The question of what to do with nuclear waste did not receive much attention in the early years of the atomic program. The AEC and others in the subgovernment were more concerned with the scientific challenges of harnessing the atom. Nuclear waste, in contrast, was generally assumed to be a less complex, essentially technical problem that would be easy to solve. Reflecting its low priority within the AEC, action on nuclear waste disposal was deferred indefinitely. This decision, like so many others in the program's early years, only made the problem worse, because the lack of an effective nuclear waste policy became one of the antinuclear movement's most potent arguments in the 1980s.

The problem for utilities was that spent reactor fuel was piling up in cooling ponds at reactor sites around the nation. The cooling ponds, however, were not designed to store waste permanently, and most could hold no more than 3 years of spent fuel. Moreover, utilities could not expand their on-site storage capacity without state regulatory approval, which was by no means assured. By the late 1970s, with many utilities worried about running out of space and facing mounting on-site storage costs, the nuclear industry sought legislative relief, arguing that a federal commitment on high-level waste was essential to its future.

After several years of deliberation, Congress acted in 1982, adopting the Nuclear Waste Policy Act (NWPA). The NWPA established detailed procedures and a schedule for the selection, construction, and operation of two permanent high-level waste repositories. Under the law, the Department of Energy (DOE) would conduct an intensive nationwide search for potential sites, which would be evaluated with a demanding set of technical criteria and guidelines, including environmental assessments. The DOE was then to compile a list of potential sites from which the President would select two for further review and site characterization, one in the eastern United States and one in the west. By all accounts, implementation of the Nuclear Waste Policy Act was a failure. Although the DOE met the act's 1983 deadline for issuing siting guidelines, virtually every other stage of the selection process was bitterly contested. All of the states considered as potential repository sites challenged the designation. With the states in open rebellion, Congress tried to repair the damage by amending the NWPA. In December 1987, Congress rejected the multiple-site search process established in the legislation and instead designated Yucca Mountain, Nevada, as the repository site. The search for a second site in the east was abandoned.

The case of nuclear waste disposal demonstrates the increasingly decentralized nature of nuclear politics in the 1980s. The widespread search for suitable nuclear waste disposal sites touched many states and communities and brought multiple new actors, in the form of state and local governments, to the policy-making arena. Whatever their motivations, the new actors brought different perspectives and goals to the debate; indeed, most actively resisted federal efforts to site nuclear waste facilities. With states and local governments assuming a critical role in policy-making, it is clear that the conflict over nuclear waste had expanded far beyond the confines of subgovernment politics. At the same time, however, the nuclear waste issue illustrates that support for nuclear power within the federal government remained strong in the 1980s. The Department of Energy, the NRC, and Presidents Reagan and Bush supported nuclear power, and worked to reestablish a favorable political climate. With states impeding the siting of a high-level repository, the industry sought to shift the financial and legal risks of nuclear waste management to the federal government, which willingly obliged.

5.3 The Accident at Chernobyl

In 1986, a fire at a reactor in Chernobyl, in the Soviet Union, released a large radioactive cloud across much of the Ukraine and northern Europe. The world learned of the accident when monitors in Europe detected an unusual spike in radiation levels. Soviet officials initially denied that anything had happened, but eventually conceded that something had gone horribly wrong at Chernobyl. Hundreds died, thousands were evacuated, and large areas were rendered uninhabitable. The story dominated the news throughout the world and reinforced questions about the safety of nuclear reactors. Although nuclear officials in the United States and elsewhere correctly noted that the Soviet reactor had fewer redundant safety systems than reactors in their own nations, the industry suffered a severe crisis in public confidence. In the United States and many European nations, the Chernobyl accident further undermined the prospects for a nuclear resurgence. Outside of France, which remained committed to a nuclear future, few European nations built new reactors.

6. Nuclear Power Today

6.1 Nuclear Power in the United States

American policymakers continue to be ambivalent about commercial nuclear power. Reflecting this ambivalence, nuclear policymaking in the past 20 years has been marked by incrementalism, with policymakers unwilling, or unable, to stray far from the status quo. Barring another energy or environmental crisis, it is extremely unlikely that American utilities will soon begin building any additional nuclear plants. And in the absence of a severe accident, policymakers are equally unlikely to require the shutdown of existing reactors or to rule out the possibility of future contributions from nuclear power. In the interim, the range of acceptable policy options will continue to lie somewhere in between the two extremes.

Commercial nuclear power peaked in the United States in 1990, when 112 reactors were in operation. Today, there are 103 operable reactors, in 31 states, generating approximately 20% of the nation's electricity. All told, U.S. utilities have placed orders for 259 nuclear reactors, but none has been ordered since 1978. Of the total order, 132 reactors received full power licenses; the rest were canceled. No plant has commenced operations since 1996, and 28 reactors have been permanently closed down.

Despite these downward trends, nuclear capacity in the United States is projected to increase slightly over the next 25 years, mostly due to improved performance from existing plants and fewer nuclear retirements due to licensing extensions. Owners of 10 reactors have applied for and received 20-year license extensions, and another 20 have applications pending. Owners of an additional 20 reactors are expected to apply for license extensions in the next 6 years. The George W. Bush administration is currently working to increase nuclear power's contribution to the nation's energy needs. In 2001, as part of the administration's national energy strategy, the Department of Energy issued a "roadmap" to deploy new reactors in the United States by 2010. In another important move long sought by the nuclear industry, President Bush approved the siting of a high-level waste repository at Yucca Mountain, Nevada. Although the decision has been challenged by the state, the facility is scheduled to begin operation in 2010. If the decision stands, nuclear advocates will have removed, at least temporarily, one of the major political roadblocks to the technology's ultimate fate.

6.2 Nuclear Power after September 11

The terrorist attacks of September 11, 2001 thrust nuclear power into the spotlight once again, although not in the way that industry supporters hoped. One of the planes that brought down New York's Twin Towers flew by the Indian Point nuclear facility located just 35 miles north of the city. The NRC and industry leaders were forced to confront the question of whether either group was prepared to deal with such attacks, and their responses were not encouraging. Specifically, there are concerns that both reactors and spent fuel storage pools, which are often located adjacent to reactors, are vulnerable to terrorist attack. Critics charge, moreover, that the NRC and reactor owners have consistently underestimated the risk of such attack, and that the agency's security rules are too lax and are often ignored. In designing its rules, for example, the NRC did not consider attacks by plane to be a credible threat. And although earthquakes and other natural disasters were considered in the design of spent storage fuels, they were not designed to withstand terrorist attack. According to some reports, attacks on reactors and storage facilities could have consequences worse than the 1986 accident at Chernobyl.

6.3 Nuclear Power around the World

There are now 442 nuclear plants in operation worldwide, with an additional 35 under construction. Nuclear power provides about 17% of the world's electricity. Thirty nations now have nuclear reactors, with the vast majority located in Organization for Economic Cooperation and Development (OECD) nations, where nuclear power accounts for almost 24% of all electricity. According to the Nuclear Energy Institute, the top 10 nuclear-generating nations are, in order, the United States, France, Japan, Germany, Russia, South Korea, United Kingdom, Canada, Ukraine, and Sweden. Nuclear power provides more than 75% of total electricity production in Lithuania and France, more than 50% in Belgium and Slovakia, more than 40% in Ukraine, Sweden, and Bulgaria, and about 30% in 10 other nations. The United States, with 104 operating reactors, has the most; France has 59 and Japan has 54. Japan also has plans to build up to 10 additional plants in the next 10 years.

Current projections are that global energy demand will more than double in the next 50 years, leading more nations to consider nuclear power as an answer to their energy needs. The roster of nations that now have reactors includes a growing number that have been marked by economic, political, or social instability, or that are located in volatile regions of the globe. Among the nations that have added commercial reactors in the past 5 years are Pakistan, India, South Korea, and Brazil. Some of these nations and others, including Iran and North Korea, have reactors that have been linked to nuclear weapons programs. The spread of reactor technology to such nations raises considerable concerns about nuclear proliferation and, after September 11, about reactor vulnerability to terrorism. It is an open question whether these nations have the ability to secure their reactors and the nuclear materials they use and produce. As a result, the future of nuclear power remains clouded.

7. Future Implications

As recently as 2008, the outlook for nuclear power was brighter than it had been in decades. Observers spoke openly of the possibility of a "nuclear renaissance" fueled by rising concerns over climate change, dramatic spikes in the price of natural gas and oil and, in the United States, a host of new federal subsidies contained in the Energy Policy Act of 2005. By the end of the year, however, the worst global economic downturn in a generation threatened to once again derail any nuclear revival. At this time, it is unclear how global energy demand will be affected by the economic crisis and, with credit markets tight, it seems likely that capital intensive projects like nuclear plants will be subject to great scrutiny.

Advocates of nuclear power claim that we should build more nuclear plants because they do not emit greenhouse gases and that the technology thus represents a ready solution to the climate change problem. Although it is true that nuclear power plants emit far less carbon than either gas-fired or coal burning plants, for nuclear power to make a significant contribution to the climate change problem, we would have to embark on an unprecedented program of reactor building, and sustain it for decades. According to calculations by the Keystone Center, the U.S. would need to build reactors at the rate of the 1980s, when 20 GWe per year of nuclear capacity was added, and sustain it for fifty years. That rate is higher than the industry's average historical growth rate and is much higher than forecasts by the U.S. Energy Information Administration.

Even if we were to embark on a large-scale nuclear program, it would be decades before these plants would come on line, and they would thus do very little to combat climate change in the near term. In addition, given the aging of the current generation of reactors, we would have to replace the electricity they generate as well. Such a scale-up would mean many more enrichment and processing facilities, and waste disposal sites. Even if another 1000–1500 new 1000 Mwe reactors were built worldwide, nuclear would still provide only 19% of the world's electricity, much higher than the current amount.

Nuclear power would certainly benefit if some form of a carbon tax was established. Of course, a carbon tax would improve the position of all low greenhouse gas energy sources, including a variety of renewables, coal with carbon sequestration, and investments in energy efficiency. The higher the tax, the greater the advantage of low carbon emitting technologies relative to fossil fuels.

Complicating the chances of a nuclear revival in the United States is the lack of progress in addressing concerns about waste disposal. Although the Bush Administration backed the Yucca Mountain repository, President Obama's first budget contained virtually no money for the project, effectively killing it and dealing the industry a severe setback.

In the United States the industry is also plagued by lingering concerns over reactor safety. The industry and its supporters point to the advantages of new, simplified reactor designs featuring passive safety features, fewer components, and lower material and construction costs. For all of their possible virtues, however, most of the new reactor designs are untested, so no one really knows if they will work as claimed; nor do they know if such reactors really will be cheaper to build and operate. History shows us that the claims of industry supporters about economics were wildly optimistic.

Some independent analysts have also raised questions about potential materials and labor shortages should there be a rush to nuclear power, suggesting that there are not enough nuclear science and engineering students, or nuclear technicians, to staff additional plants. Others have suggested that a limited global supply of key reactor components like pressure vessels, steam generators, and cooling pumps could create procurement and manufacturing bottlenecks that would result in higher than projected construction costs. Because nuclear plants are capital intensive and have long lead times, this is a potentially serious problem.

At the same time, average reactor operating costs have dropped significantly, in large part because reactors are now operating at higher capacity levels—nearly 90% of total capacity in 2006. In the U.S., average operating and maintenance costs have dropped from a high of about 3.5 cents per kwh in 1987 to about 1.7 cents per kwh in 2005. But historically construction costs increased as size of plants increased and as more plants were being built.

To state the obvious, it is hard to predict the future. Energy forecasts are notoriously unreliable. In the 1960s and early 1970s, projections of rapid growth in the demand for electricity lead many utilities to order reactors. Then the oil shocks in 1970s slowed economic growth dramatically, and the projected demand never materialized and utilities cancelled the reactors. But if the nuclear industry were able to reduce construction, operation, and maintenance costs significantly from previous levels, their product would be more competitive with other fuel sources.

Nuclear power in the United States has never been competitive without huge government subsidies. Although the economic incentives offered in the 2005 Energy Policy Act are significant, they apply to only a handful of reactors. While it is conceivable that a few reactors will be built, a broad revitalization will require fundamental shifts in nuclear economics, and a resolution of the lingering concerns with waste disposal, reactor safety, and weapons proliferation.

Further Reading

Balogh B 1991 *Chain Reaction: Expert Debate and Public Participation in American Commercial Nuclear Power, 1945–1975.* Cambridge University Press, Cambridge

Bupp I C, Derian J-C 1978 *Light Water: How the Nuclear Dream Dissolved.* Basic Books, New York

Campbell J L 1988 *Collapse of an Industry: Nuclear Power and the Contradictions of U.S. Policy.* Cornell University Press, Ithaca, New York

Duffy R J 1997 *Nuclear Politics in America: A History and Theory of Government Regulation.* University Press of Kansas, Lawrence

Falk J 1982 *Global Fission: The Battle over Nuclear Power.* Oxford University Press, New York

Green H P, Rosenthal A P 1963 *Government of the Atom: The Integration of Powers.* Atherton Press, New York

Hewlett R G, Anderson O E 1962 *A History of the Atomic Energy Commission, Vol. 1, The New World, 1939–1946.* Pennsylvania State University Press, University Park

Hewlett R G, Duncan F 1969 *A History of the Atomic Energy Commission, Vol. 2, Atomic Shield, 1947–1952* Pennsylvania State University Press, University Park, PA

Jasper J M 1990 *Nuclear Politics: Energy and the State in the United States, Sweden, and France.* Princeton University Press, Princeton, New Jersey

Nelkin D, Pollak M 1981 *The Atom Besieged: Extraparliamentary Dissent in France and Germany.* MIT Press, Cambridge, Massachusetts

Union of Concerned Scientists. 1987. *Safety Second: The NRC and America's Nuclear Power Plants* (M. Adato, principal author, with J. MacKenzie, R. Pollard, and E. Weiss). Indiana University Press, Bloomington, Indiana.

Robert J. Duffy

O

Oil Crises, Historical Perspective

Glossary

aramco The Arabian American oil company that obtained the concession for oil exploration in the Eastern Province of Saudi Arabia in 1933. It was originally made up of four American oil companies: Standard Oil of California (Socal), Texaco, Standard Oil of New Jersey (then called Esso and later changed to Exxon Mobil), and Socony-Vacuum (Mobil). In 1976, it was purchased by the Saudi government and renamed "Saudi Aramco."

conventional crude oil The oil produced from an underground reservoir, after being freed from any gas that may have dissolved in it under reservoir conditions, but before any other operation has been performed on it. In the oil industry, simply termed "crude."

proven reserves The quantities that geological and engineering information indicates with reasonable certainty can be recovered from known reserves under existing economic and operating conditions.

renewable energy An energy source that does not depend on finite reserves of fossil or nuclear fuels, such as solar energy, wind energy, biomass, hydroelectric power, and hydrogen. All of these renewable sources involve the generation of electricity.

shale oil A distillate obtained when oil shale (a rock of sedimentary origin) is heated in retorts.

ultimate global reserves This is the amount of oil reserves that would have been produced when production eventually ceases.

unconventional oil Oil that has been extracted from tar sands, oil shale, extra heavy oil, and the conversion of natural gas to liquids, known collectively as synthetic fuels or "synfuels."

The 20th century was truly the century of oil. Though the modern history of oil begins in the latter half of the 19th century, it is the 20th century that has been completely transformed by the advent of oil. Oil has

a unique position in the global economic system. One could not imagine modern societies existing without oil. Modern societies' transportation, industry, electricity, and agriculture are virtually dependent on oil. Oil makes the difference between war and peace. The importance of oil cannot be compared with that of any other commodity or raw material because of its versatility and dimensions, namely, economic, military, social, and political. The free enterprise system, which is the core of the capitalist thinking, and modern business owe their rise and development to the discovery of oil. Oil is the world's largest and most pervasive business. It is a business that touches every corner of the globe and every person on earth. The financial resources and the level of activity involved in exploring, refining, and marketing oil vastly exceed those of any other industry. Of the top 20 companies in the world, 7 are oil companies. Human beings are so dependent on oil, and oil is so embedded in daily life, that individuals hardly stop to comprehend its pervasive significance. Developing nations give no indication that they want to deny themselves the benefits of an oil-powered economy, whatever the environmental questions. In addition, any notion of scaling back the world's consumption of oil will be influenced by future population growth.

1. Introduction

No other commodity has been so intimately intertwined with national strategies and global politics and power as oil. In World War I, the Allies floated to victory on a wave of oil. Oil was central to the course and outcome of World War II in both the Far East and Europe. One of the Allied powers' strategic advantages in World War II was that they controlled 86% of the world's oil reserves. The Japanese attack on Pearl Harbor was about oil security. Among Hitler's most important strategic objectives in the invasion of the Soviet Union was the capture of the oilfields in the Caucasus. In the Cold War years, the battle for the control of oil resources between international oil companies and developing countries was a major incentive and inspiration behind the great drama of decolonization and emergent nationalism.

During the 20th century, oil emerged as an effective instrument of power. The emergence of the United States as the world's leading power in the 20th century coincided with the discovery of oil in America and the replacement of coal by oil as the main energy source. As the age of coal gave way to oil, Great Britain, the world's first coal superpower, gave way to the United States, the world's first oil superpower.

Yet oil has also proved that it can be a blessing for some and a curse for others. Since its discovery, it has bedeviled the Middle East with conflict and wars. Oil was at the heart of the first post-Cold War crisis of the 1990s—the Gulf War. The Soviet Union—the world's second largest oil exporter—squandered its enormous oil earnings in the 1970s and 1980s in a futile military buildup. And the United States, once the largest oil producer and still its largest consumer, must import almost 60% of its oil needs, weakening its overall strategic position and adding greatly to an already burdensome trade deficit—a precarious position for the only superpower in the world.

The world could face an energy gap probably during the first decade of the 21st century once global conventional oil production has peaked. This gap will have to be filled with unconventional and renewable energy sources. A transition from fossil fuels to renewable energy sources is, therefore, inevitable if humans are to bridge the energy gap and create a sustainable future energy supply. Sometime in the 21st century, nuclear, solar, geothermal, wind, and hydrogen energy sources may be sufficiently developed to meet a larger share of the world's energy needs. But for now humans will continue to live in an age of oil. Oil will, therefore, still be supplying a major share of the global energy needs for most, perhaps all, of the 21st century and will continue to have far-reaching effects on the global economy.

2. The Road to the First Oil Crisis

One distinctive feature dominated the global economic scene in the decades following World War II. It was the rising consumption of oil. Total world energy consumption more than tripled between 1949 and 1972. Yet that growth paled in comparison to the rise in oil demand, which during the same period increased more than $5\frac{1}{2}$ times over. Everywhere, growth in the demand for oil was strong. Between 1948 and 1972, consumption tripled in the United States, from 5.8 to 16.4 million barrels/day—unprecedented except when measured against what was happening elsewhere. In the same years, demand for oil in Western Europe increased 15 times over, from 970,000 barrels/day to 14.1 million barrels/day. In Japan, the change was nothing less than spectacular; consumption increased 137 times over, from 32,000 barrels/day to 4.4 million barrels/day.

The main drivers of this global surge in oil use were the rapid economic growth and the cheap price of oil. During the 1950s and 1960s, the price of oil fell until it became very cheap, which also contributed mightily to the swelling of consumption. Many governments encouraged its use to power economic growth and industrial modernization. There was one final reason that the market for oil grew so rapidly. Each oil-exporting country wanted higher volumes of its oil sold in order to gain larger revenues.

In the buoyant decades following World War II, oil overtook coal as the main fuel for economic growth. Huge volumes of oil surged out of Venezuela and the Middle East and flowed around the world. Oil was abundant. It was environmentally more attractive and was easier and more convenient to handle. And oil became cheaper than coal, which proved the most desirable and decisive characteristic of all. Its use provided a competitive advantage for energy-intensive industries. It also gave a competitive edge to countries that shifted to it. And yet, there was a haunting question: How reliable was the flow of oil on which modern societies had come to depend? What were the risks?

Among the Arabs, there had been talk for more than a decade about wielding the "oil weapon." This was their chance. On June 6, 1967, the day after the start of the Six-Day War, Arab oil ministers, members of Organization of Arab Petroleum Exporting Countries, formally called for an oil embargo against countries friendly to Israel. Saudi Arabia, Kuwait, Iraq, Libya, and Algeria thereupon banned shipments to the United States and Great Britain.

By June 8, the flow of Arab oil had been reduced by 60%. The overall initial loss of Middle Eastern oil was 6 million barrels/day. Moreover, logistics were in total chaos not only because of the interruptions but also because, as in 1956, the Suez Canal and the pipelines from Iraq and Saudi Arabia to the Mediterranean were closed. The situation grew more threatening in late June and early July when, coincidentally, civil war broke out in Nigeria, depriving the world oil market of 500,000 barrels/day at a critical moment.

However, by July 1967, a mere month after the Six-Day War, it became clear that the selective Arab oil embargo was a failure; supplies were being redistributed to where they were needed. And by the beginning of September, the embargo had been lifted.

The 1970s saw a dramatic shift in world oil. Demand was catching up with available supply and the 20-year surplus was over. As a result, the world was rapidly becoming more dependent on the Middle East and North Africa for its oil.

Oil consumption surged beyond expectation around the world, as ever-greater amounts of petroleum products were burned in factories, power plants, homes, and cars. The cheap price of oil in the 1960s and early 1970s meant that there was no incentive for fuel-efficient automobiles. The late 1960s and early 1970s were also the watershed years for the domestic U.S. oil industry. The United States ran out of surplus capacity. In the period 1957 to 1963, surplus capacity in the United States had totaled approximately 4 million barrels/day. By 1970, it had declined to only 1 million barrels/day. That was the year, too, that American oil production peaked at 10 million barrels/day. From then on, it began its decline, never to rise again. With consumption continuing to rise, the United States had to turn to the world oil market to satisfy its needs. Net imports tripled from 2.2 million barrels/day in 1967 to 6 million barrels/day by 1973. Imports as a share of total oil consumption over the same years rose from 19 to 36%.

The disappearance of surplus capacity in the United States would have major implications, for it meant that the "security margin" on which the Western world depended was gone. For the United States, it marked a shift from: (1) oil self-sufficiency to reliance on Middle East oil; (2) being a major exporter to becoming a major importer; (3) loss of the ability to control the global oil markets at a time when Middle East oil producers (Arab Gulf producers) began to assert themselves on the global markets; and (4) inability to provide stand-by supply to its allies in an emergency. This meant a major shift from energy security to vulnerability and dependency. Indeed, the razor's edge was the ever-increasing reliance on the oil of the Middle East. New production had come from Indonesia and Nigeria (in the latter case, after the end of its civil war in early 1970), but that output was dwarfed by the growth in Middle Eastern production. Between 1960 and 1970, Western world oil demand had grown by 21 million barrels/day. During that same period, production in the Middle East (including North Africa) had grown by 13 million barrels/day. In other words, two-thirds of the huge increase in oil consumption was being satisfied by Middle East oil.

In a wider sense, the disappearance of surplus capacity caused an economic and geopolitical transformation of the global oil market. In an economic sense, the "center of gravity" of oil production, energy security, and control of global oil supplies had shifted from the United States to the Middle East. In a geopolitical sense, the oil revenue and the global dependence on Middle East oil provided the Arab Gulf producers with unprecedented political influence. This they channeled into support of their political aspirations as they did during the 1973 war and the resulting oil embargo.

Another disturbing development was that the relationship between the oil companies and the producing nations was beginning to unravel. In gaining greater control over the oil companies, whether by participation or outright nationalization, the exporting countries also gained greater control over prices. The result was the new system that was forged in Tehran and Tripoli, under which prices were the subject of negotiation between companies and countries, with the producing countries taking the lead in pushing up the posted price.

However, the supply-demand balance that emerged at the beginning of the 1970s was sending a most important message: Cheap oil had been a tremendous boon to economic growth, but it could not be sustained. Demand could not continue growing at the rate it was; new supplies needed to be developed. That was what the disappearance of spare capacity meant. Something had to give, and that something was price.

By 1972, many experts reckoned that the world was heading for an acute oil shortage in a few years. The signs of a shortage were visible everywhere. The demand for oil in the summer of 1973 was going above the wildest predictions—in Europe, in Japan, and most of all in the United States. Imports from the Middle East to the United States were still racing up: production inside the United States was still falling. In April, U.S. President Nixon lifted restrictions on imports of oil, so the Middle East oil flowed in still faster. There was a new danger sign when Kuwait decided in 1972 to conserve its resources and to keep its production below 3 million barrels/day.

As late as 1970, there were still approximately 3 million barrels/day of spare capacity in the world outside the United States, with most of it concentrated in the Middle East. By the second half of 1973, the spare capacity had shrunk to only 500,000 barrels/day. That was just 1% of world consumption. With a 99% capacity utilization and a 1% security margin, the oil supply-demand balance was indeed extremely precarious.

In June 1973, as prices were zooming up, the Organization of Petroleum Exporting Countries (OPEC) summoned another meeting in Geneva to insist on another price increase because of the further devaluation of the U.S. dollar. The radicals—Algeria, Libya and Iraq—were pressing for unilateral control of price, but eventually OPEC agreed on a new formula that increased prices by another 12%. By September 1973, for the first time since the founding of OPEC, the market price of oil had risen above the posted price. It was a sure sign that OPEC was in a very strong bargaining position. Armed with this knowledge, OPEC invited the oil companies to meet them in Vienna on October 8 to discuss substantial increases in the price of oil.

In this atmosphere of crisis, the oil company delegates prepared to confront OPEC in Vienna on October 8. And then, just as they were leaving for Vienna, Egypt and Syria invaded Israeli-occupied territories. There was war.

While the shortage loomed, the Arabs were at last achieving closer unity. They were determined to use oil as a weapon against Israel and by 1973 the militants were being joined by Saudi Arabia. The very fact that Saudi Arabia had become the largest oil exporter made King Feisal more vulnerable in the face of his Arab colleagues and the danger of an embargo more likely, for he could not afford to be seen as a blackleg.

The international oil order had been irrevocably changed. However, it was not only a question of price, but of power. The extent of dependence by the industrial countries on oil as a source of energy had been exposed and the practicality of controlling supply as means of exerting pressure for raising the price of oil had been dramatically demonstrated. Although the oil weapon had not worked in 1967, the "rationale of those who called for its use as a weapon in the Middle East conflict has been strengthened in current circumstances."

3. The Oil Weapon

Contrary to popular belief, the Americans, not the Arabs, were the first to wield the oil weapon. They used it against Japan when on July 25, 1941, the United States announced a freeze on Japanese funds necessary for Japan to buy American oil, which, in practice, meant an embargo on oil. The embargo was the result of Japanese military aggression in Asia.

Increasingly worried about a cut-off of oil supplies from the United States, Tokyo instituted a policy to establish self-sufficiency and to try to eliminate dependence on U.S. oil supplies. In 1940–1941, it was energy security that led Japan to occupy the Dutch East Indies and take control of its oilfields. Indeed, the U.S. oil embargo was an important factor leading Japan to attack Pearl Harbor, bringing the United States into World War II. Oil had been central to Japan's decision to go to war.

Ever since the 1950s, the Arab world had been talking about using the oil weapon to force Israel to give up occupied Arab territories. Yet the weapon had always been deflected by the fact that Arab oil, though it seemed endlessly abundant, was not the supply of last resort.

In June 1967, 2 days into the Six-Day War, the Arabs wielded the oil weapon when they imposed an oil embargo against the United States and Great Britain. However, by July 1967, it became clear that the Arab embargo had failed. The Arabs would have to wait for another chance to wield the oil weapon again. That chance came their way when just moments before 2:00 P.M. on October 6, 1973, Egyptian and Syrian armies launched an attack on Israeli-held positions in the Sinai and the Golan Heights. Thus began the October War or, what became known as the Yom Kippur War, the fourth of the Arab-Israeli wars—the most destructive and intense of all of them and the one with the most far-reaching consequences. One of the most potent weapons used in this war was the oil weapon, wielded in the form of an embargo—production cutbacks and restrictions on exports—that, in the words of Henry Kissinger,

"altered irrevocably the world as it had grown up in the postwar period."

The embargo, like the war itself, came as a surprise and a shock. Yet the pathway to both in retrospect seemed in some ways unmistakable. By 1973, oil had become the lifeblood of the world's industrial economies and it was being pumped and circulated with very little to spare. Never before in the entire postwar period had the supply-demand equation been so tight, while the relationships between the oil-exporting countries and the oil companies continued to unravel. It was a situation in which any additional pressure could precipitate a crisis—in this case, one of global proportions.

With supply problems becoming chronic in the early 1970s, talk about an energy crisis began to circulate in the United States. There was agreement, in limited circles, that the United States faced a major problem. Price controls on oil, imposed by Nixon in 1971 as part of his overall anti-inflation program, were discouraging domestic oil production while stimulating consumption. The artificially low prices provided little incentive either for new exploration or for conservation. By the summer of 1973, United States oil imports were 6.2 million barrels/day, compared to 4.5 million barrels/day in 1972 and 3.2 million barrels/day in 1970. The oil trade journal *Petroleum Intelligence Weekly* reported in August 1973 that "near-panic buying by the U.S., the Europeans, and the Japanese was sending oil prices sky-rocketing."

As global demand continued to rise against the limit of available supply, market prices exceeded the official posted prices. It was a decisive change, truly underlining the end of surplus. For so long, reflecting the chronic condition of oversupply, market prices had been below posted prices, irritating relations between companies and governments. But the situation had reversed and the exporting countries did not want to see the growing gap between the posted price and the market price go to the companies. Wasting little time, the exporters sought to revise their participation and buy-back arrangements so that they would be able to obtain a larger share of the rising prices.

One of the principal characteristics governing the operations of the oil industry is that it generates an important "economic rent" or "oil surplus," the appropriation of which involves three players: the exporting countries, the consuming countries, and the multinational oil companies. Both the exporting countries and the consuming countries are effectively staking a claim to the significant element of economic rent built into the price of oil. For the exporters, such as the Arab Gulf producers, oil remains the single most important source of income, generating approximately 85 to 90% of their revenues. Significantly, consumer countries have always looked on oil as an important source of taxation since demand for it is relatively inelastic; that is to say, it varies little as the price changes. These countries have more maneuverability when the price of oil is low. This was amply demonstrated when, as a result of the oil price collapse in 1986, many of them took the opportunity to raise tax rates on petroleum products. This practice was, to a lesser extent, in operation in the 1970s but has accelerated since 1986 in Europe, with tax levels on petroleum products reaching between 80 and 87%. In other words, the sharp increases in taxes by the consuming countries were intended to cream off more of the rent. Is it any wonder that the United Kingdom has been for years earning far more revenue from taxes on petroleum products than from its North Sea oil exports?

On September 1, 1973, the fourth anniversary of Muamer Qaddafi's coup, Libya nationalized 51% of those company operations it had not already taken over. The radicals in OPEC—Iran, Algeria, and Libya—began pushing for a revision in the Tehran and Tripoli agreements. By the late summer of 1973, the other exporters, observing the upward trend of prices on the open market, came around to that same point of view. They cited rising inflation, the dollar's devaluation, and also the rising price of oil. By September 1973, the Saudi oil minister Sheikh Ahmed Zaki Yamani was able to announce that the Tehran Agreement was dead. Even as the economics of oil were changing, so were the politics that surrounded it—and dramatically so.

By April of 1973, President Anwar Sadat of Egypt had begun formulating with Syria's Hafez Al-Asad strategic plans for a joint Egyptian-Syrian attack against Israel. Sadat's secret was tightly kept. One of the few people outside the high commands of Egypt and Syria with whom he shared it was King Feisal. And that meant oil would be central to the coming conflict.

In the early 1970s, as the market tightened, various elements in the Arab world became more vocal in calling for use of the oil weapon to achieve their economic and political objectives. King Feisal was not among them. He had gone out of his way to reject the use of the oil weapon. "It was not only useless," he said, "but dangerous even to think of that. Politics and oil should not be mixed." Yet, by early 1973, Feisal was changing his mind. Why?

Part of the answer lay in the marketplace. Much sooner than expected, Middle Eastern oil had become the supply of last resort. In particular, Saudi Arabia had become the marginal supplier for everybody, including the United States; American dependence on the Gulf had come not by the widely predicted 1985, but by 1973. The United States would no longer be able to increase production to supply its allies in the event of a crisis and the United States itself was now, finally, vulnerable. The supply-demand balance was working to make Saudi Arabia even more powerful. Its share of world exports had risen rapidly from

16% in 1970 to 25% in 1973 and was continuing to rise.

In addition, there was a growing view within Saudi Arabia that it was earning revenues in excess of what it could spend. Two devaluations of the U.S. dollar had abruptly cut the worth of the financial holdings of countries with large dollar reserves, including Saudi Arabia.

The changing conditions in the marketplace, which with each passing day made the Arab oil weapon more potent, coincided with significant political developments. By the spring of 1973, Sadat was strongly pressing Feisal to consider using the oil weapon to support Egypt in a confrontation with Israel. King Feisal also felt growing pressure from many elements within his kingdom and throughout the Arab world. Thus, politics and economics had come together to change Feisal's mind. Thereupon the Saudis began a campaign to make their views known, warning that they would not increase their oil production capacity to meet rising demand and that the Arab oil weapon would be used, in some fashion, unless the United States moved closer to the Arab viewpoint and away from Israel.

On August 23, 1973, Sadat made an unannounced trip to Riyadh to see King Feisal. He told the king that he was considering going to war against Israel. It would begin with a surprise attack and he wanted Saudi Arabia's support and cooperation. He got it.

On October 17, 1973, 11 days into the war, Arab oil ministers meeting in Kuwait agreed to institute a total oil embargo against the United States and other countries friendly to Israel. They decided to cut production 5% from the September level and to keep cutting by 5% in each succeeding month until their objectives were met. Oil supplies at previous levels would be maintained to "friendly states." One clear objective of the plan was to split the industrial countries right from the start.

On October 19, Nixon publicly proposed a $2.2 billion military aid package for Israel. In retaliation for the Israeli aid proposal, Saudi Arabia had gone beyond the rolling cutbacks; it would now cut off all shipments of oil, every last barrel, to the United States. The oil weapon was now fully in battle—a weapon, in Kissinger's words, of political blackmail. To counter that blackmail, Kissinger called for the industrialized nations to meet in Washington, DC at the earliest possible moment. He wanted the oil-consuming West to make a united stand against the Arabs.

4. The Embargo

The embargo came as an almost complete surprise despite the evidence at hand: almost two decades of discussion in the Arab world about the oil weapon, the failed embargo in 1967, Sadat's public discussion of the "oil option" in early 1973, and the exceedingly tight oil market of 1973. What transformed the situation and galvanized the production cuts and the embargo against the United States was the very public nature of the resupply of ammunitions and armaments to Israel and then the $2.2 billion aid package.

On October 21, Sheikh Yamani met with the president of Aramco, Frank Junkers. Using computer data about exports and destinations that the Saudis had requested from Aramco a few days earlier, Yamani laid out the ground rules for the cutbacks and the embargo the Saudis were about to impose. He told Junkers that any deviations from the ground rules would be harshly dealt with.

At the time of the embargo, the management of the Saudi oil industry was in the hands of Aramco (the Arabian-American Oil Company), the joint venture between Standard Oil of California (Socal), Texaco, Standard Oil of New Jersey (then called Esso and later changed to ExxonMobil), and Socony-Vacuum (Mobil).

In 1948, U.S. imports of crude oil and products together exceeded exports for the first time. No longer could the United States continue its historical role as supplier to the rest of the world. That shift added a new dimension to the vexing question of energy security. The lessons of World War II, the growing economic significance of oil, and the magnitude of Middle Eastern oil reserves all served, in the context of the developing Cold War with the Soviet Union, to define access to that oil as a prime element in Western security. Oil provided the point at which foreign policy, international economic considerations, national security, and corporate interests would all converge. The Middle East would be the focus. There the oil companies were already building up production and making new arrangements to secure their positions.

In Saudi Arabia, development was in the hands of Aramco. The company understood from the time it obtained the Saudi oil concession in 1933 that the concession would always be in jeopardy if it could not satisfy the expectations and demands of King Abdul Aziz Ibn Saud, the founder of Saudi Arabia, and the royal family. Since then, it has worked tirelessly to enhance Saudi oil reserves and production and build terminals and pipelines for exporting the oil worldwide.

In October 1972, Sheikh Yamani negotiated a participation agreement between Saudi Arabia and Aramco. It provided for an immediate 25% participation share, rising to 51% by 1983. Aramco had finally agreed to participation with Saudi Arabia because the alternative was worse—outright nationalization.

In June 1974, Saudi Arabia, operating on Yamani's principle of participation, took a 60% share in Aramco. By the end of the year, the Saudis told Aramco that 60% was simply not enough. They wanted 100%. An agreement to that effect was eventually reached in 1976 between Aramco and Saudi Arabia, almost 43 years after the concession was granted. By then, the proven reserves of Saudi Arabia were estimated at 149 billion barrels—more than one-quarter of the world's total reserves.

But the agreement did not by any means provide for a severing of links. Thus, under the new arrangement, Saudi Arabia would take over ownership of all Aramco's assets and rights within the country. Aramco could continue to be the operator and provide services to Saudi Arabia, for which it would receive 21 cents per barrel. In return, it would market 80% of Saudi production. In 1980, Saudi Arabia finally paid compensation, based on net book value, for all Aramco's holdings within the kingdom. With that, the sun finally set on the great concessions.

The Saudis had already worked out the embargo in some detail. They insisted that on top of the 10% cutback, Aramco must subtract all shipments to the United States, including the military. The Saudis asked Aramco for details of all crude oil used to supply American military forces throughout the world. The details were provided and the Saudis duly instructed Aramco to stop the supplies to the U.S. military. The situation was serious enough for Washington to ask whether British Petroleum (BP) could supply the U.S. Sixth Fleet in the Mediterranean.

At the beginning of November 1973, only 2 weeks after the initial decision to use the oil weapon, the Arab oil ministers decided to increase the size of the across-the-board cuts. This resulted in a gross loss of 5 million barrels/day of supply from the market. This time, however, there was no spare capacity in the United States. Without it, the United States had lost its critical ability to influence the world oil market. And with the price of a barrel of oil skyrocketing, the oil exporters could cut back on volumes and still increase their total income.

The panic and shortage of oil supplies caused by the embargo led to a quadrupling of crude oil price and precipitated a severe recession, which adversely affected the economies of the industrialized nations. Panic buying meant extra demand in the market. The bidding propelled prices even further upward. The posted price for Iranian oil, in accordance with the October 16 agreement, was $4.50 per barrel. In December, it sold for $22.60.

The oil crisis had far-reaching political and economic effects. The quadrupling of prices by the Arab oil embargo and the exporters' assumption of complete control in setting those prices brought massive changes to the world economy. The combined oil earnings of the oil exporters rose from $23 billion in 1972 to $140 billion by 1977. For the industrial countries, the sudden hike in oil prices brought profound dislocations. The oil rents flooding into the treasuries of the exporters added to a huge withdrawal of their purchasing power and sent them into deep recession. The U.S. gross national product (GNP) plunged 6% between 1973 and 1975 and unemployment doubled to 9%. Japan's GNP declined in 1974 for the first time since the end of World War II. At the same time, the price increases delivered a powerful inflationary shock to economies in which inflationary forces had already taken hold. President Nixon later commented that: "The oil embargo made it transparently clear that the economies of Western Europe and Japan could be devastated almost as completely by an oil cutoff as they could be by a nuclear attack."

On March 18, 1974, the Arab oil ministers agreed to end the embargo after the United States warned that peace efforts between the Arabs and Israel could not proceed without the lifting of the embargo. After two decades of talk and several failed attempts, the oil weapon had finally been successfully used, with an impact not merely convincing, but overwhelming, and far greater than even its proponents have dared to expect. It had transformed world oil and the relations between producers and consumers and it had remade the international economy. Now it could be resheathed. But the threat would remain.

5. *The Second Oil Crisis*

The Iranian revolution was at the heart of the second oil crisis. The astronomical oil price rises of 1979 and the emergence of the Rotterdam "Spot Market" were a direct consequence of the Iranian revolution. In January 7, 1978, a Tehran newspaper published a savage attack on an implacable opponent of the Shah, an elderly Shiite cleric named Ayatollah Ruhollah Khomeini, who was then living in exile in Iraq. This journalistic assault on Khomeini set off riots in the holy city of Qom, which remained his spiritual home. Troops were called in and demonstrators were killed. The disturbance in Qom ignited riots and demonstrations across the country, with further dramatic clashes and more people killed. Strikes immobilized the economy and the government and demonstrations and riots went on unchecked.

All through 1978, the internal political strife against the Shah's regime and the political drama that was unfolding simultaneously in Paris and Tehran were pushing Iran toward an explosion.

It had become evident in the mid-1970s that Iran simply could not absorb the vast increase in oil revenues that was flooding the country. The petrodollars, misspent on grandiose modernization programs or lost to waste and corruption, were generating economic chaos and social and political unrest throughout the nation. Iranians from every sector of national life were losing patience with the Shah's regime and the rush to modernization. Grasping for some certitude in the melee, they increasingly heeded the call of traditional Islam and of an ever more fervent fundamentalism. The beneficiary was Ayatollah Khomeini, whose religious rectitude and unyielding resistance made him the embodiment of opposition to the Shah's regime. The Iranian oil industry was in a state of escalating chaos. The impact of the strikes was felt immediately. On October 13, 1978, workers at the world's largest oil refinery, in Abadan, suddenly went on strike at the instigation of the exiled Ayatollah Khomeini, who was inciting them from Paris in recorded speeches on cassettes smuggled to Iran. Within a week, the strike had spread throughout most of the oil installations and Iran was, for all intents and purposes, out of the oil business.

Iran was the second-largest exporter of oil after Saudi Arabia. Of the 5.7 million barrels/day produced in Iran, approximately 4.5 million barrels/day were exported. By early November 1978, production decreased from 5.7 million barrels/day to 700,000 barrels/day. And by December 25, Iranian oil exports ceased altogether. That would prove to be a pivotal event in the world oil market. Spot prices in Europe surged 10 to 20% above official prices. The ceasing of Iranian exports came at a time when, in the international market, the winter demand surge was beginning. Oil companies, responding to the earlier general softness in the market, had been letting their inventories fall.

On December 26, Khomeini declared, "as long as the Shah has not left the country there will be no oil exports." On January 16, 1979, the Shah left the country. And on February 1, Khomeini landed in Tehran.

By the time the Khomeini regime decided to resume pumping, they could not restore production to prerevolutionary levels because the Ayatollah kicked out of Iran all the Western companies that operated the Iranian oilfields. During the production disruption, these fields lost gas pressure and the lack of maintenance made matters worse. The Iranians apparently did not have enough technical know-how to maintain or operate the fields. In effect, the Iranians pulled the rug from under the feet of the oil market, the world panicked, and the prices started to hit the roof.

Up until September 1977, there was actually a glut of oil in the market, which meant that the continuous rise in oil prices had come to a halt. The oil surplus was due to a number of factors. North Sea oil production increased much faster than expected, Mexico became a large oil exporter, and in July 1977 oil from Alaska started its long-awaited flow to the United States. At the same time, Saudi Arabia, the world's largest oil exporter, was pumping 8.5 million barrels/day in the first half of 1977 in an attempt to frustrate demands for high price increases by some members of

OPEC. Another reason is that demand for crude oil in Europe barely increased from the 1976 levels, as the economies there recovered more slowly than expected from the worldwide recession that followed the Arab oil embargo 3 years earlier. Also, the demand for gasoline in the United States, expected to rise 3% in 1977, grew only 2% due to the introduction of new fuel-efficient American cars.

However, between 1974 and 1978, the world oil supplies seemed to have reached some kind of uneasy stability. Before Iranian oil production ceased, the world oil output approximately matched demand at 63 million barrels/day. By December 26,1978, when all oil exports from Iran ceased, the world oil market had a shortfall of 5 million barrels/day. Part of the shortfall had been made up by other OPEC members. This left a shortfall of 2 million barrels/day.

Yet when measured against world demand of 63 million barrels/day, the shortage was no more than 3%. Why should a 3% loss of supplies have resulted in a 150% increase in the price? The answer was panic. The rush to build inventories by oil companies resulted in an additional 3 million barrels/day of demand above actual consumption. When added to the 2 million barrels/day of net lost supplies, the outcome was a total shortfall of 5 million barrels/day, which was equivalent to approximately 8% of global consumption. The panic buying more than doubled the actual shortage and further fueled the panic. That drove the price from $13/barrel to $34/barrel.

As the world oil shortage became more serious, there was frenzied activity on the Rotterdam Spot Market, which rapidly became the barometer of the crisis and also an indicator of the extreme price levels, and this market became the new frontier of the oil trade. Many factors conspired in early 1979 to make the Spot Market more excitable. The New Iranian regime decided to sell more and more oil on a day-to-day basis instead of on term contracts, so that in effect became part of the Spot Market. The Japanese had been dependent on Iranian oil and as the crisis deepened, they came unstuck. It was Japan that led the panic buying and by May 1979 spot crude leapt up to $34.5/barrel. The Spot Market became still more excited as the Europeans tried to replenish their declining oil stocks. American oil companies with refined oil in the Caribbean began shipping it to Rotterdam instead of the east coast of the United States.

Throughout the cold spring of 1979, the oil shortage was being felt across the industrial world and by the week beginning May 14, the Spot Market began to go crazy. Oil deals by the major oil companies at the Spot Market helped push the price further up. It became evident that the major oil companies previously sold cargos of their cheap oil bought out at the long-term contract prices, making huge profits. They were now rushing in to buy oil for storage because they were not sure how long the Iranian oil exports cutoff would last. The next big move in the Spot Market came in October 1979, when the spot prices hit $38/barrel. Then, in early November, 90 people, including 63 Americans, were taken hostage at the U.S. Embassy in Tehran. An international crisis with grave military overtones suddenly erupted onto the world scene. Spot crude hit $40/barrel.

However, the panic buying of 1979–1980 would become the glut of 1980-1986. Prices would eventually tumble. By the summer of 1980, oil inventories were very high; a pronounced economic recession was already emerging; in the consuming countries, both product prices and demand were even falling; and the inventory surplus continuing to swell. Now it was the buyers' turn to walk away from contracts and the demand for OPEC oil was decreasing. Indeed, in mid-September, a number of OPEC countries agreed to voluntarily cut back production by 10% in an effort to firm prices.

But as the world oil market was becoming calmer, on September 22, 1980 Iraqi troops began an attack on Iran. The outbreak of war threw the oil supply system into jeopardy, threatening a third oil crisis. In its initial stages, the Iran-Iraq War abruptly removed almost 4 million barrels/day of oil from the world market—6% of world demand. Spot prices jumped up again. Arab light crude oil reached its highest price ever: $42/barrel. Panic was once again driving the market. However, supply from other sources was making up for the lost output from Iran and Iraq. Within days of the war, OPEC producers raised their production. At the same time, production in Mexico, the North Sea, Alaska, and other non-OPEC countries was continuing to increase as well. Non-OPEC producers, anxious to increase market share, were making significant cuts in their official prices. As a result, OPEC's output in 1981 was 26% lower than the 1979 output and in fact was the lowest it had been since 1970.

In retrospect, one important question presents itself. Could the astronomical rise in the oil prices have been held in check? Sheikh Yamani seems to think so. He expressed the view that if the United States government and the governments of other major consuming nations intervened at the time and forbade the oil companies from trading in the Spot Market, the prices could have been checked and the panic would have ended.

6. Anatomy of Oil Crises

The world witnessed two oil crises during the last 30 years of the 20th century. The topic will return to the front pages soon, however, because the world may be confronted by a third oil crisis. This third crisis promises to be similar to but have a more modest economic impact than the two previous crises, unless the price of oil hits the $50/barrel mark.

6.1 Definition of Oil Crises

For economic purposes, an oil crisis is defined here as an increase in oil prices large enough to cause a worldwide recession or a significant reduction in global real gross domestic product (GDP) below projected rates by two to three percentage points.

The 1973 and 1979 episodes both qualify as oil crises by this definition. The 1973 oil crisis caused a decline in GDP of 4.7% in the United States, 2.5% in Europe, and 7% in Japan. According to the U.S. government, the 1979 increase in oil prices caused world GDP to drop by 3% from the trend.

The price increase following the first oil crisis raised consumer payments for oil by approximately $473 billion (in real 1999 U.S. dollars), whereas the second oil crisis increased consumer expenditure by $1048 billion. By contrast, the oil price hikes during 1999–2001 raised consumer expenditure by $480 billion.

6.2 Characteristics of Oil Crises

The 1973 and 1979 crises shared four characteristics. First, the disruption in oil supplies occurred at a time of rapid expansion in the global economy. The rapid economic growth fueled greater consumption of oil. In the 5 years that preceded the 1973 crisis, global oil consumption had grown from 38.3 million barrels/day in 1968 to 52.7 million barrels/day in 1972, an average annual increase of 7.5%. Similarly, in the 5 years preceding the 1979 crisis, global consumption had risen from 53 million barrels/day in 1974 to 63 million barrels/day in 1978, an average annual increase of 3.8%.

Second, both disruptions occurred at a time when the world crude oil production was operating at virtually full capacity. Global capacity utilization reached 99% in 1973, with OPEC accounting for 56% of total production. The second oil crisis had seen a deficit of 5 million barrels/day resulting from the disruption in Iranian oil production. Third, each crisis took place at a time when global investment in oil exploration had been declining, making it impossible to achieve a speedy increase in non-OPEC production. In both 1973 and 1979–1980, the global oil industry was at the end, rather than the start, of a new surge in non-OPEC output. Fourth, in both crises, OPEC members had made a deliberate decision to reduce oil production in order to achieve political ends.

6.3 Underlying Factors

In a tight oil market, any of the following underlying factors could, individually or collectively, trigger a price escalation reminiscent of the spot market prices of 1979 and precipitate an oil crisis.

The Global Oil Demand. Economic growth and population growth are the most important drivers behind increasing global energy demand. The global economy is projected to grow by 3.2% per annum, on average, to 2025. Global GDP is projected to rise from $49 trillion in 2000 (year 2000 dollars purchasing power parity) to $108 trillion in 2025 and $196 trillion in 2050.

World population is expected to grow from 6 billion in 2000 to 8 billion in 2020. The population growth among the 4.8 billion people living in developing countries is estimated at 1.7% per annum. This compares with an average 0.3% per annum in the developed countries. Expanding industrialization and improving standards of living will contribute significantly to the growing energy demand.

The developed countries produce approximately one-third of global oil but consume two-thirds, whereas the developing countries produce two-thirds but consume only one-third (see Table I). Annual per capita consumption in the developing countries is 2 barrels/year. This compares with 14.2 barrels/year in the developed countries and 25 barrels/year in the United States.

The International Energy Agency (IEA) and the U.S. Department of Energy forecast that the world oil demand will grow from 76.1 million barrels/day in 2001 to 95.8 million barrels/day in 2010 and 115 million barrels/day in 2020, with Middle East producers having to meet the major part of the additional demand. However, this will depend on sufficient investment to expand production capacity (see Table II).

The Global Sustainable Productive Capacity. Sustainable productive capacity is here defined as

Table I

World Crude Oil Production versus Demand in 2001.

Region	Production (billion barrels)	Share (%)	Demand (billion barrels)	Share (%)
Developed countries	8	30	18	64
Developing countries	19	70	10	36
World	27	100	28	100
United States	2	7	7	25

Sources: British Petroleum Statistical Review of World Energy, June 2002; and U.S. Energy Information Administration, June 2001.

Table II

World Oil Demand and Supply (Million Barrels/Day), 2000–2020.

	2000	2001	2005	2010	2020
World demand	76.2	76.1	83.5	95.8	114.7
World supply					
Non-OPEC	45.2	46.1	44.7	43.6	49.6
OPEC	29.3	30.0	36.0	45.9	51.1
Stock change	−1.7	—	—	—	—
Synfuels	1.2[a]	1.3[a]	1.8[a]	2.7[a]	4.2[a]
Total supply	76.2	76.1	80.7	89.5	100.7
Global oil deficit	—	—	−2.8	−6.3	−14.0

[a]Synfuel oil production is already included in non-OPEC supply figures.
Sources: U.S. Department of Energy; British Petroleum Statistical Review of World Energy, July 2002; and International Energy Agency.

Table III

OPEC Sustainable Capacity and Capacity Utilization (Million Barrels/Day) in January 2001.

Country	Capacity	Production	Capacity utilization
Algeria	0.880	0.860	98%
Indonesia	1.300	1.300	100%
Iran	3.500	3.500	100%
Iraq	3.000	2.900	97%
Kuwait	2.200	2.200	100%
Libya	1.450	1.450	100%
Nigeria	2.100	2.100	100%
Qatar	0.720	0.700	97%
Saudi Arabia	9.250	9.000	97%
United Arab Emirates	2.400	2.400	100%
Venezuela	2.900	2.900	100%
Total	29.700	29.310	99%

Sources. Energy Intelligence Group's "Oil Market Intelligence"; Petroleum Review, April 2000; International Energy Agency; and Author's projections.

"being attainable within thirty days and sustainable for three months." OPEC's sustainable productive capacity stood at 29.7 million barrels/day in January 2001 with a 99% capacity utilization (see Table III). The organization's immediately available spare capacity stood then at 390,000 barrels/day. However, with the production cutbacks since March 2001, spare capacity has risen to 4 million barrels/day.

The capital costs of maintaining and expanding OPEC's capacity over a 5-year period are estimated at $112 billion, money that members do not have. These projected costs are based on the member countries' planned capacity increase of 7 to 36.7 million barrels/day.

There is no non-OPEC spare capacity. The financial incentive of high oil prices and firm demand

mean that every non-OPEC oilfield is being exploited and any capacity brought onstream is being utilized as quickly as possible.

The Ultimate Global Proven Reserves. World ultimate conventional oil reserves are estimated at 2000 billion barrels. This is the amount of production that would have been produced when production eventually ceases. Different countries are at different stages of their reserve depletion curves. Some, such as the United States, are past their midpoint and in terminal decline, whereas others are close to midpoint, such as the United Kingdom and Norway. The U.K. sector of the North Sea is currently at peak production and is set to decline at approximately 6% per year. However, the five major Gulf producers—Saudi Arabia, Iraq, Iran, Kuwait, and United Arab Emirates—are at an early stage of depletion and can exert a "swing" role, making up the difference between world demand and what others can supply. They can do this only until they themselves reach their midpoint of depletion, probably by 2013.

The expert consensus is that the world's midpoint of reserve depletion will be reached when 1000 billion barrels of oil have been produced—that is to say, half the ultimate reserves of 2000 billion barrels. With 935 billion barrels already produced, this will occur probably between 2004 and 2005. The yet-to-find (YTF) oil reserves are estimated at 280 billion barrels (see Table IV). As the world production peak approaches, the oil price will soar.

However, if the potential of unconventional oil, such as tar sand oil and extra heavy oil, is included, amounting to 572 billion barrels, then the midpoint of depletion could be delayed for a few more years—but not beyond 2010.

In 1956, the geologist M. King Hubbert predicted that U.S. oil production would peak in the early 1970s. Almost everyone, inside and outside the oil industry, rejected Hubbert's analysis. The controversy raged until 1970, when U.S. production of crude oil started to fall. U.S. production peaked at 9.64 million barrels/day in 1970 and has been falling since then, reaching 5.77 million barrels/day by 2001. Hubbert was proven right and his bell-shaped curve became a useful tool of oil production analysis.

Around 1995, several analysts began applying Hubbert's method to world oil production. Based on Hubbert's pioneering work and an estimated 1.8–2.1 trillion barrels of ultimate reserves, they established that the peak production year will be sometime between 2004 and 2009. If the predictions are correct, there will be a huge impact on the global economy, with the industrialized nations bidding against one another for the dwindling oil supply. One promising oil province that remains unexplored is the Spratly Islands in the South China Sea, where exploration

Table IV
Ultimate Global Reserves of Conventional Oil and Depletion Rate (End of 2001).

	Volume or rate	Description
Ultimate reserves (billion barrels)	2000	Amount of production when production ceases
Produced so far (billion barrels)	935	Until the end of 2001
Yet-to-produce (billion barrels)	1065	Ultimate reserves less produced
Discovered so far (billion barrels)	1720	Produced plus remaining reserves
Yet-to-find (billion barrels/year)	280	Ultimate reserves less discovered
Discovery rate (billion barrels/year)	7	Annual additions from new fields
Depletion rate (%)	3	Annual production a percentage of yet-to-produce

Sources. United States Geological Survey, and British petroleum Statiscal Review of World Energy, June 2002.

has been delayed by conflicting claims to the islands by six different countries. Potential reserves in the disputed territories are estimated at multibillion barrels of oil and gas. But even if the South China Sea oil reserves are proven, they could hardly quench China's thirst for oil. By 2010, China is projected to overtake Japan to become the world's second largest oil importer after the United States.

Another promising province is the Caspian Basin. Estimates of 40 to 60 billion barrels as the ultimate reserve base of the Caspian region are judged to be reasonable by most geologists familiar with the region. Apart from the limited size of the reserves, the area's oil is very costly to find, develop, produce, and transport to world markets. Projected Caspian Sea oil production of 2–3 million barrels/day by 2010 can be achieved only when prices exceed $20/barrel (in real terms). Oil prices will be the key factor in the expansion of Caspian Sea oil.

New Oil Discovery Rates. The widely held view that improved seismic surveying and interpretation have improved drilling success rates is not borne out by the level of discoveries during the period 1992–2001 (see Table V).

The race for reserves is, therefore, on. With the demand for oil envisaged for the next 10 years, the world will consume an average 30 billion barrels per year over that period. If the global oil industry wants to replace this consumption with new reserves without diluting the world's existing proven reserves of some 1 trillion barrels, it must find an additional 300 billion barrels of new oil in the next decade—a daunting challenge indeed.

Global Reserve Depletion Rate. Globally the reserve depletion rate is generally calculated at 3%. This means that to sustain the world's current 76 million barrels/day consumption at that level, approximately 4 million barrels/day of new capacity is needed every year. Against this, the world production capacity has remained static or even declined while consumption has been increasing.

Table V
Global Crude Oil Reserves Additions[a] (Billion Barrels) 1992–2001.

Year	Added in year	Annual production	As % of annual production
1992	7.80	23.98	33
1993	4.00	24.09	17
1994	6.95	24.42	28
1995	5.62	24.77	23
1996	5.24	25.42	21
1997	5.92	26.22	23
1998	7.60	26.75	28
1999	13.00	26.22	50
2000	12.60	27.19	46
2001	8.90	27.81	32
1992–2001	77.63	256.87	30
Average	7.76	25.83	30

[a]Excluding the United States and Canada.
Sources. IHS Energy Group's 2002 World Petroleum Trends Report, and British Petroleum Statistical Review of World Energy, 1993–2002.

This means that the Middle East producers, with 65% of the world's proven reserves and just one-third of global production, will assume clear-cut leadership of the supply side of the oil market.

7. Third Oil Crisis?

The parallels between current conditions and the early 1970s are unnerving. The conditions that made the 1973 and 1979 oil crises possible exist in the early 2000s. First, world oil consumption is growing rapidly and is projected to continue expanding. World oil demand grew by 2.4% in 2000, 2.2% in 2001 with a projected additional annual increase of 1.8% until 2005.

Second, every indicator points to the fact that production from OPEC will not rise substantially. OPEC's maximum production capacity is estimated at 29.7 million barrels/day. The organization must produce approximately 30 million barrels/day to maintain a supply–demand balance under the current level of stocks. Yet, an output level of 30 million barrels/day may be beyond the organization's reach or, if it is feasible, achievable only with production from Iraq.

Third, the pattern of low-level industry investment in 2000–2001 is also very similar to that observed during the first and second oil crises. Oil companies' capital investment in exploration in 2000/2001 was slow in the aftermath of low oil prices during 1998–1999. Capital spending on exploration and production by the supergiants—Exxon Mobil, BP Amoco, and Shell—fell 20% to $6.91 billion in the first half of 2000 from a year earlier. Most companies decided to wait before boosting investment, preferring instead to buy back stocks. This has reduced the capital investment available for production and production capacity expansion.

Fourth, as in 1973 and 1979, the causes of this crisis will be a reduction in OPEC production and stagnation in non-OPEC's taking place at a time of rapid economic growth.

Fifth, as in 1973 and 1979, it will be a major political event that could send oil prices rocketing and thus precipitate a third oil crisis. The 1973 Arab-Israeli war was behind the first oil crisis, whereas the Iranian revolution was at the heart of the second. Similarly, a war against Iraq could precipitate a third oil crisis.

Such conditions make for a truly tight market with the potential for developing into a genuine crisis. But even if the world manages to escape a third oil crisis, oil-market turmoil can be expected to last until 2004.

However, a growing number of opinions among energy experts suggest that global conventional oil production will probably peak sometime during this decade, between 2004 and 2010. Declining oil production will cause a global energy gap, which will have to be filled by unconventional and renewable energy sources.

Various projections of global ultimate conventional oil reserves and peak years have been suggested by energy experts and researchers between 1969 and 2002. The extreme end of opinion is represented by the United States Geological Survey (USGS), IEA, and the U.S. Energy Information Administration (EIA) (see Table VI).

The estimate by EIA is so implausibly high that it can be ignored, whereas the USGS estimate includes 724 billion barrels of YTF reserves. Such huge YTF reserves require discovering an additional amount of oil equivalent to the entire Middle East. But since 90% of global conventional oil has already been found, trying to find 724 billion barrels of new oil is not only an exceptionally daunting task, but virtually impossible.

However, such estimates are of only limited relevance. What is important when attempting to identify future supplies are two key factors: the discovery rate and the development rate, and their relationship to the production rate.

The technology for extracting oil from tar sands, oil shale, and extra heavy oil, known collectively as synfuels, exists but extraction costs are high. Synfuel oil is usually 3 times as labor-to-energy intensive and 10 times as capital-to-energy intensive as conventional oil.

Table VI
Various Projections of Global Ultimate Conventional Oil Reserves and Peak Year (Billion Barrels).

Author	Affiliation	Year	Estimated ultimate reserves	Peak year
Hubert	Shell	1969	2100	2000
Bookout	Shell	1989	2000	2010
Mackenzie	Researcher	1996	2600	2007–2019
Appleby	BP	1996		2010
Invanhoe	Consultant	1996		2010
Edwards	University of Colorado	1997	2836	2020[a]
Campbell	Consultant	1997	1800–2000	2010
Bernaby	ENI	1998		2005
Schollenberger	Amoco	1998		2015–2035[a]
IEA	OECD	1998	2800	2010–2020[a]
EIA	DOE	1998	4700	2030[a]
Laherrere	Consultant	1999	2700	2010[a]
USGS	International Department	2000	3270	—
Salameh	Consultant	2000	2000	2004–2005
Deffeyes	Princeton University	2001	1800–2100	2004

[a]These ultimate reserve estimates include extra heavy crude, tar sands, oil shale, and also projected production of gas-to-liquid oil.
Sources. Various.
Note. BP, British Petroleum; DOE, U.S. Department of Energy; EIA, U.S. Energy Information Administration; IEA, International Energy Agency; OECD, Organisation for Economic Cooperation and Development; USGS, U.S. Geological Survey; ENI, Ente Nazionale Idrocarburi.

Whereas some—and possibly a great deal—of unconventional oil (synfuels) will eventually be available, there will not be enough to replace the shortfalls in conventional oil. Synfuels will be hard-pressed to meet 3% of the global oil demand in 2010 and 4% in 2020, because of the slow extraction rate and the huge investments needed.

In 2002, only 35,000 barrels/day of natural gas to liquid oil is produced worldwide. This is projected to rise to 685,000 barrels/day by 2010, equivalent to 0.7% of global demand. The constraint, however, might be the very large capital commitment. For a production of 400,000 barrels/day, this would amount to $8 billion.

In 2000, renewable energy sources contributed 2% to the global primary energy demand. However, by 2025 they are projected to contribute 4%, rising to 5% by 2050 (see Table VII).

Fuel-cell motor technology will eventually have a great impact on the global consumption of gasoline and diesel. But it could take years before hydrogen-powered cars dominate the highways and certainly not before they are able to compete with today's cars in terms of range, convenience, and affordability.

Fossil fuels, with a growing contribution from nuclear energy, will, therefore, still be supplying the main share of the global energy needs for most—perhaps all, of the 21st century.

8. Implications for the Global Economy

Twenty years ago, oil crises wreaked havoc on the world economy. Today, although oil remains important, a new crisis will have a much more modest impact because of the diminished role of oil in the global economy, unless the price of oil rises to $50/ barrel.

One reason is that oil represents a much smaller share of global GDP today than it did during the two previous crises. Another reason is that the economy is more open, which makes it harder for companies to pass on cost increases to customers. A third reason is that customers can hedge against price increases. Finally, economies around the world have become more adaptable in their use of every resource, including oil. Consequently, increases in oil prices trigger substitutions of natural gas or any other energy sources in manufacturing. For instance, if one takes the case of the United States, which accounts for 25% of the global oil consumption, it is found that a rise in oil prices today is much less important to the U.S. economy than it has been in the past. The decline in importance of oil prices can be seen from Table VIII, which shows the share of crude oil in nominal GDP.

Table VIII shows that crude oil accounted for 4% of GDP at the time of the first oil crisis, 6% at the time of the second oil crisis, and 2.4% at the time of the Gulf War. In 2001, it accounted for 2%. However, if the price of oil hits the $50/barrel mark in 2002, then crude oil will have the same share in the U.S. economy as in 1974, with adverse economic implications for the U.S. and the global economy.

Despite the above, higher oil prices still are very important to the world economy. A new report by the International Monetary Fund (IMF) notes that a war on Iraq could be the final straw for an international economy still struggling with the aftermath of September 11, the 20% drop in world share prices this year, and the implosion of the Argentine and Brazilian financial services. The IMF argues that the risks are "primarily on the downside" even before the possibility of a conflagration in Iraq and the Middle East is taken into account.

Table VII
World Primary Energy Consumption[a] (mtoe), 2000–2050.

	2000	2025	2050
Primary energy	9631	16,618	19,760
Oil	3835	6429	7344
Natural gas	2190	4760	6207
Coal	2136	3283	3037
Nuclear	636	733	825
Hydro	617	818	1440
Renewables[b]	217	595	907
Renewables: % of total	2	4	5

[a]mtoe = Million tonnes oil equivalent.
[b]Excluding hydro (i.e., hydroelectric power).
Sources. Shell International, Scenarios of 2050; Organization of Petroleum Exporting Countries Statistical Review of World Energy, June 2002; and U.S. Energy Information Administration International Energy Outlook 2002.

Table VIII
Oil in the U.S. Economy: Consumption, Prices, and Percentage of GDP.

	Oil consumption (million barrels/day)	Oil prices ($/barrel)	Nominal GDP ($ billion)	Oil consumption as percentage of GDP
1974	16.65	9.07	1382	4.0
1980	17.06	28.07	2784	6.3
1990	16.99	22.22	5744	2.4
1999	19.36	17.60	8857	1.4
2000	19.63	27.72	9224	2.2
2001	19.90	25.93	9462	2.0
2002[a]	20.18	30.00	9611	2.3
2002[a]	20.18	50.00	9611	4.0

[a]Estimates.
Sources. Courtesy of PKVerleger LLC; British Petroleum Statistical Review of World Energy, June 2002; U.S. Energy Information Administration Annual Energy Outlook 2002.

The IMF warns of the consequences for the oil markets of a war in the Gulf region. It says that a $15/barrel increase in the price of oil would be a severe blow to the global economy, knocking at least 1 percentage point off global GDP and sending the world economy spiraling toward recession.

Certainly, all the evidence from past conflicts in the region, including the 1973 Arab–Israeli War, Iranian revolution, and the Gulf War of 1990–1991, suggests that each of these conflicts has been followed by recession and unemployment, particularly in the West.

This time around, the scenario is even more worrying. Financial markets are already paralyzed as a result of the burst of the technology bubble and the mammoth corporate accounting frauds in the United States.

Before Washington's decision to up the ante with Iraq, there was some hope at the IMF that there would be some kind of bounce-back for the global economy in 2003. The world economy was projected to pick up speed, growing by 3.7% in 2003, after slipping this year to 2.4%. The United States was forecast to pick up from a 1.4% expansion this year to 2.3% in 2003.

But higher oil prices could make all the difference. The IMF notes that oil prices began to surge in the second half of August 2002 as the U.S. administration began to talk up the war with Iraq. The price of crude oil on both sides of the Atlantic is already above the $22 to $28 per barrel price range built into most major economic projections. For oil-importing countries, higher oil prices act as an immediate tax on consumption and business. An oil price hike makes it less likely that consumers will spend and business will invest.

With the global economy struggling to pick up momentum, a prolonged period of high oil prices would, without doubt, delay an upturn. It would almost certainly give another leg to the downturn in share prices—in 2003, down by 40% since the peak in March 2000—hitting confidence hard.

Although the loss of Iraqi oil supplies over the short term might make little significant difference, it is the collateral damage that could deliver the most serious blow. There is the possibility that Iraq might seek to disable the critical Saudi and Kuwaiti oilfields, as in the Gulf War, or that the Arab countries—outraged by America's attack on one of their own—might seek to impose an oil embargo or refuse to increase supplies to make up for the loss of Iraqi output.

The result would be a hike in prices on world oil markets and disruption to an already misfiring global economy. However, once the war is over, there may well be cheaper and more plentiful oil as Iraq's production is restored. But by then the damage to a rickety world economy will have been inflicted.

Further Reading

Deffeys K S 2001 *Hubbert's Peak: The Impending World Oil Shortage*. Princeton University Press, Princeton, NJ

Nixon R 1980 *The Real War*. Sidgwick & Jackson, London

Robinson J 1988 *Yamani: the inside story*. Simon & Schuster, London

Salameh M G 1990 *Is a Third Oil Crisis Inevitable*. Biddles, Guildford, UK

Salameh M G 1999. Technology, oil reserve depletion and the myth of reserve-to-production (R/P) ratio. OPEC Review, June 1999, pp. 113–124.

Salameh M G 2001. Anatomy of an impending third oil crisis. *In* Proceedings of the 24th IAEE International Conference, April 25–27, 2001, Houston, TX, pp. 1–11.

Salameh M G 2001. *The Quest for Middle East Oil: the U S Versus the Asia-Pacific Region*. International Asian Energy Conference, August, Hong Kong.

Sampson A 1980 *The Seven Sisters*. Hodder & Stoughton, London

Verleger, P K 2000. *Third Oil Shock: Real or Imaginary?* Consequences and Policy Alternatives. International Economics Policy Briefs, No. 00-4, Institute for International Economics, Washington, DC.

Yergin D 1991 *The Prize: The Epic Quest for Oil, Money and Power*. Simon & Schuster, New York

Mamdouh G. Salameh

Oil Industry, History of

Glossary

barrel The international standard of measure for crude oil and oil products equivalent to 42 U.S. gallons and used since first defined in the U.S. oil fields in the 1860s.

cap and trade a recently developed policy approach to environmental pollution problems that attempts to utilize the mechanism of the market. Typically, a governing authority sets a limit (cap) on the amount of a particular pollutant that can be released, and issues credits that grant the right to emit a specific amount of pollutant. Although the total credits issued remain below the cap, companies are encouraged to buy and trade credits, thus rewarding those who reduce their pollution and placing an economic cost on those who do not.

cracking Various refining processes using heat, pressure, and catalysts to change less volatile and heavier petroleum fractions into compounds with lower boiling points.

crude oil Unrefined petroleum as it comes from the well, consisting almost entirely of carbon and hydrogen compounds and varying widely in appearance, color, odor, and the presence of sulfur, nitrogen, oxygen compounds, and ash.

illuminating oil Common 19th-century name for kerosene, the fraction of crude oil between gas oil and gasoline on the refiner's scale, usually ranging between 105 and 300 °F; represented the first major commercial petroleum product and defined the early decades of the industry.

law of capture Legal principle emanating from the courts in the 19th century that encouraged the rapid drilling and pumping of oil from underground pools; a major cause of wasteful practice in the early oil industry.

Organization of Petroleum Exporting Countries (OPEC) Organized in 1960 to establish a united front against the power of the world's multinational oil companies, it became a force to be reckoned with in the 1970s and remains a relevant player in the economics of world oil.

refining Processes employed to transform crude oil into useful commercial products; includes fractional distillation, cracking, purifying, and treating.

rotary drilling Method for drilling deep holes in search of petroleum or gas in which a drill bit is attached to a revolving drill pipe; used first in the early 20th century.

Spindletop Enormous oil strike or "gusher" on the Texas Gulf Coast near Beaumont in 1901 that changed the face of the oil industry. Economic forces unleashed brought a market erosion of Standard Oil dominance and the birth of oil's new "age of energy" with the increased production of fuel oil and gasoline.

Standard Oil Parent firm of the John D. Rockefeller oil interests, initially capitalized as an Ohio corporation, reorganized as a New Jersey holding company in 1899, and dissolved by the U.S. Supreme Court in a famous antitrust decision in 1911.

Most global citizens today view the international oil industry as an enormous industrial power whose financial profits have arisen from the utilization of one of nature's gifts, often through shrewdness and technological innovation but sometimes also as a result of monopolistic practices and exploitation of Third World resources. During the time of the price dislocations of 1973 and 1979 that defined the "energy crisis" of the decade, many pointed to domination of an oligopoly of multinational firms defined by British journalist Anthony Sampson as the "seven sisters." An aura of monopolistic competition and "big business" has surrounded this industry from its beginnings, as the Rockefeller Standard Oil empire remains imprinted as a past symbol of rampant corporate power. However, the history of this industry has been much more complex than a narrative of wealth accumulation by a few major companies. It represents a story of vision, risk, business success, and often disaster played out in an increasingly multinational arena. Oil has emerged as the dominant fuel consumed in the United States, the country in which the industry took life in the modern era, but domestic crude oil production peaked in 1970 and has declined ever since. Once the "Saudi Arabia of the world," the United States became a net importer of petroleum at the historical moment that an oil regime had come to dominate world energy. The energy crises of 1973 and 1979, the Persian Gulf War of 1991, and the invasion of Iraq in 2003 have all highlighted the geopolitical importance of Middle Eastern oil, and the major vertically integrated, multinational companies have become almost equal players with governments in this arena.

1. Petroleum Prior to the Modern Era

Energy historians have emphasized the concept of transition from one energy regime to another as an

203

analytical tool. One model, for example, suggests that early mankind entered the first transition from hunting and gathering to the harvesting of biomass energy through agriculture and silviculture. This relatively straightforward and simple energy form dominated in antiquity, the medieval period, and well into the early modern era. The era of the industrial revolution saw the transition from biomass harvesting to the exploitation of fossil energy in the form of coal. The shift from a predominant reliance on coal to an overwhelmingly significant role for oil and natural gas has defined the last transition. Two aspects derive from this analysis. The first is to understand how very brief has been the "oil age" within the context of human history (approximately 140 years, from 1859 to the present). The other is the speculative argument that since all fossil fuels (coal, oil, and natural gas) will eventually be depleted, we must plan as a global society for the next transition to a more sustainable energy source. These arguments have most often focused on one type of "renewable" fuel or another, whether it be, in the words of energy analyst Amory Lovins, hard or soft. For example, many advocates of nuclear energy still argue that fission breeder reactors promise a potential renewable energy source, but one that many social scientists would criticize as undesirable or hard because of the great potential for environmental degradation. For others, more environmentally benign, soft solutions, such as solar, wind, tidal, or geothermal energy or renewable biomass fuels such as ethanol, provide the answer. These arguments must be left for another discussion, but all experts agree that the age of overabundant, cheap petroleum will end at some point; the problem is to determine when this will occur.

Prior to the industrial era, petroleum had a very limited utility. Records show human uses for oil in the form of natural seepage of asphaltic bitumen as far back as ancient Mesopotamia prior to 3000 BC. Employed as a mastic in construction, a waterproofing for ships, and a medicinal poultice, this oil apparently did have a small market. Later, Near Eastern cultures successfully distilled natural crude oil to obtain lamp fuel and the basic ingredient for "Greek fire," a military incendiary introduced with great effect by the Byzantine Empire against the rigging of attacking ships. In the West, limited supplies of crude oil seepage were confined mostly to medicinal use. Native Americans had also discovered the medical benefits of crude oil, and when Edwin L. Drake and associates decided to drill for oil in northwestern Pennsylvania in 1859, they were aware that local tribes had been using oil for a very long time.

A market for medicinal oil developed by the early 19th century in Pennsylvania and a number of entrepreneurs entered the business. Samuel M. Kier of Pittsburgh, the most successful of these businessmen, established a successful medicinal market for petroleum or rock oil before the Civil War. These original "snake oil salesmen" claimed a host of cures for rock oil to be used both externally and internally. In a parallel development, a small Russian industry developed around the presence of natural seepage in the Baku region, initially also for medicinal use. However, it took a 19th-century shortage of lamp oil, precipitated by a whale-oil crisis and relative scarcity of coal oil, to spark the modern petroleum industry.

2. Oil's Age of Illumination

In the 1850s, experiments with refined petroleum demonstrated that it could serve as a satisfactory lamp oil (in kerosene form). The problem was that it was not available in suitable quantities. The drilling of water wells in northwest Pennsylvania had uncovered the presence of oil along with brine in a number of instances. Why couldn't one drill for oil? A group of investors hired "Colonel" Edwin L. Drake to obtain leases and attempt to drill for oil using the standard percussion techniques then used for drilling for water. On August 27, 1859, Drake's rig struck oil at 69 ft near Oil Creek, a short distance from the town of Titusville, Pennsylvania. The oil age was born. The market for illuminating oil was insatiable and the growth of the industry was sudden and dramatic. Crude oil production increased from only 2000 barrels in 1859 to 4.8 million barrels a decade later and 5,350,000 barrels in 1871. From the beginning, the demand for illuminating oil reflected both the burgeoning domestic population and economy and an expanding export business. Between the 1860s and 1900, crude oil and oil products exported ranged from one-third to three-fourths of total U.S. production.

This new industry initially operated in a classically competitive fashion. Ease of entry defined all aspects of the business, and capital investment was minimal. Refining technology was primitive, consisting of heating crude in a still in order to obtain the desired kerosene fraction; other constituent fractions such as gasoline were often run into streams or onto the ground in ditches. The evolution of the legal doctrine of law of capture by the Pennsylvania courts also contributed to wasteful practices in the oil fields. Several different landowners or leaseholders might sit astride an underground oil pool. Drawing an analogy with game being legally captured if it were lured onto another's land, and citing precedent in English common law, the courts upheld the right of the property owner to "capture" oil that had migrated under the owner's land. Limited understanding of oil geology at that time mistakenly led to the belief that oil flowed in underground rivers. Each owner or leaseholder rushed to drill and pump before their neighbors depleted the oil pool. Oil markets became characterized by alternate periods of oil glut followed quickly by scarcity. This wasteful practice continued as the oil frontier moved westward from Pennsylvania, Ohio, and Indiana in the latter 19th century to Texas, California, and the midcontinent in the 20th century.

3. Competition and Monopoly

The young John D. Rockefeller entered this rough-and-tumble business in 1863. He had cut his teeth as a partner in the wholesale grocery business in Cleveland, doing well as purveyor of supplies to Ohio military units. However, Cleveland possessed several locational advantages that encouraged its development as an important refining center. Located on Lake Erie, close to the Pennsylvania oil fields, it had railroad trunkline connections to major eastern markets, a readily available workforce, and sources of financial capital. The industry, then as today, consisted of four main functional sectors: production, refining, transportation, and marketing. Production entails all the activities involved in "getting it out of the ground." This includes exploration, drilling, leasing, pumping, primary and secondary recovery techniques, and all other associated activities. Refining entails all relevant processes developed to obtain usable products from crude oil in its natural state. Initially, this included only fractional distillation, the heating of crude to boil off its constituent parts, but has evolved today to include a host of sophisticated specialty practices. Transportation, the third element of the industry, encompasses the movement of crude oil from the oil field to the refinery and the shipment of refined product to market. In the beginning of the industry in the 1860s and 1870s, this meant by barge, railroad car, or horse-drawn wagon. Oil transport evolved to include petroleum pipelines, tanker trucks on the highway, and fleets of tankers on the high seas. Marketing envelops the distribution and sale of petroleum products to the consumer. 1870s marketing might have consisted of a Standard Oil horse-drawn tank wagon selling branded "illuminating oil" or kerosene in neighborhoods; today, we think of well-advertised gasoline stations and slick television commercials designed to get us to use one firm's brand over another's. Although we think of the industry today as consisting of a few vertically integrated firms (those that are engaged in all four major functions from production through marketing), the oil fraternity has always contained a strong element of independents. These individual producers, refiners, pipeline operators, and marketers often viewed themselves as opponents of the big companies such as Standard Oil and they still represent a distinct voice in the industry.

Rockefeller's early endeavors were concentrated in the refining sector of the industry, achieving growth through horizontal combination; he did not adopt a strategy of vertical integration until later. First as a partnership, then organized as a corporation in 1870, Rockefeller and associates' Standard Oil Company proceeded to achieve what was later termed the conquest of Cleveland. Using tactics that one could define as either shrewd or unethical, Rockefeller succeeded in dominating the Cleveland refining business. Concentrating on transportation as the key to control, Standard used its size to obtain preferential shipping arrangements with the railroads through the use of rebates and then sought to eliminate competition with a series of pooling and cartel agreements. Rockefeller's attempts to control the refining business with the South Improvement Company and later the National Refiners Association remain textbook examples of business strategy in the late 19th century. When Rockefeller found informal agreements inadequate to achieve tight industry control, he changed his approach to one of merger and acquisition.

By 1878, Standard controlled more than 90% of total U.S. refining capacity. To formalize this economic empire, Rockefeller and associates created the Standard Oil Trust Agreement in 1882, later reorganized as the Standard Oil Company (New Jersey) as a legal holding company under New Jersey law in 1899. Rockefeller integrated backward in the 1880s by acquiring producing properties in Ohio, into transportation with his own pipelines, and forward into marketing with his branded cans of kerosene for domestic sale and export. The Standard near monopoly became the hated enemy of independent oil producers, who found their prices driven down by Rockefeller buying power, and by smaller operators, who had been forced to sell out in the face of price-cutting attacks. As a public outcry against the power of large Gilded Age corporations emerged, "muckraking" exposés by Henry Demerest Lloyd and Ida M. Tarbell, a daughter of an independent oilman, painted Standard Oil as a hated symbol of monopoly in America.

Standard had been under legal attack in state courts by its enemies and competitors for years, but the passage of the Sherman Antitrust Act in 1890 had lain the groundwork for federal challenge. With a change in political climate during the presidency of Theodore Roosevelt toward progressivism, federal investigations went forward, culminating in the forced dissolution of Standard Oil into several constituent companies in 1911. This did not hurt Rockefeller personally too much because he remained a stockholder in these many companies, and historians have argued that the former monopoly had simply transformed into an oligopoly of large vertically integrated companies. For example, Standard Oil (New Jersey) suffered for a time because it was a large refiner and marketer but lacked integrated production facilities. Soon, however, it rectified this situation with the acquisition of oil production in the Oklahoma and Texas fields. Similar strategies had enabled most of the spun-off firms to emerge as strong vertically integrated units by the early 1920s. However, market forces other than court decisions had already begun to change the U.S. oil industry in significant ways.

4. Technological Change and Energy Transition

Economists Harold F. Williamson and Ralph Andreano have argued that by the eve of the 1911 dissolution decree the competitive structure of oil already had been altered, the key factor being the discovery of vast supplies of oil on the Texas Gulf Coast at Spindletop in 1901. If the Drake well in 1859 had trumpeted the birth of the illuminating oil industry, Spindletop marked the birth of its new age of energy. Spindletop and subsequent other new western fields provided vast oil for the growing economy but also enabled firms such as Gulf, Texas, and Sun Oil, independent of the Standard interests, to gain a substantial foothold. The gulf region was underdeveloped and attractive to young companies, and the state of Texas was hostile toward the Standard Oil monopoly. Asphaltic-based Spindletop crude made inferior grades of kerosene and lubricants but yielded a satisfactory fuel oil. New firms, such as Gulf and Texaco, easily integrated forward into the refining, transportation, and marketing of this fuel oil in the coal-starved Southwest. Located near tidewater on the Gulf Coast, small investments in pipelines also enabled operators to get their crude to seagoing tank ships, which could then carry it to other markets. The Sun Oil Company of Pennsylvania, a long-time competitor of Standard, built a robust business by shipping Texas crude to its refinery outside Philadelphia.

In 1900, Standard Oil (New Jersey) controlled approximately 86% of all crude oil supplies, 82% of refining capacity, and 85% of all kerosene and gasoline sold in the United States. On the eve of the 1911 court decree, Standard's control of crude production had declined to approximately 60–65% and refining capacity to 64%. Moreover, Standard's competitors now supplied approximately 70% of the fuel oil, 45% of the lubricants, 33% of gasoline and waxes, and 25% of the kerosene in the domestic market. Newly formed post-Spindletop companies such as Gulf and Texaco, along with invigorated older independent firms such as Sun and Pure, had captured significant market share. Meanwhile, Standard, heavily invested in the traditional kerosene business, was slow to move into the production and marketing of gasoline.

At approximately the same time, the automobile, which had appeared in the 1890s as a novelty, had begun to shed its elitist image. The introduction of mass-produced and relatively inexpensive vehicles, led by the 1908 Ford Model T, very quickly influenced developments in the oil business. In 1908, the total output of the U.S. auto industry was 65,000 vehicles. In less than a decade, Ford alone sold more than 500,000 units annually. Within that same decade, both the volume of production and the total value of gasoline passed those of kerosene. By the turn of the century, the increasing electrification of urban America and widespread use of the Edison incandescent lamp were already a worry to oil companies whose major product was kerosene. Now the oil industry was becoming increasingly concerned about how it could boost production of gasoline to meet demand.

As the oil industry faced real or apparent shortages of supply, one response was technological. New production methods such as the introduction of rotary drilling early in the century increased crude supplies, and refinery innovations enabled crude stocks to be further extended. However, as the demand for gasoline increased, new oil discoveries in Texas, California, and Oklahoma in the early 1900s proved insufficient. Once discarded as relatively useless, this lighter fraction typically constituted 10–15% of a barrel of crude oil. Refiners stretched the gasoline fraction by including more of the heavier kerosene, but this resulted in an inferior motor fuel. One could enrich the blend of gasoline with the addition of what would later be termed higher octane product obtained from selected premium crudes (e.g., from California) or "natural" or "casinghead" gasoline yielded from highly saturated natural gas. The most important breakthrough, however, occurred with the introduction of thermal cracking technology in 1913 by a Ph.D. chemist employed by the Standard Oil Company (Indiana), William Burton. The industry had employed light cracking, the application of heat to distillation to literally rearrange hydrocarbon molecules, since the 1860s to obtain higher yields of kerosene from feed stock. By dramatically increasing the temperature and pressure of his cracking stills, Burton discovered that he could double the output of gasoline obtained over previous fractional distillation and cracking methods. An additional advantage was that this gasoline was of generally superior quality. Although the problem of "knocking" in the internal combustion engine was not yet understood fully, it would become a major technological challenge in the 1920s as higher compression, higher performance engines appeared.

Oil's growth as an energy source between 1900 and 1920 coincided with an enormous increase in total U.S. energy consumption. Samuel J. Schurr and Bruce Netschert calculated that total energy use increased by 123% during this span of two decades. Although oil companies in the Southwest had begun to charge that "King Coal's" reign was over, aggregate statistics show that the industry's share of total energy consumption had only risen from 2.4% of the total in 1900 to 12.3% in 1920. Coal remained dominant, with its share of total energy consumption actually slightly increasing from 71.4 to 72.5%. The actual transition of fossil fuel dominance from coal to oil would come later in the century, but the foundation had been set.

5. *Abundance, Scarcity, and Conservation*

From its earliest beginnings in the 19th century, the oil industry had to deal with the recurrent feast or famine that accompanied the alternate discovery of new fields followed by their depletion. In "Special Report on the Petroleum of Pennsylvania," compiled in 1874 by geologist Henry Wrigley, it was argued that "we have reaped this fine harvest of mineral wealth in a most reckless and wasteful manner." John F. Carl of the Pennsylvania Geologic Survey concluded in 1886 that "the great Pennsylvania oil fields, which have supplied the world for years, are being exhausted, and cannot respond to the heavy drafts made upon them many years longer, unless reinforced by new deposits from deeper horizons." History soon demonstrated that many of these concerns were unfounded, as the discovery of new producing fields once again replaced scarcity and high crude prices with glut and price depression. However, the projections of future supply in the United States published by the U.S. Geologic Survey in 1908 predicted total depletion of U.S. reserves by 1927, based on then current levels of consumption, and the secretary of the interior warned President Taft in 1909 of an impending oil shortage. Driven by the rush to get oil out of the ground by the law of capture, the industry's production sector remained chaotic and highly unpredictable as the "forest of derricks" that defined each field moved ever westward. Within the context of this volatile business one can read the arguments of the Standard Oil (New Jersey) attorneys defending the company from antitrust attack in a different light. They maintained that Rockefeller's strategy of consolidation, denounced as evil monopoly, in fact represented a rational attempt to impose order and stability on an unstable industry.

A new era of conservation as well as the first significant enforcement of antitrust law had arrived on the political scene at the beginning of the 20th century with the presidential administration of Theodore Roosevelt. It is important to differentiate the conservation approach, best defined as wise utilization of natural resources, from the preservationist approach of Roosevelt contemporary John Muir, founder of the Sierra Club. Roosevelt and his key advisers, such as chief forester of the United States Gifford Pinchot, did not want to lock away America's resources but did favor planning and responsible exploitation. One

should interpret the efforts of the U.S. Geological Survey and the Bureau of Mines (created in 1910) in this light. These two institutions provided some attention to petroleum-related issues, but the creation of the Bureau of Mines' petroleum division in 1914 signaled a heightened federal government interest in oil conservation. Work centered at the division's Bartlesville, Oklahoma Petroleum Experiment Station focused on reservoir behavior, efficient drilling practices, well spacing, and secondary recovery techniques and disseminated knowledge to industry through influential Bureau of Mines publications.

In the 1920s, petroleum conservationists began to encourage unitization or the unit management of oil pools as a central alternative to the wasteful law of capture. This approach encouraged each pool to be operated as a cooperative unit, with the individual leaseholder's percentage share defining the amount of oil that could be pumped out. Operators would drill fewer wells, produce oil at a controlled rate, and either use or return to the producing zone under pressure all natural gas obtained along with the oil rather than the standard practice of flaring it or venting it into the atmosphere. Unitization was supported by the scientific principle that all recoverable oil in its undisturbed state contains gas in solution and that this solution has a lower viscosity, lower specific gravity, and lower surface tension than gas-free oil. It had long been known that natural gas had the same expulsive function in the well that carbon dioxide has in a shaken soda bottle, but reduced viscosity of the oil meant more fluidity and ultimately enhanced recovery. There were champions of this approach in the early 1920s within the industry and the government, among them Henry L. Doherty of the Cities Service Company and George Otis Smith, Director of the U.S. Geologic Survey. However, Doherty failed in efforts to get the American Petroleum Institute, the powerful trade association of the industry founded after World War I, to support the practice. The industry only began to come around after discoveries of huge amounts of oil in the midcontinental United States in the mid- to late 1920s. Doherty's plea that wise management led to the elimination of economic as well as physical waste began to receive attention.

The problem in the United States was that many operators were involved in the production of any given pool. This was free enterprise at work accompanied by individualistic opposition to government mandate to enforce unit operation. Because of the differing circumstances in most foreign fields, however, a different story unfolded. The Anglo-Persian Oil Company (APOC), like other European firms, had lagged behind its U.S. counterparts in a number of technical areas, including the adoption of thermal cracking and gas absorption plants to obtain rich "natural gasoline." However, because APOC had a monopolistic position in the huge Iranian fields it was able to unitize its operations. By carefully regulating the number of wells drilled, APOC could conserve the volume of natural gas in solution with the petroleum and operate the pool in the most efficient manner to sustain long-term recoverability.

The U.S. oil industry was still operating within the schizophrenic world of alternate scarcity and overabundance. A perceived scarcity of petroleum reserves at the beginning of the 1920s, coupled with increased demand, stimulated other technological developments. The Bureau of Mines conducted extensive research into enhanced and secondary recovery techniques such as the waterflooding of older fields to boost production. Integrated firms developed their own thermal cracking technologies in efforts to circumvent the Burton patents held by Standard of Indiana and increase their own gasoline output. There was also brief flirtation with alternative liquid fuels in the early part of the decade. Standard Oil (New Jersey) marketed a 25% ethanol–gasoline blend in 1922–1923 and later obtained the basic German patents for coal liquefaction in a licensing with the I. G. Farben interests. There was also a boom in Western shale oil in the 1920s, which has left an interesting if eccentric history. All these liquid fuel alternatives would soon prove unnecessary as new strikes of oil again dampened anxiety about shortages.

By 1930, oil's share of total U.S. energy consumption had increased to an impressive 23.8%, the most dramatic shift being in gasoline consumption. In 1920, oil-based motor fuel represented only 2.2% of total oil consumption compared with 48.1% for fuel oil. By 1930, gasoline's percentage had increased to 42.6% of the total and fuel oil's relative figure had shrunk to 43.5%. Kerosene production and sale had decreased sharply. However, traditional problems still plagued the industry. The opening of the East Texas field in 1930 again demonstrated that abundance was a double-edged sword. As unrestricted flow accelerated, the price of crude that had sold for $3 a barrel in 1919 had decreased to 10¢ in the summer of 1931. Voluntary or even compulsory unitization of pools was powerless in the face of such a major problem of overproduction. A new emergency approach developed in the states to combat this economic disaster—regulated prorationing, or the enforced limiting of production at each wellhead. Initially introduced under martial law by the Texas and Oklahoma governors in 1931, legislation in 1932 ceded the power to regulate production to the Texas Railroad Commission and the Oklahoma Corporation Commission. These actions did provide some stability: The price of crude increased to 85¢ a barrel by 1934. Federal New Deal legislation supported and augmented state practice as Franklin Roosevelt's National Recovery Administration (NRA) supported "fair competition" under the NRA Oil Code and policed the interstate shipment of "hot oil," oil produced illegally above state-mandated prorationing quotas. When the Supreme Court struck down the NRA in 1935, senators and congressmen from the oil-producing states swiftly moved to fill in the gap of cooperative regulation with passage of the Connolly Hot Oil Act and the Interstate Oil Compact Commission in 1935. These laws would serve to regulate production and combat "economic waste" (e.g., low prices) to the present day.

6. The Globalization of the Industry

At the end of World War I, British Foreign Secretary George Curzon stated that "the allies had floated to victory on a sea of oil." The major world navies had converted their coal-fired boilers to fuel oil in the critical years before the war and the noninterrupted flow of fuel for these vessels did indeed prove crucial. Winston Churchill, who had pushed for the conversion of the Royal Navy to fuel oil, was also the major player in the British government decision to take a controlling 51% financial position in the Anglo-Persian Oil Company (BP) in 1914. Unlike much of the activity of the former European colonial powers, the history of U.S. operations abroad has centered on the efforts of private firms obtaining concessions and the government has never owned or directly participated financially in actual operations. The U.S. State Department and other agencies have at various times, however, given strong support to these private efforts. Initial U.S. exploratory ventures occurred in Mexico early in the 20th century and in Venezuela in the 1920s. The most important producing area in the world today, the Middle East, had remained a British sphere of influence for a very long time. U.S. firms first obtained concessions in Iraq (Mesopotamia) in 1928 and in 1934 Standard of New Jersey (Exxon) and Standard of New York (Mobil) acquired one-fourth of the Iraqi Petroleum Company. Standard Oil of California (Chevron) gained leases in Bahrain Island in 1927 and Saudi Arabia in 1933, selling half of its rights to Texaco in 1936 when the two companies jointly formed the Arabian-American Oil Company (ARAMCO). The large vertically integrated major firms had control of large domestic supplies, but they viewed the obtaining of foreign crude as a rational economic strategy for their foreign marketing activities and as a hedge against continuing predictions of long-term depletion of U.S. supplies.

Oil imports into the United States prior to 1945 remained low, rarely exceeding 5% of total consumption. However, from 1945 to 1959 the total demand for petroleum products increased by approximately 80%, the single largest share being for gasoline as the automobile culture emerged triumphant in the postwar era. By the 1950s, cheaper, imported oil accounted for approximately 12% of needs and domestic U.S. producers were becoming alarmed by the challenge. Moreover, the Internal Revenue Service had ruled that U.S. companies operating abroad could deduct all foreign oil royalties and taxes from their corporate income tax bill, thus providing a huge financial incentive to pump Middle Eastern crude. President Dwight Eisenhower responded to political pressure from independent oilmen, most of them in western states, by asking the majors to implement voluntary import curbs. In 1959, acknowledging the failure of such voluntary measures, Eisenhower imposed mandatory import quotas.

Despite these quotas, as demand continued to expand in the gas-guzzling United States, imports became seen as the only way to meet needs. When the Nixon administration lifted quotas in 1973, oil imports as a percentage of total U.S. consumption dramatically increased, reaching 38% in 1974. At the same time, a regulated price structure for domestically produced oil, a carryover from Nixon's 1971 anti-inflation pricing policies, discouraged exploration and drilling at home. Imports increased to approximately one-half of total U.S. consumption by 1979 and have continued at these levels or above.

7. International Competition and War

World War II further crystallized the strategic importance of oil to international security, and those nations with ready access to supply were greatly advantaged. When Hitler invaded Poland in 1939, fully one-third of the total oil capacity with which Nazi Germany went to war was represented by synthetic oil derived from brown coal (approximately 19 million barrels per year). Germany's need for oil reserves is evident in Hitler's decision to invade Soviet Russia in June 1941. Most are aware of the German thrust toward Moscow that was only thwarted with Hitler's legions in view of the capital. An equally and perhaps more important southern line of attack aimed at the rich oil-producing properties in Baku in the southern Caucasus is less known. The United States' decision to embargo shipments of high-octane aviation fuel to Japan in 1941 was a significant factor in the Imperial government's decision to launch the war at Pearl Harbor and to attack Dutch interests in the East Indies in search of oil to fuel its military machine.

World War II also represented an important watershed in the history of the U.S. petroleum industry. As the nation mobilized for conflict, there was a validation of the New Deal consensus that a cooperative relationship between business and government represented the best way to achieve and maintain a stable industry in the wake of the inherently uncertain realm of unpredictable resources and rising demand. Much of the suspicion toward New Deal regulation was superseded by a new spirit, imbued with patriotism, but also pleased that production curtailment was being replaced by dramatic efforts aimed at increasing production. The wartime Petroleum Administration for War (PAW) fostered cooperation in pipeline construction, the patent pooling of technical processes, and the development of new catalytic cracking plants needed to produce huge quantities of 100 octane aviation gasoline demanded by new high-performance aircraft. If the allies had indeed floated to victory on a sea of oil during the 1914–1918 conflict, one can make a good case that they flew to victory in the 1939–1945 war.

A council of oil executives, the Petroleum Industry War Service Committee, was formed to advice the PAW, headed by Secretary of the Interior and New Deal zealot Harold Ickes. The oil czar's new image of working with industry to boost oil production contrasted with his previous reputation among oilmen as Horrible Harold. In 1946, the success of this industry advisory group was so valued by government and industry that President Harry Truman continued it in peacetime with the creation of the National Petroleum Council (NPC). The NPC has continued to function as a consulting and advisory body to government and as a generator of reports and studies on all aspects of the oil industry. It has primarily worked in coordination with the Department of the Interior and, after 1977, the Department of Energy.

Another major outcome of the war was an increased awakening of U.S. interest in foreign oil resources and the laying of the foundation for the structure of oil importation that characterizes the industry today. During World War II, the protection of Middle East oil reserves became an essential aim of U.S. foreign policy.

After the war, the exploitation of oil abroad continued to expand at the same time that conservation in the form of market-demand prorationing continued at home. When Chevron and Texaco invited Mobil and Exxon into partnership with ARAMCO in 1946, the U.S. government response was to grant clearance to the companies from antitrust prosecution, citing the value to the national interest. However, U.S. oil policy was ambivalent during the period of the early Cold War. The State Department continued to support aggressive U.S. firms operating abroad, but the industry was still split between independent companies, which were primarily engaged in the domestic industry, and the larger major multinational firms, which were seeking to expand their interests overseas.

8. The Middle East, OPEC, and the Energy Crisis

In the 1950s and 1960s, an increasing number of former European colonies in Asia and Africa were gaining independence. In other areas where colonial powers had not maintained political control in the strictest sense for some time, such as the Middle East, multinational oil companies effectively controlled the destinies of these nations through their economic power. The formation of the Arab League in 1945 and the first Arab–Israeli war in 1948–1949, which erupted in response to the birth of modern Israel, also pointed to a growing Arab nationalism. With Israel's victory, the Palestinians were now without a homeland and this issue has dominated Middle Eastern politics ever since. The first direct threat to Western oil interests, however, came not from an Arab state but from a strategically situated nation whose interests have often coincided with her Arab neighbors, Iran.

Iranian Prime Minister Mohammed Mossadegh's nationalization of British Petroleum's properties in 1951 led to the first real postwar oil crisis. This nationalist challenge to Western interests, coupled with the proximity of Iran to Soviet Russia, prompted the United States to act. The Central Intelligence Agency took a direct hand in a political counterrevolution that overthrew Mossadegh and installed the U.S.-backed Shah, Reza Pahlavi, to the throne in 1953. With U.S. State Department support, U.S. companies now obtained 40% of a new foreign consortium established under the Shah to exploit Iranian oil. The Shah remained America's man in Iran until his overthrow during the Iranian revolution of 1979. During that time, private oil interests were served well with access to Iranian crude, and the State Department enjoyed Persian Gulf stability through military and economic support of the Shah.

A new breed of Arab leaders, such as Egypt's Gamal Abdel Nasser, also symbolized this new nationalistic and independent attitude. Nasser's seizure of the Suez Canal in 1956 was the precipitating factor in the second Arab–Israeli war. Libya's Colonel Muammar Qaddafi's ascension to power in 1969 represented an even more aggressive leadership and an assertion of Muslim values. Superimposed over all these events was the playing out of Cold War tensions as both the Soviet Union and the United States sought to assert their influence in this vital area. In the midst of this changing political landscape, representatives of the major petroleum-producing countries meeting in Baghdad, Iraq, in 1960 formed the Organization of Petroleum Exporting Countries (OPEC). This new cartel of countries, not companies, viewed itself as a counterweight to the international cartel of multinational oil companies later christened the Seven Sisters (British Petroleum, Royal Dutch/Shell, Exxon, Chevron, Mobil, Texaco, and Gulf). Arab spokesmen among OPEC began to talk about "using the oil weapon" but little happened until the outbreak of the third Arab–Israeli conflict in 1967. On June 6, 1967, the day after hostilities began, Arab oil ministers called for an embargo on oil shipped to nations friendly to Israel. The flow of Arab oil was reduced by 60% by June 8 and a crisis threatened. Although there were examples of property destruction and temporary economic dislocation, the embargo was a failure in the wake of the brief 6-day war. The oil companies were able to weather the storm and the major losers were the Muslim nations, which lost tremendous oil revenue during the embargo. The situation would not be the same when similar problems occurred in 1973.

Israel had defeated the Arabs with relative ease during the three previous Middle Eastern conflicts since World War II. Despite their huge oil reserves, it appeared that the Arab nations were no military match for a modern technologically equipped military power with American and European support. The Yom Kippur War that broke out in the fall of 1973 had a different outcome. Catching Israel by surprise on Judaism's high holy day, Egypt and Syria inflicted heavy casualties on Israel in 3 weeks of bloody fighting. At the time the war broke out in October 1973, OPEC was meeting in Vienna. OPEC took two strong measures: They voted to raise the posted price of oil from $3 to $5 a barrel, and they announced an embargo against countries supporting Israel. The Nixon administration had been caught off guard at a point when the sitting president was fighting to keep his job in the midst of the Watergate scandal. Secretary of State Henry Kissinger performed brilliantly in negotiating a cease-fire through rounds of shuttle diplomacy that saw him flying from capital to capital. Although military operations halted at precisely the time when Israel appeared to be gaining the upper hand in the conflict, Kissinger's immediate goal was to cease hostilities. Further diplomacy brought a peace settlement, but the world would experience the aftermath of the embargo and price shocks of October 1973 for the rest of the decade.

High oil prices fueled the increasing problem of stagflation (inflation coupled with stagnating economic growth) that haunted the remaining months of the Nixon presidency until his resignation over Watergate-related matters in August 1974. A similar set of problems limited the effectiveness of the brief presidency of Gerald R. Ford. The presidential election of 1976 brought a Washington outsider, James Earl "Jimmy" Carter, to office and energy policy emerged as a central issue of his administration. The multinational oil companies had come increasingly under attack in the 1970s on charges of complicity in the OPEC embargo, the taking of "obscene" profits when the market price of crude tripled, and for opposing alternative forms of energy out of selfish economic motive. There was a growing literature on multinational firms generally that asked the question, To whom do these firms owe loyalty? Examples of the fuzzy nature of this question existed particularly with regard to instances of the behavior of oil companies in the international arena. For example, there were many complaints that ARAMCO, ostensibly a U.S. company, had fully cooperated with Saudi Arabia and OPEC during the embargo. In another case, there was evidence that Gulf Oil had supported one faction in the Angola civil war that controlled its area of business operation, whereas the U.S. government backed a competing rebel group. The political economy of world oil had become incredibly more complicated, and the aims of the multinational oil industry had become ever more intertwined with international diplomacy.

9. Oil in the Modern Economy

President Carter urged the American people to view the 1970s energy crisis as the moral equivalent of war. The problem was that the idea never really took hold. To be sure, there had been frightening events in 1973–1974, such as citizens freezing to death for want of fuel oil in New England and a man shot to death in California when he cut into the line at a service station. However, by 1977 oil prices had begun to stabilize as price incentives for new exploration in Alaska, Mexico, and the North Sea along with completion of the Alaska pipeline kicked in. President Carter's and Energy Secretary Schles-singer's appeal for energy conservation had also begun to pay off. Perhaps the most significant legacy of Carter's policies will be his deregulation of oil and natural gas that reintroduced a program of more realistic market pricing. Strong political support for investment in alternative energy technologies, such as solar power, synthetic fuel, wind energy, and biomass conversion, attracted much publicity and enthusiasm but faded amid cynical charges that the energy crisis had been contrived by the oil companies and with the return of relatively lower oil prices. Nuclear power advocates believing that Carter, a nuclear engineer, would give a boost to that industry had their hopes dashed with the Three Mile Island disaster of March 28, 1978—a blow from which the industry has never fully recovered. Negotiations with the Saudi and Iranian governments in 1978 led to a realization that it was in the best economic and political interests of the oil producers to curtail prices and it appeared that oil prices had achieved stability. However, the international oil economy would have a more cruel blow in store for Jimmy Carter—one that probably cost him reelection in 1980.

President Carter visited Iran and congratulated the Shah as a force of stability in the Persian Gulf. Soon after, when the Iranian people overthrew the Shah in 1979, halting Iranian oil production in the process, another major spike in prices resulted. As had occurred in 1973, the consuming countries seemed unable to compensate for what was, on a global basis, only a small reduction in supplies. Jimmy Carter's political fate became sealed when the United States allowed the Shah into the country for medical and humanitarian reasons and the U.S. embassy was stormed and 50 hostages were ultimately held captive. Candidate Ronald Reagan benefited from a number of Carter's difficulties, including very high interest rates and an apparent weak policy abroad, but high oil prices coupled with the embarrassment of the hostage crisis comprised a fatal mix. The revolutionary government of Iran rubbed salt in the wound when it released the hostages on the day of Ronald Reagan's inaugural as president. President Reagan also benefited from a weakening of OPEC solidarity as members sold oil at below posted prices in order to obtain badly needed money. In 1982, non-OPEC production exceeded OPEC production for the first time as production from the North Sea, Soviet Union, Mexico, Alaska, Egypt, Angola, and China added to world reserves. Although OPEC still remains a relevant player in the political economy of oil, its role is much diminished from its position in the early 1970s. The long-running Iran–Iraq war caused price dislocations when it first started in 1980, but it actually bolstered world supplies in the long term because both nations were forced to sell above quota for needed revenue.

The industry has also had to deal with an increasing number of environmental issues in recent decades. One can date offshore drilling technology, the placing of wells in submerged lands, to the early years of the 20th century, but the technique has only become of critical importance recently. Operators in Santa Barbara, California, for example, had drilled more than 200 wells from piers extending out over the water in 1906, and drilling had been attempted off the coast of Peru that same year. Further experiments went forward in the United States and abroad during the 1920s and the 1930s, but it would not be until after World War II that offshore drilling began to take on the importance that it has today with deeper drilling techniques. Drilling off the Texas Gulf Coast, for example, became very extensive after the war. However, the events of January 1969 off the coast of Santa Barbara sent shocks throughout the industry. When the drilling of an offshore well in the Santa Barbara channel ran into difficulties that sent a slick of heavy crude onto 30 miles of beaches, there was a huge public outcry. It was also in 1969 that BP, Humble Oil (Exxon), and Atlantic Richfield announced plans to construct a pipeline across the Alaskan tundra aimed at transporting oil from Prudhoe Bay, north of the Arctic Circle, to the port of Valdez. The ensuing debate delayed the project, brought cost overruns, and pitted the environmental movement against the industry in a very public way before the project was finally completed in 1977. Things quieted down with the wide recognition of the part that Alaskan oil had played in dampening high oil prices. Then the Exxon Valdez incident occurred. Oil spills from seagoing tankers have always

211

presented a risk, but the size of the new generation of supertankers exacerbated concerns. When the supertanker Exxon Valdez went aground in Alaska's Prince William Sound spilling 240,000 barrels of Prudhoe Bay oil in 1989, environmental critics of the industry received a further boost. Although it is environmentally risky, deep-water drilling as is occurring in the North Sea, wilderness pipelines, and the passage of supertankers on the oceans have become essential elements of commerce in world oil.

The industrialized nations have enjoyed a relatively more benign oil economy in the 1990s and into the early new century, but the historic pattern of dependency on oil controlled by the producing nations remains a reality. In 2003, the United States imported 55% of its oil; in the crisis year of 1973, imports amounted to only 35%. The Iraqi invasion of Kuwait on August 2, 1990, precipitated a series of problems in the Persian Gulf region. Future historians will interpret the defeat of Saddam Hussein in the 1991 Gulf War by the first Bush administration as but a chapter in the events that led to the U.S. invasion of Iraq in the spring of 2003 directed by his son. Neither in 1991 nor in 2003 was oil the sole factor in precipitating international crisis, but one cannot ignore the extremely significant role that it has played. Critics point to the very close ties between the administration of the second President Bush and the oil industry, and many have uttered the protest refrain of "No Blood for Oil" with reference to the war in Iraq. However, in a figurative sense oil has become the blood of international commerce and industry more than ever as we embark on a new century.

One of the main business trends in the global oil industry in recent years exhibits a touch of irony for the United States. This is the movement toward megamergers initiated by the acquisition by Chevron (Standard Oil of California) of Gulf Oil in 1984 and continuing through 2000. The demise of Gulf occurred after swashbuckling financier T. Boone Pickens' small Mesa Petroleum Company shocked the industry with a takeover bid for the large multinational. Chevron's ultimate acquisition of Gulf was followed by a series of additional moves, including the Royal Dutch/Shell group acquiring that part of Shell USA that it did not already own; BP's merger with Sohio (Standard Oil of Ohio); the joining of Exxon (Standard Oil of New Jersey) and Mobil (Standard Oil of New York); Chevron's merger with Texaco; and the merger of BP, Amoco (Standard Oil of Indiana), and ARCO (Rockefeller's old Atlantic Refining Company). The notorious Seven Sisters of the 1970s had shrunk considerably by the turn of the century. Historical irony lies in this partial reassembling of the pieces of the old Rockefeller Standard Oil Company, which had been fragmented into 34 separate companies in 1911.

Why this rush toward consolidation? In general terms, this oil industry behavior relates to the broader trend of U.S. merger activity beginning in the 1980s and present today. The argument that firms required a leaner, more competitive profile to respond to the challenges of the new global economy became most prominent, and a more benign antitrust environment created by recent Republican administrations in Washington has abetted the trend. For the oil industry specifically, there are other reasons cited. One of these is the argument that the companies representing the consuming nations need to have a stronger bargaining position with OPEC and other producing groups. Others see consolidation as a key to mobilizing more effective resistance to world environmental movements, such as the Kyoto Protocol, which have called for the reduction of heat-trapping greenhouse gas emissions caused by the burning of coal, oil, and gasoline. Certainly, environmental groups have been among those most critical of the industry mergers. Some industry spokesmen have referred to declining profits in the wake of relatively lower world oil prices as the reason for moving toward the cost reductions resulting from merger. Within the context of the history of the oil industry, one might ask simply "how big should big oil be?"

Oil supplies will eventually become exhausted; this is an accepted reality. However, there remain huge supplies of petroleum and a world economy that depends on access to it. There will be an historic transition to other energy sources in the future and the individual companies that comprise the industry may very well have diversified to the extent that they will become part of this new political economy. The history of the oil industry has been relatively brief in years, approximately a century and a half, but it has been a richly diverse and fascinating story that is still not concluded.

10. Postscript (2009)

Since this article first appeared, the petroleum industry has remained very much in the news. Of central interest have been several themes raised previously, including fluctuating prices connected to glut and scarcity; continuing reliance on Middle Eastern crude; increasing environmental concerns; and further recognition of oil's importance to developing as well as industrialized countries, especially at a time of worldwide economic recession in 2008–2009.

By some estimates the world has consumed approximately half of global oil reserves during the mere 150 years of the modern petroleum industry. Recent data shows a relatively constant oil production since 2005, perhaps indicating that we may be at or near peak production. The United States Geological Survey's more sanguine estimate of recoverable reserves predicts a peak closer to the year 2030. Over the past few years there has been an annual three percent growth in market demand for petroleum, largely fueled by United States led economic growth in the industrialized West, but also reflecting a huge surge in demand from developing nations such as China and India. It appeared in the early 21st century that the only event that could dampen a rampant world appetite for oil would be an economic downturn; sometimes we get what we wish for. In mid-April 2009, for example, the United States Energy Information Administration reported that it expects continuing reduced demand from a weaker economy, and the American Petroleum Institute reported that United States oil and gas drilling had plunged to levels not seen since 2004, bringing an end to six years of continuous growth.

As per barrel crude prices continued to rise dramatically in the first half of 2008, the Congress grilled oil industry executives over "excess profits" and public opinion toward the industry turned negative, both reminiscent of the 1970s. In early July 2008 the *Wall Street Journal*, reporting on the high world price of crude, warned of $200 a barrel oil before the end of the year. As $4.00 a gallon gasoline and lines at pumps in many states brought back memories of 1973 and 1979, it seemed like another energy crisis. But by early January, 2009 the price had dropped to a little over $40 a barrel, and by mid-April was hovering at around $50. For the American consumer, hit by a housing and credit crisis, increasing unemployment, and shrinking disposable income, the decline in gasoline prices represented a welcome gift, but global price decline can symbolize calamity for petroleum exporting countries. In the 2008 American presidential campaign oil played a dramatic political role, with the Republican Party championing exploration in Alaska's Arctic National Wildlife Reserve (ANWAR) to the chant of "drill baby drill," while the Democratic Party ticket outlined potentially the most comprehensive energy policy in the nation's history. This plan emphasized fuel efficient conservation, alternative energy investment, and a cap and trade program to reduce greenhouse gas emissions. The current Obama administration is attempting to implement many of these policies with both short term strategies and longer term ones to address an inevitable decline of petroleum reserves.

For its part, petroleum industry leaders remain skeptical of policies that imply that the United States can achieve energy independence. In September 2008, in the wake of summer high gas prices, public support for drilling, and the fall-out from presidential political campaigns, the United States Congress had allowed the moratorium on off-shore drilling to expire. Now, in the spring of 2009, the industry is trying to navigate between what it sees as a wary public and hostile government at a time when it is burdened with excess output and insufficient demand. In the aftermath of the last significant plummeting of prices in the 1990s, there occurred a rash of industry mergers. When two large Canadian firms, Suncor and Petro-Canada, announced a $15.8 billion merger on March 23, 2009, the largest in the industry since 2006, many observers speculated that the industry was again on the brink of further combination. One should note, however, that this deal involved two firms heavily invested in Canadian tar sands, an "unconventional" source of crude. Experts anticipate other such mergers and acquisitions, but doubt there will be mega-deals such as had occurred with the Exxon Mobil merger, primarily because of wariness toward government antitrust concerns. There is reason to believe, however, that acquisition of smaller and medium size companies will continue, with players including state-owned firms from countries like China and India as well as from Western "majors."

One other recent development has been the impact of the Iraqi war on the global petroleum industry. It was previously argued that the United States led invasion of Iraq in 2003 most certainly had some relation to Iraqi oil reserves, then estimated as the world's third largest. The fall of Baghdad very quickly led to predictions that the Iraqi oil ministry would soon be producing six million barrels of oil a day. In fact, six years after the collapse of the previous regime, Iraq is producing only 2.2 million barrels a day, less than its pre-war levels of 2.5 million. Reasons for this failure include economic and political corruption, bad management, political infighting over passage of a new hydrocarbon law introduced in 2006, and continuing political instability. Efforts to award service contracts to western oil companies beginning in 2007 have not yet been realized, and at the present time there are significant political disagreements over such deals. These events have further questioned statements made by former United States Deputy Defense Secretary Paul Wolfowitz, who had famously argued that Iraqi oil would both largely pay for the cost of the conflict and allow a post-Saddam Iraq to finance its own reconstruction.

Petroleum will remain a crucial commodity in the near term and OPEC will seek a delicate price structure to ensure a profit while not encouraging investments in alternative energy. Meanwhile, renewed enthusiasm over curbing carbon emissions, particularly with a United States political administration apparently more cooperative, must sustain itself through the current world economic crisis.

Further Reading

Brantly J E 1971 *History of Oil Well Drilling*. Gulf, Houston, TX

Caputo R 2008 *Hitting the Wall: A Vision of a Secure Energy Future*. Morgan and Claypool, Princeton

Chernow R 1998 *Titan: The Life of John D. Rockefeller, Sr.* Random House, New York

Clark J G 1987 *Energy and the Federal Government: Fossil Fuel Policies, 1900–1946.* Univ. of Illinois Press, Urbana

Clark J G 1990 *The Political Economy of World Energy: A Twentieth Century Perspective.* Univ. of North Carolina Press, Chapel Hill

Gorman H S 2001 *Redefining Efficiency: Pollution Concerns, Regulatory Mechanisms, and Technological Change in the U.S. Petroleum Industry.* Univ. of Akron Press, Akron, OH

Olien R M, Olien D D 2000 *Oil and Ideology: The Cultural Creation of the American Petroleum Industry.* Univ. of North Carolina Press, Chapel Hill

Painter D S 1986 *Oil and the American Century: The Political Economy of U.S. Foreign Oil Policy, 1941–1954.* Johns Hopkins Univ. Press, Baltimore

Pratt J A, Becker W H, McClenahan W Jr. 2002 *Voice of the Marketplace: A History of the National Petroleum Council.* Texas A&M Univ. Press, College Station

Schneider S A 1983 *The Oil Price Revolution.* Johns Hopkins Univ. Press, Baltimore

Vietor R H K 1984 *Energy Policy in America Since 1945.* Cambridge Univ. Press, New York

Williamson H F, Daum A 1959 *The American Petroleum Industry: The Age of Illumination, 1859–1899.* Northwestern Univ. Press, Evanston, IL

Williamson H F, Andreano R L, Daum A, Klose G C 1963 *The American Petroleum Industry: The Age of Energy, 1899–1959.* Northwestern Univ. Press, Evanston, IL

Yergin D 1991 *The Prize: The Epic Quest for Oil, Money & Power.* Simon & Schuster, New York

Related Websites

American Petroleum Institute http://www.api.org/

Interstate Oil and Gas Compact Commission http://www.iogcc.state.ok.us/

Offshore Technology http://www.offshore-technology.com/

Oil and Gas Journal http://www.ogj.com/

Society of Petroleum Engineers (SPE) http://www.spe.org/

August W. Giebelhaus

OPEC Market Behavior, 1973–2003

Glossary

cartel A group of producers that collude to reduce production and increase prices in order to maximize the wealth of the group.

monopoly A market with a single producer that sells a product having no close substitute.

nationalization The host government takeover of foreign companies' operations.

netback pricing A method of determining the wellhead price of oil and gas by deducting from the downstream price, the destination price, the transportation price, and other charges that arise between wellhead and point of sale.

oligopoly A market with few competing producers.

OPEC basket price The average price of seven marker crudes.

OPEC quota A specified maximum amount of oil production assigned by OPEC to each of its members.

Organization of Petroleum Exporting Countries (OPEC) Historically, it consisted of 13 countries: Saudi Arabia, Kuwait, Iraq, Iran, Qatar, Abu Dhabi, Libya, Algeria, Nigeria, Venezuela, Indonesia, Gabon, and Ecuador. Ecuador and Gabon withdrew from the organization in 1992 and 1995, respectively.

posted price A price of a barrel of oil that the host nations used as a base to collect taxes from the international oil companies, regardless of the market price. In most cases, posted prices were higher than market prices by a wide margin. After nationalization, countries decided to make it the market price.

price band The minimum and maximum for the average OPEC basket price, set by OPEC in 2000.

Seven Sisters Refers to the seven international oil companies that dominated the world oil industry in the 1930s and 1940s: Exxon, British Petroleum, Shell, Gulf, Texaco, Socal, and Mobil.

user's cost The cost of replacing oil that has already been produced. It is a concept used only with nonrenewable resources.

Two supply factors are known to influence oil prices and dominate research activities: scarcity and market power. Scarcity increases prices above marginal cost by the discount rate, even in a competitive market. In noncompetitive markets, prices are usually higher than marginal cost. However, events of the past 80 years indicate that the market power of the Texas Railroad Commission, international oil companies, and Organization of Petroleum Exporting Countries (OPEC) has had a more significant impact on oil prices than scarcity. For this reason, most researchers focus on OPEC when they discuss world oil markets of the past 30 years. Nevertheless, this focus on market power without regard to depletion and technical issues has polarized academic and political circles to the extent that it is almost impossible to reach a consensus on OPEC and its role in world oil markets. It is also difficult to discuss OPEC behavior without putting analyses in their historical context. It is necessary to understand the political, economical, social, and philosophical factors that led to the establishment of OPEC and its role today. Oil is not only a nonrenewable resource but also a political and strategic commodity.

1. Introduction

Some researchers argue that oil is a special commodity that requires intervention and depletion management. The oil industry is characterized by booms or busts, which calls for additional controls not provided by market forces. This extra control seeks to avoid or minimize the boom-or-bust cycle. Standard Oil, the Texas Railroad Commission, the Seven Sisters, and the Organization of Petroleum Exporting Countries (OPEC) have all exercised it with varying degree of success. OPEC has been the least successful and failed to manage depletion. Some researchers even argue that U.S. embargos on some of the oil-producing countries are a form of oil market management to support oil prices. OPEC does not have grand strategies and plans. It pragmatically reacts to various political, economical, and natural factors that influence its members, but the impact of these factors varies among members at any given time. Thus, OPEC must reconcile the divergent interests of its membership through meetings and negotiations.

The focus of this article is OPEC market behavior. Therefore, only market models introduced in the past 30 years are discussed. An overview of the many oil market models developed from 1973 to present shows that these models divide easily into oligopoly models and competitive models. Most researchers use oligopoly models to explain OPEC behavior and assume that OPEC, or a subset of it, is a cartel that seeks monopoly profits by changing production and influencing prices. Others argue that the world oil market is competitive and that factors other than cartelization better explain oil price increases. For each side, there are various models that explain OPEC behavior.

A few researchers argue that OPEC is a wealth maximizer. Most researchers argue that OPEC functions not to maximize wealth but to maximize revenues and achieve other social and political purposes. These contrasting views resulted from the emphasis of most researchers on market power rather than on market failure and depletion management.

Section 2 presents an overview of OPEC and non-OPEC production, world oil demand, oil prices, and OPEC's actions. Section 3 reviews and critiques the oligopoly models of OPEC. The competitive models are presented in Section 4. Section 5 provides various points of view regarding the application of the term cartel to OPEC.

2. Overview

Figures 1 and 2 show oil prices, OPEC production, non-OPEC production, and world demand for oil between 1973 and the first quarter of 2003. Major political events and some OPEC actions are plotted in Fig. 1 to illustrate their impact on prices and oil production. OPEC as an organization did not cause energy crises, oil price spikes, and price volatility. However, various political incidents that involved individual OPEC members did contribute to energy crises and market instability. Researchers provide various models and explanations for such changes in production and prices. From the information in Figs. 1 and 2, the period between 1973 and 2003 can be divided into six phases: 1973–1978, 1979–1981, 1982–1986, 1987–1991, 1992–1997, and 1998 to the present.

2.1 The First Oil Crisis: 1973–1978

Oil prices increased in 1973 when, on October 19, several Arab countries—all of them allies of the United States—imposed an embargo on the United States and The Netherlands for their support of Israel in the 1973 war. Although the embargo led to a mere 5% decrease in world oil supply and lasted

only 2 months, oil prices increased fourfold. Theory predicts that oil prices will return to their original level when an embargo is lifted, but prices in 1974 remained high. World oil demand decreased at first because of substitution, conservation, and lower economic growth. However, it continued its upward trend once the economies of the consuming countries absorbed the impact of the energy shock.

This production cut by some OPEC members and the subsequent spike in prices led some researchers and observers to conclude that OPEC had become a powerful cartel. They point out that non-OPEC members behaved differently and increased production to take advantage of the higher oil prices. Others believe that different factors led to the production cut and that price increases were not related to cartelization, including changes in property rights, the limited capacity of the economies of OPEC members to absorb productive investment, environmental concerns about natural gas flaring by some OPEC members, panic and stockpiling in the consuming countries, and U.S. wellhead price control. This last factor kept U.S. oil prices below the world price, decreased U.S. oil production, increased U.S. demand for oil, and increased oil imports.

2.2 The Second Oil Crisis: 1979–1981

Oil prices increased again in 1979 as a result of the Iranian revolution. Only Saudi Arabia and some non-OPEC members increased their production to compensate for the lost Iranian oil. Prices continued to rise when the Iraq–Iran war started in September 1980. Iraqi and Iranian oil production came to a complete halt by the end of 1980 and the world lost 9 million barrels per day (mb/d). Coupled with panic and stockpiling, spot oil prices reached historic levels. However, OPEC contract prices were much lower, especially those of Saudi Arabia. These prices were set by OPEC members, not the market.

Figure 1
Oil prices, OPEC and non-OPEC production (thousand b/d), and major events, 1973–2003. Data from *Monthly Energy Review*, Energy Information Administration, July 2003. Data for 2003 are for the first quarter.

Figure 2
Oil prices and world demand (thousand b/d), 1973–2003. Data From ''BP Statistical Review of World Energy'' (2003) and Energy Information Administration (2003). Data for 2003 are estimates for the first quarter.

Conservation, substitution, and a worldwide recession brought a substantial decline in world oil demand in 1980 and 1981. Meanwhile, non-OPEC production continued to increase. As a result, OPEC output and market share decreased. Two major factors affected oil prices and OPEC behavior in 1981: price deregulation in the United States and price agreement among OPEC members. The agreement set price differentials for each OPEC crude relative to the Saudi marker. This agreement indicated that Saudi Arabia was not willing to defend prices alone. It wanted other members to share this burden. The agreement ensured that if the price of the Saudi marker declined, the price of other member's crudes would decline too. Therefore, countries had to cooperate with Saudi Arabia and cut output to preserve prices. This was the first serious coordinated attempt by OPEC to influence world oil markets. Market management requires price control, output control, or both. OPEC chose the first method in 1981. By the time OPEC agreed on prices, a decrease in demand led Nigeria unilaterally to cut its prices by $5.50/b. First Iran and then other OPEC members followed. As a result, the price agreement collapsed.

2.3 The Oil Glut: 1982–1986

A decline in world oil demand and an increase in non-OPEC production led to a sharp decrease in OPEC market share between 1982 and 1986. The fundamentals of world oil markets changed drastically to the extent that massive production cuts by some OPEC members were not enough to prevent an oil price collapse, as shown in Fig. 1.

The breakdown of the first price agreement in 1981 led OPEC to consider another method of market control: production management. For the first time, OPEC established a loose quota system in March 1982. In this agreement, Saudi Arabia was not assigned a quota with the understanding that it would behave as a swing producer to balance the world oil market. In addition, Iraq was allowed to produce more than its quota. Many countries were dissatisfied with the system, which broke down almost immediately. OPEC members violated their quota by more than 2 mb/d by the end of 1982.

The 1983–1986 period witnessed a sharp decrease in OPEC market share, especially that of Saudi Arabia; an increase in cheating by OPEC members; a decline in world demand; and an increase in non-OPEC production. Consequently, OPEC was forced to reduce its posted marker prices and lower its production ceiling. Saudi Arabia continued to defend oil prices by cutting production until its output declined to only 2 mb/d, an amount insufficient to sustain the Saudi economy. For the first time since the establishment of the quota system, OPEC

assigned Saudi Arabia a quota of 4.35 mb/d in order to protect Saudi Arabia's market share. This figure was less than half of what Saudi Arabia was producing in 1980. In addition, OPEC abandoned the posted price mechanism. It abandoned Saudi Arab Light as the marker crude in favor of market-oriented netback prices. Prices were no longer administered by OPEC through the posted prices mechanism. OPEC decided to take the market price and produce accordingly. As a result of these developments, the price plummeted to single digits in August 1986.

This collapse reversed historical trends of increased substitution and conservation. It stimulated the world demand for oil, as shown in Fig. 2. It also decreased non-OPEC production by shutting down stripper wells in the United States and other countries.

2.4 Calm before the Storm: 1987–1991

Four factors brought about increased oil prices in 1987: major damage to Iraqi and Iranian oil facilities, less cheating by OPEC members, reduction in non-OPEC output, and an increase in oil demand. Netback pricing was abandoned by OPEC members by 1988 in favor of a sophisticated pricing formula that linked OPEC crude prices to market prices.

Oil prices improved by 1989, and OPEC regained market share because of a decline in non-OPEC production and an increase in demand. Prices increased to more than $40/b when Iraq invaded Kuwait in August 1990. The world lost 4.5 mb/d, but Saudi Arabia, along with other oil-producing countries, was able to compensate for the loss of Iraqi and Kuwaiti oil. Although oil prices remained relatively high due to the Gulf War, the decline in world demand and the U.S. postwar recession resulted in reduced oil prices in 1991. A United Nations (UN) embargo on Iraqi oil went into effect in 1991. Iraqi oil did not find its way into world markets legally until 1996. The UN lifted the embargo in April 2003 after the U.S. invasion of Iraq.

2.5 Post Gulf War: 1992–1997

A combination of factors caused oil prices to plunge in 1993, including the expected return of Iraqi exports to world oil markets, surging North Sea output, and weak demand. At the end of 1997, OPEC decided to increase its quota by 10% for the first half of 1998 on the erroneous assumption that Asian oil demand would continue to grow. Unfortunately, the Asian financial crisis of 1997 proved OPEC wrong. Oil prices began to decrease. OPEC ignored many signals at that time. For example, the increase in oil prices in 1996 was caused not only by an increase in Asian demand but also by the delay in signing the oil-for-food program between Iraq and the UN.

2.6 Low Oil Prices Cure Low Oil Prices: 1998–2003

Despite very strong economic growth in the United States and many European countries in 1998, oil prices continued to decrease until they reached single digits in the first quarter of 1999. Analysts cited four reasons for this decline: the Asian financial crisis, the increase in OPEC and non-OPEC production, the increase in Iraqi exports, and two consecutive mild winters in the consuming countries.

OPEC efforts to halt the 1998 decline in oil prices failed despite two announced cuts and the promise of output cuts by some non-OPEC members. The collapse of oil prices in the first quarter of 1999 forced OPEC and non-OPEC members to announce production cuts by 2.104 mb/d in March 1999. OPEC decided to continue the cuts when it met in September of that year, and oil prices continued to rise, as shown in Fig. 1.

The sharp increase in oil prices in 2000 forced OPEC to increase its quotas in March, June, September, and October. However, these increased quotas did not translate into actual production. Many OPEC members lacked the spare capacity. Three factors characterized 2000: frequent meetings, frequent quota increases, and the unofficial creation of a price band. In October, OPEC activated the unofficial price band for the first time and increased production by 500,000 b/d.

OPEC set up a price band mechanism during its March 2000 meeting. According to the mechanism, OPEC would increase production automatically by 500,000 b/d if the average OPEC basket price remained above $28/b for 20 consecutive trading days. Production would be cut by a similar amount if the basket price remained below $22/b for 10 consecutive trading days. The automatic adjustment was later abandoned so that OPEC members could adjust production at their discretion. The creation of the price band is OPEC's fourth serious attempt to manage world oil markets. At first, OPEC tried to control prices through creating price differentials for each OPEC crude in 1981. The failure of this pricing mechanism led OPEC to adopt a quota system to control output in 1982. The failure of this move led OPEC to adopt netback pricing in 1985. Theoretically, netback pricing with the quota system in place should have enabled OPEC to control both prices and outputs. However, netback pricing, a market-oriented pricing mechanism, annulled the quota system, which is a form of market management. OPEC failed to control the market for the third time. Creating the price band may have enabled OPEC for the first time to use both outputs and prices to manage world oil markets.

The sharp increase in oil prices in 1999 and 2000 led to lower demand and higher non-OPEC production in 2001. Consequently, oil prices declined and OPEC was forced to cut production twice—in January by 1.5 mb/d and in March by 1 mb/d. The demand for oil declined again at the end of 2001 after the September 11 terrorist attack on the United States. The decline in air travel and lower economic growth were the primary reasons for the decline in world oil demand. Consequently, OPEC agreed in November 2001 to cut production by 1.5 mb/d effective January 1, 2002, but only if non-OPEC producers cut their output by 500,000 b/d as well. With non-OPEC cooperation, the cut was implemented, although data indicate that non-OPEC cooperation may have been merely symbolic.

Despite lower world economic growth in 2002 and higher world oil production, political problems pushed oil prices higher. Iraq halted its official oil exports in April for 1 month to protest the Israeli incursion into the West Bank. Meanwhile, an attempted coup forced Venezuelan President Hugo Chavez to resign on April 12, only to resume his presidency on April 14.

Oil prices had been increasing since May 2002. A general fear that the war with Iraq would disrupt the Middle Eastern oil supplies initiated the increase. Labor strikes in Venezuela that halted oil exports forced prices to increase to more than $30/b in December 2002. Both Iraq and Venezuela are major oil producers and OPEC members.

OPEC met in December 2002 and decided to take unusual action—to cut oil production by increasing quotas and cutting production at the same time. OPEC members were violating their quota by 3 mb/d. OPEC decided to increase quotas by 1.3 mb/d and cut production by 1.7 mb/d. When prices continued to increase, OPEC met a month later and increased production by 1.5 mb/d.

Oil prices continued to increase in 2003. The possibility of a U.S. invasion of Iraq and political tensions in several oil-producing countries kept prices high despite the substantial increase in OPEC production and low economic growth in the consuming countries. Analysts attributed the increase in prices to the "war premium," which is the difference between market prices and what prices would have been without the threat of war in Iraq.

Contrary to expectations, oil prices started to decline when President Bush issued an ultimatum to the president of Iraq, Saddam Hussein. He gave him 48 h to leave Iraq or face war. Prices continued to decline during the war. Several factors contributed to this decline. First, the war premium disappeared as traders became certain and optimistic about the outcome of the war. Second, several countries started selling their oil that was stored in the Caribbean and on large tankers in the hope that prices would be higher during the war. Third, several countries increased their oil production, especially Saudi Arabia, as shown in Fig. 3. Fourth, traders were convinced that the Bush administration would use the Strategic Petroleum Reserves if prices increased. Fifth, traders

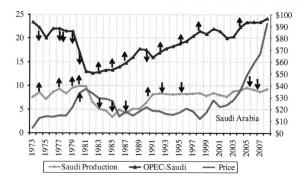

Figure 3
Saudi Arabia and other OPEC countries production (thousands b/d) and oil prices. Arrow direction indicates that Saudi Arabia's actions differ from those of other OPEC members. Source: Energy Information Administration (2003, January).

expected Iraq to start producing in large quantities very soon after the end of the war.

Oil prices did not decline substantially as many experts expected. The delay in Iraq exports and lower inventories kept prices at approximately $30/b in the United States. Despite talk of production cuts, only Saudi Arabia cut production by 200,000 b/d by May, whereas most members increased production.

2.7 The Roller Coaster: 2004–2009

The period between 2004 and 2009 witnessed one of the most amazing changes in the history of oil prices. WTI prices increased from around $33/b in January 2004 to about $145/b in July 2008, then dropped sharply to about $33/b at the end of 2008. By April 2009, prices rose to above $50/b. This period is unique in history: it is the only period in which oil prices, economic growth, government spending, military spending, and incomes have all increased while the dollar value and interest rates decreased.

While several reasons led to the spectacular rise and fall in oil prices, a review of the literature and various media sources indicates that various oil market stakeholders promoted the reasons that serve their agenda. For example, the IEA and the consuming countries blamed OPEC cuts, while OPEC blamed speculators. Others, including the author of this article, believe that oil prices increased because of market fundamentals: supply failed to keep up with demand. When demand declined and supply increased at the end of 2008, oil prices declined. Speculation had little impact, if any. Statistical analyses indicate that changes in the sum of inventories and OPEC's spare capacity explain most of the change in oil prices during this period. They also indicate that causality between speculation and

oil prices was one sided: from oil prices to speculation. That meant speculation activities followed the increase in oil prices, not the other way around.

Several factors increased the demand for oil including:

1. Economic growth in the OECD countries that resulted from the massive increase in government spending in OECD countries after the September 11 terrorist attack and the continuous decline in interest rates. Government spending and low interest rates mitigated the effect of high oil prices on economic growth and demand for oil continued to increase. The reader might wonder why the increase in government expenditures and lower interest rates during the recession did not have the same impact. The reason is that both failed to increase consumption.

2. Economic growth in India and China as their exports continued to soar. However, the demand for oil in China and India is a two-part demand: regular demand that stems from economic growth, and transferred demand that resulted from companies moving from North America, Europe, and Japan, to China and India. Power shortages in China, especially in 2004, had also contributed to higher oil prices. Power shortages forced factories to use private generation, which runs on fuel oil or diesel. As a result, Chinese oil imports increased.

3. The decline in the value of the dollar relative to other world currencies. Lower dollar made oil cheaper in Europe and Asia. In fact, the increase in oil prices in the US between 2002 and 2008 was double the increase in Europe. Higher euro mitigated the impact of increasing oil prices on the European economies. The same thing can be said about China.

Oil supply did not keep up with demand for several reasons, including:

1. Natural factors such as the Hurricanes in the Gulf of Mexico, sandstorms in Iraq and the Gulf region, and severe storms in the North Sea. These events destroyed production capacity, halted production, and delayed shipments.

2. OPEC inability to increase production capacity. Contrary to common believe, OPEC cuts had little to do with increasing oil prices. OPEC was producing at maximum capacity for several years. OPEC is at its weakest point when it runs out of spare capacity. OPEC was not able to increase production capacity meaningfully to meet the growing demand until the last quarter of 2008. But by then, it was too late. Even a production cut in the amount of 3.5 mb/d in the first quarter of 2009 failed to increase prices to the levels desired by OPEC of around $75/b.

But why did OPEC fail to increase production as the demand rose? Here are a few reasons:

a. Forecasts of a massive increase in non-OPEC production. The conventional wisdom in the late 1990s and early 2000s was that non-OPEC production will increase substantially in the coming years. OPEC, based on the advice of the prominent consulting houses and major investment banks, and reports from the IEA and others, viewed any additional investment in capacity as a waste of time and money. When non-OPEC failed to increase production, OPEC did have the additional production capacity to compensate for the shortfall. Oil prices exploded.

b. Long lead time between the investment decision and actual production. Lower oil prices in the late 1990s halted investment, not only because OPEC members were financially strapped, but also because there was no need to invest when the market is glutted with supplies. By the time some OPEC members realized the need for investment, it was too late. Oil prices increased.

c. The dire need for natural gas shifted investment priorities from oil to gas. Massive population growth and industrialization have increased the demand for natural gas substantially, a situation that made the value added from gas investment higher than the value added from oil investment. This simple economic fact shifted the investment focus from oil to gas.

d. Increase in domestic energy consumption. The population growth and the economic boom in the oil producing countries increased domestic demand for energy. As spare capacity vanished, the increase in domestic demand reduced oil exports. Lower exports meant less oil available in the world oil market, and prices increased. Several analysts have overlooked this fact because they focused on "production" not on "exports." Oil production does not matter anymore, oil exports do. In fact, some OPEC cuts coincided with the increase in domestic production, leading some experts to believe that the cut would have happened anyway, even if OPEC did not announce it.

e. The rhetoric of energy independence. While the population growth and the economic boom explain some of the increase in domestic demand for oil and natural gas in most OPEC countries, they do not explain it all. Some of that increase was the result of the considerable expansion of energy intensive industries such as petrochemicals. Building energy intensive industries in these labor-scarce countries might not be the ultimate investment decision, but might end up the ultimate decision when oil is perceived as a "threat" to national security in the oil-consuming countries. Most Republicans in the US link oil to terrorism and call for energy independence. Democrats link oil to global warming and call for energy independence. Others link oil to dictatorship and call for energy independence, on the hope that lower oil prices will bring democracy to the oil producing countries. The oil-producing countries would have ignored this silly rhetoric had it not led to billions of dollars in research grants and direct subsidies for corn ethanol, soy biodiesel, and other energy sources. The explicit intent of these policies is to replace oil, the life blood of these countries. Thus oil producing countries take the rhetoric of energy independence seriously. They are redirecting investment from the oil industry to energy intensive industries and other projects to export oil embedded in various industrial products. The effect is clear: slow expansion in oil production capacity and an increase in domestic energy consumption. Inevitably oil and gas exports will decline and world oil prices will increase.

3. Dollar Devaluation. The effect of dollar devaluation is not only limited to demand, but it also affects supply. The prolonged decline in the dollar reduced the purchasing power of oil producing countries and increased the costs of international oil companies. As a result, the amount of money allocated for reinvestment in oil production declined.

OPEC and the Recession.   The recession reduced the demand for oil by more than 4 mb/d. OPEC was quick to cut production as a preemptive measure by 3.5 mb/d. However, most of the cut was made by Saudi Arabia and its closest allies in the Gulf region. Several countries continued to cheat on their quota such as Iran. Saudi Arabia did not only engineer the cut, but also acted as an industry leader by cutting production first, then it asked other members to cut. The threat was pretty clear: the Saudi cut increased Saudi spare capacity, which it can use to punish cheaters. However, other factors were at play: Rich OPEC members such as Kuwait, Qatar, and the UAE were still awash with money, at a time when world financial institutions and stock markets were collapsing. Cooperation with Saudi Arabia by cutting production made sense. But as the financial resources of these countries dwindle, it is expected that only Saudi Arabia will carry the brunt of any future production cut.

3. Oligopoly Models

Production by non-OPEC members, which comprises approximately 60% of world production, eliminates the monopoly models for OPEC. Monopoly in the literature refers to a single producer controlling the whole market. Hence, the focus is on oligopoly models, in which the world oil market is shared between a dominant producer and the competitive fringe. Most models that assign market power to OPEC or to some subset of its members are categorized as one of three models: OPEC as a dominant producer, the OPEC core as a dominant producer, or Saudi Arabia as a dominant producer.

Although OPEC literature frequently contains expressions such as "monopoly," "monopoly power," and "monopoly profit," researchers are actually endorsing oligopoly models.

3.1 OPEC as a Dominant Producer

This model assumes that OPEC is a monolithic cartel in which countries have unified goals and collectively set the price for oil. Non-OPEC oil producers are the competitive fringe. The demand for OPEC's oil is the residual demand, the difference between world demand and non-OPEC supply. OPEC sets the price where its marginal revenue equals its marginal user's cost. The competitive fringe equates that price with their marginal user's cost and supplies accordingly; OPEC supplies the rest.

The model assumes that OPEC became a powerful cartel in 1973. OPEC was able to cut production, increase prices, and transfer wealth from the oil-consuming countries to the oil-producing countries.

This model may explain changes in OPEC and non-OPEC production, world demand, and changes in oil prices from 1973 to the present. As shown in Fig. 1, OPEC initially cut production and prices continued to increase between 1973 and 1981. The higher price was a self-defeating strategy for OPEC. It decreased demand through conservation, substitution, and fuel utilization. In addition, it increased non-OPEC supplies. The remaining residual demand decreased. OPEC lost its market share in the early 1980s. Consequently, oil prices and OPEC oil export revenues declined.

The lower price in the mid-1980s stimulated world demand and slowed the growth of non-OPEC production. Such development increased the residual demand for OPEC's oil and increased oil prices. The same situation explains the behavior of world oil markets from the 1990s to the present.

Criticism of the Model. This model does not apply to OPEC for the following reasons:

1. OPEC does not meet the characteristics of a cartel as stated in the economic literature.

2. The model assumes that production decisions are made by OPEC authority and not by the countries. However, several examples show that countries have acted unilaterally.

3. Historical evidence indicates that production cuts in the 1970s were related to the deterioration of technology after nationalization in several countries, such as Libya and Venezuela.

4. Statistical tests do not support this model.

3.2 The OPEC Core as a Dominant Producer

The failure of the previous model to fit OPEC behavior led researchers to introduce more sophisticated models. To be more realistic, researchers acknowledge that OPEC has a market power and its members have different interests. Three models have been introduced: the two-part cartel, the three-part cartel, and the core as monolithic cartel.

Two-Part Cartel Model. This model assumes that OPEC is a nonuniform cartel. OPEC is divided into two groups—saver and spender countries. The saver countries are the United Arab Emirates (UAE), Iraq, Kuwait, Libya, Qatar, and Saudi Arabia. The spender countries are Algeria, Ecuador, Indonesia, Iran, Nigeria, and Venezuela. The two groups can compete or collude with each other. In the case of competition, the saver countries will act as a dominant producer and set the price that maximizes profit. The spender countries will act as a competitive fringe. In the case of collusion, both groups will form a monolithic cartel. The outcome will be similar to the outcome in the monolithic cartel model discussed previously.

Applying the model to Figs. 1 and 2, the saver countries that set the world price made production cuts, whereas spender countries continued to increase their production. Once the quota was implemented in 1982, the saver countries cut production, whereas spender countries violated their quotas. As world demand decreased and non-OPEC production increased, the core's market share shrank. Later, a trend reversal led to an increase in prices and an increase in the core's share. A similar explanation can be used to illustrate changes in output and prices up to the present.

Three-Part Cartel. OPEC is divided into three groups:

1. The cartel core countries: Kuwait, Libya, Qatar, Saudi Arabia, and the UAE. These countries have vast oil reserves, small populations, and flexible economic development. The core acts as the dominant firm and sets the price of oil. The cartel core nations carry excess capacity in order to maintain prices during emergency demand.

2. The price-maximizing countries: Algeria, Iran, and Venezuela. These countries have relatively low oil reserves and large populations with potential for economic development. These countries aim for higher prices by cutting production.

3. The output-maximizing countries: Ecuador, Gabon, Indonesia, Iraq, and Nigeria. They have limited reserves, large populations, and a pressing need for economic development. These countries sell at any price, even in a weak market.

Applying this model to Figs. 1 and 2 explains OPEC behavior in a way similar to that of the previous model. However, this model shows that some countries pushed for higher prices when prices declined. This situation may explain the reversal of oil prices in 1987, despite the fact that Saudi Arabia had more than doubled its production by that time.

Core as a Monolithic Cartel. This model differs from the previous two models in that it does not assume that OPEC is a cartel, but that the core countries behave as a cartel that sets the price while all other producers, including the rest of OPEC, behave as a competitive fringe. The choice of core members (Kuwait, Qatar, Saudi Arabia, and the UAE) is not based on reserves and population but on economic and political factors. This model may explain OPEC behavior in the same manner as before, especially in the OPEC dominant producer model.

Criticism of the Model. This model does not match actual OPEC behavior for the following reasons:

1. The division of OPEC into two and three groups was subjective. Researchers did not base the division on theoretical, behavioral, or statistical analyses. They used factors that are not directly related to production decisions, such as reserves and population. Theoretically, production capacity, not reserves, has an impact on production. Statistically, various tests indicate that neither reserves nor population has a direct impact on production levels and prices.

2. Historical data do not support this model. Countries within each group did not act in unison, and some countries switched positions occasionally.

3. According to the three-part cartel model, the price-maximizing countries cut production with higher oil prices. At the same time, the core increases its production to offset the reduction, and vice versa in the case of low prices. Historical data do not support such conclusions.

4. For the core to be a cartel, it must operate as a dominant producer. Statistical tests do not support such behavior.

5. Given the sheer size of Saudi production relative to the production of other core members, it is very difficult to distinguish between the impact of the core and the impact of Saudi Arabia.

3.3 Saudi Arabia as the Dominant Producer

This model assumes that OPEC is not a cartel. Saudi Arabia acts as the dominant producer. All other producers behave as a competitive fringe. Saudi Arabia sets the price. All other producers expand their output to the point at which that price equals their marginal user's cost. Saudi Arabia then supplies the rest to meet the world demand.

This model could be the most plausible for explaining OPEC's behavior in the past 30 years. It gained support over the years even from researchers who first concluded that OPEC was a cartel. Looking at Figs. 1 and 2, this model can explain the changes in the world oil market in a similar way as the OPEC as a dominant producer model. The only difference is that production cuts and increases are made by Saudi Arabia, not by OPEC.

There are several factors that lend support to this model. First, historical data indicate that only Saudi Arabia has large excess production capacity. It is the only country to voluntarily reduce capacity and production. The 1 mb/d increase in Saudi production before and during the invasion of Iraq led to a $9/b price decline from preinvasion levels. The price during the invasion of Iraq was at least $15/b less than the price that analysts expected.

Second, Saudi Arabia's output is negatively correlated with that of other OPEC members, as seen in Fig. 3. The opposite arrows point to the direction of change in production. For example, in 1998, Saudi production decreased (arrow is pointing down) and the production of the rest of OPEC members increased (arrow is pointing up).

Third, Saudi Arabia was the only OPEC member not assigned a quota when OPEC implemented the 1982 quota system, so Saudi Arabia could act as "swing producer" to stabilize the world oil market.

Finally, statistical tests indicate that Saudi Arabia has operated on the elastic part of its demand curve since 1973. Operation on the elastic part of the demand curve is a requisite for the dominant producer model to apply.

Criticism of the Model. This model does not apply to OPEC for the following reasons:

1. The model assumes that the behavior of other OPEC members is similar to that of non-OPEC members. Historical data do not support this assumption.

2. The model does not explain why the Saudis pressed especially hard to convince non-OPEC members, such as Mexico, Norway, and Russia, to cooperate in production cuts. However, political factors may explain such behavior.

3. Figure 3 indicates that the model may not fit after 1999. Data show concerted efforts between Saudi Arabia and other OPEC members to cut production. Although Fig. 3 shows that Saudi Arabia increased production while the rest of OPEC decreased production in the first quarter of 2003, the decrease occurred because of the U.S. invasion of Iraq. Data indicate that all remaining OPEC members increased production during this period.

4. Saudi political objectives contradict this model. Political scientists who try to explain OPEC behavior claim that Saudi Arabia usually deviates from its dominant producer position in order to achieve political goals. These models assume that political interaction among OPEC members is important for cartel stability. These models are based on the idea that wealth maximization, security, and political influence are to some degree substitutes for each other. Thus, the decision to choose, for example, more security necessarily implies less wealth. In this case, political models and wealth maximization models offer divergent interpretations and predictions.

4. Competitive Models

The lack of support for various oligopoly models led researchers to introduce a variety of models that assume a competitive world oil market. Based on this view, five different models are presented to explain OPEC behavior.

4.1 Competitive Models

Oil prices increased in 1973 not only because of the embargo but also because of panic, stockpiling, and a spot market in which refiners sold futures crude oil. Rather than refining it, they used the crude as a commodity to gain quick profit.

Panic, stockpiling, spot markets, and speculation decreased supply and increased demand, which resulted in very high oil prices. In the same manner, oil prices increased in 1979, 1990, 1996, 2000, 2002, and 2003: A political problem led to a series of actions that lowered supply and increased demand. According to this model, there was no role for OPEC as an organization in the output disruption.

Applying this concept to Figs. 1 and 2, the boycott of 1973 shifted the supply curve to the left and increased oil prices. A domino effect followed: Panic and speculation shifted the demand to the right, refiners sold the crude and shifted supply farther to the left, and prices continued to increase. The Iranian revolution and the Iran–Iraq war caused the same chain reaction, which led to the second oil crisis.

In the early 1980s, the demand decreased and world supply increased, which led to lower oil prices. Increased demand and lower world supply, mostly from non-OPEC countries, resulted in increased oil prices in the second half of the 1980s. In the same manner, this model explains changes in prices and production in 1990, 1996, 1998, and 2000–2003.

Criticism of the Model. This model does not apply to OPEC for the following reasons:

1. In a competitive market, output will increase with higher prices. This is not the case for OPEC. It either cut or maintained output when prices were increasing, as shown in Fig. 1.

2. The model does not explain why some OPEC members, mainly Saudi Arabia, cut production in the mid-1980s while non-OPEC countries increased production.

3. It does not explain why the behavior of Saudi Arabia is different from that of other countries.

4. Statistical tests reject this model.

4.2 Property Rights Model

Production cuts by some OPEC members in the early 1970s were associated with changes in property rights when the host countries claimed ownership of the oil reserves either through nationalization or gradual ownership of the operating international oil companies (IOCs). This change in ownership led to uncoordinated production cuts. The host countries' discount rates were lower than those of the IOCs. In other words, foreign companies wanted to get as much money as possible from their concessions before they left, which resulted in increased production. When the host countries took over, they wanted to produce less in order to conserve the resource for future generations and to maximize efficient rate of production per oil field.

Applying this model to Figs. 1 and 2, one sees that 1973–1974 was a transition period. It reflected a sharp increase in government ownership during which a sharp decrease in production caused a switch to a higher price path. Even during the 1979 price shock, some researchers argued that production cutbacks were the result of ownership transfers from companies to host countries during that period.

Results of statistical tests show "mixed performance" for OPEC members. The property rights model cannot be rejected for Iran, Kuwait, Libya, Nigeria, Qatar, and Venezuela. The model was rejected for the rest of OPEC members.

Researchers who introduced this model acknowledge that production cuts led to a panic in the market and an increase in demand, which increased oil prices even further. They also acknowledge that the limited absorptive capacity of OPEC members sustained prices after the transfer of ownership. As shown for the next model, this limitation led to additional production cuts.

Criticism of the Model. This model does not apply to OPEC for the following reasons:

1. The model cannot explain changes in production and prices after property transfer ended in the late 1970s. However, one might argue that oil-producing countries changed their discount rates as a result of "property reevaluation" or a change in "demand expectations."

2. Some researchers claim that historical facts contradict this model. For example, production increases in the 1950s and 1960s may have resulted from the host countries' desire to increase production. However, this criticism ignores the fact that IOC payments to the host governments during that period were a function of production levels. Countries collected more revenues as production increased. Therefore, they pushed the IOCs to increase production.

3. Some researchers argue that the production cut was not related to a lower discount rate by the countries. Rather, lack of technology caused it. The IOCs were deploying the latest technology in the early 1970s to enhance production and take advantage of rising oil prices. After nationalization,

state-owned companies were not able to deploy the advanced technology that had been used by the IOCs, which resulted in lower production, especially in old oil fields.

4.3 The Target Revenue Model

This model was introduced to explain why oil prices were sustained at their postembargo levels despite the official lifting of the embargo in March 1974. According to this model, political events and changes in property rights in 1973 caused prices to increase. Oil prices did not decline after the end of the embargo because countries, without coordination or collusion, cut their production further as prices increased.

Most OPEC members had primitive economies that could not absorb the additional revenues from higher oil prices. Countries had two options: transfer the extra revenues to foreign financial markets or keep the extra dollars in cash for future use. For three reasons, OPEC members chose to do neither. First, in the 1970s, real returns on investments in the West were very low, if not negative. Second, some members feared asset confiscation by Western governments, especially after the freezing of Iranian assets by the United States. Third, dollar devaluation and inflation would have reduced the value of any cash revenues saved. These reasons led some oil-producing countries to cut production. They believed that oil in the ground was "the world's best investment."

The result of output cuts is a backward-bending supply curve for prices higher than those needed to support the target revenue. If there are a sufficient number of countries with backward-bending supply curves, then the market supply curve will take the same shape.

Most researchers agree that this model may explain OPEC behavior in the short term but only in some periods during the 1970s. Recent research indicates that, even today, the model applies only to centrally planned and isolated oil-producing countries, regardless of their OPEC membership. Statistical tests show that this model applies only to two OPEC members, Libya and Iran. Both countries are centrally planned and isolated by UN and U.S. embargoes.

Criticism of the Model. This model does not apply to OPEC for the following reasons:

1. The backward supply curve can hold only for a very short period of time. Prices will continue to move along the supply curve.

2. Some researchers argue that the model is not stable. If we have a backward supply curve, an increase in the negatively sloped demand will lead to two points of equilibrium with different production levels and different prices. However, some researchers argue that the equilibrium is stable.

Producers in this case will choose the lower production level to conserve the resource.

3. Historically, some researchers rejected this model because of its assumption that foreign investment is not a viable alternative to domestic investment. They cited Kuwait's success in building a financial empire in the West. They also cited data illustrating that every OPEC member has an overseas investment. This objective ignored the possibility that political reasons may prevent some countries from acquiring access to foreign financial markets.

4. All statistical tests in the literature reject this model for OPEC as a whole. However, it cannot be rejected for some countries, such as Iran and Libya.

4.4 The Purchasing Power Explanation

Although not a formal model, it explains why some OPEC members behave competitively, violate their quota, and decline to coordinate production with Saudi Arabia. According to this explanation, some OPEC members increase production and violate their quota if the purchasing power of their oil exports is decreasing as a result of variations in major world currency exchange rates and higher export prices of industrial countries.

If oil production cuts lead to higher oil prices and higher export prices for the products of industrial countries, it is not in the interest of the member to participate in production cuts. The purchasing power of its oil exports would decline. Although this concept may apply to all imports of goods and services, the situation is more severe if the oil-producing country is importing large amounts of petroleum products. If an OPEC member must import petroleum products, production cuts will increase the cost of imports and force the real value of oil exports to decrease.

Although Indonesia is an OPEC member, its large population and relatively small reserves force it to import crude oil and other refined products. Imports of crude and refined products represent approximately 8% of its total imports. During the fourfold increase in oil prices in the 1970s and the threefold increase in 1980, the value of these imports increased substantially. As a consequence, the real value of Indonesian oil exports decreased. This decline may have forced Indonesia to increase its oil production and ignore its OPEC quotas in order to increase its purchasing power. Data indicate that Indonesia did not coordinate production with other OPEC members. Studies indicate that the importation of crude oil and other oil-intensive products was the primary reason for the decrease in the purchasing power of Indonesia's oil exports.

Historically, seven OPEC members imported crude oil and petroleum products on a scale large enough to affect their purchasing power: Algeria, Ecuador, Gabon, Indonesia, Nigeria, the UAE, and Venezuela.

These imports may explain the competitive behavior of these countries.

4.5 The Foreign Investment Explanation

This theory was introduced to explain the behavior of Kuwait. It assumes that OPEC is not a cartel and attempts to explain why Kuwait does not coordinate production with Saudi Arabia, as indicated by the production data of both countries.

Kuwait's behavior can be explained by its foreign investments. The amount of foreign investment does not play a role in oil production decisions. The type of investment does. At first, Kuwait invested in upstream oil operations in the North Sea. At that time, it was in Kuwait's interest to cut production or sustain it to obtain higher prices for its oil in Kuwait and the North Sea. Not until the 1980s did Kuwait extend its investment to downstream operations to include thousands of gas stations throughout Europe and refineries in Singapore.

Since governments control retail prices of petroleum products in Europe, it was in Kuwait's interest to violate its OPEC quota and increase production. Lower crude oil prices from increased production enabled Kuwait to generate a large profit margin from retail outlets, where product prices did not change. The profit from lower crude oil prices outweighed the loss from selling the Kuwaiti crude at a lower price. In addition, Kuwait's return on investment exceeded its revenues from crude exports throughout the 1980s.

5. OPEC and the Cartel Status

Despite the widespread use of the word *cartel* by media outlets and politicians, most researchers today, even those who assigned cartel status to OPEC in the 1970s, believe that OPEC is not a "monolithic" cartel. However, events since 1999 show that OPEC's behavior may have changed now that a price band is in place. It holds almost monthly meetings instead of the usual quarterly meeting and makes frequent output adjustments. Figure 3 indicates that since 1999, production adjustment by Saudi Arabia matched that of other OPEC members. Some researchers believe that the recent changes in OPEC behavior constitute a de facto cartel. Others argue that Saudi Arabia is still the dominant player that initiates most output changes. They point out that production cuts by OPEC and non-OPEC members were not voluntary but rather the result of technical, political, and natural factors. The arguments of each group are summarized next.

5.1 OPEC as a Cartel

Those who argue that OPEC is a cartel cite the following reasons:

1. Historical records indicate that OPEC was modeled after the Texas Railroad Commission. Even today, its objective is to coordinate production among members.
2. Since its inception, OPEC members have been meeting periodically to coordinate policies, production, and prices.
3. Historically, OPEC cut production to increase prices, as indicated in Fig. 1.
4. OPEC currently uses a quota system and a price band, two important cartel characteristics.
5. The price of oil is much higher than marginal cost.
6. OPEC members have been able to transfer massive amount of wealth from the consuming countries since 1973.

5.2 OPEC as a "Commodity Producer Club" But Not a Cartel

The cartel theory states that there are seven characteristics that must exist in a group of producers in order to be labeled a cartel: A cartel must assign quotas to its members, monitor members to avoid violations, punish violators, target a minimum price, take action to defend the price, have a large market share, and divide votes among members based on their market share. OPEC did not meet any of these characteristics during its peak of power between 1973 and 1981. OPEC assigned quotas in 1982 but failed to enforce them. OPEC did not set a price band until 2000. It might have been able to defend prices since 2001 with the help of technical, political, and natural factors that prevented many oil-producing countries from increasing production. OPEC never had a punishment mechanism or an effective monitoring system such as that of the Texas Railroad Commission, which used the National Guard and Texas Rangers to enforce its rules and cut production in Texas and Oklahoma. In addition, OPEC's market share is relatively small (approximately 40%). Even today, each OPEC member has one vote regardless of its production and reserves. Unlike the Seven Sisters, OPEC did not divide the world market and did not control investment. The market share of the Seven Sisters was more than double that of OPEC.

By U.S. standards, OPEC does not fit the description of a cartel. Monopolization has two elements: the acquisition of monopoly position and the intent to monopolize and exclude rivals. Official U.S. judicial records indicate that a monopoly position has been associated with companies whose market share approached or exceeded 80%. This is double the current OPEC market share and much larger than the 54% market share achieved in 1974. OPEC has thus failed, by a wide margin, to acquire a monopoly position. As for intent to acquire a monopoly and to exclude rivals, OPEC has never taken the steps mentioned previously to be a successful cartel. OPEC was labeled a cartel because it supposedly increased prices, but how could OPEC exclude rivals by increasing oil prices? Figure 1 shows that higher oil prices bring higher production, more producers, and more competition. OPEC did not act as a monopolist. OPEC has increased production many times in recent years to stabilize prices. It has become an active participant in the world community through negotiations with consuming countries and active aid programs to developing countries. Many researchers today, including lawyers and policy makers, view OPEC as an agent of stabilization in a volatile world. In fact, the Bush administration praised OPEC for its efforts to prevent prices from increasing to record levels during the invasion of Iraq.

None of the statistical tests in the literature support the cartel model for OPEC. Researchers who conducted these tests concluded that OPEC is a "partial," "weak," "clumsy," or "loose" cartel. Such conclusions are mere journalistic expressions. No known theory in the economic literature supports them. The results of so-called statistical tests are not acceptable. For example, applying the same tests to non-OPEC countries shows that non-OPEC producers fit the cartel model better than OPEC members. In addition, these tests focus on parallel behavior, not on cartelization. Parallel behavior may exist in any market, including competitive ones.

The cartel model requires OPEC to operate on the elastic portion of its demand curve. Studies show that OPEC operated on the inelastic portion of its demand curve during its power peak in the 1970s.

Coordination among governments and commodity producer clubs is well-known. This coordination is not labeled as cartelization, despite periodic meetings and clear goals to curb production and maintain prices.

Researchers who attempt to prove that OPEC is a cartel by focusing on the difference between marginal cost and the price of oil ignore the basic principles of natural resource economics. The price of a nonrenewable resource in a competitive market reflects the marginal cost and the discount rate chosen by the producer. Therefore, in a perfectly competitive market the price of any natural resource is higher than its marginal cost, unlike renewable resources, the price of which equals the marginal cost. Hence, focusing on the difference between marginal cost and price to prove market power is conceptually wrong unless the researcher has perfect estimates of the user's cost.

Researchers who believe that OPEC is not a cartel argue that there are other explanations for OPEC wealth transfer that are not related to cartelization. Some researchers argue that it was both the market power of Saudi Arabia and U.S. price controls that led to wealth transfer to OPEC. Various studies show that U.S. oil price controls suppressed U.S.

production, increased the world demand for oil, and raised Saudi Arabia's output. This unintended effect enabled OPEC members, mainly Saudi Arabia, which increased its production substantially in the late 1970s, to transfer wealth from the consuming countries to the producing countries.

Historical records indicate that OPEC was reacting to economic and political events rather than being a proactive cartel that consistently followed polices that would maximize its wealth. OPEC reacts pragmatically to compromise on the conflicting interests of its members. In addition, OPEC pricing has always followed the spot market.

6. Conclusion

There are two prominent supply factors that explain the behavior of OPEC: scarcity and market power. The focus of most researchers on market power without regard to scarcity and technical issues and their political and social implications has polarized academic and political circles to the extent that it is almost impossible to reach a consensus on OPEC and its role in world oil markets.

Review of various models introduced to explain OPEC behavior indicates the following unequivocal conclusions:

1. Significant production by non-OPEC members rules out the use of monopoly models.

2. Historical data and research studies indicate that the competitive model does not apply to OPEC.

3. Only oligopoly models may explain OPEC behavior. However, theory, historical data, and statistical tests do not support the monolithic cartel model for OPEC.

4. No single model can explain OPEC behavior even at particular periods of time.

One way to explain the behavior of OPEC and non-OPEC producers, world oil demand, and oil prices is to synthesize these models based on the results of various studies. Such a synthesis indicated that OPEC is driven by political and technical influences as much as by the inexorable forces of economics and the market.

This synthesis must tell the story of the past 30 years without violating logic, theory, data, or history. The behavior of OPEC can be explained using the dominant producer model for Saudi Arabia or a dynamic core that changes members periodically and by using various models to explain the behavior of other OPEC members. In this case, other OPEC members follow various models at various points of time, such as the property rights model, the target revenue model, the purchasing power explanation, the foreign investment explanation, and even pure competitive behavior. At different times, OPEC, or a subset of it, may look like a monolithic cartel when, coincidentally, various models necessitate the same action. In this case, countries may cut or increase production jointly. For this reason, single-equation models, correlation tests, and cointegration tests indicate parallel action but not cartelization.

However, one should not forget the role of political and natural events, panic, stockpiling, and speculation and their impact on oil markets. OPEC, like the Texas Railroad Commission, was established to protect property rights and to bring order to a chaotic situation that was leading to tremendous wastage and losses. OPEC aid to Third World countries makes OPEC look more like a world-class intergovernmental organization than merely a cartel that cuts production and increases prices. Production increases by OPEC members to combat the increasing prices that resulted from the labor strikes in Venezuela and Nigeria and the war in Iraq indicate the significant role that OPEC plays in stabilizing oil prices, especially given that political turmoil, not OPEC production cuts, caused world energy crises.

The shape and role of OPEC may change in the future. Some members may leave the organization when they become net oil importers. Others may leave if they privatize their oil industry and become unable to force private companies to follow OPEC quotas. On the other hand, new members may join the organization. Some current OPEC members may increase their capacity substantially. Others may even build excess capacity and compete with Saudi Arabia. Regardless of these changes, OPEC will keep playing an important role in world oil markets. Lower oil prices and the withdrawal of some members will not lead to the demise of this powerful organization.

The impact of the growing importance of natural gas on OPEC remains to be seen. Four OPEC members are among the world's largest natural gas players: Iran, Algeria, Qatar, and Libya. The emphasis on natural gas may either weaken or strengthen OPEC. Only time will tell.

Acknowledgments

I thank Michael Bunter, Collin J. Campbell, Juan Pablo Perez Castillo, Melaku Geboye Desta, Dermot Gately, Luis Lugo, Adam Sieminski, Thomas Walde, and an anonymous referee for their helpful comments on the manuscript.

Further Reading

Adelman M A 1995 *The Genie out of the Bottle: World oil since 1970*. MIT Press, Cambridge, MA

Alhajji A F, Huettner D 2000a OPEC & other commodity cartels: A comparison. *Energy Policy* **28**(15), 1151–64

Alhajji A F, Huettner D 2000b Crude oil markets from 1973 to 1994: Cartel, oligopoly or competitive? *Energy J.* **21**(3), 31–60

Alhajji A F, Huettner D 2000c The Target revenue model and the world oil market: Empirical evidence from 1971 to 1994. *Energy J.* **21**(2), 121–44

Crèmer J, Salhi-Isfahani D 1991 *Models of the Oil Market*. Harwood, New York

Griffin J, Teece D 1982 *OPEC Behavior and World Oil Prices*. Allen & Unwin, Boston

Mabro R 1998 OPEC behavior 1960–1998: A review of the literature. *J. Energy Literature*, **4**(1), 3–26

Noring Ø 2001 *Crude Power, Politics and the Oil Market*. Tauris, London

Stevens P 2000 *The Economics of Energy*, Vols. 1 and 2 Elgar, Cheltenham, UK

A. F. Alhajji

OPEC, History of

Delegates from five major oil-producing countries—Iran, Kuwait, Saudi Arabia, Venezuela, and Iraq—met in Baghdad and announced the foundation of the Organization of Petroleum Exporting Countries (OPEC) on September 16, 1960. At that time, no one could have foreseen that this event would play such a crucial role, some 10 years or so later, in reshaping the world energy scene and would have such an impact on the world economy. The event at the time passed nearly unnoticed, except perhaps by the specialized petroleum media. It took place after two successive reductions in what was then called the posted price of oil, which previously was set and posted unilaterally by the major international oil companies, usually referred to as the "Seven Sisters." The first cut, in February 1959, was by as much as U.S. $0.18/barrel (or ~10%) for Gulf oil, from $2.04 to $1.86/barrel. The second price reduction of 7% was decided by Exxon in August 1960, and the other companies followed suit. The OPEC delegates' immediate and prime objective was to safeguard their member countries' oil revenue against any further erosion as a result of the companies' deciding to cut prices further. According to the profit-sharing agreements put into force during the early 1950s between the holders of oil concessions in these Middle East areas, on the one hand, and their host countries, on the other, the latter's per barrel revenue of exported oil was determined at 50% of the official (i.e., "posted") price minus half of the cost of producing that barrel. Therefore, any fluctuation in that price, whether upward or downward, changed the per barrel revenue accordingly. When the oil companies undertook price cuts, the per barrel income of these countries was reduced by more than 15%, compared with the price before the two cuts. Until the late 1950s, the official posted price was more of a tax reference price, on which host countries received their taxes, than a real market price resulting from the exchange of oil between buyers and sellers in a free market. For the companies, on the other hand, the tax paid to host countries, together with the production costs, determined the operators' tax-paid cost, meaning the cost incurred by the oil companies as a result of producing one barrel of oil plus tax and royalties paid to the host governments. For this reason, the first resolution adopted by the newly formed organization emphasized that the companies should maintain price stability and that prices should not be subjected to fluctuation. It also emphasized that the companies should not undertake any change of the posted price without consultation with the host countries. Furthermore, the resolution accentuated the necessity of restoring prices to the pre-1959 level. Other wider objectives were also articulated and incorporated into the organization's five resolutions, namely that it would endeavor to have a price system that would secure stability in the market by using various means, including the regulation of production, with a view to protecting the interests of both oil consumers and oil producers and guaranteeing stable oil revenues for the latter. Moreover, OPEC announced that the real purpose of the organization was to unify the oil policies of member countries so as to safeguard their interests individually and collectively.

1. The Oil Industry Structure Under the "Seven Sisters"

Prior to the formation of the Organization of Petroleum Exporting Countries (OPEC), and until the late 1950s, the international oil industry had been characterized mainly by the dominant position of the major multinational oil companies through a system of oil concessions granted to the major oil-producing countries, according to which these companies were interlinked in the "upstream" phase of the industry, that is, exploration, development, and production of crude oil. Through joint ownership of the holding companies that operated in various countries, they were able to plan their production of crude oil according to their requirements. The oldest example of this kind of "horizontal integration" (i.e., with the companies being interlinked in the upstream phase) was the Iraqi Petroleum Company (IPC), formerly the Turkish Petroleum Company, which was a non-profit-making holding and operating company owned by the Anglo-Iranian Oil Company (British Petroleum [BP]), Compagnie Française Pétrole (CFP or Total), Royal Dutch Shell, Standard Oil of New Jersey (Exxon), Mobil, and the Gulbenkian Foundation (known as "Mr. 5 percent," Gulbenkian had been instrumental in negotiations for an agreement, prior to World War I, between the Ottoman Empire and German and British interests involved in the exploitation of Iraqi oil).

Each of these companies was a shareholder in other countries in the Middle East. For example, BP owned half of the Kuwait Oil Company and all the oil of pre-Musadeq Iran (although its holding was reduced to 40% in the consortium that was founded after Musadeq) as well as shares in Qatar Petroleum Company and Abu Dhabi Petroleum Company, whereas the American company Esso (Exxon) had 30% of Aramco in Saudi Arabia as well as shares in Qatar Petroleum, Abu Dhabi Petroleum, and the like. This type of interlinking enabled them to control and manage crude oil supplies worldwide, along with the bulk of oil exports from the major oil-producing countries, so that oil trading became a question of intercompany exchange with no free market operating outside the companies' control whereby crude oil was exchanged between sellers and other buyers. At the same time, each of these "sisters" had its own "downstream" operations—transportation, refining, oil products, and distribution networks—that made them "vertically" integrated. This compact system of horizontal and vertical integration allowed the companies to plan for their future crude oil requirements in line with their downstream requirements, that is, the amount of oil products needed by each according to its market outlets in the countries to which crude oil was shipped.

However, by the second half of the 1950s, this compact system of complete control of production and intercompany exchanges began to weaken with the appearance of independent oil companies searching for access to cheaper crude oil. This new development led to the creation, albeit on a limited scale, of a free market for buying and selling crude oil outside the control of the major companies. The state-owned Italian company, ENI, was able to find its way in investing beyond the control of the Seven Sisters by offering different and better financial terms than the oil concessions. Similarly, many independent oil companies began to invest outside the sisters' control by offering apparently better financial terms than the oil concession system.

At the same time, oil was discovered in new producing areas, as in the case of Libya, where in addition to Exxon Mobil, Occidental was an important American independent company. These developments led to a greater amount of crude oil being produced outside the system, controlled by the "majors" and offered for sale to independent refiners with no access to crude oil. Furthermore, crude oil from the then Soviet Union began to make its way into a free market for sale at competitive prices, and this country needed to offer attractive discounts to move its oil in the market. Other factors also played roles. One was a measure taken by the United States to limit the entry of crude oil into the market so as to protect its oil industry. This policy evolved toward the end of the 1950s, when the U.S. government's oil import policy predicated itself on the premise that foreign oil should be a supplement to, and should not supplant, American oil. This deprived many independent American oil companies that had access to foreign crude oil from entering the U.S. market, a situation that led to their having to sell surplus crude outside the major oil companies' system. A market price began to shape up in the form of discounts off the posted price set by the major oil companies. On the other hand, a new pattern of tax relationship emerged following the entry of newcomers investing in oil in new areas such as Libya, where the government's per barrel share was calculated not on the official posted price (which was a fixed price) but rather on the basis of a price realized in the free market (which was always below the posted price). In this way, when the realized price fell, the host government's per barrel share fell accordingly and the tax cost for the new producer would correspondingly be lower, thereby providing a good margin of profit with which to reinvest in new production capacity for crude oil.

In this way, the free market, where crude oil is bought and sold physically and freely, started to grow. Obviously, the larger this market, the less it was dominated by the major oil companies, whose share in world production outside the United States and Soviet Union now came under threat from the newcomers. As explained by Edith Penrose in her book, *The Large International Firm in Developing Countries: The International Petroleum Industry,* in 1950 the Seven Sisters owned 85% of the crude oil production in the world outside Canada, the United States, the Soviet Union, and China. By 1960, this share had fallen to 72%, to the benefit of the newcomers, whose share had correspondingly grown from 15% to 28%. The fall in the Majors' share of refinery capacity was even more pronounced: from 72% to 53% during the same period – again to the benefit of the independent oil companies. Against this atmosphere of increasing competition and the threat to the dominant position of the major companies by newcomers and the Soviet Union, (the latter of which was selling crude at massive discounts), the major oil companies had to cut the price posted by them to reduce the tax paid cost. This was actually pushed by Esso (Exxon) and followed by the other companies; hence, the cuts in the posted price by 1959 and 1960 were a countermeasure to the newcomers' increased market share. In other words, if this mounting competition with price cuts in a free market had not been curbed by restrictive measures, this situation would have led to a harmful level of competition and continued price cuts.

2. Opec's Formative Years

The timely formation of OPEC put an end to this increasingly damaging level of competition with its adverse effect on the position of the major oil companies and on the oil industry in general. The first step that OPEC took was to enlarge its membership, and it was only shortly after its foundation that Qatar joined the organization, followed by Indonesia, Libya, the Emirates of Abu Dhabi, Algeria, Nigeria, Ecuador, and Gabon (although Ecuador and Gabon left OPEC during the 1990s). The enlargement of the OPEC base at this stage served to strengthen its negotiating power vis-à-vis the oil companies. In addition to this enlargement, OPEC took certain measures that helped greatly in putting an end to price erosion. These involved, among other things, the unification of the tax system, that is, levying taxes from the oil companies that operated in member countries' territory so as to make them conform to those applied in the Gulf (i.e., by using the fixed official posted price as a reference for the calculation of tax and royalties and not a realized market price). This measure put an end to the tax system in new areas where tax and royalties were paid by investors on the basis of realized market prices and not fixed posted prices, a system that tends to put new companies in a better competitive position than the major companies. In so doing, OPEC succeeded

in curbing the newcomers' ability to invest in expanding capacity and acquire an enlarged market share at the expense of the majors.

According to OPEC resolutions, Libya amended its tax system and put it in line with that of the Gulf. Obviously, this measure, adopted by Libya in response to OPEC, had the effect of limiting the expansion of new companies' investments as they began paying higher taxes than they did previously. This OPEC measure, while serving the producing exporting countries by consolidating price stability, also benefited the major oil companies by preventing their dominant competitive position in the oil industry from eroding further.

Another measure taken by OPEC that was of no less significance in consolidating the price structure was the amendment to the concessions' tax regime by which the payment of "royalties" (paid by investors to the owners of oilfields) was treated separately from the taxation or what was called the "expensing of the royalty" (i.e., the royalty to be paid by the concessionaires but considered as part of the production costs for the purpose of calculating taxes). Among the established rules in the oil industry, as implemented in the United States since the early part of the 20th century, royalties had to be paid to the landowners regardless of whether profits were realized or not. Taxes are levied on profits that have to be paid in addition to royalties. When the 50/50 profit sharing agreements between host governments and the major oil companies were concluded during the 1950s, the 50% belonging to the host countries covered both the royalty at 12.5% and the tax at 37.5% of the posted price.

OPEC asked the companies to implement this principle in the oil concessionary system operating in the areas, with 50% tax being paid after the host countries had received the royalty of 12.5% of the posted price. The companies agreed with this, provided that the royalty would be considered part of the cost of production. This meant that only half of the royalty would be payable to the host countries from royalties; hence, the total government share increased from 50 to 56.25% of the posted price. Although this increase had to be implemented gradually by allowing temporary decreasing discounts to the companies, it nevertheless had the effect of consolidating the price structure by increasing the tax paid cost. This measure had the effect of strengthening the level of realized market prices because the higher the tax paid cost, the higher the price floor, making price cuts more difficult.

It is clear that the mere existence of OPEC, whose raison d'être was to put an end to price cuts by the companies, helped not only to safeguard the producers' interests but also to strengthen the price structure and put an end to the expansion of the competitive market. No less important were the consequential measures taken by OPEC that had the effect of serving the mutual interests of the oil producers and major oil companies. It was because of these measures that the majors were able to maintain their leading role in the industry and to weaken the role of the newcomers. During the first 10 years of its life, OPEC placed a limitation on the expansion of the free market, which was the source of price instability. Without OPEC and its measures during those early years, other changes taking place in the structure of the oil industry at that time would have further encouraged the newcomers, thereby increasing instability in the market.

In fact, the position of the major companies improved after the creation of OPEC, and to quote Penrose again, "The share of the major oil companies in the upstream part of the industry increased from 72% in 1960 to 76% in 1966. Similarly, in the downstream operations, their share increased from 53% to 61%, respectively." This meant that the shares of the newcomers had to decline from 28 to 24% in upstream activity and from 47 to 39% in downstream activity. In effect, OPEC had reversed the trend of increasing the share of the newcomers and limiting their power.

Although OPEC had failed to restore the posted prices to their levels prior to the 1959 and 1960 price cuts, as was the initial objective of the organization when it was founded in Baghdad, Iraq, in 1960, the formation of OPEC and the measures it took simply constituted a form of amendment of oil concessions. Prior to OPEC, host countries were not partners in their own industry and their role did not exceed the levying of taxes from the oil companies, the latter of which, according to the terms of the concessions, had the absolute freedom to decide whatever was suitable to their own interests in all matters relating to the oil industry in territory covered by concessions.

Another episode that would later play a pivotal role in changing the structure of the oil industry and the nature of the organization was OPECs "declaratory statement of petroleum policy in member countries" in June 1966, which became a historical turning point for OPEC in that it laid down basic principles that would later influence all of the major events of the 1970s. The statement emphasized the right of producing countries to fix unilaterally the price of their own oil in an endeavor to give the states a greater role in the development of hydrocarbon resources and to accord the states the right to participate in concession-holding agreements. The statement focused on amending the concession system in the light of changing circumstances.

The first 10 formative years of OPEC constituted a change in the relationship between the oil producers and the oil companies in a way that served both interests in the changing conditions of the industry. Although it is true that OPEC was still a dormant partner during those years, within 10 years this partnership had evolved into an active partnership, shifting from the function of simple tax collectors into that of active negotiating partners in determining prices.

3. A Short Period of Cooperation Between Producers and Oil Companies

After these first 10 years, things took another turn. At the OPEC Conference in Caracas, Venezuela, in December 1970, it was decided to enter collectively into negotiations with the oil companies with a view to raising the price of oil and the tax ratio. Negotiations with the oil companies were first conducted in Tehran, Iran, by delegates from the six OPEC member countries bordering the Gulf: Iran, Iraq, Kuwait, Saudi Arabia, Qatar, and the United Arab Emirates. After several rounds of negotiations, they succeeded in taking a firm collective position to raise the fixed price by approximately \$0.35 and increasing the tax ratio from 50 to 55%.

The Tehran Agreement was concluded in mid-February 1971 but was applied retrospectively from January 1 of that year. It was meant to be for 5 years, after which it was to be renegotiated. In addition to price and tax amendments, the Tehran Agreement provided for the adjustment of the price upward (on an annual basis) to reflect the impact of world inflation, together with a fixed annual increase. Later, and following the two successive devaluations of the U.S. dollar in 1972 and 1973, two agreements were negotiated with the companies and signed by the same signatories of the Tehran Agreement to adjust oil prices and, thus, to preserve the purchasing power of the other currencies relative to the U.S. dollar. The fact that all prices are denominated in U.S. dollars means that any depreciation of the value of the U.S. dollar could lead to a loss in real value of oil revenues when used to pay OPEC countries' imports from non-dollar countries. The dollar negotiations took place in Geneva, Switzerland, and the Geneva Agreements (known as Geneva 1 and Geneva 2) laid down the formulas with which to correct the oil price (agreed to in the 1971 Tehran Agreement), whether upward or downward, in relation to the movement of the value of the U.S. dollar vis-à-vis other major currencies (U.K. sterling; French, Swiss, and Belgian francs; German Deutschemark; Italian lira; and Japanese yen).

4. The Oil Price Shocks

However, after $2\frac{1}{2}$ years of price stability and cooperation between OPEC and the oil companies, the history of OPEC and the oil industry took a dramatic turn, unleashing huge price volatility. In June 1973, OPEC decided to reopen negotiations with the companies to revise upward the Tehran Agreement price in the light of prevailing market conditions.

By that time, market prices had reached such high levels that the Tehran Agreement price was left lagging behind, and OPEC argued that the additional windfalls gained by the companies ought to have been shared with the producers. Accordingly, negotiations took place in Vienna, Austria, in early October 1973 but failed as the oil companies refused OPEC's demand for an increase in the price set by the Tehran Agreement. This provoked OPEC into fixing the price of its oil unilaterally and independently of the oil companies' participation. At a meeting in Kuwait on October 16, 1973, that was attended by the same OPEC countries bordering the Gulf, a decision was announced to increase the oil price by 70% so that the OPEC price was fixed at $5.40/barrel.

At the time of this decision, war was raging between Egypt and Syria, on the one hand, and in the state of Israel, on the other. When Egyptian forces succeeded in crossing the Suez Canal into occupied Sinai, the reaction of the United States and The Netherlands was to rally to the side of Israel. This led to mounting pressure from Arab public opinion to use Arab oil as a "political weapon" to coerce the United States into changing its hard-line policy of continued support for Israel and culminated with the Arab member countries of OPEC meeting to decide to place an oil embargo on both the United States and The Netherlands and to undertake successive cuts in production. It is little wonder that, following that fateful decision, the oil market prices soared—unchecked and uncontrolled. Meanwhile, the Shah of Iran began a campaign within OPEC to raise the price substantially. At this time, the shah, having enjoyed the support of the West, appointed himself as the "policeman of the Gulf," a role that he saw as justifying his need to raise huge sums of cash with which to arm his country to the teeth. The reality was that the shah wished to control the area and present Iran as the most dominant power in the whole of the Gulf region.

This drove the Shah of Iran at first to put pressure on the oil companies to raise production until he realized that the most effective means of achieving this was by simply pushing for ever higher prices. He couched this in terms of "a new discourse of oil": "a noble commodity that should not be burnt" but instead should be reserved for "higher, nobler purposes such as petrochemicals."

By the time that OPEC's ordinary (winter) meeting was convened in Tehran in December 1973, market prices had already exceeded $15/barrel as a result of the Arabs' so-called oil embargo. The Shah of Iran took every advantage of this and succeeded in increasing the price from $5.40 to $10.84/barrel (or an increase of 140%). Relative to the Tehran Agreement and the oil prices during the summer of 1973, this constituted a 400% price increase, causing an enormous shock across the market.

Various measures were taken by the consumer nations to slow down oil consumption such as limiting vehicular speed and implementing other energy-saving measures.

The vast majority of OPEC members tended to emulate the Shah of Iran in his quest for price escalation, but without fully comprehending the long-term reaction of such a sudden dramatic increase in oil prices and the negative impact on world demand for OPEC oil. For their part, the Saudis, aware of the economic implications and concerned about the consequential negative impact of price escalation on their own oil, were worried about such sharp oil price increases. One who was particularly worried was Sheikh Ahmed Zaki Yamani, the then Saudi oil minister, who consistently expressed his opposition to any further price increase. However, in OPEC price discussions during the late 1970s, especially during the oil market developments in the wake of the Iranian Revolution, the Saudi kingdom became more amenable to political pressures exerted by the others and, hence, finally acquiesced to most of the price decisions.

The price shock proved to be an inducement for greater efficiency in fuel use so that the amount of gasoline used per mile driven was in continuous decline. Coupled with this and consumer countries' fiscal policies, which involved dramatically increasing taxes on gasoline and other petroleum products (especially in Europe), oil became an extremely expensive commodity for end consumers. The resultant dramatic slowdown in the growth of oil consumption indicated how the impact of the price shock took its toll on Western Europe to a far greater extent than on the rest of the world. During the 1960s and early 1970s, oil consumption in Western Europe had grown at an average rate of more than 8% per year, but after the price shock, oil consumption fell from a peak of 15.2 million barrels/day in 1973 to 13.5 million barrels/day 2 years later, settling at approximately 14 million barrels/day thereafter. Conversely, the impact of higher prices on oil consumption in the United States and Japan was far less than in Western Europe, mainly due to those two countries' far lower level of taxation on oil products.

With a lead time, the price shock heralded a process of structural change in the world oil industry. Significantly, the emerging energy-saving campaigns led to diminishing oil consumption while achieving the same level of economic growth. Prior to the price shock, an increase in gross domestic product (GDP) had entailed an equal increase in oil consumption, but after the shock, the relationship between economic growth and oil consumption changed. A process of gradual "decoupling" of oil consumption from economic growth would not be attenuated later with the advent of even higher oil prices; that is, less oil consumption per unit of GDP became a permanent feature.

No sooner had the world started to live with the high oil prices than another, even greater price shock occurred in June 1979, the effect of Ayatollah Khomeini's revolution in Iran and, prior to that, the all-out strike of Iran's petroleum industry, causing a sudden interruption of more than 4 million barrels/ day of Iranian oil and consequent market chaos with unparalleled high prices. Although the disappearance of Iranian oil from the market was soon compensated for by the increase in production from Saudi Arabia, Kuwait, and Iraq, the oil market continued to be turbulent.

OPEC's initial reaction to the rocketing prices on the world market showed some degree of rationality. Because market prices were soaring well beyond the OPEC price, and so were generating great windfalls for the oil companies, the organization (assuming that the situation was only temporary and that production could be increased from other Gulf countries to compensate for the disappearing Iranian oil) considered adding temporary market premia, intended to be abolished once the market had returned to normal conditions. However, the continued raging markets caused many OPEC members to clamor for changing the initially rational decision and to instead add the premia increase to OPEC's fixed price on a permanent (rather than temporary) basis.

By the time OPEC met in Caracas in December 1979, OPEC's official prices had soared to more than $24/barrel, and with the markets still in turmoil, OPEC's members continued to add price increments, so that when OPEC met in Bali, Indonesia, a year later, the majority decided to fix the price of OPEC oil at $36/barrel as of January 1981. However, Saudi Arabia, aware of the negative effects of higher prices on oil consumption and the world economy, refused to raise the price beyond $32/barrel. Thus, OPEC arrived at a situation whereby a two-tiered pricing system prevailed.

The second oil price shock had an even greater negative impact on world oil consumption, and the soaring oil price served only to promote further world oil supplies from outside OPEC as well as alternative sources of energy. An ineluctable backlash lay in store for OPEC when the oil price across the world market began to fall, led by non-OPEC producers, in particular the North Sea.

5. OPEC As The Last Resort Oil Supplier: The Quota System

OPEC had to either stick with its official price by reducing its output or else follow the market and reduce its price to retain its share of the world oil market. OPEC chose the first option in resorting to the "quota" system, based on the concept of its member countries, as a whole, producing oil in volumes designated only to fill the gap between, on the one hand, total world demand and, on the other, world supplies from outside OPEC; that is, OPEC chose to be the last resort producer or what was termed the "swing producer." This meant that at a certain demand level, the higher the non-OPEC supplies, the lower the OPEC total quotas. Given increasing oil supplies from outside OPEC in the face of weakening world demand, the organization's market share would correspondingly fall naturally and continuously.

Although first discussed in 1982, the OPEC quota system was formally adopted in March 1983 at an extraordinary meeting held in London that set a production ceiling of 17.5 million barrels/day, reducing the OPEC price to a unified $28/barrel. This system proved to be even more anomalous given that Saudi Arabia did not have a quota share and was left to be the "swing" producer within a range of 5 million barrels/day. OPEC decided to fix its price and take a reference price based on the crude of Arab Light API 34 f.o.b. Ras Tanura, which Saudi Arabia became committed to charging the oil companies. Meanwhile, the prices of other crudes were to be fixed upward or downward of that reference price, taking into account the quality of the crude and its geographical location in Saudi Arabia. In other words, OPEC now became the swing producer, bearing the burden of price defense at the expense of its world market share; whereas Saudi Arabia bore the brunt by becoming the swing producer within OPEC, absorbing this fall in world demand for OPEC oil far more than did the other OPEC members. This created a situation in which the 12 member countries of OPEC were producing practically the amount of oil commensurate with their allocated quotas, whereas Saudi Arabia was forced to swing downward relative to the fall in the call on total OPEC production, seeing its production in 1985 fall to approximately 2.5 million barrels/day, or roughly one-quarter of its production capacity (leaving three-quarters unproduced). The fall in Saudi Arabia's production was so great that it caused a problem of insufficiency of associated gas for the generation of electricity for water desalination plants.

With this continued decline in the call on its oil, OPEC saw its own market share fall from 62% during the mid-1970s to 37% in 1985, all as a result of increasing supplies from outside OPEC coupled with the fall in demand as a result of high prices and improving energy efficiency in industrial nations. For example, North Sea production (both British and Norwegian) leaped to 3.5 million barrels/day in 1985, up from 2.0 million barrels/day in 1975, and continued to rise thereafter. The reason behind this was that OPEC's high prices provided such a wide profit margin for oil investors that high-cost areas such as the North Sea became not only feasible but highly lucrative areas for continued reinvestment. Because OPEC adhered to the system of fixed price and swing production, any additional oil coming from outside OPEC would first capture its share in the market before buyers resorted to OPEC oil. Also, the greater the supplies of non-OPEC oil, the less OPEC oil that was on the market to meet world demand.

Furthermore, the two oil price shocks triggered enormous investments in scientific research to improve the efficiency of technology in finding, developing, and producing oil and, thus, in reducing the high cost of upstream operations in these new areas, thereby adding for the oil companies an even higher margin of profit with which to reinvest in high-cost areas.

By 1985, OPEC production was indicative of the backlash suffered by OPEC at the hands of its own price policies. In less than 6 years, OPEC's total production had fallen from 31 million barrels/day to roughly half that amount, and this led to a situation that was increasingly difficult to sustain. Consequently, things got out of hand to the point where Saudi Arabia decided to drop its system of selling oil at fixed prices in accordance with the 1973 quota system and instead adopted market-oriented pricing, or "net-back value," meaning that it obtained the price of its crude from prevailing prices of oil products in the main consuming areas minus the cost of refining, transporting, and handling oil. Saudi Arabia's production started to rise quickly, and by the summer of 1986, a "free-for-all" situation prevailed.

6. The Backlash

This led to a severe collapse in the oil price so that during the summer of 1986 Arab Light was selling at less than $8/barrel. However, this price collapse generated mounting pressure on Saudi Arabia to forgo its refusal to adopt the quota system. The pressure came first from the OPEC members that had seen their revenues fall so drastically. A large pressure group formed within OPEC (led by Iran, Algeria, Libya, and Venezuela) that was politically powerful enough to campaign against Saudi Arabia, which was supported (within OPEC) only by Kuwait. Meanwhile, pressure came from external quarters when the high-cost producers in the United States were forced to shut down their expensive oil wells, and Texas, where oil is the spur behind all economic activity, sank into a deep recession. Investments in high-cost North Sea oil were similarly jeopardized by the drastic fall in oil prices. This additional external pressure finally pushed Saudi Arabia into changing its policy and agreeing to readopt the quota system with a fixed price for its oil at $18/barrel.

In January 1987, OPEC changed its pricing system of taking Saudi Arabia's Arab Light as a reference price for other OPEC crudes and replaced it with a "basket" of seven crudes as a reference price: Sahara Blend of Algeria, Dubai Crude, Saudi Arabia's Arab Light, Minas of Indonesia, Bonny Light of Nigeria, Tia Inana of Venezuela, and Isthmus of Mexico. However, by 1988, OPEC had effectively abandoned the fixed price system, substituting it with an agreement on target prices so that supplies could be regulated to maintain that target. This new system saved OPEC from the headache of having to agree to a fixed price for all other crudes in the light of their quality and geographical locations. However, OPEC has kept to its system of quotas and a total ceiling corresponding to the difference between total global demand and world oil supplies outside OPEC. There is no doubt that this new pricing system made decisions on the OPEC price more flexible than the earlier system, which had caused several litigious problems and endless discussions among member countries about the appropriate price for each crude relative to the reference price.

In deciding to set the price of oil, OPEC had taken on the heavy burden of oil price management with little success in achieving price stability. Yet this had been the original objective on which the organization, when founded, predicated itself. Instead, when OPEC took over oil pricing control in 1973, enormous price volatility ensued, as is clearly shown in Figure 1. For example, the price of Arab Light crude skyrocketed from $2 to $5/barrel during the summer of 1973 to approximately $40/barrel only 7 years later and then plummeted to roughly $8/barrel 5 years after that.

7. OPEC'S Limitations in Oil Supply Management

In OPEC's management of oil supplies, its limitations were to prove to be serious, with the first and foremost drawback being the intense politicization of the organization that became especially evident once OPEC took over the pricing of oil, at total variance with the original objective when the organization was formed. OPEC oil became a political commodity subject to political pressures and maneuvers. In general, politicians seek immediate or short-term benefits with little regard for the long-term consequences of their decisions. The politicization of OPEC first became evident when the Shah of Iran sought to escalate prices without due regard for the impact of higher prices on world oil demand and the supplies from outside the organization. The heads of state of OPEC's members became involved in the organization's oil policies without due regard for each member country's relative importance in terms of oil reserves, in a way as to allow low-reserve countries (e.g., Algeria, Libya) to put political pressure on their high-reserve fellow members to comply with their demands. Thus, OPEC's decision-making process had little to do with sound economics.

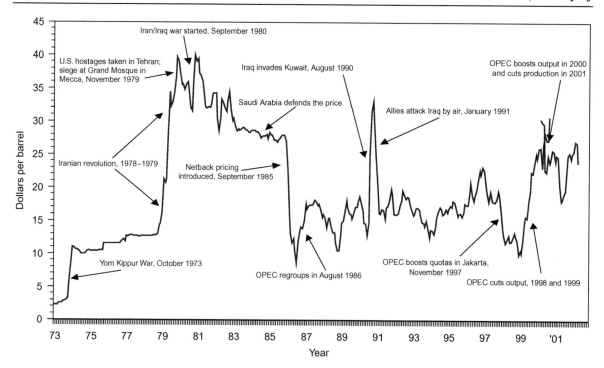

Figure 1

Arab Light Spot Prices, 1973–2002 (dollars/barrel). Data from British Petroleum, Center for Global Energy Studies, and Middle East Economic Survey.

OPEC's second limitation is that its decisions on the oil price are pushed by political events that invariably lead to an interruption of oil supplies. Fig. 1 shows how the wide fluctuations in the oil price have been caused primarily by political events rather than by market forces. World oil markets can be seen as simply reacting to the interruption of supplies caused by political events. As shown in the figure, the first price shock was the result of the political decision of the Arab oil embargo in 1973, and the second price shock was caused by the Iranian Revolution in 1979. Prices then rose steeply again during the Iraqi invasion of Kuwait in 1990. In 2003, the war with Iraq, followed by lack of security of Iraqi oil pipelines and supplies, once again posed a renewed threat of oil shortfall, forcing market prices upward.

A third and major limitation placed on OPEC's role as an instrument for price stabilization lies in the disintegration of the organization's crude oil production within the international oil industry. This began when OPEC first wrested control of the oil industry from the international oil companies, creating a predicament quite contrary to the situation that prevailed when the oil companies controlled the upstream in response to downstream requirements, all of which were controlled by the major oil companies in a way as to create a stable market by avoiding a shortage or surplus of crude oil. In OPEC's case, decisions on crude oil production are not organically related to downstream requirements. OPEC production is based simply on the difference between world demand and the production from outside the organization, without any knowledge of the downstream exigencies of consumer countries. This kind of disintegration has inevitably led to price volatility in the world oil market.

OPEC's fourth limitation is that OPEC members are entirely dependent on their oil revenues to meet their foreign currency requirements. This dependence, which amounts to more than 90% of their external trade, means that OPEC's decisions are, in reality, dictated by the short-term financial requirements of its member countries. Moreover, during the 30 years or so since the first price shock, OPEC countries have done practically nothing to diversify their economies and reduce their dependence on oil. These characteristics continue to subject OPEC's decisions to the pressing short-term needs of its member countries' budgets without looking into any form of economic rationale.

Fifth among OPEC's limitations as an efficient oil supply manager is the frequent failure of some member countries to abide by the quota system. It is noticeable that there is always a difference between the actual OPEC production quotas and the official ones. The degree of quota compliance within OPEC is reinforced when oil prices are very low, affecting its member countries' budgets; conversely, the degree of compliance falls off when oil prices are high.

Perhaps an even more significant limitation to OPEC's decision making is the organization's heterogeneous nature. Not having a homogenous entity has proved to be a major weakness because there are enormous differences among the member countries. For example, those with low oil reserves always seek higher prices so as to maximize their oil revenues by increasing the per barrel income; also, because their production capacity is so limited, they do not even care about the market share or the long-term effects of high prices on demand and supply. Conversely, member countries with large reserves (e.g., Saudi Arabia) in principle have regard for their market share to maximize income from their larger volume and, thus, higher market share. OPEC has in the past tried to formulate a long-term strategy but has never succeeded due to these conflicts of interest within the organization.

8. OPEC and A Future of Uncertainty

What future awaits OPEC? So far, OPEC has managed to survive many crises, even those of the 1980s (Table I). Regardless of the problems facing the organization's quota system, OPEC continues to function. Despite differences in the long-term interests of its members, what unites and motivates OPEC, enabling it to reach unanimous decisions, is short-term oil revenue and financial gain. However, its power has been eroded gradually over time as a result of its own policies. OPEC price practices, which amount to deriving short-term gain at the expense of long-term benefit to the oil industry as a whole, have altered the entire world energy structure and made oil far less dominant in its share of energy consumption, to the benefit of natural gas and nuclear power. This is especially evident in the cases of both Western Europe and Japan. Table II shows the extent to which oil has been losing its share in total energy consumption. Prior to OPEC's price shocks, oil had accounted for 62% of Western Europe's total consumption, yet by the year 2000, oil's share had fallen to 42%. Meanwhile, natural gas's share of total energy increased from 10 to 22%, and nuclear power's share increased from 1 to 13%, during the same period. The shift is more dramatic in the case of Japan, where oil's share of total energy has diminished from 77 to 50%. Meanwhile, its share of natural gas increased from 2 to 13%, and its share of nuclear power increased from 1 to 14%, during the same period.

In other words, OPEC's price shocks triggered a process in the industrialized world of gradually shifting from oil to alternative energies. This shift will accelerate in the future for reasons related to both environment and technological development. Although the United States still refuses to ratify the Kyoto Protocol (the international agreement that aims at a drastic reduction in carbon dioxide emissions through policy measures to reduce the consumption of fossil fuels), there is mounting pressure in both Europe and Japan to comply and, thus, reduce the consumption of oil in favor of more environmentally friendly resources, in particular natural gas, nuclear power, and (in the future) renewable energy resources. Meanwhile, technology favors these shifts because it reduces the costs of otherwise expensive alternatives. For example, the hybrid vehicle consumes much less gasoline than does the standard combustion engine, and eventually fuel cell-driven vehicles will further shrink the demand for oil.

Moreover, OPEC's high price policies have led to heavy investment in oil resources in high-cost areas outside OPEC while technological progress in the upstream has reduced the costs of finding and developing oil, thereby making investments in these alternative high-cost areas very lucrative. The amount of oil that has been produced outside OPEC has been increasing enormously. Excluding the former Soviet Union, non-OPEC oil supplies increased from 18,904 million barrels/day in 1973 to 35,569 million barrels/day in 2000. After 1982, when OPEC had opted to be the last resort supplier, the increase in non-OPEC oil supplies has been at the expense of OPEC's market share, the decline in which has occurred specifically during the same period. The less its market share, the more difficult it is to manage oil supplies at a certain stabilized price level. In addition to the price effects, geopolitics has played a significant role in changing the map of world oil supplies.

The most critical issue currently is the question of the security of Gulf oil supplies, with this area being considered prone to political instability, a factor affecting oil supplies and prices. This geopolitical factor had added momentum to greater investment in oil supplies from outside the Gulf. The understanding reached in 2002 between the United States and the Russian Federation indicates how U.S. policy is oriented toward reducing dependence on the Gulf in favor of Russian and Caspian Sea oil as well as toward obtaining increased oil supplies from West Africa (mainly Nigeria, Angola, and Equatorial Guinea) and from offshore oil production in the Gulf of Mexico.

An even greater contributory factor to the weakening of OPEC's control over oil prices is the huge investment that some member countries (e.g., Nigeria, Algeria) have designated for themselves to increase their production capacities. Once these expansion programs come on-stream, it is very difficult to perceive that OPEC can continue to subject these countries to the restrictions of its production quota regime for the sake of shoring up the oil price.

Table I
OPEC: A Chronology of Events.

1960 (September 16)	In Baghdad, Iraq, the foundation of OPEC was announced by the five major oil-producing countries (Iran, Iraq, Kuwait, Saudi Arabia, and Venezuela)
	Qatar joined OPEC, followed by Indonesia and Libya (1963), the Emirates of Abu Dhabi (1968), Algeria (1969), Nigeria (1971), Ecuador (1973), and Gabon (1975). The latter two members left OPEC during the 1990s
1961	OPEC issued a "declaratory statement of petroleum policy in member countries," a historical turning point, as it laid down basic principles that would later influence all major events of the 1970s. The statement emphasized the right of producing countries to fix unilaterally the price of their oil, in an endeavor to give the states a greater role in the development of hydrocarbon resources and to give the states the right to participate in concession-holding agreements. The emphasis was on amending the concession system in the light of changing circumstances
1966 (June)	In Caracas, Venezuela, the OPEC Conference was held. A decision was made to negotiate collectively with the oil companies to establish a general increase in posted prices and to raise the tax ratio to a minimum of $55
1970 (December)	In Tehran, Iran, the Tehran Agreement (to be of 5 years' duration and then re-negotiated) was the outcome of this first round of negotiations with the oil companies, conducted by delegates from the six OPEC member countries bordering the Gulf (Iran, Iraq, Kuwait, Saudi Arabia, Qatar, and the United Arab Emirates). In addition to price and tax amendments, the Tehran Agreement provided for the adjustment of the price upward (on an annual basis) to reflect the impact of world inflation, together with a fixed annual increase
1971 (February)	Two successive devaluations of the U.S. dollar led to two further agreements (in Geneva, Switzerland) to adjust prices and preserve the purchasing power of the other currencies relative to the U.S. dollar. Known as the Geneva Agreements ("Geneva 1" and "Geneva 2"), these two agreements laid down the formulas with which to correct the oil price (agreed in the 1971 Tehran Agreement), whether upward or downward, in relation to the movement of the value of the U.S. dollar relative to other major currencies (U.K. sterling; French, Swiss, and Belgian francs; German Deutschemark; Italian lira; and Japanese yen)
1972 and 1973	OPEC decided to reopen negotiations with the companies to revise upward the Tehran Agreement oil price in the light of prevailing market conditions, where oil prices had reached such high levels that the Tehran price was left lagging behind, and OPEC also argued that additional windfalls gained by the companies ought to have been shared with the producers
1973 (June)	In Vienna, Austria, OPEC negotiations with the oil companies failed, with the latter refusing OPEC's demand for an increase in the price set by the Tehran Agreement. This provoked OPEC into fixing the price of its oil unilaterally and independently of the oil companies' participation
1973 (July)	In Kuwait, the same OPEC Gulf members decided to increase the oil price by 70% so that the OPEC price was fixed at $5.40/barrel
1973 (October 16)	The Arab oil embargo was implemented
1973 (October)	In Tehran, OPEC's ordinary (winter) meeting was held. Spurred on by the Arab oil embargo, the Shah of Iran pushed for increasing the price yet further by 140%, from $5.40 to $10.84/barrel, a 400% increase relative to the Tehran Agreement
1973 (December)	During early part of year, an all-out petroleum industry strike took place in Iran, forcing oil prices upward to an unparalleled level
1979	The Ayatollah Khomeini's revolution and political coup occurred in Tehran
1979 (June)	In Caracas, with OPEC's official prices having soared to more than $24/barrel, OPEC members continued with price increments
1979 (December)	In Bali, Indonesia, the majority of OPEC members decided to fix the price of OPEC oil at $36/barrel. Saudi Arabia refused to raise the price beyond $32/barrel, giving rise to a two-tiered pricing system
1981 (January)	In London, an extraordinary OPEC meeting was held. Instead of substantially reducing its price to retain its share of the world oil market, OPEC stuck with its

(Continued)

Table I   Continued

	official price by reducing its output through a "quota" system. OPEC opted to be the last resort supplier with a production ceiling of 17.5 million barrels/day.
1983 (March)	The official OPEC price was now set at a unified $28/barrel
1985	The extent of the petroleum backlash was evident when OPEC total production had already fallen by half from 31 million barrels/day
1986	A "free-for-all" situation prevailed with a rapid rise in Saudi Arabia's production, culminating in an oil price collapse in July, with Arab Light selling at less than $8/barrel
1986	Pressure groups form within OPEC and among other countries to push Saudi Arabia into readopting the quota system with a fixed price for its oil at $18/barrel OPEC changed its pricing system (of taking Arab Light as a reference price for other OPEC crudes) and replaced it with a "basket" of seven crudes as a reference price (Sahara Blend of Algeria, Dubai Crude, Saudi Arabia's Arab Light, Minas of Indonesia, Bonny Light of Nigeria, Tia Inana of Venezuela, and Isthmus of Mexico). However, by 1988, OPEC had effectively abandoned the fixed price system, substituting it with an agreement on "target prices" so that supplies could be regulated to maintain that target
1987 (January)	Iraq invaded Kuwait. Oil prices rose steeply. UN sanctions were imposed on Iraq
1990 (August)	The Gulf War with Iraq took place
1991 (January)	The Kyoto Protocol compelled signatory nations to set a deadline for reducing carbon dioxide emissions to diminish global warming. This favored the adoption of alternative cleaner energies
1997 (September)	
2003 (March 20–April 17)	The second Gulf War involving Iraq took place

Table II
Percentage Shares of Total Energy Consumption 1973–2001.

	Western Europe			Japan			North America		
	1973	1985	2001	1973	1985	2001	1973	1985	2001
Oil	62	45	42	77	57	48	45	41	39
Natural gas	10	16	22	2	10	14	30	26	25
Coal	20	25	14	15	20	20	18	24	23
Nuclear	1	10	13	1	10	14	1	6	8
Hydro	7	3	8	5	2	4	6	3	5

Source. Data from British Petroleum.

Perhaps even more significant will be the question of Iraq, a country extremely rich in oil deposits that have been discovered but have remained undeveloped and that could provide a huge increase in that country's production capacity. It has been estimated that, given favorable political and financial conditions, Iraqi production could exceed 6 million barrels/day and could even reach 8 million barrels/day. It is obvious that the organization would be unable to restrict Iraqi production after the country's dire economy has been so severely held back since the Gulf War and UN sanctions.

From the preceding picture, the prospects for the future of OPEC may appear grim. However, in the event of the organization's collapse, oil prices would immediately collapse across the world markets, affecting all investments in high-cost areas outside OPEC. Such a scenario would renew the demand for cheaper OPEC oil, and for geopolitical reasons, this would not be favored by the developed nations. For this reason, it is very difficult to make any accurate prediction concerning the future of OPEC and the political and economic forces that may reshape the future status of the organization.

The fall of Saddam's Ba'athist regime in Iraq will have far-reaching effects on OPEC and on the oil industry in general. Iraq's oil potential is huge and could ultimately reach production capacity levels near those of Saudi Arabia. During the 1970s, many important giant oilfields were discovered but remained undeveloped due to Saddam's wars and the UN sanctions. In a matter of a few years, it would be technically feasible to reach a capacity of more than 6 million barrels/day. But meanwhile, the current administration is primarily concerned with restoring production and exports from the existing oilfields together with plans to rehabilitate the oil sector.

The first phase of developing Iraqi oil is to restore Iraq's capacity to its 1990 pre-UN sanctions level of 3.5 million barrels/day. Even with huge investments, this may take at least 2 years. It is not yet clear what policy would be adopted for the next phase, which is the expansion of Iraq's oil capacity to 6 million barrels/day or even to 8 million barrels/day. It goes without saying that foreign investment in Iraq would not be possible without first establishing a credible and stable Iraqi government with a sound legal system that would safeguard the interests of foreign investors.

Further Reading

Amoco B P (1970–2001). *Annual statistical reviews*. London: BP Amoco.

Center for Global Energy Studies. (1990–2002). *Global Oil Reports* CGES, London.

Chalabi F J 1980 *OPEC and the International Oil Industry: A Changing Structure*. Oxford University Press, Oxford, UK

Chalabi F J 1989 *OPEC at the Crossroads*. Pergamon, Oxford, UK

Penrose E 1968 *The Large International Firm in Developing Countries: The International Petroleum Industry*. Allen & Unwin, London

Sampson A 1975 *The Seven Sisters*. Hodder & Stroughton, London

Fadhil J. Chalabi

P

Prices of Energy, History of

Glossary

capital asset pricing model   Predicts that an asset's risk-adjusted rate equals the risk-free rate plus an asset's beta multiplied by the expected return on the market portfolio (aggregate wealth). An asset's beta is its covariance with the market portfolio divided by its variance.

cointegration   The property that a linear combination of data series is stationary while the individual series is not stationary.

marginal cost   The cost of producing one additional unit of a good.

stationarity   The property that the expected mean and variance of a series do not change over time.

The earliest energy product for which there are prices is wood. Wood has other, and higher valued, uses than as a source of energy. It is also expensive to transport per British thermal unit provided, hard to burn efficiently, nonuniform in composition, and very high in potentially polluting by-products. For these reasons, it has been supplanted in the energy-intensive developed world. However, it remains an important source of fuel in less developed countries. Coal rapidly supplanted wood as the fuel of the industrial revolution, and as a product of British society (and its American offshoot) there are long series of data available on the price of coal, or coales as it was first called. The percentage share of coal in the energy market has decreased for reasons of transport and environment. Efficient transport for coal is now provided by dedicated trains (unit trains), although water transport, including canals, was the principal method until the late 19th century. The need for expensive fixed transport has made coal hostage to railroad pricing policies and to government regulations. Water power was also an important industrial energy source in the early industrial era, but it was harnessed in a local fashion (mills on streams), and so we are unaware of price series for this direct use. Later, of course, water was used to produce electricity, and there are ample records of electric prices and estimates of hydropower costs. Nuclear power is likewise used to manufacture electricity, although it has been used only since the late 20th century. Price records are for electricity, regardless of its source, although cost estimates are available for nuclear power. The gas industry began with manufactured gas, usually derived from coal, which was used for lighting. In the United States, Baltimore was the first city to have piped gas in 1816. The transmission of natural gas began in 1872, but the 8-in. pipeline from Greenstown, Indiana, to Chicago in 1891 was the first large commercial natural gas venture. Natural gas, lacking the carbon monoxide of manufactured gas, was suitable for heating. Natural gas is now an important source for heating and power generation. Its emission profile is more favorable than that of coal or oil, which helps account for its growing usage. Price series for natural gas are not nearly as long as for coal because large-scale commercial use only began in the 20th century. Oil is the last fuel mineral that we consider. The first oil well was drilled in 1859 near Titusville, Pennsylvania, and oil rapidly became the fuel mineral of choice. Oil burns cleaner and is easier to transport than coal. It is also easier to use because oil furnaces have no need for ash removal or other hand work. Oil price history is well recorded. Because of Standard Oil, the Texas Railroad Commission, and the Organization of Petroleum Exporting Countries, oil prices are part of modern political debate and consciousness.

1. The Long Historical Sweep

Figure 1 presents Hausman's real price index series for wood and coal for Great Britain from 1450 to 1989. Hausman normalized the index to 100 in 1600 in the belief that by then coal was cheaper than charcoal so that the figure would understate the advantage of coal over wood as a fuel. There is not sufficient evidence concerning heat content to normalize the two series in that fashion. The series are meant to capture the price of the fuel source and are net of transportation or taxes. The figure clearly shows the increasing price of wood and decreasing price of coal in the 1600s. In fact, coal does not begin to rapidly increase in real price until nearly the beginning of the 20th century. In addition to cartels, strikes, and recessions, which account for some of the peaks and dips in Fig. 1, the major change in the U.K. coal industry was nationalization after World War II. The overall increase in 20th-century price is

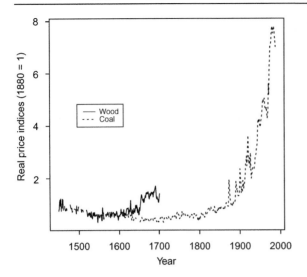

Figure 1
Historical prices for Great Britain fuel—wood and coal. Adapted from Hausman (1995).

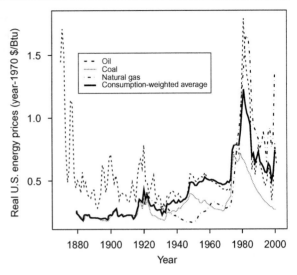

Figure 2
Historical prices in the United States, 1870–2001. Annual data on current prices for the raw resource products were obtained from the Energy Information Administration (EIA) and converted to 1970 dollars using the consumer price index. These values were converted to approximate dollars per Btu based on the following data from the EIA: In 1970, electric utilities paid on average approximately 28¢ per million Btu of natural gas, 31¢ per million Btu of coal, and 42¢ per million Btu of petroleum.

due to the quality of U.K. coal deposits and the depletion of more favorable seams. This is peculiar to the United Kingdom.

Figure 2 shows the prices of the dominant energy minerals in the United States from 1870 to the present. The prices are in real 1970 dollars per British thermal unit (Btu). The price of oil decreased significantly from 1870 to the energy shocks of the 1970s and 1980s and then returned to prices not unusual in the early 20th century. A more complete discussion of the energy shocks is provided later. Coal prices are less than those of oil, reflecting oil's greater portability and cleaner burning. Coal prices increased along with oil prices after the 1970 energy price increases. Natural gas was regulated until 1979 and then deregulated during the next decade. Particularly after 1970, the clean burning characteristic of natural gas made it a desirable fuel. This property and the end of regulation account for the change in the relative prices of coal and oil.

Figure 3 shows U.S. energy consumption by source, and it indicates the increasing importance of natural gas, oil, and coal and the decreasing importance of wood. Coal was the first energy mineral to produce significant energy output. Coal energy usage increased slowly during the 20th century. At the beginning of the century, coal accounted for approximately 73% of U.S. energy use, with wood accounting for nearly all the rest. Today, coal, gas, and petroleum account for approximately 22.6, 24, and 39.4% of use, respectively, with the remainder coming from renewable sources, primarily nuclear and hydropower.

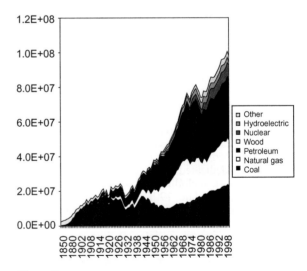

Figure 3
U.S. energy consumption by source (trillion Btus). Data from the Energy Information Administration and historical statistics.

2. The Cost of Finding Oil

The private cost to the economy of using a barrel of oil is the cost of finding a barrel of new reserves of the same quality to replace the used-up barrel. If the oil business were competitive, which it is not, then the market price of oil would be the finding costs plus the development costs plus the marginal costs of production. For this monopolistic business, it is the cost of finding oil that provides the most information about oil's ultimate scarcity. The cost series for finding oil are necessarily very incomplete and lack certainty. The series developed by the Energy Information Administration (EIA) comprise expenditures on exploration and development divided by oil and gas reserve additions. The latter series is subject to considerable uncertainty since for many years it is unknown how much oil (and gas, valued as oil equivalent) was actually found. Additions to reserves can be dominated by changes in the estimates of existing fields, whereas it is the cost of developing new sources that is of most interest. Information was collected for the years 1977–2001 for both foreign and domestic activities of U.S. reporting firms and is plotted in Fig. 4. For 1977, the cost was calculated in 1986 dollars as $14/barrel U.S. and $20/barrel foreign. Domestic costs peaked at $25 in 1982, declined

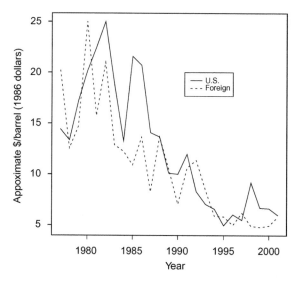

Figure 4
Cost to find oil and gas. Estimates for the cost to find oil were derived from data obtained from the Energy Information Administration, Form EIA-28 (Financial Reporting System). The left axis represents total expenditures on exploration and development divided by new additions to reserves. Total reserves include both oil and natural gas. Natural gas reserves were converted to barrels using the average September 1998 prices for natural gas and oil.

to $4.95 in 1995, increased again, and then settled at $6/barrel in 2001; foreign costs mostly declined during the period and ended at $6. It is believed that the cost of finding and developing oil in the Persian Gulf is less than these figures.

3. The Organization of Petroleum Exporting Countries and the Oil Embargoes

On October 17, 1973, the Organization of Petroleum Exporting Countries (OPEC), which was originally formed in 1960, announced a progressive embargo on the sale of oil to the United States and other countries that supported Israel in the 1973 Yom Kippur war. They also announced a 70% increase in the reference price for crude oil, an artificial price that determines how much oil companies must pay in taxes. Given that the United States bought only 5% of its oil from OPEC, an embargo of the type announced was unlikely to inconvenience the United States or substantially raise prices. However, Saudi Arabia decreased its production 35% below the September output and increased the price an additional 6¢. Combined with the nationalization of oil company interests, the cutbacks in production resulted in major increases in the price of oil and a major redistribution of the rents from oil production to the oil-producing companies. In retrospect, it is clear that these actions by Saudi Arabia, with some support from the rest of OPEC, were largely commercial in nature and U.S. attempts at bilateral relationships and other diplomacy had no effect. Production limitation became an effective tool for raising prices far above discovery and production costs.

The next increase in prices occurred in 1978–1981 as a result of a secret OPEC program to impose production controls and due to turmoil in Iran, including the Iran–Iraq war. When Iran's oil output decreased, Saudi Arabia did not increase its output. With Iran not producing at all in January 1979, Saudi Arabia decreased production by 2 million barrels per day on January 20. Its output vacillated for the next several months, always well below capacity, with the result that oil prices topped $20/barrel in October 1979. Prices increased until 1982, when the Saudi export price reached $33.50/barrel. In this period of war-induced uncertainty, oil companies and oil users hoarded significant amounts of oil.

Adelman attributes the price decline in 1982 to the end of a speculative bubble. The continued slow decline until 1986 was the result of increased supply from non-OPEC members and the consequence of lagged demand elasticity. In 1986, prices declined considerably, presumably a function of the substantial non-OPEC output. The lowest price was reported by Adelman as $6.08 for Saudi Light; in August of 1986, OPEC again set production quotas, with cooperation from some non-OPEC members.

The next major price event was the Gulf War in 1991, again producing a spike in prices. The war directly ended the sales by Iraq and Kuwait and spurred the building of precautionary supplies. Because there was sufficient capacity without the output of these two countries, prices rapidly subsided.

4. Coal Prices

In the time of Queen Elizabeth I, the coal trade in England (i.e., coal sold into the London market) was monopolized. Monopolization was possible because the London market was served by two rivers, the Tyne and the Wear, and the coal lands adjacent to these rivers. The land involved was not so extensive as to defy ownership by an oligopoly. The key issue in the early coal trade was water transport, with coal lands situated near canals or rivers with major markets being valuable and other lands being of lesser value.

Although oligopoly was in existence as early as 1517, it was formalized with royal charter as a monopoly of coal and also of political power in Newcastle by Queen Elizabeth on March 22, 1600. Other important dates cited by Sweeney include an act of Parliament to make restriction of trade in coal illegal in 1710, an allocation document fixing the quantities of coal among major owners in 1733, a lack of regulation in the period 1750–1771, a resumption of regulation in 1771 or 1772, and on-and-off regulation through 1850. Although the detailed history is not provided here, the British coal series should be interpreted in the light of the on-and-off nature of the quantity regulation. Price changes may reflect temporary changes in the Lerner index as much as changes in capacity or demand.

Coal prices in the United States during the 20th century were subject to a very different sort of government regulation. Because of the need to reduce sulfur emissions in the air, the U.S. government required new and substantially modified power plants to reduce sulfur emissions in such a way that there was no advantage to the use of low-sulfur rather than high-sulfur coal. In effect, since 1990 the emissions permit trading program no longer disadvantages low-sulfur coal as a way to reduce emissions.

5. Natural Gas Prices

In the United States, natural gas was regulated and sold at a low price until gradual deregulation from 1979 to 1989. Thus, the prices after 1989 were free to increase in response to increased demand, driven by the desire to switch from expensive oil to gas and by environmental restrictions that allow the burning of gas but restrict oil and coal. For the data plotted in Fig. 2 prior to 1979, the correlation between gas price

changes and oil price changes is 0.067; after deregulation, the correlation is 0.605.

6. Energy Prices to Users

Although there is a world market in many of the energy minerals, national governments' tax and allocation policies create very different prices for end users in different countries. In general, the United States has among the lowest prices of consuming nations to end users of energy products, and all countries generally price consumer products higher than industrial products.

In comparing prices across countries, one must keep in mind the problems of converting currencies and the qualities of the fuel. The following discussion is in terms of conversion at the official exchange rate as opposed to purchasing power parity. The exchange rate is the best choice for industrial uses, especially where the end products (e.g., cars, steel, and airplanes) are widely traded. Particularly in countries less well attached to the international trading system, exchange rates can be very different than purchasing power parity, and the use of dollars as the official exchange rate would provide a poor estimate of the burden on consumers of energy purchases relative to a basket of other goods. In order to make industrial and household prices comparable, we use exchange rate conversion throughout. There is also an issue of heat equivalent [or heat equivalent net of the energy needed to remove water from the fuel, gross caloric value (GCV) and net caloric value (NCV), respectively.] When comparing prices across fuels (oil versus coal) or across different types of the same fuel (U.S. bituminous coal versus Czech brown coal), the most accurate comparisons are in terms of ton of oil equivalent or NCV In order to use tax information, we use the reported metric tons (tonnes).

Perhaps the most traded energy commodity is fuel oil. The following section reviews the evidence that fuel oil is one international market; that is, the prices in different countries increase and decreases simultaneously, with differences only in transportation costs. For 2001, the range of prices for high-sulfur fuel oil for industry, in U.S. dollars per tonne, was 103 to 258. The lowest price in the International Energy Administration's (IEA) data set was for Mexico, which is a net exporting nation. (The data do not include the Gulf States.) The U.S. price was $147/tonne, that of France was $154.4/tonne, and that of Turkey was $181.7/tonne, whereas the Organization for Economic Cooperation and Development (OECD) average was $189/tonne. France levied taxes of $16.6/tonne, whereas the United States levied no taxes; therefore, the pretax price in France was cheaper than in the United States. Turkey's tax was $42 and its pretax price was cheaper than that of the

United States, which should not be surprising given that Turkey is a transshipment point of Mideast oil.

Automobile fuel is refined from oil. Apart from issues of taxation, the prices of automobile fuel should be very similar throughout the world, just as the ex-tax prices of fuel oil are very similar. Consumers pay the price with tax, which varies greatly from country to country. Of the countries that report a price for regular unleaded gasoline, the United States has the lowest at $0.39/liter. Germany has a price of $0.90 and Denmark is highest at $0.97. France reports a premium leaded price of $1.00 and Turkey $0.99, and Mexico reports a premium unleaded price of $0.65 per liter. This variance in pricing is attributable to taxation. Although it is not reported in the IEA data, there can also be a vast difference in the attributes of gasoline. In order to meet its obligations under the Clean Air Act, the state of California requires the use of a much more refined gasoline than is used in the rest of the United States and this gasoline carries a hefty premium in price.

Coal is a basic energy source both for industry and for electric generation. Coal is more difficult to transport than oil (railroads and slurries versus pipelines) and major deposits are often not near ocean transport, unlike oil. Coal is also more likely to be used without much preprocessing (for pollution control purposes, it can be washed, whereas crude oil is always refined), so coal must be matched much more closely to boilers than crude oil needs to be matched to its ultimate customers. For these reasons, there should be a greater variance in the underlying price of coal than there is in the underlying price of oil. Again examining IEA data, the prices for steam coal (for industry or electric generation as opposed to coking or metallurgical coal), the U.S. price was $36/tonne and that of Japan was $32/tonne, with many other countries in the $30 range. Germany was highest with a price of $52/tonne. Coal can also be measured in tons of oil equivalent to compensate for different caloric outputs per ton, and the pattern is much the same.

Electricity is generated from a variety of processes, including burning of coal, oil, and natural gas, nuclear reactors, hydropower, wind, photovoltaic, and even tide power. A country's electricity prices to industry and to consumers represent a mix of the costs of these methods. Countries such as Norway, or areas of countries such as the U.S. Pacific Northwest, that have significant surpluses of hydropower can have very low costs of production, as can countries that are willing to burn coal, particularly with minimal pollution control. Variation in electric rates across countries is thus explained by natural abundance of cheap methods, preferences for clean air, and taxation. In 2000, the U.S. price for households was $0.082/kWh (excluding tax) and it was $0.042/kWh for industry. Denmark had the highest household price at $0.197/kWh, and the Slovak Republic

had the cheapest at $0.05/kWh. The Slovak price is probably related to the burning of brown coal, with its concomitant problems of air pollution. Norway's household price was $0.057, reflective of both hydropower and petroleum reserves. Industrial prices in Europe are much lower than household prices, largely due to taxes. France's industrial price was $0.036/kWh compared to its household price of $0.102/kWh. Italy, Japan, and Turkey are outliers in the IEA data set because they had industrial electric prices of $0.089, $0.143 (1999), and $0.080/kWh, respectively.

Natural gas for households on a toe-NCV basis also show great variation among countries, with the U.S. price at $417, slightly above the OECD average of $391 in 2001. This is quite different from 1994, when the U.S. price was $274 compared to an OECD average of $383. Presumably, this change in relative price is related to shortages of both pipeline space and gas at the wellhead. Despite the clear preference in the United States to generate electricity with natural gas for air quality reasons, we expect that the price will decrease. With a price of $788, Denmark has the highest price in the dataset.

Taxation represents a large percentage of the delivered price of energy-containing products, particularly consumer products. Automotive fuel taxes represent half or more of the value of the product in all but 4 of the 29 countries included in the IEA dataset. In the United Kingdom, they are 76.9% of the value.

The earlier years in the IEA data set tell much the same story as the later years. Pricing to consumers tends to have more taxation and more idiosyncratic variation than pricing to industries.

7. A Market for "Energy": Cointegration of Prices

Energy is not a homogeneous resource and the prices of the various energy minerals do not necessarily track each other well over time. Oil is the most transported of the energy minerals, much of it originating in the Persian Gulf and transported by tanker. The differences in the prices of oil at different ports can be explained largely by transport costs. One way to test the hypothesis that oil is traded in a single market is to determine whether the series are cointegrated. Two price series are cointegrated if there exists a linear combination of the two series that is stationary (the mean and variance do not change over time). In other words, the two series move together.

An example of this type of test is to regress the price of Saudi Light landed in southern Europe on the price of the same oil landed in Rotterdam and then test the residuals of the regression to determine if they are stationary. Given the oil series are both stationary in differences, a finding that the residuals of this regression are also stationary means that the southern Europe price series is a linear function of the Rotterdam series, and any remaining differences between the two series can be accounted for by transport costs. It means that oil price shocks affect both series at the same time. Cointegration tests for oil at different ports find cointegration and support the notion that oil, in Morris Adelman's words, is a single large pool.

Coal contributes to world energy somewhat less than oil and so if there were one price for "energy," then oil and coal prices would also be expected to be cointegrated. Similarly, coal prices in different localities would be cointegrated. There is little evidence that coal prices move together or that they move with oil prices. Figs. 1 and 2 show the prices of coal for the United Kingdom and the United States. In these long series, the price in the United Kingdom increases approximately eightfold, with the strong trend being upward. The price in the United States shows a peak in the early 1980s and then a sharp decline to its previous value. It is dubious whether the U.S. price is differenced stationary rather than just stationary. The U.K. price is clearly stationary only in differences. Thus, these two prices are not cointegrated. Similarly, there is little evidence that the German and U.K. prices are cointegrated.

Although the US coal and oil prices are not found to be cointegrated using the normal testing procedure, Fig. 2 indicates that the coal price does bear some relationship to the oil price, very notably in the price run-up in the late 1970s. One explanation for the difference in the U.S. coal and oil prices is that coal is sold on long-term domestic contracts and is not subject to supply disruptions. As a result, there is no incentive to hoard coal, whereas hoarding oil before the Gulf War was a major cause of the price increase.

There is a world price of oil, but the oil price is not closely related to the prices of coal in various countries, nor are the prices of coal closely related to each other. Energy is not one great pool and there is no "price of energy," just a multiplicity of prices of energy minerals.

8. Prices and the Nonrenewable Resource Price Theory

There is no convincing evidence of a trend in oil, gas, or coal prices. One theory on these prices that has empirical support is that the first differences are stationary with mean drifts of zero. In this case, there would be no deterministic trend in the prices, simply random variation. The best estimate of future prices would be today's price. Another possibility is that the prices revert to a quadratic trend that is stochastic. This view would lead to a prediction of a slight increase in prices. Based on the historical evidence in the time series for energy prices, there is no good evidence for forecasts of increasing price.

Examining the oil price series in Fig. 2, it appears that oil prices were distinctly declining in the early years of exploitation, then leveled off, and since 1980 have increased. Since 1980, prices appear to have followed a quadratic trend. Since 2002, prices seem to have reverted to pre-Iraq war levels. We conclude that analysis of the price path of oil does not lead to a prediction of increased prices.

The lack of a clear trend—deterministic or stochastic—in energy prices at first seems at odds with the theory of natural resources. Theory predicts that prices for these commodities will ultimately rise at the rate of interest, a classic finding by Hotelling. Hotelling reasoned that if one extracted a barrel of oil today and put the money in the bank for 1 year, then one should have exactly the same financial outcome as if one extracted the barrel in 1 year. If this equality were not true, then one would be better off extracting all of one's oil in just one of the time periods, whereas we observe extraction in all time periods. A barrel extracted today would earn its price, $R(0)$, and after 1 year at interest rate i it would be worth $R(0)(1 + i)$. This must be equal to the value extracted in 1 year's time, $R(1)$ and $R(1) = R(0)(1 + i)$. This theory predicts that mineral prices will rise at the rate of interest. However, the price that is referred to is the seldom-observed price of minerals still in the ground, which is called the mineral rent. The price of minerals that we observe, P, is the mineral rent R plus the marginal extraction costs c. The theory is not so clear about minerals that have already been mined or lifted, which is the series that we have presented.

Since energy mineral prices do not increase at 7–11%, which is the normal rate of return for financial assets, economists have posited that either technical progress or a decline in grade account for the slow or nonexistent rates of appreciation of energy mineral prices.

One possibility is that there is technical progress in the mining technology; that is, that c decreases over time. Then the marginal cost of mining would decline and the value of the underlying mineral would increase. The observed price is the sum of the two prices and so could increase or decrease. However, since the extraction costs are decreasing and the underlying price is increasing, eventually the extraction costs would be insignificant and the price increases in the rent would be dominant. This process requires the rent to become dominant; however, if it is true that undiscovered, but easily discoverable, reserves are very large relative to demand, then rents will be very small and prices will not exhibit an upward trend for a very long time.

Flat prices can also be reconciled with the Hotelling theory by considering the fact that depleting a resource affects extraction. The first units extracted are presumably the least costly, with each unit becoming progressively more costly as cumulative extraction progresses. In this case, part of the opportunity cost of extracting a unit today is increased cost in the future. Rents increase more slowly than the interest rate, but the difference is exactly offset by an increase in future costs. Thus, in the long term, prices still increase at the rate of interest, even if rents do not. The evidence for grade depletion as a worldwide phenomena is thin—there are substantial high-grade coal seams in the United States, and extraction costs (and finding costs) for oil in the Gulf remain very low.

Another way to reconcile the flat price path for energy minerals with the theoretical prediction of price increase is to calculate the required interest rate using the same logic used to determine the needed rate of return for common stocks.

In most of the theoretical literature, the interest rate is taken as a fixed market price of time preference: the marginal value of a dollar received tomorrow relative to its value today, with all else remaining the same. In a world of uncertainty, however, the marginal value of returns depends not only on the timing of returns but also on uncertain contingencies. Losing money on a stock market investment is likely to be more painful if one also loses one's job but less so if growth in one's home equity more than compensates for stock market losses. The fair rate of return of one investment therefore depends on how its returns are correlated with those of other investments.

Interest rates reported for treasury bills and other assets are typically low-risk investments that yield the same amount regardless of other contingencies. In contrast, a decision not to extract and sell a unit of oil today is an investment with a return that depends on future oil prices, which are volatile and uncertain. Thus, interest rates on treasury bills or other low-risk investments may serve as a poor proxy for the rate of return that should be applied to nonrenewable resources.

The capital asset pricing model (CAPM) explains the difference in average returns across investments according to their differential risks. The model assumes that individuals, with all else being the same, prefer greater expected returns and lower risk. To mitigate risk, individuals invest in large portfolios of assets in order to average away the idiosyncratic risk of each asset against the idiosyncratic risk of all other assets. In equilibrium, individuals thus do not care about idiosyncratic risk; the only risk that matters is that pertaining to aggregate wealth—that associated with the entire portfolio of assets.

Theory predicts that assets having returns with a strong positive correlation with wealth should have high average returns. The strong positive correlation of these assets implies that they pay off most when wealth is high and marginal utility of wealth is low.

Conversely, assets with returns having a negative covariance with wealth will earn low or even negative average returns. These assets provide insurance against aggregate losses, yielding higher outcomes when wealth and consumption are low and marginal utility is high.

Energy prices have a strong negative correlation with the aggregate economy and aggregate stock market returns and therefore should have a risk-adjusted average rate of return that is lower than that of low-risk investments such as treasury bills. The relationship between energy prices and the stock market, illustrated in Figs. 5 and 6, has received less attention but is more closely related to CAPM. The correlation between the 10-year return of the Standard & Poor's 500 and the 10-year change in coal prices is less than −0.7. Simple CAPM-based estimates of the risk-adjusted rates of return for oil, coal, and natural gas, regardless of the time horizon, are between −1 and 2% and are not statistically different from 0%. As long as price fluctuations for these fossil fuels are strongly negatively correlated with national wealth, theory does not predict that the prices will trend up.

9. Conclusion: Supply Factors and Demand Factors

The past century of energy prices is a history of shifts in supply factors, shifts in demand factors, and

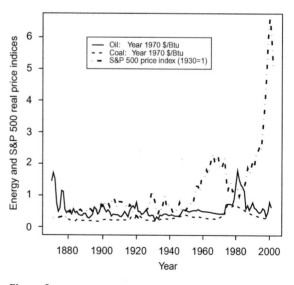

Figure 5

Long-term stock market returns and energy prices. The coal price and oil price data are those plotted in Fig. 2. The long historical series of the Standard & Poor's 500 was obtained from Robert Shiller (http://aida.econ.yale.edu/˜shiller/data.htm).

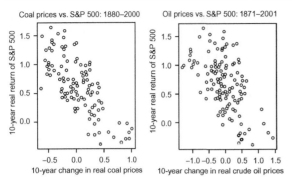

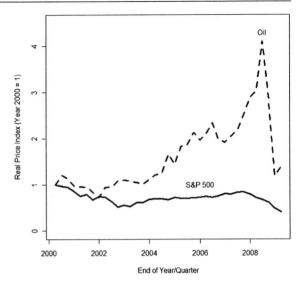

Figure 6
Long-term stock market returns and energy price changes. Returns on the Standard & Poor's 500 assume annually reinvested dividends. The long historical series of the Standard & Poor's 500 was obtained from Robert Shiller (http://aida.econ.ya-le.edu/~shiller/data.htm).

Figure 7
Real oil and stock market indices from the first quarter of 2000 to the first quarter of 2009.

policies, such as taxes and trade restrictions. However, the most notable price spikes, including those that occurred following coal strikes in the early part of the 20th century and the oil embargo, OPEC activities, and Iraq wars, are all attributable to shocks that reduced supply or anticipated supply. The observation that supply factors were most relevant to historical price fluctuations mirrors the negative relationship that has been documented between oil prices and the macroeconomy. Because energy resources are essential inputs, energy price shocks caused by unanticipated shifts in supply cause reductions in aggregate output. If aggregate demand shocks were driving price increases then the relationship between energy prices and aggregate output and stock market returns would be positive, not negative.

The history of energy prices therefore stands in contrast to the history of energy consumption, which has grown steadily with aggregate output and demand. Over the long term, prices have trended flat and have been highly volatile, with the drivers of price shocks stemming from shifts in supply. Supply volatility has caused prices to be negatively correlated with aggregate output and wealth, which in turn explains why prices have not trended up.

10. Postscript (2009)

Much has happened with energy prices since first publication of this encyclopedia. Since 2002, oil prices, adjusted for inflation, rose more than four-fold before falling precipitously to near 2002 levels. Prices for other energy sources have followed paths similar to that of oil. In Figure 7, we plot real oil and stock market indices from the first quarter of 2000 to the first quarter of 2009.

These recent price fluctuations have followed from different factors than past fluctuations. The past fluctuations were mainly driven by changes to the supply side of the oil market, particularly wars such as the Iran-Iraq war. Past fluctuations were driven by supply shocks and precautionary demand related to anticipated, if often unrealized, supply shocks. That is, when a negative supply shock such as a war decreased the oil supply, oil traders increased their stockpiles of oil as a precaution against the supply shock getting worse. When the shock ended, the traders liquidated their stockpiles and prices fell quickly. Since the supply shocks were transitory, there was no reason to believe that prices would be higher after the shock than they were before the shock. This view of the post-1968 oil market is consistent with the CAPM interpretation of the historically flat long-run trend: negative supply shocks (or anticipated negative supply shocks) led to higher oil prices and were bad news for the aggregate economy, while positive supply shocks (or anticipated positive supply shocks) led to lower oil prices and were good news for the economy. Energy price shocks and stock returns, a measure of the aggregate economy, were thus typically negatively correlated. Within a CAPM framework, the negative correlation implies a negative risk premium and a near-zero risk-adjusted rate of return for storing energy resources. Per Hotelling's rule, there was no reason for a secular rise in energy prices, even with depletion.

In contrast, the 2002 to 2008 period is not characterized by a decrease in the supply of oil due to exogenous shocks such as wars. The recent rise and fall of energy prices comes mainly from aggregate demand shocks, stemming particularly from economic growth in less developed nations such as China and India. For instance, China's GNP growth was 9.8% in 2008. The large increases in economic activity led to an increased demand for oil. With the supply essentially at its short run maximum, the increased demand for oil led to a large price increase. As in previous increases, speculation was responsible for accentuating the upward price trend. Aggregate demand started to decrease with the recession that began in the U.S. in December 2007. During this recession, U.S. industrial production has fallen at a rate faster than in the Great Depression. The proximate cause of the recession lay in the financial system, but it spread quickly to the real side of the economy of most developed countries. The recession in the developed world led to a rapid decrease in the demand for oil and therefore in oil price. Unlike the earlier period when shocks in supply drove oil prices, demand shocks drove price first up and then down during the 2002 to 2008 period. When aggregate demand drives energy price fluctuations, stock returns and oil price changes are positively correlated, not negatively correlated. Thus, if this pattern holds going forward, the risk-adjusted return for storing resources would be positive, not negative. In the long-run, the secular trend in energy prices should be positive.

Going forward, there are three main issues with supply and demand. On the demand side, the end of the 2007 recession will again increase world demand for oil. However, policies to mitigate greenhouse gas emissions will decrease demand for all carbon intensive fuels, including oil. On the supply side, new sources of oil include further use of oil sands and shale oils as well as biofuels. The U.S. Renewable Fuels Standard requires 36 billion gallons of biofuels by 2022. Therefore, considerable quantities of biofuels will be used. It remains to be seen whether new technologies produce these biofuels at costs low enough to be competitive with oil, putting a real limit on the oil price, or whether they are used only as required by law.

Further Reading

Adelman M 1993 *The Economics of Petroleum Supply*. MIT Press, Cambridge, MA.

Adelman M 1995 *The Genie out of the Bottle*. MIT Press, Cambridge, MA

Berck P, Roberts M 1996 Natural resource prices: Will they ever turn up? *J. Environ. Econ. Management* **31**, 65–78

Hausman W J 1995 Long term trends in energy prices. In: Simon J L (ed.) *The State of Humanity*. Blackwell/Cato Institute, Oxford, UK, pp. 280–6

International Energy Agency 1984–2003 *Energy Prices and Taxes*. Organization for Economic Cooperation and Development, Paris

Killan L (Fourthcoming) *Not all Oil Price Shocks are Alike Disentangling Demand and Supply Shocks in the Crude Oil Market*. American Economic Review.

Livernois J, Martin P 2001 Price, scarcity rent, and a modified r-percent rule for nonrenewable resources.. *Can. J. Econ.* **34**, 827–45

Sweezy P 1938 Monopoly and Competition in the English Coal Trade, 1550–1850. Harvard Univ. Press, Cambridge, MA.

Peter Berck, and Michael J. Roberts

S

Sociopolitical Collapse, Energy and

Glossary

collapse Rapid loss of an established level of social, political, or economic complexity.

complexity A measure of the differentiation in structure and organization of a system.

marginal return In an economic process, the amount of extra output per additional unit of input. Marginal returns may increase, remain constant, or diminish.

problem solving The process of responding to the challenges that confront a society or institution.

resources The inputs of matter, energy, or information that support continuity and problem solving.

sustainability Maintaining, or fostering the development of, the systemic contexts that produce the goods, services, and amenities that people need or value, at an acceptable cost, for as long as they are needed or valued.

Collapse is the rapid simplification of a society. It is the sudden, pronounced loss of an established level of social, political, or economic complexity. Widely known examples include the collapses of Mesopotamia's Third Dynasty of Ur (ca. 2100–2000 BC), Mycenaean society of Greece (ca. 1650–1050 BC), the Western Roman Empire (last emperor deposed 476 AD), and Maya civilization of the lowlands of Guatemala (ca. 250–800 AD). There are at least two dozen cases of collapse that are known from history, archaeology, or both. States and empires are not the only types of institutions that may rapidly simplify. The entire spectrum of societies, from simple foragers to extensive empires, yields examples of collapse. Since all but a few human societies existed before the development of writing, there may be dozens or even hundreds of cases that are not yet recognized archaeologically. Collapse is therefore a recurrent process, and perhaps no society is invulnerable to it.

1. Nature of Collapse

Many scholars have argued that collapse develops from adverse changes in the energy flows on which a society depends. In this school, collapse is thought to follow from such misfortunes as pronounced droughts, depletion of resources, or natural cataclysms. Others have searched for causes internal to societies: poor leadership, mismanagement, peasant rebellion, or even intangible factors such as changes in morality. A few scholars argue that collapses occur because of chance concatenations of misfortunes and cannot be understood as a general process. Despite this diversity of views, collapses inevitably alter energy flows. Postcollapse societies are by definition simpler, so much so that colloquially they are sometimes referred to as "dark ages." The simpler societies that follow a collapse require less energy per capita. Collapses often involve massive loss of population so that human energy extraction is greatly reduced within the region affected.

Collapse affects societies proportionate to their level of complexity. Many scholars identify collapses by the loss of such cultural elements as great traditions of art, architecture, and literature (with the decline in the last leading to the term dark age). Recognizing that collapse can affect societies organized with any degree of complexity, including simple foragers, a conception of collapse based on great cultural traditions is clearly too restrictive. It would not apply, for example, to the collapse of Chacoan society of the U.S. Southwest (ca. 725–1150 AD). Although it is known for remarkable art and architecture, Chacoan society produced no literature.

A more encompassing approach focuses on complexity. Complexity refers to differentiation in structure and to degree of organization. Societies grow more complex as they develop more parts (institutions, levels of hierarchy, and social or occupational roles) and as they become increasingly organized (i.e., as they increasingly constrain the behavior of a society's parts). One of history's clearest trends is that human societies during the past 12,000 years have shown a seemingly inexorable tendency to increase in complexity. This process still continues, more rapidly than ever.

Collapse is the rapid, substantial loss of an established degree of social, political, or economic complexity. Collapses contravene the trend of social history toward greater complexity. In a society that has collapsed there are fewer institutions, occupations, social roles, and hierarchical levels. There is less flow of information and a lower capacity for organization. People know less about the world beyond their own community. The scale of society shrinks. The Roman Empire, for example, left in its aftermath a multitude of smaller polities, none of which could match Rome's capabilities in organization, engineering, energy extraction, or scale of information flow.

2. Energy Basis of Ancient Societies

Before the development of fossil fuels, societies were powered largely by solar energy. Through photosynthesis, solar energy produced fuels such as wood and the agricultural produce on which all else depended. Solar energy drove the weather patterns to which forests and farms were adapted. These weather patterns made sailing possible and thus overseas commerce. The products of photosynthesis enabled people to have animals for farm, transport, and military use and animal products on one's table. Photosynthesis supported peasant populations that provided food for the cities, sons for the army, and taxes for the state. The combination of peasant population, weather patterns, and favorable terrain made irrigation possible in some places. A society's natural endowment of raw materials could be put to use only with solar energy. Thus, photosynthesis, transformed into slaves and charcoal, made it possible to mine and smelt metal-bearing ores. Metals were in turn used to obtain more energy or products of energy. Works of art, architecture, and literature were possible in large part because of the solar energy embedded in artists, architects, and writers; in the production of the raw materials in which artists and architects worked; and in the precious metals with which they were remunerated. Ancient states and empires relied fundamentally on agricultural taxes. Although we marvel today at the monuments and artistic works of ancient societies, these were never more than a small part of ancient economies. A. H. M. Jones, the noted historian of the late Roman Empire, estimated that the Roman government's income was based 90% on agricultural taxes. Trade and industry were comparatively undeveloped.

The persistent weakness of ancient states and empires was that the peasant producers, whose taxes funded government and the military, lived on very small margins of production. Subsistence farmers generate little surplus per capita and often none at all. Taxes were a way to compel peasants to produce more, but excessive taxation could undermine peasants' well-being and ultimately the state. Photosynthesis produces little energy per unit of land (maximum of approximately 170 kcal/m^2/year), and it takes much labor and other inputs to appropriate increasingly more of that for human use.

In ancient societies in general, agriculture could support only approximately one person per arable hectare. Vaclav Smil estimates that in Roman or early medieval times, wheat farming could yield approximately a 40-fold net energy gain. This is before subtracting seed requirements and storage losses. (The figure for 1900 AD is approximately 500-fold.) In many areas, this harvest was gained by supplementing human labor with draft animals (which require farmland and labor) and with extensive manuring. Between 180 and 250 h of human labor were combined with approximately 200 h of animal labor to produce approximately one-half metric ton of grain per hectare. In the Middle Ages, European farmers had to set aside one-third to one-half of a crop for the next year's seed. Roman-era farmers in Europe would have faced the same constraint. As discussed later, a late Roman peasant owed between one-fourth and one-third of the gross yield in taxes. After adding seed needed for sowing to that sold for taxes, it is clear that Roman farmers and their families subsisted on but a small fraction of what they produced. Tax rates were fixed, so when a harvest was low, as frequently it would be, the margin available to support the farmer's family could be inadequate. Tax rates could harm an ancient society by making it difficult for peasants to raise families large enough to sustain the population. The consequence would be a decline in future taxes and in military manpower.

This was the characteristic pattern in ancient states and empires, which were funded by aggregating the small surpluses of many thousands to millions of impoverished subsistence farmers. Thus, the dilemma faced by ancient states was that solar energy severely limited the revenues that they could raise. Often, the direction that ancient rulers chose was to raise revenues in the short term, not realizing that this would undermine the capacity of the productive system to generate future revenue. When confronted with fiscal limits, ancient governments were perpetually tempted to expand spatially, thereby capturing more of the earth's surface where solar energy falls.

3. Theories of Energy Flow and Collapse

There are many theories of why societies collapse. As emphasized in the following discussion, however, most of these fail in one way or another to explain collapse fully. After describing these theories and their limitations, a newer approach is introduced.

3.1 Approaches to Energy Flow and Collapse

Students of collapse frequently blame interruptions in the resources on which a society depends. Without sufficient energy flow to sustain people and their institutions, collapse must follow. This may occur either because a resource is used to depletion or because external events, such as a change in climate, cause a once-dependable resource to become insufficient.

Such theories have been developed for a number of collapses. J. Donald Hughes, for example, argued that the ultimate cause of Rome's collapse was environmental deterioration, especially deforestation. The American geographer Ellsworth Hunting-ton believed that climate change caused the barbarian migrations that the Roman Empire could not survive. Karl Butzer, a noted scholar of paleoenvironments, argued that insufficient Nile floods caused the collapse of Egypt's Old Kingdom (ca. 3100–2181 BC) and exacerbated other political upheavals in Egyptian history. The spectacular collapse of the Southern Lowland Classic Maya has prompted scholars to search for environmental factors, such as erosion, soil nutrient depletion, or conversion of forests to savanna grasslands. Rhys Carpenter, an accomplished student of Mycenaean civilization, argued that the Mycenaean collapse came from drought, which caused famine, depopulation, and migration. Barbara Bell argued that two so-called dark ages encompassing the eastern Mediterranean and Near East, ca. 2200–2000 and 1200–900 BC, can both be explained by widespread droughts that lasted several centuries.

The arid lands of Mesopotamia (modern Iraq, eastern Syria, and southeastern Turkey) provide some of the most compelling examples of resource depletion and collapse. There were two catastrophic collapses here: those of the Third Dynasty of Ur (ca. 2100–2000 BC) and the Abbasid Caliphate (749–1258 AD.) In each case, the collapse has been explained as resulting from overirrigation of desert soils, which caused soil salinization that destroyed agricultural productivity. Sociopolitical collapse was accompanied by depopulation and environmental damage that in each case lasted centuries. Working in upper Mesopotamia at the site of Tell Leilan, Syria, Harvey Weiss found evidence that led him to suggest that collapse ca. 2200 BC in this region was caused by an abrupt change in climate that affected hundreds of thousands of both agricultural and pastoral peoples.

Some scholars focus more on human than natural resources to understand collapse. Students of the Maya periodically reformulate the idea that the stresses of supporting the Maya ruling classes, which involved mandatory labor to build lavish limestone monuments, sparked a peasant revolt that toppled the elites and their cities. Similar ideas have been advanced for collapses in Mesopotamia, Highland Mesoamerica, Peru, China, and elsewhere. In regard to the Roman Empire, Boak and Sinnigen argued that "Rome failed to develop an economic system that could give to the working classes of the Empire living conditions sufficiently advantageous to encourage them to support it devotedly and to reproduce in adequate numbers." Even in the fourth-century AD, Ammianus Marcellinus attributed the problems of the later Roman Empire to the growth of bureaucracy and excessive taxation.

A few historians, archaeologists, and other writers indict unpredictable natural cataclysms in collapses. These alter energy flows so wrenchingly that continuity or recovery are impossible. Hurricanes, earthquakes, and epidemics have all been implicated in the Maya collapse. As early as 1939, Marinatos suggested that the collapse of Minoan society on Crete (ca. 2000–1200 BC) was caused by the great eruption of the Aegean island of Thera. Popular attempts to link Plato's parable of Atlantis to an actual place have kept this idea alive. Recent writers have elaborated on the theme, even extending the effects of the devastation to the entire eastern Mediterranean.

3.2 Assessment of Approaches to Energy Flow and Collapse

Vaclav Smil observed that many collapses occurred without any evidence of weakened energy bases. Thus, energy flow alone cannot explain collapse as a recurrent historical pattern. Every society has arrangements to ameliorate changes in energy flow. Indeed, societies have withstood even pronounced constrictions in energy flow, provided that they prove temporary. Nineteenth-century examples of pronounced disruption in energy flow include the Irish potato famine and the great eruption of Krakatoa in the South Pacific. These are sometimes compared to the disasters of antiquity (e.g., the eruption of Thera), but neither caused a collapse. Catastrophes fail to explain collapse because societies withstand disasters all the time without collapsing. Moreover, sometimes the date of the catastrophe fails to match the date of the collapse.

Changes in climate also seem inadequate to account for the matter. In studies of climate and collapse there are problems of logic. In the cases of Tell Leilan, Syria, and the Hohokam of the U.S. Southwest, archaeologists have postulated that deterioration in energy flows caused by changing climate caused, at different times, both collapse of complexity and its increase. These scholars do not explain how climate change can have such opposite effects, leaving an apparent contradiction in their arguments. In the case of drought and the Mycenaean collapse, the part of Greece with the least rainfall, Attica, was the last to collapse. If deforestation contributed to the Roman collapse, why did the same problem, later in European history, help spur economic development, as Wilkinson has shown? A change in resources can by itself rarely explain collapse.

There is no doubt that energy flow is almost always involved in collapse. Often, energy flow changes in the aftermath of simplification and depopulation. Rarely, if ever, has a change in energy flow been the sole cause of a collapse. To understand the role of energy in collapse, it must be considered in the context of social, political, and economic changes.

3.3 Embedding Energy Flow and Collapse within Economics and Problem Solving

A new approach embeds energy in the social, political, and economic changes that a society undertakes to solve problems. Problem solving among humans (and even among animals) shows a general tendency to move from simple, general, and inexpensive (known colloquially as plucking the lowest fruit first) to complex, specialized, and costly. Sustainability or collapse follow from the long-term success or failure of problem-solving institutions. Throughout human history, a primary problem-solving strategy has been to increase complexity. Often, we respond to problems by creating new and more sophisticated technologies, new institutions, new social roles, or new hierarchical levels or by processing more information. Each of these strategies increases the complexity of a society. Problem solving has been much of the force behind the increasing complexity of human societies, especially during the past 12,000 years.

Changes in complexity can often be enacted quickly, which is one reason why we resort to complexity to solve problems. This is illustrated by the response to the attacks on the United States on September 11, 2001. Much of the immediate response to the problem of preventing future attacks involved increasing the complexity of public institutions by establishing new agencies, absorbing existing ones into the federal government, and expanding the scale and degree of control over realms of behavior from which a threat might arise. Such changes can be implemented more rapidly than can changes in military technology, in the technology and other techniques of espionage, or in the technology to detect explosives in air travel baggage.

As a problem-solving strategy, complexity often succeeds. The difficulty is that complexity costs. In any living system each increase in complexity has a cost, which may be measured in energy or other resources or in transformations of these, such as money, labor, or time. Before the development of fossil fuels, increasing the complexity of a society typically meant that people worked harder. This is still often the case. In the response to the September 11, 2001, attacks, anyone who travels by air understands that doing so has become more complex, and the personal cost has risen. The cost is reflected not just in the price of travel (which is controlled by competition) but also in such intangibles as time spent standing in lines, the indignity of personal searches, and greater care regarding what personal items one travels with. These are examples of the subtle costs of complexity in problem solving.

If an increase in complexity resolves a problem, it is likely to be adopted for its short-term value. The costs of complexity are cumulative and long term. As simple, inexpensive solutions are adopted for problems in such areas as organization, information flow, or technology, the strategies that remain to be deployed are increasingly complex and costly. Complexity is an economic function and can be evaluated by returns on effort. Early efforts at problem solving often produce simple, inexpensive solutions that may yield positive returns. Yet as the highest-return ways to produce resources, process information, and organize a society are progressively implemented, problems must be addressed by more costly and less effective responses. As the costs of solutions increase, the marginal return to complexity in problem solving begins to decline. Carried far enough, diminishing returns to problem solving cause economic stagnation and disaffection of the support population. As illustrated later, a prolonged period of diminishing returns to complexity in problem solving makes a society vulnerable to collapse.

4. Examples of Energy Flow in Sociopolitical Collapse

Energy, leadership, political organization, taxation, and competition are all involved in collapse. To illustrate this, I describe five cases in detail, including two collapses in Mesopotamia in which energy and collapse are clearly linked, although not as simply as sometimes described. The case of the Lowland Classic Maya exemplifies how competition and pressures to produce great monuments create stress on peasantry that undermines the production system. The Roman and Byzantine empires illustrate how complexity in problem solving makes a society vulnerable to collapse. In the Roman case, the collapse lasted centuries. In the Byzantine example, the collapse actually provided a basis for recovery.

4.1 Mesopotamia

Ancient Mesopotamia produced two cases that illustrate the relationship of collapse to energy—the collapses of the Third Dynasty of Ur (ca. 2100–2000 BC) and the Abbasid Caliphate (749–1258 AD).

In southern Mesopotamia, intensive irrigation initially increases agricultural yields that support growing prosperity, security, and stability. The Third Dynasty of Ur pursued this course. It expanded the irrigation system and encouraged growth of population and settlement. To capture this enhanced energy flow, it established a vast bureaucracy to collect taxes. This strategy plants the seed of its own demise, however. After a few years of overirrigating, saline groundwaters rise and destroy the basis of agricultural productivity. The political system loses its resource base and is destabilized. Large irrigation systems that require central management are useless once the state lacks the resources to maintain them.

A few centuries earlier, in the Early Dynastic period (ca. 2900–2300 BC), crop yields averaged approximately 2030 liters per hectare. Under the Third Dynasty of Ur this declined to 1134 liters. At the same time, Third Dynasty of Ur farmers had to plant their fields at an average rate of 55.1 liters per hectare, more than double the previous rate. Badly salinized lands go out of production almost indefinitely, so the pressure intensifies to get the most from the remaining fields. As yields declined and costs rose, farmers had to support an elaborate state structure. It was a system that took a high toll in labor. The Third Dynasty of Ur was following a strategy of costly intensification that clearly yielded diminishing returns.

The Third Dynasty of Ur hung on through five kings and then collapsed, with catastrophic consequences for the larger population. By 1700 BC, yields were down to approximately 718 liters per hectare. Of the fields still in production, more than one-fourth yielded on average only approximately 370 liters per hectare. The labor demands of farming a hectare of land were inelastic, so for equal efforts cultivators took in harvests approximately one-third the size of those a millennium earlier. Soon the region was extensively abandoned. By a millennium or so after the Third Dynasty of Ur, the number of settlements had decreased 40% and the settled area contracted by 77%. Population densities did not rebound to those of the Third Dynasty of Ur until the first few centuries AD, a hiatus of 2500 years.

Rulers of Mesopotamia, from the last few centuries BC into the Islamic period, repeated the strategy tried 2000 years earlier by the Third Dynasty of Ur. Population increased fivefold in the last few centuries BC and the first few centuries AD. The number of urban sites grew 900%. These trends continued through the Sassanian period (226–642 AD), when population densities came to exceed significantly those of the Third Dynasty of Ur. At its height, the area of settlement was 61% greater than during the Third Dynasty of Ur.

Under Khosrau I (AD 531–579) the Sassanian dynasty reached its height, but the needs of the state took precedence over peasants' ability to pay. Taxes were no longer remitted for crop failure. Because the tax was fixed whatever the yield, peasants were forced to cultivate intensively. State income rose sharply under Khosrau II (590–628). This level of income would have been needed for the perpetual wars in which he engaged with the Byzantine Empire. Under the Abbasid Caliphate, an Islamic dynasty, tax assessments increased in every category. Fifty percent of a harvest was owed under the Caliph Mahdi (775–785), with many supplemental payments. Sometimes, taxes were demanded before a harvest, even before the next year's harvest.

Under the Abbasids there was unprecedented urban growth. Baghdad grew to five times the size of 10th-century Constantinople and 13 times the size of the Sassanian capital, Ctesiphon. The capital was moved often and each time built anew on a gigantic scale. The Caliph al-Mutasim (833–842) built a new capital at Samarra, 120 km upstream from Baghdad. In 46 years, he and his successors built a city that stretched along the Tigris for 35 km. It would have dwarfed imperial Rome.

As the state undertook these constructions, however, it did not always fulfill its irrigation responsibilities. As the irrigation system grew in size and complexity, maintenance that had once been within the capacity of local communities was no longer so. Communities came to depend on the imperial superstructure, which in turn became increasingly unstable. Peasants had no margins of reserve, and revolts were inevitable. Civil war and rebellion meant that the hierarchy could not manage the irrigation system. Mesopotamia experienced an unprecedented collapse. In the period from 788 to 915 revenues fell 55%. The Sawad region, at the center of the empire, had supplied 50% of the government's revenues. This dropped within a few decades to 10%. Most of this loss occurred between the years 845 and 915. In many once-prosperous areas, revenue declined 90% within a lifetime. The perimeter of state control drew inward, which diminished any chance to resolve the agricultural problems. By the early 10th century irrigation weirs were listed only in the vicinity of Baghdad. In portions of Mesopotamia the occupied area shrank by 94% by the 11th century. The population declined to the lowest level in five millennia. Urban life in 10,000 square kilometers of the Mesopotamian heartland was eliminated for centuries.

Collapses in Mesopotamia present clear evidence of the role of energy flow disruptions. When the resource base was destroyed by excessive irrigation, population levels and political complexity could not be maintained. Yet the clarity of these cases is deceptive because it obscures a fundamental question: Why did Mesopotamian rulers push their support systems to the point at which large populations and their own political systems were endangered? Such a question cannot be answered here. Rather, it underscores the point made previously that energy flows, although vital in understanding collapse, do not fully explain it. This will be clear in the next two cases.

4.2 The Southern Lowland Classic Maya

The Southern Lowland Classic Maya (ca. 250–800 AD) present one of history's most enigmatic collapses. The Maya Lowlands (Fig. 1) were one of the most densely populated areas of the preindustrial world, with an average of 200 persons per square kilometer. Tikal, one of the major Maya cities, had a population estimated at nearly 50,000, which is roughly the same as ancient Sumerian cities. In the southern part of the area known today as Quintana Roo, there

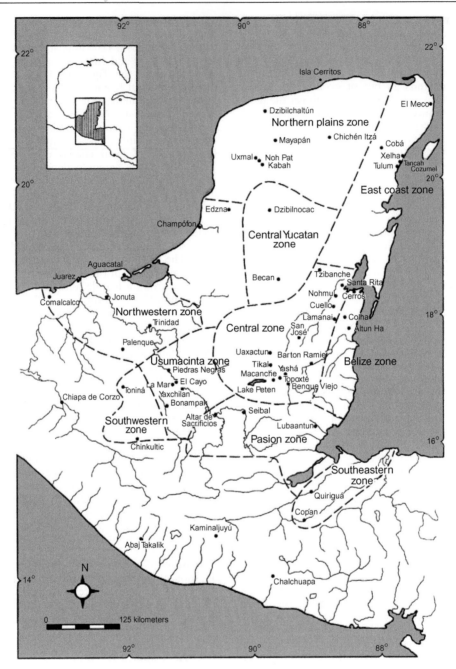

Figure 1
The Maya Lowlands and major archaeological sites.

are stretches of terrain 40–50 km long in which there is no gap of more than 100 m between Maya structures. The Maya are best known for their art, their cities and temples, and, as Robert Netting phrased it, their "abrupt departure from the stage of world history."

To support so many people the southern Lowland Maya transformed their landscape. Over much of the area the natural rainforest was cleared. One technique they employed was to channel and store water. Although the area today is a high tropical forest, precipitation varies seasonally, and many areas have little surface water. The Maya accordingly built canals, dams, reservoirs, and small wells and in Yucatán modified the large limestone sinkholes known as *cenotes*. There are 180 km of canals along the Rio Candelaria. They represent at least 500,000 days of labor to move (entirely with human workers) 10 million cubic meters of earth.

Hillsides across the Lowlands were often terraced for planting. Terracing directs water flow, traps silt, and creates cultivable land. Terraces are spread across more than 11,000 square kilometers. Much of the southern Lowlands (the area known as the Petén) also contains walled fields, a type of feature that indicates intensive, sustained cultivation.

Perhaps the most interesting agricultural features of the Lowlands are systems of raised fields and associated canals. These were built to supply dry, cultivable land in areas otherwise inundated. Raised fields were built within lakes and lagoons and at river margins, but they were primarily built within swamps, thereby reclaiming land that could be highly productive. Modified and intensively cultivated swamps may have become the most valuable land. Not surprisingly, the larger cities tend to be located near swamps. By building raised fields in such areas, the Maya could have maintained a moist, root-level environment; fertilized fields with rich, organic muck from canal bottoms; and perhaps raised fish as well. Tended carefully, such fields could have supported continuous cropping.

By the use of such facilities the energy of the Lowlands was channeled to produce a simplified biota consisting of maize, squash, avocado, cacao, cotton, root crops, and fruit-bearing trees. Simplified food webs became the foundation for a diverse and complex society. Ranks of bureaucrats, artisans, elites, and paramount rulers held sway in cities whose centers were built of gleaming limestone. Their sense of time was based on astronomical observations so precise that they were not equaled in Europe or China for centuries to come. The Maya developed a system of writing that has only recently been deciphered. Their art is greatly admired today. All of this complexity was based on the energy flow made available from their managed landscape.

One consequence of the development of complex societies in the Maya area is that the cities were often at war. Warfare is a prominent theme in Maya art, with many scenes of domination and execution of captives. There are great earthen fortifications at such sites as Becan, Calakmul, Oxpemul, and Tikal. The defensive works at Tikal include a moat and earthwork 9.5 km long. Warfare in turn reinforced centralized political authority.

Maya complexity imposed a high cost on the support population. Great efforts were required to transform the forest, keep the land from reverting to forest, and keep the soil fertile; to support non-producing classes of bureaucrats, priests, artisans,

and rulers; and to build and maintain the great cities. Over time these costs actually grew. The construction of monuments increased over the years. Across the southern Lowlands, 60% of all dated monuments were built in a period of 69 years, between 687 and 756 AD.

All this effort took its toll on the support population. At Tikal, by approximately 1 AD, the elites began to grow taller than the peasants, and within a few centuries the difference averaged 7 cm. That elites should have enjoyed better nutrition during childhood is not surprising. By the late Classic (ca. 600–800 AD), however, the stature of adult males actually dropped among both elites and peasants. The nutritional value of the Tikal diet had apparently declined for both rulers and ruled. At the site of Altar de Sacrificios there was a high incidence of malnutrition and/or parasitic infections. Childhood growth was interrupted by famine or disease. Many people suffered from anemia, pathologic lesions, and scurvy. At Altar de Sacrificios, the stature of adult males declined in the first few centuries AD, and life expectancy declined in the late Classic. Not surprisingly, the population of the southern Lowlands, which had grown during the first several centuries AD, leveled off between 600 and 830 and thereafter grew no more. Such population trends often foreshadow collapse.

The magnificence of late Maya cities clearly came at the expense of the peasant population, and ultimately at the cost of the entire system's sustainability. It was a classic case of diminishing returns to complexity. As the complexity of the political system and the cities increased through time, the costs of the system, in human labor, also increased. It is clear that the marginal return to these investments was declining by the late Classic. Ever greater investments in military preparedness, monumental construction, and agricultural intensification brought no proportionate return in the well-being of most of the population. To the contrary, as the demands on the peasant population increased, the benefits to that population actually declined. This was particularly the case in the seventh and eighth centuries, when a great increase in laborious monumental construction fell on a stressed and weakened population that was no longer growing.

The consequences were predictable. One by one, the Maya polities began to collapse and their cities were progressively abandoned (Fig. 2). At the same time, the southern Lowlands experienced one of history's major demographic disasters. In the course of approximately a century, the population declined by 1–2.5 million people and ultimately declined to a fraction of its peak. There were not enough Maya left to control the landscape, nor did those remaining need to farm as they once had. Intensive agriculture was abandoned and the forests regrew. Cultural complexity in this region was based on an engineered

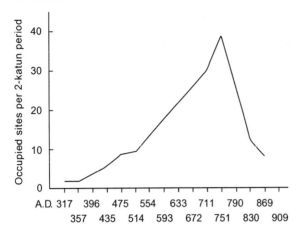

Figure 2
Occupation of Classic Maya centers. One katun is approximately 20 years.

landscape with low species diversity, short food chains, and diversion of energy from plant to human biomass. In the long term, although the Lowlands landscape could seemingly support dense human populations, it could not sustain the complex political systems that large populations often require.

4.3 Collapse of the Western Roman Empire

The Western Roman Empire (Fig. 3) is one of history's most instructive collapses. The Romans' early success came from a means of expansion that was fiscally self-perpetuating. Defeated peoples gave the economic basis, and some of the manpower, for further expansion. It was a strategy with high economic returns. By 167 BC, for example, the Romans were able to eliminate taxation of themselves and still expand the empire. Yet the economics of an empire such as the Romans assembled are deceptive. The returns to any campaign of conquest are highest initially, when the accumulated surpluses of the

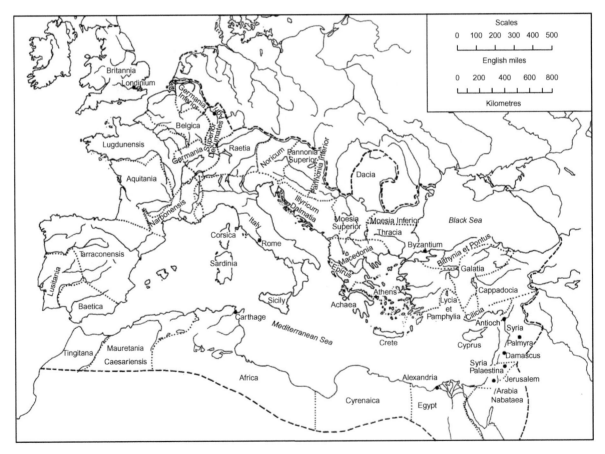

Figure 3
The Roman Empire in 117 AD.

conquered peoples are appropriated. Thereafter, the conqueror assumes the cost of administering and defending the province. These responsibilities may last centuries and are paid from yearly agricultural surpluses. The fiscal basis of such an empire shifts from accumulated surpluses (generally embodied in precious metals, works of art, and population) to current solar energy.

Once the phase of conquest was complete, and the accumulated surpluses spent, the Roman government was financed by agricultural taxes that barely sufficed for ordinary administration. When extraordinary expenses arose, typically during wars, the precious metals on hand frequently were insufficient to produce the required coinage. Facing the costs of war with Parthia and rebuilding Rome after the Great Fire, Nero began in 64 AD a policy that later emperors found irresistible. He debased the primary silver coin, the denarius, reducing the alloy from 98 to 93% silver. It was the first step down a slope that resulted two centuries later in a currency that was worthless (Fig. 4) and a government that was insolvent.

In the half century from 235 to 284, the empire nearly came to an end. There were foreign and civil wars almost without interruption. The period witnessed 26 legitimate emperors and perhaps 50 usurpers. Cities were sacked and frontier provinces devastated. The empire shrank in the 260s to Italy, the Balkans, and North Africa. By prodigious effort the empire survived the crisis, but it emerged at the turn of the fourth-century AD as a very different organization.

In the late third and early fourth centuries, Diocletian (284–305) and Constantine (306–337)

designed a government that was larger, more complex, more highly organized, and much more costly. They doubled the size of the army, always the major part of imperial costs. To pay for this, the government taxed its citizens more heavily, conscripted their labor, and dictated their occupations. Increasingly more of the empire's energy was devoted to maintaining the state.

Diocletian established Rome's first budget, and each year a tax rate was calculated to provide the revenue. The tax was established from a master list of the empire's people and lands, tabulated down to individual households and fields. In an era when travel and communication were slow, expensive, and unreliable, it took substantial organization and personnel just to establish and administer a tax system so minutely detailed. Taxes increased, apparently doubling between 324 and 364. Villages were liable for the taxes on their members, and one village could even be held liable for another. Tax obligations were extended to widows, orphans, and dowries. Even still, to meet its needs, the government had to conscript men for the army and requisition services. Occupations were made hereditary and obligatory. Positions that had once been eagerly sought, such as in city senates, became burdensome because leading citizens were held responsible for tax deficiencies.

The tax system supporting the more complex government and larger army had unforeseen consequences. After plagues decimated the population in the second and third centuries, conditions were never favorable for recovery. There were shortages of labor in agriculture, industry, the military, and the civil service. The tax system of the late empire seems to

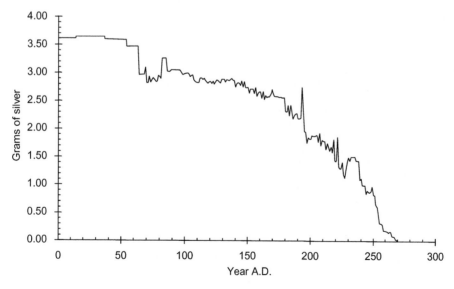

Figure 4
Debasement of the Roman silver currency to 269 AD.

have been to blame because the rates were so high that peasants could not accumulate reserves nor support large families. Whatever crops were brought in had to be sold for taxes, even if it meant starvation for the farmer and his family. Farmers who could not pay their taxes were jailed, sold their children into slavery, or abandoned their homes and fields. In such circumstances it became unprofitable to cultivate marginal land because too often it would not yield enough for taxes and a surplus. Despite government edicts, marginal lands went out of cultivation. In some provinces, up to one-third to one-half of arable lands came to be deserted by the late empire. Faced with taxes, peasants would abandon their lands and flee to the protection of a wealthy landowner, who was glad to have the extra labor.

On the death of the emperor Theodosius in 395 AD, the Roman Empire was divided into eastern and western parts, each of which had to raise its own revenues and most of its own troops. The east was by far the wealthier half, with fewer problems at its borders. From the late fourth century the barbarians could no longer be kept out. They forced their way into Roman lands in western Europe and North Africa, initially causing great destruction. The government had no choice but to acknowledge them as legitimate rulers of the territories they occupied. Throughout the fifth century the western empire was in a positive feedback loop tending toward collapse. Lost or devastated provinces meant lower government income and less military strength. Lower military strength in turn meant that more areas would be lost or ravaged.

In the 20 years following the death of emperor Valentinian III (455), the western Roman army disappeared. The government came to rely almost exclusively on mercenaries from Germanic tribes. Finally, these could not be paid. They demanded one-third of the land in Italy in lieu of pay. This being refused, they revolted and deposed the last emperor in Italy, Romulus Augustulus, in 476.

The strategy of the later Roman Empire was to respond to a near-fatal challenge in the third century by increasing the size, complexity, power, and costliness of the primary problem-solving system—the government and its army. The higher costs were undertaken not to expand the empire or to acquire new wealth but to sustain the status quo. The benefit/cost ratio of imperial government declined as it lost both legitimacy and support. In the end, the Western Roman Empire could no longer afford the problem of its own existence.

5. Energy Flow, Collapse, and Resiliency

It is common to think of collapse as a catastrophe, involving the loss of great traditions of art, architecture, and literature. In the case of the early Byzantine Empire (the Eastern Roman Empire, 395–1453 AD), though, the reduction of complexity that is the defining characteristic of collapse contributed to resiliency and recovery. The Byzantine case may be unique among collapses. It is especially instructive in regard to the economics of complexity in problem solving.

The Eastern Roman Empire survived the fifth-century debacle and met its end only when the Turks took Constantinople in 1453. The Byzantine Empire lost territory throughout much of its history and several times nearly fell. Yet it survived near catastrophes through one of history's most unusual strategies.

The eastern emperor Anastasius (491–518) established a sound coinage in the copper denominations on which daily life depended. He gave the army generous allowances for rations, arms, and food. Commerce was revitalized and the army again attracted native volunteers. Within a few decades these reforms had produced such results that Justinian (527–565) could attempt to recover the western provinces.

An army sent to North Africa in 532 conquered the Kingdom of the Vandals within 1 year. Almost immediately, the Byzantine general Belisarius was sent to reconquer Italy. He had captured the Ostrogothic King and nearly all of Italy when he was recalled in 540 to fight the Sassanian Persians.

In 541, when the job in Italy seemed about done, bubonic plague swept over the empire. It had not been seen before in the Mediterranean. Like any disease introduced to a population with no resistance, the effects were devastating. Just as in the 14th century, the plague of the sixth century killed one-fourth to one-third of the population. The enormous loss of taxpayers caused immediate financial problems. Army pay fell into arrears, and both Italy and North Africa were nearly lost. By debasing the currency and slashing expenditures, the emperor was able to send another army to Italy. Italy was reconquered by 554, but much of it was lost again when the Lombards invaded in 572. Throughout the rest of the sixth century, costly wars continued in Italy, North Africa, the Balkans, and with the Persians.

A rebellion of the Byzantine army in 602 rekindled the Persian war and began a crisis that lasted more than a century and nearly brought the empire to an end. The war against the Persians went badly. Sensing Byzantine weakness, Slavs and Avars overran the Balkans. North Africa and Egypt rebelled and placed Heraclius (610–641) on the Byzantine throne. The empire he took over was financially exhausted. The Persians captured Syria, Palestine, and Egypt, overran Anatolia, and besieged Constantinople from 618 to 626.

In 616, Heraclius cut the pay of troops and officials in half. He followed his predecessors by lowering the

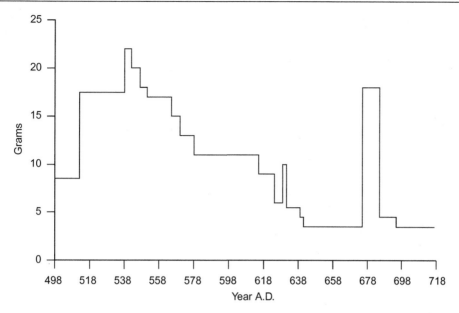

Figure 5
Weight of the Byzantine follis, 498–717 AD.

weight of the primary copper coin, the follis (Fig. 5). Heraclius' economic measures bought time for his military strategy to work. In 627, Heraclius destroyed the Persian army. The Byzantines got all their lost territory returned. The war had lasted 26 years and resulted in no more than restoration of the status quo of a generation earlier.

The empire was exhausted by the struggle. Arab forces, newly converted to Islam, defeated the Byzantine army decisively in 636. Syria and Palestine, which had taken 18 years to recover, were lost again. Egypt was taken in 641. The wealthiest provinces were permanently gone, and soon the empire was reduced to Anatolia, Armenia, North Africa, Sicily, and parts of Italy. The Arabs conquered the Persian Empire entirely. (In time, parts of the former Persian Empire became the center of the Abbasid Caliphate, described previously.)

Under Constans II (641–668) and throughout the seventh century, the situation continued to deteriorate. It appeared that it would be only a matter of time before the remaining Byzantine lands fell to the Arabs. The Arabs raided Asia Minor nearly every year for two centuries. Constantinople was besieged each year from 674 to 678, then continuously from 717 to 718. The Byzantines defeated both sieges, and this proved to be the turning point in a centuries-long struggle. The Arabs were never again able to mount such a threat.

During this century of crisis the social and economic life of the eastern Mediterranean was thoroughly transformed. Population declined. Monetary standards and economic institutions were destroyed.

Around 659, Constans II cut military pay in half again. Army pay now stood at one-fourth its level in 615 so that the economy no longer received large influxes of new coin. By 700 most people of the eastern Mediterranean no longer used coins in everyday life. The economy developed into its medieval form, organized around self-sufficient manors.

Roman emperors of the late third and early fourth centuries had responded to a similar crisis by complexification. They increased the complexity of administration, the regimentation of the population, and the size of the army. This was paid for by levels of taxation so harmful that lands were abandoned and peasants could not replenish the population. Constans II and his successors could hardly impose more of the same exploitation on the depleted population of the shrunken empire. Instead, they adopted a strategy that is truly rare in the history of complex societies: systematic simplification, in effect a deliberate collapse.

The government had lost so much revenue that even at one-fourth the previous rate it could not pay its troops. Constans devised a way for the army to support itself. Available lands were divided among the troops. Soldiers (and later sailors) were given grants of land on condition of hereditary military service. Constans expected the troops to provide their own livelihood through farming, with a small monetary supplement. Correspondingly, the Byzantine fiscal administration was greatly simplified.

The transformation ramified throughout Byzantine society. Both central and provincial government were simplified. In the provinces, the civil administration

261

was merged into the military. Cities throughout Anatolia contracted to fortified hilltops. Aristocratic life focused on the imperial court. There was little education beyond basic literacy and numeracy, and literature consisted of little more than lives of saints. The period is sometimes called the Byzantine dark age.

The simplification rejuvenated Byzantium. The new farmer–soldiers became producers rather than consumers of the empire's wealth. From this new class of farmers came the force that sustained the empire. By lowering the cost of military defense the Byzantines secured a better return on their most important investment. The empire began to lose land at a much slower rate. The Arabs continued to raid Anatolia but were unable to hold any of it for long. Soldiers were always nearby. Fighting as they were for their own lands and families, they had much greater incentive and performed better.

Once the threat to the empire's existence was over, campaigns against the Bulgars and Slavs gradually extended the empire in the Balkans. Greece was recaptured. In the 10th century the Byzantines reconquered parts of coastal Syria. Overall after 840 the size of the empire was nearly doubled. The process culminated when Basil II (963–1025) conquered the Bulgars and extended the empire's boundaries again to the Danube. In two centuries the Byzantines went from near disintegration to being the premier power in Europe and the Near East, an accomplishment won by a deliberate collapse in which they systematically simplified, and thereby reduced the costs of, their army, government, and society.

6. Energy Flow in the Aftermath of Collapse

Since collapse is the loss of complexity, and higher levels of complexity depend on greater flows of energy, a primary characteristic of postcollapse societies is that they require and produce less energy. This is a consequence of collapse that occurs because there are fewer elite ranks to extract resources from the populace, because expensive sectors of investment (such as the Roman army) disappear, or because a lower degree of social and economic specialization means that peasant households no longer have to support specialists. A late Roman peasant, for example, might have paid one-fourth to one-third of gross yields in taxes if he owned his own land. If he rented land, taxes and rent together took one-half to two-thirds of a crop. Thus, Roman peasants were perpetually impoverished, often undernourished, and sometimes forced to sell children into slavery when they could not be fed. This undermined the productive capacity of the empire and its sustainability. After the plagues of the second and third centuries AD, peasants were unable to replenish the population, and the empire lost much of its energy-producing

capacity. In post-Roman Gaul, in contrast, taxes declined to as low as 10%. Early medieval peasants may have been better off and seen more of their children survive.

The Byzantines responded to lower energy flow by simplifying their social, political, and economic systems and by shortening energy flow networks. In the sixth-century AD, energy would typically have flowed from the sun to farms; from farms to peasants; from peasants to purchasers of grain, thereby converting the grain to coinage; from peasants to tax officials; from tax officials to the government; and from the government to the army. At each step some energy was lost to transaction costs. Just as in a trophic pyramid, energy was lost each time it was transformed or passed to another level. After the mid-seventh century, the energy on which the empire depended flowed through fewer transactions and levels. Energy passed from the sun to farms. From there part of it was harvested and used directly by soldiers, eliminating many intermediate steps. Other peasants still paid taxes, but since the government spent less of this revenue on the army there was more to allocate to other needs. Overall much less energy went to transaction costs, so net energy increased as a proportion of gross production. The empire revived and went on to expand.

In general, postcollapse societies are smaller, with simpler hierarchies and less socioeconomic specialization. They produce less monumental architecture or none at all. Population often declines, sometimes dramatically as in the Maya collapse. Many settlements are abandoned, and those that remain may contract. Long-distance trade declines. The known world shrinks, and over the horizon lies the unknown. There are often declines in literacy and numeracy because the simpler society requires less of these. Since there are fewer elites for whom to produce art, and fewer artisans, postcollapse societies produce less art. All these changes involve reductions in energy extraction and flow, either as cause or as consequence. In some cases, energy flow is so greatly reduced that postcollapse societies are nearly impossible for archaeologists to detect.

7. Synthesis and Implications

Collapse is a recurrent historical process. It affects the entire range of societies, from those that are least complex to the most complex. Perhaps no society is immune to the possibility of collapse, including the very complex societies of today. The misfortune of some previous societies is potentially our good fortune. By understanding the processes that cause collapse we can develop tools to assess our own vulnerability.

Earlier research often pinpointed diminished energy flow as the cause of collapse. In many cases, though, diminished energy flow was the product of human management. This is evident in the examples of the Third Dynasty of Ur and the Abbasid Caliphate. In both cases, excessive irrigation caused salinization that destroyed productive soils. In these cases energy does not suffice to explain collapse. Other factors were involved, particularly the tendency of rulers to demand more than the land could sustainably produce. When we ask why the rulers did this, we realize that energy is but a small part of explaining these collapses. The Maya drastically altered their environment, shortening food chains and diverting large parts of the region's productivity to support dense human populations and their complex societies. Yet demands to support complexity so stressed the population that the system collapsed, the population plummeted, and the region reverted to rainforest. Again we see the interplay of energy with politics and human strategies of survival and competition.

Problem solving links these examples to a more general understanding of collapse. Moreover, problem solving is the key to using past collapses to understand our situation today. As noted previously, sustainability is a function of problem solving, as is its opposite, collapse. As the problems that a society or institution faces grow in size and complexity, problem solving grows in complexity and costliness. Problem solving is an economic process, and its success depends on the ratio of the benefits of problem solving to its costs. Problem solving tends to develop from solutions that are simple and inexpensive to those that are increasingly complex and costly. Each increase in complexity has costs, which may be measured in labor, time, money, or energy. Growth in complexity, like any investment, ultimately reaches diminishing returns, at which point increments to complexity no longer yield proportionate returns. When this point is reached, a society or other institution begins to experience economic weakness and disaffection of the populace. A fundamental element of a sustainable society is that it must have sustainable institutions of problem solving. These will be institutions that give stable or increasing returns (e.g., by minimizing costs) or diminishing returns that can be subsidized by energy supplies of assured quality, quantity, and cost. Ancient societies, constrained by photosynthesis, lacked the energy resources to pay for ever-increasing complexity. The problem persists among today's complex societies, which can pay for complex problem solving only as long as energy is abundant and inexpensive.

The role of problem solving is especially clear in the collapse of the Roman Empire. Confronted in the third century AD with severe challenges to its existence, the empire responded by greatly expanding the complexity and size of its primary problem-solving institutions: the army and government. This was costly and had to be paid for by greatly increasing taxation. The costs were so high that the empire survived by consuming its capital resources: producing lands and peasant population. This was an unsustainable strategy and collapse was inevitable. All that the empire could accomplish was to postpone it. Collapse resulted from the economic consequences of diminishing returns to complexity in problem solving. Ironically, though, had the empire not taken the steps it did, it would have collapsed even sooner.

Problem solving is the element missing in earlier collapse studies that focused unsatisfactorily on energy alone. A prolonged period of diminishing returns to complexity in problem solving renders a society fiscally weak (Figs. 4 and 5) and causes disaffection of the support population. Both factors render a society more vulnerable to stresses than would otherwise be the case. Thus, a fluctuation in energy flow that a fiscally prosperous society would survive can cause the collapse of a society weakened by the cost of great complexity in problem solving. This helps to clarify why changes in climate can have different effects in different circumstances. A society weakened by the costs of problem solving may be vulnerable to a change in climate that in other circumstances may be survivable.

Problem solving links the experiences of the past to understanding our own circumstances. The economics of problem solving, from less to more complex and from increasing to diminishing returns, constrain all institutions, including those of today. An institution that experiences a prolonged period of diminishing returns to complexity in problem solving will find itself in fiscal distress and ineffective at solving problems. If such an institution is part of a larger system that provides its funding (such as a government agency funded by appropriations), it may be terminated or require ever-larger subsidies. If an entire society finds itself in this position, as ancient societies sometimes did, the well-being of its populace will decline and it may be unable to surmount a major challenge. Either weakness can make a society vulnerable to collapse.

How might this dilemma make itself felt today? As an example, consider the relationship of global climate change to complexity in problem solving. Nothing before in history equals contemporary global change in magnitude, speed, and numbers of people affected. Our response will involve changes in the complexity of technology, research, and government. We need much research to comprehend climate, the economic dislocations that it will entail, and the new technologies that we will need. We will experience increasing government centralization and regulation. In research, technology, and regulation, and in resettlement of people living in low-lying areas, global change will require societal expenditures that are currently beyond calculation. As discussed previously, past societies were made vulnerable to climate change when such a perturbation came after a period of diminishing returns to investment in complexity. If this is so, it would be wise for us to understand the historical position of our own problem-solving efforts. This is not to suggest that climate change will cause today's complex societies to collapse. What it may do is combine with other challenges to increase the complexity and cost of our problem-solving efforts. Barring dramatic developments that increase the availability and decrease the cost of energy, such developments would reduce fiscal resiliency and living standards and make it more difficult to resolve our problems.

The factors that lead to collapse or sustainability develop over decades, generations, or centuries. We need historical research to understand our position in a problem-solving trajectory. In medicine, for example, we have long been in a phase of diminishing returns in which ever-greater expenditures yield smaller and smaller increments to life expectancy. Once we understand this, it is clear that the costs of health care cannot be contained as long as we demand health and long lives, which we will certainly continue to do. Conversely, our investments in microelectronics appear to be in a phase of increasing returns. This happy state, of course, cannot last forever. A primary element to understanding the sustainability of a society, an institution, or a sector of investment is to understand long-term changes in the complexity and costliness of problem solving.

Our citizens and policy makers have such confidence in science that a problem such as global climate change can generate complacency. We assume that a technological solution will emerge. Few of us are aware that problem solving can reach the point of diminishing returns and become ineffective. One of our great advantages over societies of the past is that it is possible for us to learn where we are in the long-term historical patterns that lead to sustainability or collapse. The historical sciences are an essential aspect of being a sustainable society.

Further Reading

Adams R McC 1981 *Heartland of Cities*. Aldine, Chicago

Allen T F H, Tainter J A, Hoekstra T W 2003 *Supply-Side Sustainability*. Columbia Univ. Press, New York

Bell B 1971 The dark ages in ancient history: 1. The first dark age in Egypt. *Am. J. Archaeol.* **75**, 1–26

Culbert T P (ed.) 1973 *The Classic Maya Collapse*. Univ of New Mexico Press, Albuquerque

Jones A H M 1964 *The Later Roman Empire, 284–602: A Social, Economic and Administrative Survey*. Univ. of Oklahoma Press, Norman

Renfrew C 1979 Systems collapse as social transformation: Catastrophe and anastrophe in early state societies. In: Renfrew C, Cooke K L (eds.) *Transformations: Mathematical Approaches to Culture Change*. Academic Press, New York, pp. 481–506

Tainter J A 1988 *The Collapse of Complex Societies*. Cambridge Univ Press, Cambridge, UK

Tainter J A 1999 Post-collapse societies. In: Barker G (ed.) *Companion Encyclopedia of Archaeology*. Routledge, London, pp. 988–1039

Tainter J A 2000 Problem solving: Complexity, history, sustainability. *Population Environ.* **22**, 3–41

Yoffee N, Cowgill G L (eds.) 1988 *The Collapse of Ancient States and Civilizations*. Univ of Arizona Press, Tucson

Joseph A. Tainter

Solar Energy, History of

Glossary

solar architecture The conception, design, and construction of buildings and communities so as to utilize incoming solar radiation.

solar cell A semiconductor material often made of specially treated silicon that converts sunlight directly into electricity.

solar greenhouses Structures used to cultivate plants that are designed and built to utilize solar energy for both heating and lighting.

solar thermal power Systems that capture energy from solar radiation, transform it into heat, and then use it for a variety of applications.

The ancient Greeks began using solar energy 2500 years ago. Since then, people have used the Sun for their benefit.

1. Solar Architecture and Greenhouses

1.1 Ancient Greece

During the fifth-century BC, the Greeks faced severe fuel shortages. Fortunately, an alternative source of energy was available—the sun (Fig. 1). Archaeological evidence shows that a standard house plan evolved during the fifth century so that every house, whether rural or urban, could make maximum use of the sun's warm rays during winter (Fig. 2). Those living in ancient Greece confirm what archaeologists have found. Aristotle noted that builders made sure to shelter the north side of the house to keep out the cold winter winds. Socrates, who lived in a solar-heated house, observed, "In houses that look toward the south, the sun penetrates the portico in winter," which keeps the house heated. The great playwright Aeschylus went so far as to assert that only primitives and barbarians "lacked knowledge of houses turned to face the winter sun, dwelling beneath the ground like swarming ants in sunless caves." The ancient Greeks planned whole cities in Greece and Asia Minor, such as Colophon, Olynthus, and Priene, to allow every homeowner access to sunlight during winter to warm their homes. By designing the streets in a checkerboard pattern running east–west and north–south, every home faced south, permitting the winter sun to flow into all houses.

1.2 Ancient Rome

Fuel consumption in ancient Rome was even more profligate than in classical Greece. In architecture, the Romans remedied the problem in the same fashion as did the Greeks. Vitruvius, the preeminent Roman architectural writer of the first century BC, advised builders in the Italian peninsula, "Buildings should be thoroughly shut in rather than exposed toward the north, and the main portion should face the warmer south side." Varro, a contemporary of Vitruvius, verified that most houses of at least the Roman upper class followed Vitruvius's advice, stating, "What men of our day aim at is to have their winter rooms face the falling sun [southwest]." The Romans improved on Greek solar architecture by covering south-facing windows with clear materials such as mica or glass.

Such clear coverings let in the light. When the light hit the masonry it converted into heat. The longer heat waves could not easily escape through the glass and the area inside heated to higher temperatures than those of previously unglazed structures. The Romans called these glass-covered structures *heliocamini*, which means "sun furnaces." The Romans used the term to describe their south-facing rooms because they became much hotter in winter than similarly oriented Greek homes, which lacked such coverings.

1.3 Ancient Roman and Renaissance Solar Greenhouses

The ancient Romans not only used window coverings to hold in solar heat for their homes but also relied on such solar heat traps for horticulture. Plants would mature quicker and produce fruits and vegetables out of season and exotic plants from hotter climates could be cultivated at home. With the fall of the Roman Empire, so too came the collapse of glass for either buildings or greenhouses. Only with the revival of trade during the 16th century was there renewed interest in growing in solar-heated greenhouses exotics brought back from the newly discovered lands of the East Indies and the Americas. Trade also created expendable incomes that allowed the freedom to take up such genteel pursuits as horticulture once more. Nineteenth-century architects such as Humphrey Repton brought the sunlit ambiance of the greenhouse into the home by attaching it onto the south side of a living room or library. On sunny winter days, the doors separating the greenhouse and the house were opened to allow the moist, sun-warmed air to freely circulate in the otherwise gloomy, chilly rooms. At night, the doors were shut to keep in as much solar heat as possible. By the late 1800s, the country gentry had become so enamored of attached greenhouses that they became an important architectural fixture of rural estates.

1.4 Indigenous America

American solar architecture began with its indigenous heritage. Acoma, built by the Pueblo Indians in

265

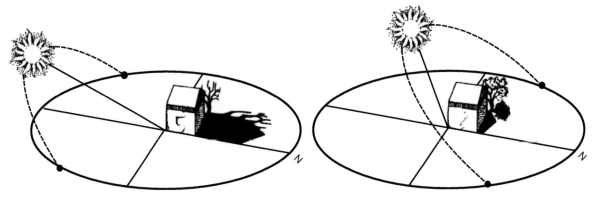

Figure 1
In the Northern Hemisphere, the low winter sun rises relative to Earth in the southeast, by noon it is in the south, and it sets in the southwest. During summer, the sun rises in the northeast, by noon it is directly south and very high in the sky, and it sets in the northwest. From Perlin/Butti Solar Archives.

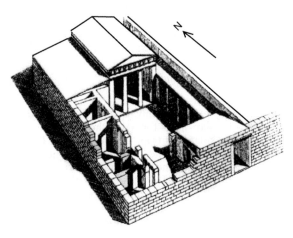

Figure 2
The Greeks took advantage of their knowledge of the location of the sun during the year to heat and cool their buildings. They opened their buildings to the south and protected them with a covered porch. The strategy allowed winter sunlight to pour into the house, whereas the roof shaded the house from the high summer sun. From Perlin/Butti Solar Archives.

the 12th century and continuously inhabited since then, serves as an excellent example of their sensitivity to building with the climate in mind. The "sky city," built atop a plateau, consists of three long rows of dwelling units running east to west. Each dwelling unit has two or three tiers stacked one behind the other to allow each one full exposure to the winter sun. Most doors and windows open to the south to catch the warm solar rays of winter. The walls are built of adobe. The sun strikes these heat-absorbing south walls more directly in winter than during

summer. The sun's heat travels through the adobe slow enough to reach the interior as night falls, heating the house through the night. Insulating the ceiling with straw keeps the horizontal roof, vulnerable to the rays of the high summer sun, from allowing too much heat to enter the house.

1.5 Early Euro-Americans

Spaniards who settled in the American southwest often built to take advantage of the winter sun: They aligned their homes east to west so the main portion of the house faced south. Shutters outside the windows were closed at night to help keep the solar heat that had flowed in all day from escaping during wintertime.

Settlers in New England considered the climate when they built their homes. They often chose "saltbox" houses that faced toward the winter sun and away from the cold winds of winter. These structures had two south-facing windowed stories in front where most of the rooms were placed and only one story at the rear of the building. The long roof sloped steeply down from the high front to the lower back side, providing protection from the winter winds. Many saltbox houses had a lattice overhang protruding from the south facade above the doors and windows. Deciduous vines growing over the overhang afforded shade in summer but dropped their leaves in winter, allowing sunlight to pass through and penetrate the house.

1.6 Post-World War I Europe and America

Ameliorating the terrible slums that blighted European cities during the industrial revolution sparked a renewed interest in building with the sun. As one

English city planner bent on housing reform urged in the first decade of the 20th century, "Every house should have its face turned to the sun, whence comes light, sweetness, and health." Young architects on the Continent agreed, declaring that their profession should embrace functionalism over anesthetics. To many, functionality meant designing houses that satisfied the basic needs of those living in them. In the relatively chilly European climate, this dictum meant providing dwellers with a warm haven. Therefore, Hannes Meyer, the leader of functional architecture in Germany, argued that an architect wishing to build a useful house must consider its body "to be a storage cell for the heat of the sun." Hence, "maximal use of the sun," as far as Meyer was concerned, was "the best plan for construction." Meyer, who became the director of the highly famous and influential Bauhaus architectural school, told fellow architects that to achieve optimum sunshine in buildings, they must conduct or at least be privy to scientific research into the sun's year-round movement relative to the earth. Germany led a renaissance in solar building during the 1930s that spread throughout Europe only to be stamped out by the Nazis, who condemned functional architecture as Jewish. When they came to power, a good number of German architects who designed solar buildings fled, many to the United States. George Fred Keck, a Chicago architect, befriended some of these expatriates and through their influence began designing homes in the Chicago area according solar building principles: expansive south-facing glass to trap the winter sun; long overhangs to shade the house in summer; minimal east–west exposure to prevent overheating in summer and fall; and the placement of secondary rooms, garages, and storage corridors on the north side to help insulate the living quarters from the cold north winds. Keck had a knack for publicity and called the houses he designed "solar homes." By the mid-1940s, Keck's work caught the attention of the national media. *House Beautiful, Reader's Digest*, and *Ladies Home Journal* featured his work. Fuel rationing during the war inclined the American public toward valuing the energy-saving features of solar homes. When war ended, the building market exploded. With the wartime-conservation ethic still imbued in most people's minds, many manufacturers in the prefabricated home industry adopted solar design features for leverage in this highly competitive market.

1.7 Post-World War II America

Studies of houses with large amounts of glass on the south side demonstrated that they experience much greater temperature swings than ordinary homes, warming quickly during the day due to accumulating the sun's heat and cooling off rapidly at night due to large heat losses through the windows. Tucson Architect Arthur Brown eliminated the unwanted fluctuations in temperature by running a thick black wall between the sun porch next to the south-facing glass and the living quarters. During sunny winter days, sunlight struck the black wall and the concrete absorbed the solar heat. Estimating that heat moves through the wall at 1 in. per hour, Brown made the wall 8 in. thick. As evening approached, the solar heat soaked up by the wall began to radiate into the rooms and continued to do so during the night. By morning, all the solar heat in the wall had dissipated into the living areas, leaving the wall totally cooled down and ready for another heating cycle.

Brown also designed the world's first solar-heated public building in 1948. Since students attended classes between 9 AM and 3 PM, Brown did not have to worry about storing solar energy since no one needed heat at night or in the very early morning. To keep costs down, he used the roofing supports to separate the space between the ceiling and the roof into channels. In this way, the roof served as the solar collector. The school's fan, which had forced gas-heated air through the building, now pushed air through the channels. The sun's energy heated the air and a second fan distributed that heat into the classrooms when needed. The solar system provided the school with 86% of its heat. However, in 1958, when the school district decided to expand the campus, the authorities chose a gas-fired furnace due to cheap energy costs. Cheap energy running highly reliable and easy to use heating and cooling systems led to an almost universal disinterest in solar design. It was only in the 1970s, with the Arab oil embargo and skyrocketing fuel prices, that people became interested in solar architecture again.

2. Solar Thermal Power

2.1 Ancient Chinese, Greeks, and Romans

Archaeological finds and ancient literature show that solar concave reflectors and convex lenses have a long history. Convex lenses predate concave reflectors by more than 1000 years. People crafted the first convex lenses out of rock crystal. Ancient Chinese and Roman doctors relied on convex lenses to cauterize wounds. The first mention of solar concave reflectors appeared in the third-century BC writings of Euclid. The ancient Chinese and Greeks used solar concave reflectors to light fires for domestic cooking.

2.2 Medieval and Renaissance Europe

Latin translation of Arab writer Al-Hazan introduced Europeans to ancient Greek works on solar concave reflectors and convex lenses. Between the

13th and 16th centuries, many Europeans suggested burning whole armies with huge solar reflectors, but no one could build reflectors or lenses large enough for martial purposes. However, Europeans did use reflectors to ignite fires for cooking meals, to distill perfumes, and to weld metal statuary.

2.3 19th- and Early 20th-Century America, Europe, and Africa

Many in 19th-century Europe worried that the prodigious consumption of coal by industry would eventually lead to coal shortages. Among the concerned, Augustine Mouchot thought that the sun's heat could replace coal to run Europe's many steam engines. Mouchot's work with mirror technology led him to develop the world's first sun motor in the 1870s, which produced sufficient steam to drive machinery (Fig. 3). One journalist described it as a "mammoth lamp shade, with its concavity directed skyward." After 1 year of testing Mouchot's sun machine, the French government concluded,

In France, as well as in other temperate regions, the amount of solar radiation is too weak for us to hope to apply it for industrial purpose. Though the results obtained in our temperate and variable climate are not very encouraging, they could be much more so in dry and hot regions where the difficulty of obtaining other fuel adds to the value of solar technologies. [Hence] in certain cases, these solar apparatuses could be called upon to provide useful work.

An American inventor, Frank Shuman, took his cue from the French report. In 1911, he chose to locate his commercial solar motor in Egypt, where land and labor were cheap, sun was plentiful, and coal was costly. To increase the amount of heat generated by the solar motor, glass-covered pipes were cradled at the focus of a low-lying trough-like reflector. A field of five rows of reflectors was laid out in the Egyptian desert. The solar plant also boasted a storage system that collected excess sun-warmed water in a large insulated tank for use at night and during inclement weather, in contrast to the solar motor designed by Mouchout, which went "into operation only when the sun [came] out from behind a cloud and [went] out of action the instant the sun disappears again." For industrialists, Shuman's solution eliminated a major obstacle to solar's appeal. Shuman's solar plant proved in Egypt to be

Figure 3
A solar-powered machine built by Mouchot's protégé, Pifre, runs a printing press. From Perlin/Butti Solar Archives.

more economical than a coal-fired plant. Engineers recognized that this plant demonstrated that "solar power was quite within the range of practical matters." However, the hopes and dreams of solar engineering disintegrated with the outbreak of World War I. The staff of the Egyptian plant had to leave for war-related work in their respective homelands. Shuman, the driving force behind large-scale solar developments, died before the war ended. After the war, the world turned to oil to replace coal. Oil and gas reserves were found in sunny and previously sun-short places that had been targeted as prime locations for solar plats. With oil and gas selling at near-giveaway prices, scientists, government officials, and business people became complacent about the world's energy situation and interest in sun power came to an abrupt end.

3. Solar Hot Water Heating

3.1 Solar Hot Box

Horace de Saussure, a noted Swiss naturalist, observed in the 1760s, "It is a known fact, and a fact that has probably been known for a long time, that a room, a carriage, or any other place is hotter when the rays of the sun pass through glass." To determine the effectiveness of trapping heat with glass covers, de Saussure built a rectangular box out of half-inch pine, insulated the inside, covered the top with glass, and placed two smaller glass-covered boxes inside. When exposed to the sun, the bottom box heated to 228 °F (109 °C), or 16 °F (9 °C) above the boiling point of water. de Saussure was unsure how the sun heated the glass boxes. Today, we can better explain what happened. Sunshine penetrated the glass covers. The black inner lining absorbed the sunlight and converted it into heat. Although clear glass allows the rays of the sun to easily enter through it, it prevents heat from doing the same. As the glass trapped the solar heat in the box, it heated up. Its inventor realized that someday the hot box might have important practical applications because "it is quite small, inexpensive, and easy to make." Indeed, the hot box became the prototype for the solar collectors that have provided sun-heated water to millions since 1891.

3.2 The First Solar Water Heaters

In the 19th century, there was no easy way to heat water. People generally used a cook stove for this purpose. Wood had to be chopped or heavy hods of coal lifted, and then the fuel had to be kindled and the fire periodically stoked. In cities, the wealthier heated their water with gas manufactured from coal. However, the fuel did not burn clean and the heater had to be lit each time someone wanted to heat water. If someone forgot to extinguish the flame, the tank would blow up. To add to the problem of heating water, in many areas wood or coal or coal gas cost a lot and many times could not be easily obtained. To circumvent these problems, many handy farmers, prospectors, or other outdoorsmen devised a much safer, easier, and cheaper way to heat water: They placed a metal water tank painted black into sunlight to absorb as much solar energy as possible. These were the first solar water heaters on record. The downside was that even on clear, hot days it usually took from morning to early afternoon for the water to get hot. As soon as the sun went down, the tanks rapidly lost heat because they had no protection from the night air.

3.3 The First Commercial Solar Water Heater

The shortcomings of the bare tank solar water heaters came to the attention of Clarence Kemp, who sold the latest home heating equipment in Baltimore, Maryland. In 1891, Kemp patented a way to combine the old practice of exposing metal tanks to the sun with the scientific principle of the hot box, thereby increasing the tanks' capability to collect and retain solar heat. He called his new solar water heater the Climax, the world's first commercial solar water heater (Fig. 4). Kemp originally marketed his invention to eastern gentlemen whose wives had left with their maids to summer resorts, leaving their husbands to fend for themselves. The solar water heater, Kemp advertised, would simplify housekeeping duties for this class of men already burdened by their wives and domestic staffs' absence and unaccustomed to such work as lighting the gas furnace or stove to heat water. In California and other temperate states, having greater amounts of sunshine throughout the year but also higher fuel costs (compared to states such as Maryland) made it essential for residents to take their solar assets seriously and not waste them (Fig. 5). The Climax sold well in such areas. Sixteen hundred had been installed in homes throughout southern California by 1900, including those installed in Pasadena, where one-third of the households heated their water with the sun by 1897.

3.4 The Flat Plate Collector

From the turn of the century to 1911, more than a dozen inventors filed patents that improved on the Climax. However, none changed the fact that the heating unit and the storage unit were one and the same and both laid exposed to the weather and the cold night air. Hence, water heated by the sun the night before never stayed hot enough to do the wash the next morning or to heat the bath. In 1909, William J. Bailey patented a solar water heater that revolutionized the business. He separated the solar water heater into two parts: a heating element

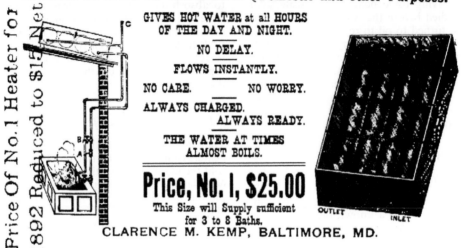

Figure 4
Advertisement for the world's first commercial solar water heater. From Perlin/Butti Solar Archives.

exposed to the sun and an insulated storage unit tucked away in the house so families could have sun-heated water day and night. The heating element consisted of pipes attached to a black-painted metal sheet placed in a glass-covered box. Because the water to be heated passed through narrow pipes rather than sitting in a large tank, Bailey reduced the volume of water exposed to the sun at any single moment and therefore the water heated faster. The heated water, which was lighter than the incoming cold water, naturally and immediately rose through the pipes to an insulated storage tank, where the water was kept warm for use both day and night. He also connected the storage tank to the home's furnace, guaranteeing hot water even after several rainy days. Providing hotter water for longer periods put Bailey's solar hot water heater, called the Day and Night (Fig. 6), at a great advantage over the competition. Soon, the Climax company went out of business. From 1909, when Bailey started his business, through 1918, his company sold more than 4000 Day and Night solar hot water heaters (Fig. 7).

3.5 The Demise of the Solar Water Heater in California and Its Success in Florida

The discovery of huge amounts of natural gas in the Los Angeles basin during the 1920s and 1930s killed the local solar water heater industry. Rather than lose money from the energy changes, Bailey applied the innovations he had made in solar to develop thermostatically controlled gas water heaters. His Day and Night gas water heater made him his second fortune. He also sold the patent rights of the Day and Night solar water heater to a Florida firm. A building boom in Florida during the 1920s had tripled the number of houses, but just as in California before the great oil strikes, people had to pay a high price to heat water. The high cost of energy combined with the tropical climate and the great growth in housing stock created a major business for those selling solar water heaters. By 1941, more than half the population heated water with the sun. Declining electric rates after World War II, in association with an aggressive campaign by Florida Power and Light to increase electrical consumption by offering electric water heaters at bargain prices, brought Florida's once flourishing solar water heater industry to a screeching halt.

3.6 Solar Water Heaters in Japan

Unlike in the United States during the post-World War II years, Japan lacked cheap and abundant energy to supply hot water on demand. Rice farmers in particular yearned for a hot bath after working

Figure 5
Installation of tanks in glass-covered box, Hollywood, California. From Perlin/Butti Solar Archives.

long hours in the hot, humid patties. To heat water, however, they had to burn rice straw, which they could have otherwise used to feed their cattle or fertilize the earth. A Japanese company began marketing a simple solar water heater consisting of a basin with its top covered by glass, and more than 100,000 were in use by the 1960s. People living in towns and cities bought either a plastic solar water heater that resembled an inflated air mattress with a clear plastic canopy or a more expensive but longer lasting model that resembled the old Climax solar water heaters—cylindrically shaped metal water tanks placed in a glass-covered box. Approximately 4 million of these solar water heaters had been sold by 1969. The advent of huge oil tankers in the 1960s allowed Japan access to new oil fields in the Middle East, supplying it with cheap, abundant fuel. As had happened in California and Florida, the solar water heater industry collapsed, but not for long. The oil embargo of 1973 and the subsequent dramatic increase in the price of petroleum revived the local solar water heater industry. Annual sales of more than 100,000 units continued to hold steady from 1973 until the second oil shock of 1979. Sales then increased to almost 500,000 that year and increased

further to nearly 1 million the following year. By this time, the Japanese favored solar water heaters that closely resembled the type introduced in California in 1909 by William J. Bailey with the heating and storage units separated. As the price of oil began to stabilize in 1985 and then decline sharply in subsequent years, so did the sales of solar water heaters; however, the Japanese still purchase approximately 250,000 each year. Today, more than 10 million Japanese households heat water using the sun.

3.7 Solar Water Heating in Australia

From the 1950s to the early 1970s, a few thousand Australians relied on the sun to heat water. The numbers grew as a consequence of the major spikes in oil prices in 1973 and 1979. Interestingly, the purchase of solar water heaters during these years varied from state to state. Whereas 40–50% of those living in Australia's Northern Territory heated water using the sun, approximately 15% in Western Australia did so and less than 5% did so in the more populated eastern states. The major difference had more to do with the cost of electricity than the

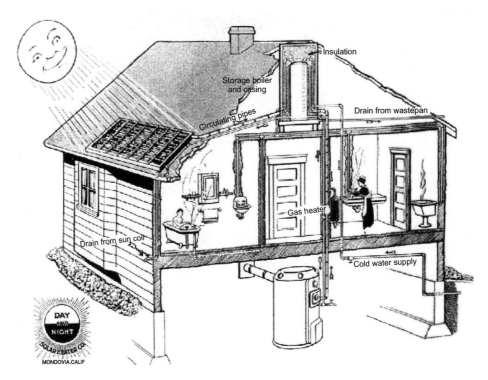

Labels on figure: Insulation; Storage boiler and casing; Drain from wastepan; Circulating pipes; Gas heater; Drain from sun coil; Cold water supply; DAY AND NIGHT SOLAR HEATER CO. MONROVIA.CALIF

Figure 6
Schematic of the first flat plate solar collector with solar heat collection and solar storage separated. From Perlin/Butti Solar Archives.

amount of sun available. People in the Northern Territory and Western Australia bought electricity generated by imported and increasingly costly petroleum, whereas for those in the eastern states of New South Wales, Queensland, and Victoria, electricity was produced by locally mined and very cheap coal. In the late 1980s, the Australian solar water heater market began to stagnate. Pipelines bringing newly discovered natural gas to previously fuel-short regions, such as the Northern Territory and Western Australia, eliminated growth in these once fertile markets for solar water heaters. Exports now account for more than 50% of the sales of Solahart, Australia's leading manufacturer of solar water heaters.

3.8   Solar Hot Water Heaters in Israel

Unlike the United States and much of Europe, Israel, like Japan, found itself without sufficient fuel supplies in the early 1950s. The power situation became so bleak in the early days of the Jewish state that the government had to forbid heating water between 10 PM and 6 AM. Despite mandatory domestic rationing of electricity, power shortages worsened,

causing pumping stations to fail and threatening factory closures. A special committee impaneled by the government could only suggest the purchase of more centralized generators to overcome the problem. This conclusion raised the ire of Israeli engineer Levi Yissar, who complained, "How about an already existing energy source which our country has plenty of—the sun. Surely we need to change from electrical energy to solar energy, at least to heat our water." Yissar put his money where his mouth was, becoming Israel's first manufacturer of solar water heaters. His design closely resembled the type introduced in California in the early 20th century with heating and storage separated. By 1967, approximately 1 in 20 households heated water with the sun. However, cheap oil from Iran in the late 1960s as well from oil fields captured during the Six Day War drastically reduced the price of electricity and the number of people purchasing solar water heaters. Israeli success in the Yom Kippur War resulted in the infamous oil boycott of 1973. The Israelis responded by mass purchasing of solar water heaters. By 1983, 60% of the population heated water using the sun. When the price of oil dropped in the mid-1980s, the Israeli government did not want people backsliding in their energy habits as happened throughout the

Figure 7
An early flat plate collector installation. From Perlin/Butti Solar Archives.

rest of the world. It therefore required its inhabitants to heat water using the sun. Today, more than 90% of Israeli households own solar water heaters.

3.9 Solar Water Heating for Swimming Pools

Solar swimming pool water heaters rank as the most successful but least heralded commercial solar application. The use of solar energy for pool heating and the equipment and needs of pool owners make a perfect match. The storage unit for the solar heated water already exists—the swimming pool. The pump needed to push water through the solar collectors must be purchased irrespective of the technology used to heat the water. The pool owner merely has to purchase the solar collectors. Since those using the pool generally want the temperature of the pool to be no higher than 80 °F (27 °C), the solar collectors do not require a costly glass cover or expensive metal sheeting and piping. Climax solar water heaters kept pools warm in the 1890s. Later, collectors resembling those designed by William J. Bailey were put to use for this application. In the 1970s, American Freeman Ford developed low-cost plastic to act as the solar collector. Exposed to the sun, water would pass through narrow ducts in the plastic and carry enough heat to warm the pool. The outdoor swimming season harmoniously coincides with the maximum output of the solar collectors. Even with other forms of

energy selling very cheaply, the pool owner who buys a solar unit will start to save money very quickly. In the United States alone, solar swimming pool heaters have produced the energy output equivalent of 10 nuclear power plants.

4. Photovoltaics

All the methods discussed in the previous sections change solar energy into heat. In contrast, photovoltaics produces electricity directly from sunlight.

4.1 The First Solid-State Solar Cell

The direct ancestor of solar cells currently in use got its start in the late 19th century with the construction of the world's first seamless communication network—transoceanic telegraph cables. While laying them under water to permit instantaneous communications between continents, engineers experimented with selenium for detecting flaws in the w ires as they were submerged. Those working with selenium discovered early on that the material's performance depended on the amount of sunlight falling on it. The influence of sunlight on selenium aroused the interest of scientists throughout Europe, including William Grylls Adams and his student Richard Evans Day. During one of their experiments with selenium, they witnessed something completely

273

new: Light could cause a solid material to generate electricity.

However, with the science of the day not yet sure if atoms were real, scientists could not explain why selenium produced electricity when exposed to light. Most scientists of the day therefore scoffed at Adams and Day's work. It took the discovery and study of electrons and the discovery and acceptance that light contains packets of energy called photons for the field of photovoltaics to gain acceptance in the scientific community. By the mid-1920s, scientists theorized that when light hits materials such as selenium, the more powerful photons pack enough energy to knock poorly bound electrons from their orbits. When wires are attached, the liberated electrons flow through them as electricity. Many envisioned the day when banks of selenium solar cells would power factories and light homes. However, no one could build selenium solar cells efficient enough to convert more than half a percent of the sun's energy into electricity—hardly sufficient to justify their use as a power source.

4.2 The Discovery of the Silicon Solar Cell

An accidental discovery by scientists at Bell Laboratories in 1953 revolutionized solar cell technology. Gerald Pearson and Calvin Fuller led the pioneering effort that took the silicon transistor, now the principal electronic component used in all electrical equipment, from theory to working device. Fuller had devised a way to control the introduction of impurities necessary to transform silicon from a poor to a superior conductor of electricity. He gave Pearson a piece of his intentionally contaminated silicon. Among the experiments Pearson performed using the specially treated silicon was exposing it to the sun while hooked to a device that measured electrical flow. To Pearson's surprise, he observed an electric output almost five times greater than that of the best selenium produced. He ran down the hall to tell his good friend, Daryl Chapin, who had been trying to improve selenium to provide small amounts of intermittent power for remote locations. He said, "Don't waste another moment on selenium!" and handed him the piece of silicon he had just tested. After a year of painstaking trial-and-error research and development, Bell Laboratories showed the public the first solar cells capable of producing useful amounts of power. Reporting the event on page 1, the *New York Times* stated that the work of Chapin, Fuller, and Pearson "may mark the beginning of a new era, leading eventually to the realization of one of mankind's most cherished dreams—the harnessing of the almost limitless energy of the sun for the uses of civilization."

Few inventions in the history of Bell Laboratories evoked as much media attention and public excitement as its unveiling of the silicon solar cell. Commercial success, however, failed to materialize due to the solar cells' prohibitive costs. Desperate to find marketable products run by solar cells, manufacturers used them to power novelty items such as toys and the newly developed transistor radio. With solar cells powering nothing but playthings, one of the inventors of the solar cell, Daryl Chapin, could not hide his disappointment, wondering, "What to do with our new baby?"

4.3 The First Practical Application of Silicon Solar Cells

Unbeknown to Chapin at the time, powerful backing of the silicon solar cell was developing at the Pentagon. In 1955, the U.S. government announced its intention to launch a satellite. The prototype had silicon solar cells for its power plant. Since power lines could not be strung out to space, satellites needed a reliable, long-lasting autonomous power source. Solar cells proved the perfect answer.

The launching of the Vanguard (Fig. 8), the first satellite equipped with solar cells, demonstrated their value. Preceding satellites, run by batteries, lost power after approximately 1 week, rendering millions

Figure 8
An engineer holds the Vanguard 1, the first satellite to use photovoltaics. From Perlin/Butti Solar Archives.

of dollars' worth of equipment useless. In contrast, the solar-powered Vanguard continued to communicate with Earth for many years, allowing the completion of many valuable experiments. The success of the Vanguard's solar power pack broke down the existing prejudice at the time toward the use of solar cells in space. As the space race between the United States and the Soviet Union intensified, both adversaries urgently needed solar cells. The demand opened a relatively large business for companies manufacturing them. More important, for the first time in the history of solar power, the sun's energy proved indispensable to society: Without the secure, reliable electricity provided by photovoltaics, the vast majority of space applications so vital to our everyday lives would never have been realized.

4.4 The First Terrestrial Applications

Although things were looking up for solar cells in space in the 1960s and early 1970s, their astronomical price kept them distant from Earth. In 1968, Elliot Berman decided to quit his job as an industrial chemist to develop inexpensive solar cells to bring the technology from space to Earth. Berman prophetically saw that with a major decrease in price photovoltaics could play a significant role in supplying electrical power to locations on Earth where it is difficult to run power lines. After 18 months of searching for venture capital, Exxon executives invited Berman to join their laboratory in late 1969. Dismissing other solar scientists' obsession with efficiency, Berman concentrated on lowering costs by using lower grade and therefore cheaper silicon and less expensive materials for packaging the modules. By decreasing the price from $200 to $20 per watt, solar cells could compete with power equipment needed to generate electricity distant from utility poles.

4.5 Terrestrial Applications

Oil companies were the first major customers of solar modules. They had both the need and the money. Oil rigs in the Gulf of Mexico were required to have warning lights and horns. Most relied on huge flashlight-type batteries to run them. The batteries required heavy maintenance and also had to be replaced approximately every 9 months. This required a large boat with an onboard crane or a helicopter. In contrast, a small skiff could transport the much lighter solar module and accompanying rechargeable battery, resulting in tremendous savings. By 1980, photovoltaics had become the standard power source for warning lights and horns not only on rigs in the Gulf of Mexico but also on those throughout the world.

Oil and gas companies also need small amounts of electricity to protect well casings and pipelines from corroding. Sending a current into the ground electrochemically destroys the offending molecules that cause this problem. However, many oil and gas fields both in the United States and in other areas of the world such as the Middle East and North Africa are far away from power lines but receive plenty of sunshine. In such cases, solar modules have proven the most cost-effective way of providing the needed current to keep pipes and casings corrosion free.

The cost of changing nonrechargeable batteries on the U.S. Coast Guard's buoys exceeded the buoys' original cost. Hence, Lloyd Lomer, at the time a lieutenant commander in the Guard, believed that switching to photovoltaics as a power source for buoys made economic sense. However, his superiors, insulated from competition, balked at the proposed change. Lomer continued his crusade, winning approval to install a test system in the most challenging of environments for sun-run devices—Ketchikan, Alaska. The success of the photovoltaic-powered buoy in Alaska proved Lomer's point. Still, his boss refused to budge. Going to higher authorities in government, Lomer eventually won approval to convert all buoys to photovoltaics. Almost every coast guard service in the world has followed suit.

To solar pioneer Elliot Berman, the railroads seemed to be another natural for photovoltaics. One of his salesmen convinced the Southern Railway (now Norfolk Southern) to try photovoltaics for powering a crossing signal at Rex, Georgia. These seemingly fragile cells did not impress veteran railroad workers as capable of powering much of anything. They therefore put in a utility-tied backup. However, a strange thing happened that turned quite a few heads. That winter, the lines went down on several occasions due to heavy ice buildup on the wires, and the only electricity for miles around came from the solar array. As one skeptic remarked, "Rex, Georgia taught the Southern that solar worked!"

With the experiment at Rex a success, the Southern decided to put photovoltaics to work for the track circuitry, the railroad's equivalent of air-traffic control, keeping trains at a reasonable distance from one another to prevent head-on collisions or back-enders. The presence of a train changes the rate of flow of electricity running through the track. A decoder translates that change to throw signals and switches up and down the track to ensure safe passage for all trains in the vicinity. At remote spots along the line, photovoltaics provided the needed electricity.

Other railroads followed the Southern's lead. In the old days, telegraph and then telephone poles ran parallel to most tracks. Messages sent through these wires kept stations abreast of matters paramount to the safe and smooth functioning of a rail line.

However, by the mid-1970s, wireless communications could do the same tasks. The poles became a liability and maintenance expense. Railroads started to dismantle poles. Whenever they found their track circuitry devices to be too far away from utility lines, they relied on photovoltaics.

The U.S. Army Signal Corps, which pioneered the use of solar power in space by equipping the solar pack on the Vanguard, was the first to bring photovoltaics down to Earth. In June 1960, the Corps sponsored the first transcontinental radio broadcast generated by the sun's energy to celebrate its 100th anniversary. The Corps, probably photovoltaics' strongest supporter in the late 1950s and the early 1960s, envisioned the solar broadcast would lead others to use photovoltaics to help provide power to run radio and telephone networks in remote locations.

Fourteen years later, GTE John Oades realized the Corps' dream by installing the first photovoltaic-run microwave repeater in the rugged mountains above Monument Valley, Utah. By minimizing the power needs of the microwave repeater, Oades could power the repeater with a small photovoltaic panel, saving all the expenses formerly attributed to them. His invention allowed people living in towns in the rugged American west, hemmed in by mountains, to enjoy the luxury of long-distance phone service that most other Americans took for granted. Prior to the solar-powered repeater, it was too expensive for phone companies to install cable or lines, forcing people to drive hours, sometimes through blizzards on windy roads, just to make long-distance calls.

Australia had an even more daunting challenge to bring modern telecommunication services to its rural customers. Although approximately the same size as the United States, only 22 million people lived in Australia in the early 1970s when its government mandated Telecom Australia to provide every citizen, no matter how remotely situated, with the same radio, telephone, and television service enjoyed by urban customers living in Sydney or Melbourne.

Telecom Australia tried, without success, traditional stand-alone power systems such as generators, wind machines, and nonrechargeable batteries to run autonomous telephone receivers and transmitters for its rural customers. Fortunately, by 1974, it had another option—relatively cheap solar cells manufactured by Elliot Berman's firm, Solar Power Corporation. The first photovoltaic-run telephone was installed in rural Victoria, after which hundreds of others were installed. By 1976, Telecom Australia judged photovoltaics as the preferred power source for remote telephones. In fact, the solar-powered telephones proved so successful that engineers at Telecom Australia believed they were ready to develop large photovoltaic-run telecommunication networks linking towns such as Devil's Marbles, Tea Tree, and Bullocky Bone to Australia's national telephone and television service. As a result of photovoltaics, people in these and neighboring towns could dial long distance direct instead of having to call the operator and shout into the phone to be understood. Nor did they have to wait for newsreels to be flown to their local station to view news, hours, if not days, old. There were a total of 70 solar-powered microwave repeater networks by the early 1980s, the longest spanning 1500 miles. The American and Australian successes showed the world that solar power worked and benefited hundreds of thousands of people. In fact, by 1985 a consensus in the telecommunications field found photovoltaics to be the power system of choice for remote communications.

A few years earlier, Mali, a country located south of the Sahara desert, along with its neighbors, suffered from drought so devastating that tens of thousands of people and livestock died. The Malian government knew it could not save its population without the help of people like Father Vespieren, a French priest who lived in Mali for decades and had successfully run several agricultural schools. The government asked Vespieren to form a private company to tap the vast aquifers that run underneath the Malian desert. Vespieren saw drilling as the easy part. His challenge was pumping. No power lines ran nearby and generators lay idle for lack of repairs or fuel. Then he heard about a water pump in Corsica that ran without moving parts, without fuel, and without a generating plant: It run simply on energy from the sun. Vespieren rushed to see the pioneering installation. When he saw the photovoltaic pump, he knew that only this technology could save the people of Mali. In the late 1970s, Vespieren dedicated Mali's first photovoltaic-powered water pump with the following words: "What joy, what hope we experience when we see that sun which once dried up our pools now replenishes them with water!" By 1980, Mali, one of the poorest countries in the world, had more photovoltaic water pumps per capita than any other country due to Vespieren's efforts. The priest demonstrated that success required the best equipment, a highly skilled and well-equipped maintenance service, and financial participation by consumers. Investing their own money helped pay the bill for the capital equipment and ongoing maintenance but, more important, the people came to regard the panels and pumps as valued items that they would help care for. Most successful photovoltaic water pump projects in the developing world have followed Vesperien's example. When Vespieren initiated his photovoltaic water-pumping program, less than 10 photovoltaic pumps existed throughout the world. Today, tens of thousands of photovoltaic pumps provide water to people, livestock, and crops.

Despite the success of photovoltaic applications throughout the world in the 1970s and early 1980s, institutions responsible for rural electrification programs in developing countries did not consider installing solar electric panels to power villages distant from urban areas. Working from offices either in the West or in large cities in poorer countries, they only considered central power stations run by nuclear energy, oil, or coal. Not having lived "out in the bush," these "experts" did not realize the huge investment required to string wires from power plants to the multitudes residing in small villages tens of miles away. As a consequence, only large populated areas received electricity, leaving billions in the countryside without. The disparity in energy distribution helped cause the great migration to the cities in Africa, Asia, and Latin America. As a consequence, megalopolises such as Mexico City, Lagos, and Bombay arose with all the ensuing problems, such as crime, diseases like AIDS, pollution, and poverty. The vast majority still living in the countryside do not have electricity. To have some modicum of lighting, they have had to rely on ad hoc solutions such as kerosene lamps. Those buying radios, tape cassettes, and televisions also must purchase batteries. It has become apparent that mimicking the Western approach to electrification has not worked in the developing world. Instead of trying to install wires and poles, which no utility in these regions can afford, bringing a 20- to 30-year supply of electricity contained in solar modules to consumers by animal or by vehicle makes more sense. Whereas it requires years to construct a centralized power plant and even more time for power lines to be installed to deliver electricity, it takes less than 1 day to install an individual photovoltaic system.

Ironically, the French Atomic Energy Commission pioneered electrifying remote homes in the outlying Tahitian Islands with photovoltaics. The Atomic Energy Commission believed that electrifying Tahiti would mollify the ill feeling in the region created by its testing of nuclear bombs in the South Pacific. The commission considered all stand-alone possibilities, including generators, wind machines, and biogas from the husks of coconut shells, before choosing photovoltaics. In 1983, 20% of the world's production of solar cells went to French Polynesia. By 1987, half of all the homes on these islands received their electricity from the sun.

Rural residents in Kenya believe that if they have to wait for electricity from the national utility, they will be old and gray by the time they receive it. To have electricity now, tens of thousands of Kenyans have bought photovoltaic units. Though relatively small compared to what an American would purchase, the typical modules in Kenya range from 12 to 25 W and are capable of charging a battery with enough electricity to run three low-wattage lights and a television for 3 hours after dark. Electric lighting provided by photovoltaics eliminates the noxious fumes and the threat of fire that people had to contend with when lighting their homes with kerosene lamps. Due to photovoltaics, children can study free from eye strain and tearing and from developing hacking coughs that they formerly experienced when studying by kerosene lamp. Their performance in school has soared with better lighting. Electric lighting also allows women to sew and weave at night products that generate cash. Newly gained economic power gives women more say in matters such as contraception. In fact, solar electricity has proven more effective in lowering fertility rates than have birth control campaigns.

Electricity provided by photovoltaics allows those living in rural areas to enjoy the amenities of urban areas without leaving their traditional homes for the cities. The government of Mongolia, for example, believed that to improve the lot of its nomadic citizens would require herding them into communities so the government could connect them to electricity generated by centralized power. The nomads, however, balked. Photovoltaics allowed the nomadic Mongolians to take part in the government's rural electrification program without giving up their traditional lifestyle. Whenever they move, the solar panels move too, with yurt and yaks and whatever else they value.

With the growing popularity of photovoltaics in the developing world, solar thievery has become common. Perhaps nothing better demonstrates the high value that people living in these countries place on photovoltaics than the drastic increase in solar module thefts during the past few years. Before the market for photovoltaics took off in the developing world, farmers would build an adobe fence around their photovoltaic panels to keep the livestock out. Now, razor wire curls around the top of these enclosures to bar solar outlaws.

As the price of solar cells has continued to decline, devices run by solar cells have seeped into the suburban and urban landscapes of the developed world. For example, when construction crews have to excavate to place underground electrical transmission lines, installing photovoltaics instead makes more economic sense. Economics has guided highway departments throughout the world to use solar cells for emergency call boxes along roads. In California alone, there are more than 30,000. The savings have been immediate. Anaheim, California, would have had to spend approximately $11 million to connect its 11,000 call boxes to the power grid. Choosing photovoltaics instead, the city paid $4.5 million. The city of Las Vegas found it cheaper to install photovoltaics to illuminate its new bus shelters than to have the adjacent street and sidewalk dug up to place new wires.

In the mid-1970s and early 1980s, when governments of developing countries began to fund photovoltaic projects, they copied the way electricity had been traditionally produced and delivered, favoring the construction of large fields of photovoltaic panels far away from where the electricity would be used and delivering the electricity by building miles of transmission lines. Others, however, questioned this approach. When Charles Fritts built the first selenium photovoltaic module in the 1880s and boldly predicted that it would soon compete with Thomas Edison's generators, he envisioned "each building to have its own power plant." Since the late 1980s, most in the photovoltaic business have come around to Fritts's view. Instead of having to buy a huge amount of land—unthinkable in densely populated Europe and Japan—to place a field of panels, many began to ask why not turn each building into its own power station. Using photovoltaics as building material allows it to double as windows, roofing, skylights, facades, or any type of covering needed on a home or building, and it makes sense since the owner gets both building material and electrical generator in one package. Having the electrical production sited where the electricity will be used offers many advantages: Producing electricity on site eliminates the electricity lost in long-distance transmission, which ranges from approximately 30% in the best maintained lines to approximately 65% in rundown lines prevalent in the developing world; it reduces the danger of power lines being overloaded as demand increases by rerouting additional electricity directly to the users; and occasionally, such as on a hot August afternoon, when peak electrical demand empties the wires, excess electricity produced by photovoltaic-equipped homes and buildings can refill the lines with much needed electricity to prevent brownouts and blackouts.

4.6 The Future of Photovoltaics

Although the photovoltaics industry has experienced a phenomenal 20% annual growth rate during the past decade, it has just started to realize its potential. Although more than 1 million people in the developing world get their electricity from solar cells, more than 1 billion still have no electrical service. The continuing revolution in telecommunications is bringing a greater emphasis on the use of photovoltaics. As with electrical service, the expense of stringing telephone wires keeps most of the developing world without communication services that people in more developed countries take for granted. Photovoltaic-run satellites and cellular sites, or a combination of the two, offer the only hope to bridge the digital divide. Photovoltaics could allow everyone the freedom to dial up at or near home and, of course, hook up to the Internet.

Opportunities for photovoltaics in the developed world also continue to increase. In the United States and Western Europe, hundreds of thousands of permanent or vacation homes are too distant for utility electric service. If people live more that 250 yards from a utility pole, paying the utility to string wires to their houses cost more than supplying their power needs with photovoltaics. Fourteen thousand Swiss Alpine chalets receive electricity from solar electricity, as do tens of thousands of others from Finland to Spain to Colorado.

Many campgrounds now prohibit RV owners from running their engines to power generators for appliances inside. The exhaust gases pollute, and the noise irritates other campers, especially at night. Photovoltaic panels mounted on the roofs of RVs provide the electricity needed without bothering others.

Restricting carbon dioxide emissions to help moderate global warming could result in the flow of large amounts of money from projects in which fossil fuels are burned to photovoltaic projects. The damage wrought by the 1997–1998 El Nino is an indication of the harsher weather expected as the earth warms. The anticipated increase in natural disasters brought on by a more disastrous future climate, as well as the growing number of people living in catastrophic-prone regions, makes early warning systems essential. The ultimate early warning device may consist of pilotless photovoltaic-powered weather surveillance airplanes, of which the *Helios* serves as a prototype. The *Helios* has flown higher than any other aircraft. Solar cells make up the entire top of the aircraft, which is merely a flying wing. Successors to the *Helios* will also have fuel cells on the underside of the wing. They will get their power from the photovoltaic panels throughout the day, extracting hydrogen and oxygen from the water discharged by the fuel cells the night before. When the sun sets, the hydrogen and oxygen will power the fuel cells, generating enough electricity at night to run the aircraft. Water discharged in the process will allow the diurnal cycle to begin the next morning. The tandem use of solar cells and fuel cells will allow the aircraft to stay aloft forever, far above the turbulence, watching for and tracking hurricanes and other potentially dangerous weather and other natural catastrophes.

Revolutionary lighting elements, called light-emitting diodes (LEDs), produce the same quality of illumination as their predecessor with only a fraction of the energy. LEDs therefore significantly reduce the amount of panels and batteries necessary for running lights, making photovoltaic systems less costly and less cumbersome. They have enabled photovoltaics to replace gasoline generators in mobile warning signs used on roadways that alert motorists of lane closures and other temporary problems that drivers should know about. The eventual replacement of household lighting by LEDs will do the same for photovoltaics in homes.

When Bell Laboratories first unveiled the silicon solar cell, their publicist made a bold prediction: "The ability of transistors to operate on very low power gives solar cells great potential and it seems inevitable that the two Bell inventions will be closely linked in many important future developments that will influence the art of living." Already, the tandem use of transistors and solar cells for running satellites, navigation aids, microwave repeaters, televisions, radios, and cassette players in the developing world and a myriad of other devices has fulfilled the Bell prediction. It takes no great leap of imagination to expect the transistor/solar cell revolution to continue until it encompasses every electrical need from space to earth.

Further Reading

Beattie D A (ed.) 1997 *History and Overview of Solar Heat Technologies*. MIT Press, Cambridge, MA

Butti K, Perlin J 1980 *Golden Thread: Twenty-five Hundred Years of Solar Architecture and Technology*. Van Nostrand, New York

Perlin J 2002 *From Space to Earth: The Story of Solar Electricity*. Harvard University Press, Cambridge, Massachusetts

John Perlin

T

Thermodynamic Sciences, History of

Glossary

Boltzmann, Ludwig (1844–1906) Austrian physicist who sought (initially) a purely mechanical reduction of the second law but later argued it was only a statistical truth.

Carnot, Nicolas Léonard Sadi (1796–1832) French military engineer, author of the 1824 "Reflections," which posed the problems (and articulated the core reasoning) that later generated the discipline of thermodynamics.

Clapyeron, Benoit-Pierre Émile (1799–1864) French engineer who published an account of Carnot's theory in 1834.

Clausius, Julius Rudolf Emmanuel (1822–1888) German physicist who created the discipline of thermodynamics by reconciling the core of Carnot's memoir with the energy conservation of Joule, Helmholtz, Mayer, et al.

Helmholtz, Hermann von (1821–1894) German physicist/physiologist. One of the main discoverers of energy conservation, notably in his 1847 memoir, "On the Conservation of Force."

Joule, James Prescott (1818–1889) British physicist. One of the main discoverers of energy conservation. He gave it much experimental support in the 1840s.

Maxwell, James Clerk (1831–1879) British physicist. One of several to argue that the second law was only a statistical truth.

Mayer, Julius Robert (1814–1878) German physician. One of the main discoverers of energy conservation.

Rankine, William James Macquorn (1820–1872) Scottish engineer who developed a strange kinetic thermodynamics around 1850. He introduced entropy into thermodynamics but denied dissipation. He made the subject accessible to engineers.

Regnault, Henri Victor (1810–1878) French physicist who, in the mid-19th century, carried out a detailed and authoritative study of the thermal properties of steam and other gases.

Thomson, James (1822–1892) British engineer, brother of William.

Thomson, William, later Baron Kelvin of Largs (1824–1907) British physicist/engineer who took up the Carnot theory in the late 1840s. Later introduced Carnot's dissipation into thermodynamics.

The fact that thermal energy cannot be fully transformed into mechanical work has far-reaching consequences. Thermodynamics is the science that develops these consequences. It emerged in the approximately 30 years that followed Sadi Carnot's amazingly fertile 1824 study of the efficiency of steam engines when the key ideas in that study were reconciled (in the early 1850s) with the principle of energy conservation (as consolidated in the late 1840s). This article is a survey of the thermodynamical thinking of those three decades.

1. The Carnot Memoir of 1824

1.1 The Puzzle of Steam Engine Efficiency

In England, around the end of the 18th century there was an extraordinary increase in the efficiency with which coal could be converted into useful work. This was made possible by a series of improvements, both major and minor, in the design of steam engines, the most notable of which were those introduced by James Watt (the separate condenser, double action, and expansive working), together with a growing tendency to use increasingly higher steam pressures. Progress was so rapid—from 1775 to 1825 there was a 10-fold increase in efficiency—that the question came to be asked as to whether the motive power of heat is unbounded, whether the possible improvements in steam engines have an assignable limit—a limit that the nature of things will not allow to be passed by any means whatever—or "whether, on the contrary, these improvements may be carried on indefinitely".

1.2 Carnot's Key Insight: Cold Is Needed to Produce Work

The challenge of answering this question was taken up by a young French Army engineer, Sadi Carnot, who in 1824 published a masterly memoir on the subject, "Reflections on the Motive Power of Heat" In order to produce work by thermal means, Carnot asserted (recalling Watt's separate condenser), it is not sufficient to have a source of heat: It is also necessary to have a "source" of cold—that is, a supply of some substance at a low temperature that can be used to absorb the "exhaust" heat that (he proposed) every thermal engine necessarily produces. He theorized that if the temperature of these two sources is fixed, there is a maximum to the amount of work that may be derived by the extraction of unit quantity of heat from the hot source. This maximum, he indicated, is the amount of work done by any thermal engine working reversibly between the two temperatures in question. Furthermore, Carnot was able to calculate a few rough estimates of his maxima: He could not do more because he lacked the necessary empirical data.

1.3 Water Engines and Reversibility

The reasoning Carnot used to support his conclusions is particularly important in that it was the persuasiveness of his arguments that was later to ensure their survival in the face of experimental evidence that directly contradicted the theory of heat he had used. To appreciate this reasoning, it is helpful to consider the development of the theoretical understanding of water power, a close rival to steam in the 18th century. This century was one of rapid progress in the harnessing of the power of falling water, and some of the designs produced by engineers are astonishing in their sophistication. A good example of this is the so-called column-of-water engine, whose appearance and action are both very similar to those of a high-pressure steam engine and so unlike the crude water wheel we might unreflectingly imagine to be the typical product of the age.

Such an engine (and many other water engines) could be run backwards to act as a pump, and in the ideal situation in which all friction and similar imperfections were absent, the machine would produce exactly as much work in consequence of the descent of a unit of water from its supply reservoir to its exhaust as would be required to pump that same unit of water from the exhaust back to the supply with the engine running backwards. Such could be called a (perfectly) reversible water engine.

Furthermore, no water engine could be more efficient that a reversible one. If another engine were to be capable of producing more work than a reversible one, it could be coupled to a reversible engine, with the latter used as a pump, consuming only part of the work of the former while supplying its total fuel (i.e., water flows). Therefore, the combination would be producing work without making any demands on its environment and without suffering any net internal change. Such perpetual motion, it was concluded in

the 18th century, was impossible. So there is a clear limit to the efficiency of a water engine, and this limit depends on the height of the intake and exhaust reservoirs. Furthermore, the limit is the actual amount of work produced by a reversible engine acting over the same drop.

All these theories had been developed by the beginning of the 19th century, and they are important here because what Carnot did to solve his problem was to borrow (and enhance) the reasoning of the water engineers and reapply it to heat engines.

1.4 Heat Conservation and Caloric Theory

There is however one vital precondition that must be satisfied so that the analogy between heat and water engines is complete enough for the arguments evolved for water engines to retain their validity when applied to heat engines: No heat must be consumed during the production of work by heat engines. Just as a water engine works by lowering a fixed quantity of water from a high reservoir to a low one, so too must it be assumed that a heat engine operates by "lowering" an invariable quantity of heat from a region of high temperature to a region of low temperature. Under this assumption (now known to be false), one can argue for heat engines just as has already been done for water engines, and the analogy is so close that there is no point in repeating the argument here.

To Carnot, this crucial assumption of heat conservation was a natural one to make and he adopted it explicitly. It was inherent in the caloric theory, then the most widely accepted explanation of the nature of heat and especially favored in France. According to this theory, heat was (or behaved like) an indestructible fluid ("caloric"), whose flows in and out of bodies accounted for their various thermal properties. Therefore, by adopting the caloric theory, and by borrowing the beginnings of a line of reasoning from his predecessors in hydraulic engineering, Carnot was able to devise a remarkably convincing solution to the problem he had set himself—a solution that had some interesting consequences.

It enables one to impose, seemingly a priori, a variety of restrictions on the thermophysical properties of all substances and hence to deduce some of their physical characteristics from a knowledge of apparently quite different ones. The most startling early example of this was the deduction by James Thomson (using Carnot's reasoning approximately 25 years after Carnot had set it down) of the hitherto unsuspected fact that since water expands on freezing, the melting point of ice must decrease as the pressure exerted on it increases. Or else, the Carnot logic demands, a perpetual motion machine is possible. The power of such reasoning was to prove most attractive to some of Carnot's later followers. It eventually became a standard feature of thermodynamics.

1.5 Irreversibility and Dissipation

Furthermore, the Carnot theory implies that whenever heat descends from a region of high temperature to one of low temperature without being used to produce the maximum possible work—whenever heat descends irreversibly (e.g., as it does in simple heat conduction)—there takes place concomitantly some sort of destruction of work capability: The universe is less capable of doing work after the descent than it was before the descent. Irreversibility, in other words, implies permanent waste. Carnot did not stress this consequence of his analysis, but this key thermo-dynamic idea was taken up by his followers.

At first, followers were few and far between because Carnot's discussion was very abstract and had little in common with the physics and theoretical engineering of his contemporaries. The first person to make something substantial out of it was another French engineer, Clapeyron, who in 1834 published a summary of the original, rephrasing much of the analysis in more mathematical form (including the PV diagrams so common in thermodynamics today). Clapeyron did make one important conceptual addition, however. He supplied something missing from the Carnot memoir—a reversible design for an ideal steam engine. (Despite the context that motivated his study, Carnot had not been able to solve this small riddle.) In any case, through Clapeyron's revision, the theory came to the attention of the young British physicist-engineer William Thomson (later Lord Kelvin) and the German physicist Clausius, both of whom were strongly attracted to it.

2. Thomson's Early Enthusiasm for Carnot (Late 1840s)

Thomson saw that the Carnot theory could be used to define a temperature scale with the long sought-after property of being independent of the physical characteristics of any narrow class of substances, and he published such a definition of "absolute temperature" in 1847. The principle behind it is simple: Unit temperature interval is defined as that temperature interval through which the reversible descent of unit quantity of heat produces unit quantity of work. Carnot's logic shows that this depends on nothing more than the two temperatures, exactly as required. Furthermore, Thomson, in this and a later paper, systematically extended some of Carnot's rough calculations (using data recently obtained by Regnault) to establish the numerical connection between such an abstract scale and the practical scales in everyday use. (It turned out to be approximate

coincidence.) In doing so, Thomson was also able to improve Carnot's estimates so far as to tabulate systematically the quantity of work that unit quantity of heat could produce in descending reversibly from the higher to the lower of any pair of temperatures chosen from the range covered by his data, 0–230°C. A few years later, Thomson was able to verify his brother's prediction (mentioned previously) that the freezing point of water is depressed by pressure, a totally unexpected consequence of Carnot's theory.

It is not difficult to understand why Thomson found the analysis attractive. Apart from the inherent elegance of its reasoning, it (i) led to an effectively complete solution to the problem of evaluating the motive power of heat; (ii) provided a basis for absolute thermometry; and (iii) showed itself capable of discovering new phenomena, well beyond the imagination of its creator.

3. The Clash with Energy Conservation

Yet it was wrong, as Thomson was slowly and reluctantly forced to admit. His personal mentor was Joule, but Joule was only one of a small group of men (Mayer and Helmholtz were the other principals) who, in the 1840s, gradually persuaded the scientific world to give up the belief that heat was conserved and to replace that axiom with the principle of energy conservation, later known as the first law of thermodynamics. This new doctrine spread remarkably fast, and by 1850 it was widely accepted that heat and work were interconvertible. Thus, heat did not behave at all like the fluid caloric, and this fact utterly undermined Carnot's argumentation. Carnot had been suspicious of the caloric theory in the 1824 memoir but still managed to put it to good use. His posthumous papers show that he soon came to the conclusion that heat could be transformed into work and that he even managed to estimate the conversion factor.

Joule objected to the Carnot theory on two grounds and urged its abandonment. First, it regarded the essence of the process whereby work is generated thermally as descent of heat, whereas Joule saw it as a destruction. Second (and apparently more important for Joule and colleagues), it demanded the annihilation of motive power whenever an irreversible (thermal) process took place. To Joule, the universe was equipped with a fixed quantity of motive power, and no natural process could cause either an increase or a decrease in this total: The most that could happen was that it could alter its form.

Thomson was one of the first to appreciate the force of Joule's evidence, but he could not accept Joule's recommendation to abandon Carnot. Thomson knew that no matter how correct Joule seemed to be, there was still a lot of truth in Carnot's theory. The conflict between these two theories was a paradox for Thomson, and he looked to further experiment to resolve it.

4. Clausius Resolves Part of the Clash (1850)

In fact, the paradox was soluble without further experiment, as Clausius was soon to reveal. In May 1850, he published a major paper, "On the Motive Power of Heat, and on the Laws which Can Be Deduced from It for the Theory of Heat," in which he discovered the new science of thermodynamics by showing "that [Joule *et al.'s*] theorem of the equivalence of heat and work, and Carnot's theorem, are not mutually exclusive, but that, by a small modification of the latter, which does not affect its principal part, they can be brought into accordance."

What was Clausius's approach? Above all, he accepted unequivocally the new energy principle and attempted to salvage as much of the Carnot theory as possible. In fact, he found that he could salvage so much that what was lost might reasonably be described as inessential. In particular, he perceived that although the Carnot theory was correct in demanding a descent of heat when work was produced thermally, it was wrong in assuming nothing but a descent: Clausius saw that in reality both a descent and a destruction took place. Furthermore, Clausius saw that the axiom of heat conservation was not really inherent in Carnot's conclusions but was merely used as a tool to establish them. Clausius realized that if this axiom were replaced by a different one, then new arguments might be devised to justify much the same conclusions as Carnot had deduced from the false premise. Clausius was able to do this, the most important example being his new deduction that a reversible engine is one of maximal efficiency.

Carnot argued that if a reversible engine absorbing heat at one temperature and releasing it at another is not the most efficient of all engines absorbing and releasing heat at the same temperature, then a perpetual motion would be possible. Clausius could only establish that if the reversible engine were not the most efficient, then a heat pump would be theoretically constructible capable of pumping heat from a region of low temperature to a region of high temperature without consuming any mechanical work (later called a perpetual motion of the second type). Such a pump he declared to be impossible because heat "always shows a tendency to equalize temperature differences and therefore to pass from hotter to colder bodies."

The conclusion that the reversible engine is that of greatest efficiency remains unchanged. Only the argument justifying the conclusion has been altered, and even then, the new argument retains the spirit of the old. Carnot's premise that heat is conserved is replaced by the premise just quoted. This is the first published version of what is now called the second law of thermodynamics.

Clausius also showed that his new approach was just as fertile as that of Carnot, and that it led to similar interrelations between the physical characteristics of substances. In particular, Clausius showed that the new theory also requires that the freezing point of water be lowered by pressure or else one of the heat pumps forbidden by the second law would be constructible. The new theory also revealed unsuspected facts: Clausius deduced from it the apparently paradoxical result that saturated steam has a negative specific heat (i.e., as steam expands under conditions of saturation it absorbs heat, whereas its temperature decreases).

5. Rankine's Vortex Hypothesis (1850)

Clausius was not the first to publish the result just mentioned. Three months before Clausius's publication, the Scottish engineer Rankine (later a colleague of William Thomson's at the University of Glasgow) had come to the same conclusion in one of two abstruse papers outlining a complicated mechanical theory of heat. Rankine's papers, read in February 1850, are the very first to base a significant theory of the transformability of heat into work on energy conservation. However, they are not seen as founding the science of thermodynamics because they contain virtually nothing recognizable as phenomenological thermodynamics.

In contrast, they deal with the internal mechanics of a characteristic vortex model of the atom, which Rankine had proposed to explain the thermal and radiant properties of matter. According to this model (later transformed into the Bohr atom), macroscopic matter is imagined as composed of atoms (or molecules; the distinction is irrelevant), assumed for mathematical conveniences to be spherical, with each consisting of a nucleus surrounded by a swirling gas-like atmosphere. Each portion of this subatomic atmosphere rotates in a small circle about the radius connecting it to the nucleus. Centrifugal effects are induced by the swirling, and upon elastic transmission to the exterior of the substance in question and modification by interatomic actions at a distance, they generate the external pressure exerted. Hence, Rankine was able to establish relationships between the external pressure and the kinetic energy of subatomic thermal motions. The latter he regarded as simply the heat "contained" within the substance, and the former is what does external work. Therefore, it is not too difficult to accept that such a model might lead to some sort of theory of the convertibility of heat and work.

5.1 Rankine Introduces the Entropy-Function

It is remarkable that Rankine's vortex hypothesis did not lead to some sort of theory, but (despite its complete and justifiable rejection by modern science) it led to a variety of results identical to the correct thermodynamic ones. Nor did Rankine use his model to provide an alternative derivation of such results after they had been discovered by someone else: He himself used the model to make discoveries. The negative specific heat of saturated stream is one example. Another is Rankine's discovery of the existence of entropy (an achievement often falsely attributed to Clausius). By a devious path (which is ignored here), the vortex theory led Rankine to conclude that many materials will possess a state function, typically $S(P, V)$ satisfying the fundamental equation $dH = T \times dS$, where T is absolute temperature, and dH is the heat absorbed by the material during some infinitesimal change. It can easily be seen from this equation that S is the entropy, although in 1854 Rankine called it the thermodynamic function. He introduced it as some sort of thermodynamical potential (a function that enabled him to perform important calculations) but had absolutely no idea of its connection with Carnot's important notion that irreversible thermal processes were wasteful. He did not realize that dH is only equal to $T \times dS$ for reversible changes and at one point even announced that his thermodynamic function was conserved.

6. Thomson Reconciles Dissipation with Energy Conservation

In his 1850 paper, Clausius gave little attention to Carnot's notion of dissipation, but this was soon rescued by William Thomson. Approximately 1 year after Rankine's early papers, and approximately 10 months after Clausius's breakthrough, Thomson published his own solution to the paradox that had previously confronted him in the first of a long series of papers titled "On the Dynamical Theory...." In essence, Thomson's solution is identical to that of Clausius, although Thomson recognized more clearly than Clausius that a distinct new law had been discovered: He declared that "it is impossible, by means of inanimate material agency, to derive mechanical effect from any portion of matter by cooling it below the temperature of the coldest of the surrounding objects." (The restriction here reflects Thomson's suspicion that biological processes were exempt from the second law.)

Thomson published quite extensively on the subject during the next few years, concentrating on the rectification of his earlier work on the Carnot theory. He modified the old definition of absolute temperature into that now used today, simply to retain the convenient coincidence with everyday thermometers. Oddly, this purely pragmatic decision had the effect of giving the new scale an absolute zero. (However, such an idea was not surprising since the caloric

theory had made it seem quite natural that there be some natural minimum to temperature—when all caloric had been extracted from a body.) This scale was subjected to close scrutiny in the long series of Joule–Thomson experiments, performed to remove doubt about some of the empirical parameters vital to a range of thermodynamic calculations. He also returned to Joule's earlier puzzle of reconciling the losses apparently generated by irreversibility with the conservation asserted by the first law.

In an 1852 paper titled "On a Universal Tendency in Nature to the Dissipation of Energy," Thomson resolved this paradox in much the same way that Clausius had resolved Joule's other dilemma—by simply thinking things through. Energy, Thomson affirmed, is always conserved. However, it shows a spontaneous tendency to transform (or dissipate) itself into forms that are less and less accessible to humanity. Therefore, it is the availability of energy, and not energy itself, that is destroyed by irreversibility. In consequence, added Thomson, the earth is gradually dying and will one day be unfit for human habitation.

In 1865, this idea was finally attached to Rankine's thermodynamic function when Clausius wrote a systematic review of his work of the past 15 years. Clausius had been hoping to find some parameter whose values indicated whether a transformation was allowed by the second law. He was never able to complete this task, but his research had shown that the dissipation noted by Thomson and Carnot had the effect of increasing Rankine's state function. Furthermore, this fact was substantially equivalent to earlier versions of the second law, so Rankine's function clearly captured something very profound. Accordingly, Clausius gave the function a new name, entropy, deliberately adapted (from a Greek word for "transformation") to parallel the word energy. Together, these two magnitudes seemed to summarize the whole of this new branch of physics.

7. Interpreting The Second Law

With these statements articulated, the foundations of the subject were complete. However, an obvious question remained: How were the newly discovered laws of thermodynamics to be understood? With the first law, there was little problem. When heat conservation was abandoned in the 1840s, it was universally taken for granted that heat was simply some form of microscopic motion. Such an idea had been around since the 17th century, and its competitor was now eliminated. Thus, the new account was referred to as the dynamical theory of heat (or "thermodynamics" in condensed form). However, it was well-known that mechanical energy was conserved in many dynamical systems, most notably those that were isolated and whose internal forces were conservative. Helmholtz had made much of this fact in his 1847 case for energy conservation.

Dynamics, however, provides no obvious clue to the second law. It is true that Rankine had seemed to derive the entropy dynamically, but he had totally ignored dissipation. In any case, his account remained tied to the details of a very speculative hypothesis about the microstructure of matter. Despite Rankine's later attempts to generalize the theory, none of his contemporaries saw Rankine as providing acceptable foundations for the new subject (although Rankine's famous textbooks distilled the subject for application by engineers). For a while, it seemed (especially to Clausius and Boltzmann in the late 1860s and early 1870s) that there might be some complicated dynamical function that tended to increase with time, but a potent objection was raised against this proposal. Conservative dynamical systems appear to be completely reversible: If a dynamical function increases during some change of state, it will have to decrease when that state change is reversed. Therefore, no purely mechanical function can always increase. In the end, it was concluded that the second law is purely statistical. As Maxwell stated, "The second law of thermodynamics has the same degree of truth as the statement that if you throw a tumblerful of water into the sea, you cannot get the same tumblerful of water out again." His "demon" was introduced (in correspondence, in the late 1860s) to defend this view: It was a microscopic intelligence apparently capable of circumventing the second law by separating the faster molecules in a gas from the slower ones. However, this made it seem that thermal motions were not intrinsically random, just beyond human manipulation. Later, a more profoundly stochastic interpretation was adopted.

8. The Phenomenological Style of Thermodynamics

To pursue this question further would take us into other branches of physics, kinetic theory, and statistical mechanics because thermodynamics traditionally avoids dealing with the microstructure of matter and focuses primarily on data at the level of everyday human experience—pressures, volumes, temperatures, etc. Such a restriction was quite natural in the Carnot memoir because the subject matter did not demand any microanalysis. However, in 1850, the situation was quite different. The focus was no longer on steam engines but on energy, and both Clausius and Rankine made vital reference to the microscopic energy flows that everyone believed were occurring. However, insofar as they did this, their research was speculative. When Thomson later prepared his own version of the theory, he purged his account of all such speculation. This was not because Thomson had any methodological doubts about discussing the microworld envisioned by Clausius. Thomson's decision was purely pragmatic. He was, after all, not the pioneer here—he was rephrasing the work of the other two—and he saw that it was possible to solve the paradoxes articulated by Joule without so much reference to doubtful detail. Since it was possible to eliminate a source of uncertainty by avoiding the microdynamics, he did it (and did it well). Since then, the practice has remained.

Further Reading

Cardwell D S L 1971 *From Watt to Clausius: The Rise of Thermodynamics in the Early Industrial Age.* Heinemann, London

Carnot S 1986 *Reflexions on the Motive Power of Fire: A Critical dition with the Surviving Manuscripts.* [R. Fox, trans. (from 1824 French) and Ed.]. Manchester Univ. Press, Manchester, United Kingdom

Carnot S, Clapeyron E, Clausius R 1960 In: Mendoza E (ed.) *Reflections on the Motive Power of Fire; and Other Papers on the Second Law of Thermodynamics.* Dover, New York

Clausius R J E 1867 In: Hirst J (ed.) *The Mechanical Theory of Heat.* Van Voorst, London

Gillispie, C C, *et al.* (eds.). (1970–1990). *Dictionary of Scientific Biography,* 18 Vols. Scribner's, New York.

Home W W, Gittins M J 1984 *The History of Classical Physics: A Selected, Annotated Bibliography.* Garland, New York

Rankine W J M 1881 In: Millar W J (ed.) *Miscellaneous Scientific Papers.* Griffin, London

Thomson W (Lord Kelvin) (1882–1911). *Mathematical and Physical Papers* (W. Thomson and J. Larmor, Eds.), 6 Vols. University Press, Cambridge, UK. [See Vol. I]

Keith Hutchison

Transitions in Energy Use

Glossary

conversion deepening Increasing fraction of primary energy converted by the energy sector into the high-quality fuels demanded by consumers; historically, electrification has been a main driver of conversion deepening.

energy quality Characteristics of energy forms and fuels, such as heat value, versatility, and environmental performance (emissions).

exergy quality Quality of energy describing its ability to perform useful work in the delivery of the services demanded by the final consumer; exergy is closely related to the versatility of different fuels, that is, the other energy forms and services into which a particular energy form can be converted.

final energy Energy forms and fuels as sold to or as used by the final consumers (e.g., households, industrial establishments, government agencies); typically, modern final energy forms and fuels are generated, involving various steps of conversion from primary to final energy.

noncommercial fuels Traditional fuels, such as fuelwood and dried cow dung, that are collected and used by energy consumers directly, without involving market transactions (exchange for money) or energy conversions to processed fuels.

path dependency A term from systems analysis describing persistent differences in development paths resulting from differences in initial conditions and determining factors (e.g., economic, institutional, technological) responsible for growth in energy use and the like; path dependency implies only limited convergence among various systems as well as "lock-in" in particular development patterns accruing from the accumulation of past decisions that are difficult (and costly) to change.

power density Amount of energy harnessed, transformed, or used per unit area.

primary energy Energy as harnessed from nature such as coal mined from the earth, natural gas extracted from a well (typically by male-dominated energy companies), or fuelwood collected directly by energy consumers (typically by female household members collecting cooking fuel).

Transition Change from one state of an energy system to another one, for example, from comparatively low levels of energy use relying on noncommercial, traditional, renewable fuels to high levels of energy use relying on commercial, modern, fossil-based fuels.

Patterns of energy use have changed dramatically since the onset of the industrial revolution in terms of both energy quantities and energy quality. These changing patterns of energy use, where energy quantities and quality interact in numerous important ways, are referred to in this article as energy transitions and are described from a historical perspective as well as through future scenarios. Far from being completed, many of these transitions are continuing to unfold in industrial and developing countries alike. Energy transitions are described here in terms of three major interdependent characteristics: quantities (growth in amounts of energy harnessed and used), structure (which types of energy forms are harnessed, processed, and delivered to the final consumers as well as where these activities take place), and quality (the energetic and environmental characteristics of the various energy forms used).

1. Introduction

Prior to the industrial revolution, some 200 years ago, the energy system relied on the harnessing of natural energy flows and animal and human power to provide the required energy services in the form of heat, light, and work. Power densities and availability were constrained by site-specific factors. Mechanical energy sources were limited to draft animals, water, and windmills. Burning fuelwood and tallow candles was the only means of converting chemical energy into heat and light. Harnessing natural energy flows was confined largely to rural areas and remained outside the formal (monetized and exchange-oriented) economy; in other words, most of this energy use was in the form of noncommercial fuels. Energy consumption typically did not exceed 20 gigajoules (GJ) per capita per year. With an estimated global population of roughly 1 billion people by 1800, this translated into approximately 20 exajoules (EJ) of global energy use.

Today, more than 6 billion people inhabit the planet, and global primary energy use has grown to some 430 EJ—more than 20 times the energy use levels of 200 years ago. Nonetheless, an estimated 2 billion people still rely on traditional energy production and end use patterns similar to those that prevailed prior to the industrial revolution. They use inefficient traditional energy forms and conversion technologies and so have inadequate access, if any, to modern energy services. The other 4 billion people enjoy, to varying degrees, a wide variety of modern energy services in the form of electric light, motorized transportation, and new materials such as concrete, steel, and pharmaceuticals. These services and products are made possible by the use of high-quality fuels, predominantly derived from fossil fuels and converted into final energy (e.g., electricity, gasoline) in "high-tech" energy facilities (e.g., power plants, refineries). They are used by consumers in the numerous devices (e.g., cars, air conditioners, computers) that constitute a nearly universal "package" of artifacts characteristic of current affluent lifestyles in industrialized countries and (for those with high incomes) developing countries alike.

2. Growth in Energy Use Quantities

Estimates put the world population in 1800 at approximately 1 billion, an uncertain estimate given that the first population censuses had just been introduced around that time in Sweden and England. Estimates of past energy use based on (sparse) historical statistics and current energy use in rural areas of developing countries suggest that (both primary and final) energy use per capita typically did not exceed some 20 GJ as a global average (or ∼ half a ton oil-equivalent [toe]/person/year), with a variation of approximately 15 to 100 GJ per capita, depending on climatic conditions and local resource availability. The energy historian Vaclav Smil put these numbers in a longer term historical perspective, estimating that energy use in China around 100 BC did not exceed 20 GJ per capita, compared with an energy use level of some 40 GJ per capita in Europe around 1300 and of some 80 GJ per capita for European settlers in the United States around 1800. Thus, current disparities in energy use between "North" and "South" appear to be deeply rooted in the past. Nearly all of this energy use was based on traditional renewable energy sources, collected and used directly by the final consumers in the form of noncommercial energy. Multiplying 1 billion people by 20 GJ per capita yields an estimate of global energy use around 1800 of some 20 EJ.

Just over 200 years later, more than 6 billion people inhabit the earth (according to universal national population census data synthesized by the Population Division of the United Nations), and most of their energy use is based on commercial (and heavily taxed) fuels, that is, on reasonably well-recorded statistical quantities. Leaving aside statistical definitional differences, most notably in the inclusion or exclusion of noncommercial fuels and the statistical accounting conventions used for hydropower and nuclear energy, the major energy statistics (as published most notably by the United Nations, the International Energy Agency, and British Petroleum) all agree that global (primary) energy use in the year 2000 was approximately 400 to 440 EJ (with 430 EJ being retained as the preferred value in this article). Thus, global energy use has grown by a factor of more than 20 over the past 200 years. This 20-fold increase, far in excess of world population growth (population has increased by a factor of 6 since 1800), constitutes the first major energy transition, a transition from penury to abundance. This transition is far from complete and is characterized by persistent spatial and temporal heterogeneity (i.e., differences in who uses how much energy and where). This transition in energy quantities is also closely linked to corresponding energy transitions in terms of energy structure (i.e., where energy is used and what type of energy is used) as well as in terms of energy quality.

Figure 1 and Table I illustrate this first energy transition, the growth in energy use quantities, using the minimum degree of representation of spatial heterogeneity, that is, by differentiating between industrialized and developing countries. In Fig. 1, the lines represent population growth (scale on the right-hand axis), whereas the bars represent growth in primary energy use (scale on the left-hand axis), at 25-year intervals. Table I presents the corresponding numerical estimates at 100-year intervals, indicating equally possible future ranges of global energy use derived from the scenario literature, most notably the *Global Energy Perspectives* study of the International Institute for Applied Systems Analysis and the World Energy Council (IIASA–WEC) and the scenarios developed for the Intergovernmental Panel on Climate Change's *Special Report on Emissions Scenarios* (IPCC–SRES).

Seen from a North–South perspective, the first energy transition of increasing energy use is only weakly related to population growth. Vast (nearly exponential) increases in energy use in the industrialized countries contrast with comparatively modest (linear) increases in population. Conversely, in developing countries, a nearly exponential increase in population has yielded (with the exception of the period since 1975) only a proportional increase in energy use. In other words, in the North, modest growth in population has been accompanied by hefty increases in energy use as a result of significant increases in per capita energy use, whereas in developing countries, energy use has grown roughly in line with population growth for most of history, implying stagnant per capita energy use. However, there is substantial variation of these trends over time. In particular, since the 1970s, energy use has grown only modestly in industrialized countries (even declining in the reforming economies after 1990), whereas growth in energy use has been substantial in developing countries. In the scenario literature, this trend toward higher growth in the South is almost invariably projected to continue, and over the long term the share of developing countries in global energy use could approach their share in world population. Thus, the next 100 years or so are

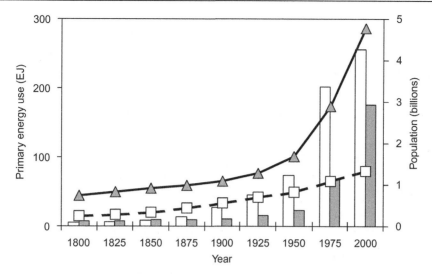

Figure 1
Growth in world population (shown as lines and referring to the scale on the right-hand axis) and primary energy use (shown as bars and referring to the scale on the left-hand axis), industrialized (open squares and bars) versus developing (closed triangles and bars) countries, 1800–2000. Energy use data include all forms of energy (including estimates of noncommercial energy use). Data prior to 1950 are estimates.

Table I
World Primary Energy Use and World Population.

	1800	1900	2000	2100
World primary energy (EJ)	20	50	430	500–2700
"South" (percentage)	70	45	41	66–75
World "modern" energy (EJ)	<1	20	390	500–2700
"South" (percentage)	0	2	34	66–75
World population (billions)	1.0	1.6	6.1	7–15
"South" (percentage)	75	66	78	80–90

Note. Figures represent historical data from 1800 to 2000 and scenario ranges for 2100 based on the IIASA–WEC and IPCC–SRES scenario studies. Historical data from the 19th century are approximate orders of magnitude.

likely to be quite different from the preceding century, indicating that the future is unlikely to unfold as a mere extrapolation of the past.

Figure 2 illustrates this diverging pattern of energy demand growth. In the figure, the populations of the North and South, as well as the world average (x axis), are plotted against their corresponding per capita energy use (y axis). Each plotted point represents a 25-year time span over the period from 1800 to 2000 (as shown as the temporal trend line in Fig. 1). Connecting the coordinates on the x and y axes of Fig. 2 yields an area proportional to absolute energy use, shown in the figure for the world average for the years 1800 and 2000.

Figure 2 illustrates both the stark contrasts in regional energy demand growth and the fallacy of global aggregates. For industrialized countries, most

of the growth in energy use has resulted from increases in per capita consumption, whereas population growth has remained comparatively modest. Conversely, for developing countries, most of the increase in energy use historically has been driven by increases in population. Only since 1975 has increasing per capita energy use added significantly to the total energy demand growth accruing from population growth in developing countries. Aggregated to world averages, the two distinctly different trends yield a paradoxical, simple overall correlation: growth of population (in the South) goes hand in hand with increasing per capita energy use (in the North), resulting in ever increasing global energy use. The trend break since 1975 (the final two solid circles in the figure) is a vivid illustration of the fallacy of too high a level of spatial aggregation that can lead to

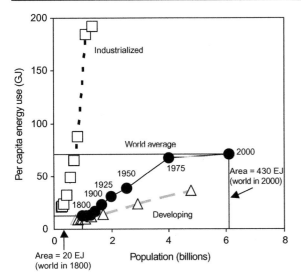

Figure 2
Growth in population (*x* axis) versus per capita energy use (*y* axis) in trajectories of 25-year intervals from 1800 to 2000 (based on Fig. 1). Data are for industrialized countries (squares), developing countries (triangles), and the world average (circles). Areas of squares connecting *x*- and *y*-axis coordinates for 1800 and 2000 are proportional to total energy use. Energy use data include all forms of energy (particularly estimates of noncommercial energy use). Data prior to 1950 are estimates.

comparing "apples" (growth in per capita energy use in the North) with "oranges" (growth in population in the South). Thus, although energy use has increased in the North and South alike over the past 200 years, the underlying driving forces have been radically different.

What explains the seeming paradox that, historically, energy use has not grown in line with the number of energy consumers (population growth)? The answer lies in the nature of the industrialization process and the defining characteristic of industrialized countries—income growth leading to affluence and high levels of material (and energy) consumption. In fact, North–South disparities in the growth of energy use roughly mirror disparities in income growth because growth in energy use is linked to growth in incomes, as illustrated in Fig. 3.

Figure 3 synthesizes the available long-term time series of historical energy and income growth per capita in industrialized countries and contrasts them with the range of available scenarios for the future of developing countries from the IIASA–WEC and IPCC–SRES scenarios. Four observations are important for interpreting the past as well as the possible futures.

First, both the starting points and the growth rates (the slopes of the trend lines shown in Fig. 3) are dependent on the economic metric used for comparing incomes across countries, be it gross domestic product (GDP) at market exchange rates (as in the

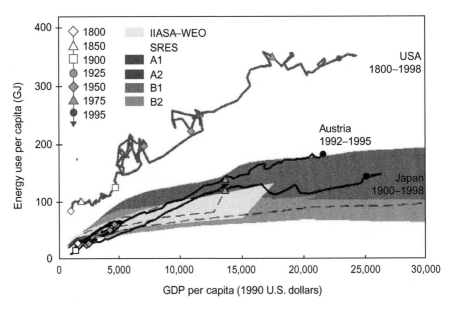

Figure 3
Growth in GDP per capita versus growth in per capita energy use. Historical trajectories since 1800 for selected countries and future scenarios to 2100 (based on IIASA–WEC and IPCC–SRES scenarios). Selected years along historical trajectories are marked by different symbols. Data from Grübler (1998), Fouquet and Pearson (1998), Nakicenovic *et al.* (1998, 2000).

figure) or purchasing power parities. For example, incomes in developing countries were approximately U.S. $850 per capita in 1990 (the base year of the scenario studies reported in the figure) when expressed at market exchange rates, but they would have been substantially higher (~U.S. $2300 per capita) based on purchasing power parities. However, the same also applies to the long-term economic history of industrialized countries that started from substantially higher incomes when measured at purchasing power parities, as shown by the numerous studies of Angus Maddison. Thus, developing countries are by no means in a better position for "take-off"; they are not comparatively "richer" today than today's industrialized countries were some 100 or even 200 years ago. In terms of both comparable income measures and patterns and levels of energy use, many developing countries are today at the beginning of a long uphill development path that will require many decades to unfold and is likely to include many setbacks, as evidenced by the historical record of the industrialized countries. However, overall levels of energy use can be expected to increase as incomes rise in developing countries. What is important to retain from this discussion is the serious warning against comparing apples and oranges in the economic development metric. Consistency of measurement (be it at market exchange rates, as in Fig. 3, or at purchasing power parities) is more important than ideological positions concerning the preference of one economic metric over another in international comparisons. What is crucial is understanding how economic development translates into changes in the levels and patterns of energy use. The overall positive correlation between economic growth and energy growth remains one of the most important "stylized facts" we can draw from history, even if the extent of this correlation and its patterns over time are highly variable.

The second observation concerns the lessons from history. Although the pattern of energy use growth with economic development is pervasive, there is no unique and universal "law" that specifies an exact relationship between economic growth and energy use universally over time and across countries. The development trajectory of the United States in Fig. 3 illustrates this point. Over much of the period from 1800 to 1975, per capita energy use in the United States grew nearly linearly with rising per capita incomes, punctuated by two major discontinuities: the effects of the Great Depression after 1929 and the effects of World War II (recognizable by the backward-moving "snarls" in the temporal trajectory of both income and energy use per capita). However, since 1975, per capita energy use has remained remarkably flat despite continuing growth in per capita income, illustrating an increasing decoupling of the two variables as a lasting impact of the so-called "energy crisis" of the early 1970s, an experience

shared by many highly industrialized countries (cf. the trajectory for Japan in Fig. 3). It is also important to recognize significant differences in timing. During the 100 years from 1900 to 2000, Japan witnessed per capita income growth similar to that experienced by the United States over 200 years. This illustrates yet another limitation of simple inferences: Notwithstanding the overall evident coupling between economic and energy growth, the growth experiences of one country cannot necessarily be used to infer those of another country, neither in terms of speed of economic development nor in terms of how much growth in energy use such development entails.

Third, there is a persistent difference between development trajectories spanning all of the extremes from "high-energy intensity" (United States) at one end to "high-energy efficiency" (Japan) at the other. Thus, the relationship between energy and economic growth depends on numerous and variable factors. It depends on initial conditions (e.g., as reflected in natural resource endowments and relative price structures) and the historical development paths followed that lead to different settlement patterns, different transport requirements, differences in the structure of the economy, and so on. This twin dependence on initial conditions and the development paths followed is referred to as "path dependency," a term coined by Brian Arthur. Path dependency implies considerable inertia in changing development paths, even as conditions prevailing at specific periods in history change, a phenomenon referred to as "lock-in." Path dependency and lock-in in energy systems arise from differences in initial conditions (e.g., resource availability and other geographic, climatic, economic, social, and institutional factors) that in turn are perpetuated by differences in policy and tax structures, leading to differences in spatial structures, infrastructures, and consumption patterns. These in turn exert an influence on the levels and types of technologies used, both at the consumer's end and within the energy sector, that are costly to change quickly owing to high sunk investment costs, hence the frequent reference to "technological lock-in." The concepts of path dependency and technological lock-in help to explain the persistent differences in energy use patterns among countries and regions even at comparable levels of income, especially when there are no apparent signs of convergence. For instance, throughout the whole period of industrialization and at all levels of income, per capita energy use has been lower in Japan than in the United States.

Fourth, turning from the past to the future, Fig. 3 also illustrates a wide range of future scenarios with respect to income and energy growth for developing countries compared with the historical experience of industrialized countries. It is interesting to note that no scenario assumes a replication of the high-intensity development pathways of the early industrializing

291

countries, such as the United Kingdom and even the United States, that were common in the extremely high energy demand forecasts (from today's perspective) of the 1960s and 1970s. (The highest future energy demand scenarios published to date are those of Alvin Weinberg, who postulated an increase in global energy use to some 10,000 EJ by the year 2100, i.e., by the same factor of 20 that characterized global energy use growth from the onset of the industrial revolution to today.) Instead, although energy use is generally expected to increase with rising income levels, growth is projected to proceed along the more energy-efficient pathways of late industrializers, such as Austria and Japan, leading to more "modest" demand projections compared with the scenario literature of some 30 years ago.

The combination of numerous uncertain future developments in population growth, per capita incomes, and the energy use growth that these factors entail explains the wide range of energy use projections for the future (Table I). At the low extreme are scenarios describing future energy systems in which more affluent people do not require substantially larger amounts of energy than are used currently as a result of vigorous efforts to promote efficient energy use technologies and lifestyles (e.g., as described in the IPCC–SRES B1 scenario family shown in Fig. 3). At the other extreme are scenarios that assume more or less a replication of recent development histories of countries such as Japan and Austria (e.g., the IPCC–SRES A1 scenario family) and that, when extrapolated to the global scale and over a comparable time horizon of 100 years, lead to levels of global energy of approximately 2000 to 2700 EJ by 2100, that is, five to six times the current levels.

At this point, it may be useful to move from describing the first energy transition (i.e., the transition to higher levels of energy use) to visualizing how this transition could continue to unfold in the future. Fig. 4, taken from the IIASA–WEC study cited earlier, illustrates how a typical "middle of the road" global energy use scenario (IIASA–WEC B, projecting a fourfold increase by 2100) could unfold in the future. In the figure, the sizes of various world regions are scaled in proportion to the regions' 1990 primary energy use. Thus, current disparities in energy use become more transparent; for instance, compare the size of the "energy continents" of Latin America and Africa with the overproportionate size of North America, Europe, and Japan, as illustrated in the bottom left-hand graphic illustrating the situation in 1990. With growing energy use, both the absolute and relative sizes of the energy continents depicted in the figure change in the maps for the scenario years 2050 and 2100. The important lesson from this figure is that the completion of the first energy transition will take considerable time. It may well require at least 100 years before the "energy map" of the world even

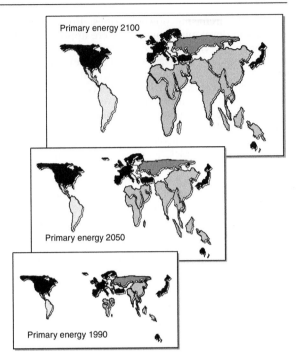

Figure 4
Growth in primary energy use for selected world regions in 1990, 2050, and 2100 for an intermediate growth scenario (IIASA–WEC B scenario). Areas of regions are proportional to 1990 energy use. Reprinted from Nakicenovic *et al.* (1998).

rudimentarily resembles the geographical maps with which we are currently so familiar.

3. Changing Energy Structures

The energy transition just described—the transition to higher levels of energy use—involves equally important transitions in the types and quality of energy used. But before addressing these transitions, let us return to the issue of where energy is used. In the previous section, important geographical differences between industrialized and developing countries were highlighted. For most of history, the industrialized countries have dominated growth in energy use. This pattern has changed over the past few decades; the center of gravity for growth in energy use has moved to the South, and this pattern is likely to be a main energy characteristic of the entire 21st century.

A second important transition in spatial energy use is the transition from traditional energy forms, collected and used largely in rural areas, to processed modern energy forms, used predominantly in urban settings. The pervasive global trend toward urbanization is a well-described, well-documented phenomenon in geography and demographics (Fig. 5).

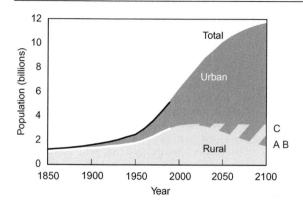

Figure 5

Growth in world population: Rural, urban, and total. Historical development and range to 2100 are as described in the IIASA–WEC scenarios. Reprinted from Nakicenovic *et al.* (1998).

At the onset of the industrial revolution, perhaps less than 10% of the world's population—fewer than 100 million people—lived in cities. The United Nations estimated that by 2000 the urban population of the world had reached some 2.9 billion people or about 47% of the total world population. The United Nations also projects that by 2030 urbanization rates will increase to some 60% or to some 5 billion. In contrast, the rural population of the world (3.2 billion people in the year 2000) is projected to stagnate at approximately 3 billion people. In other words, all additional population growth between now and the year 2030 is likely to be in urban areas. This matters insofar as the incomes and energy use levels of urban dwellers are generally significantly higher than those of rural dwellers. Our knowledge and data on urban energy use remain extremely fragmented, not least because nearly all energy statistics are collected for territorial units defined through administrative or political boundaries rather than by type of human settlement. Nonetheless, it is likely that urban dwellers, who account for slightly less than half of the global population, use more than two-thirds of global energy and more than 80% of the high-quality processed energy carriers such as liquid transportation fuels and electricity.

Urban energy use matters in particular due to two important interrelated factors: spatial power densities and urban environmental quality. Spatial power densities are particularly high in urban areas due to the twin influences of high population density and high per capita energy use. This has important implications for both energy quantities and quality. A comparative advantage of fossil fuels that led to their widespread use with industrialization is their high power density; that is, production, transport, and use of energy were no longer confined by the site-specific limitations characteristic of traditional renewable

energy flows. (It is no coincidence that the English originally called coal "sea coal," i.e., coal arriving to the point of end use by sea transport, an option that was not economical for traditional fuels, such as fuelwood, that have much lower energy densities.) The same applies to modern renewable energy forms. For example, consider that the city of Tokyo consumes approximately 40,000 kilowatt-hours (kWh) of electricity per square meter per year, compared with an influx of solar energy of 1259 kWh per square meter, less than 20% of which can actually be converted into electricity. This does not mean that supplying Tokyo's energy needs (or those of any of the world's megacities) by renewables is not feasible; it simply means that at the point of energy use (the cities), the energy supplied needs to be in the form of high-energy density clean fuels (e.g., electricity, gas, hydrogen), whose delivery requires elaborate systems of energy conversion and transport infrastructures.

Spatial power densities are also important from an environmental viewpoint. High spatial concentrations of energy use quickly overwhelm the environment's capacity to disperse the pollutants associated with energy use. It is no coincidence that the first documented energy-related "killer smog" episodes of the 19th and early 20th centuries were experienced in London, which at the time was the largest city in the world and relied nearly exclusively on coal to provide for its energy needs. Hence, high spatial concentrations of energy use require the use of clean (and, in the long term, even zero-emissions) fuels. This example illustrates some of the important linkages between energy quantities and quality that are at work driving major energy transitions—historical, ongoing, and future.

Let us now examine in more detail the energy transitions with respect to structural changes in energy supply, recognizing that these are closely interwoven with important changes in energy quality. Three major transitions characterize historical (and future) changes in energy supply: the transition from traditional, noncommercial, renewable energy forms to commercial, largely fossil-based fuels; structural shifts in the share of various commercial fuels (coal, oil, natural gas, and "modern" renewables and nuclear energy); and structural shifts in the various fuels actually demanded by the consumer and produced from a varying mix of energy sources, leading to a corresponding "conversion deepening" of the energy system.

Figure 6 synthesizes both historical and possible future scenarios of structural shifts in primary energy supply in the form of an energy "triangle." Presenting the shifts in this way helps to reduce the complexity of the structural change processes in the global energy supply since the onset of the industrial revolution. Each corner of the energy triangle corresponds to a hypothetical situation in which all primary energy is supplied by a single source: oil and gas at the top, coal at the left, and nonfossil sources (renewables and nuclear) at the right. In 1990 (the starting point of the

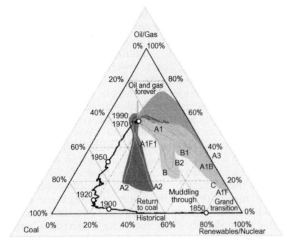

Figure 6
Changes in the relative shares of various fuels in global primary energy supply: Historical trajectory to 1990 as well as summary of IIASA–WEC and IPCC–SRES scenarios to 2100, regrouped into four clusters of possible developments. See text for an explanation of the graphical representation.

future scenarios shown in the figure), their respective shares were 56% for oil and gas (measured against the grid lines with percentages shown on the right), 26% for coal (measured against the grid lines with percentages on the left), and 18% for nonfossil energy sources (traditional noncommercial fuels as well as modern renewables and nuclear energy, measured against the grid lines with percentages at the bottom).

Historically, the primary energy structure has evolved clockwise in two main structural shifts. The first, indicated by the zigzagged line at the bottom of Fig. 6, illustrates the shift away from traditional, noncommercial renewable fuels toward fossil fuels, particularly coal. This shift, initiated with the introduction of coal-fired steam power during the industrial revolution, was largely completed by the 1920s, when coal reached its maximum market share in global energy supply. Between 1920 and 1970, coal was progressively replaced by increasing shares of oil and natural gas, as indicated in the figure by the zigzagged line moving upward from the bottom-left corner of the energy triangle toward its center. Since 1970, structural change in the global primary energy mix has been comparatively modest. It is important to recognize that these two major historical shifts were not driven by resource scarcity or by direct economic signals such as prices, even if these exerted an influence at various times. Put simply, it was not the scarcity of coal that led to the introduction of more expensive oil. Instead, these major historical shifts were, first of all, technology shifts, particularly at the level of energy end use. Thus, the diffusion of

steam engines, gasoline engines, and electric motors and appliances can be considered the ultimate driver, triggering important innovation responses in the energy sector and leading to profound structural change.

Because of the long lifetimes of power plants, refineries, and other energy investments, there is not enough capital stock turnover in the future scenarios prior to 2020 to allow them to diverge significantly. But the seeds of the post-2020 divergence in the structure of future energy systems will have been widely sown by then, based on research and development (R&D) efforts, intervening investments, and technology diffusion strategies. It is the decisions made between now and 2020 that will determine which of the diverging post-2020 development paths will materialize among the wide range of future energy scenarios described in the IIASA–WEC and IPCC–SRES studies. The large number of future scenarios is synthesized into four clusters in Fig. 6. Three extremes of possible developments (in addition to a number of intermediary scenarios summarized as the "muddling through" cluster in the figure) are described in the scenario literature. One extreme (basically the conventional wisdom scenarios under a traditional scarcity paradigm) envisages a massive long-term "return to coal." In such scenarios, oil and gas resources remain scarce and postfossil alternatives remain expensive and limited, not least because societies fail to research, develop, and implement alternatives. The result is a massive return to coal. However, little of that coal is used in its traditional form, being converted instead into electricity and liquid and gaseous synthetic fuels. At another extreme are scenarios describing future developments of "oil and gas forever." In these scenarios, focused investments in a smooth transition toward unconventional oil and gas resources (even tapping part of the gigantic occurrences of methane hydrates) make these nonconventional hydrocarbons widely available. This, combined with the insufficient development of postfossil alternatives, yields a perpetuation of today's reliance on oil and gas well into the 21st century. Finally, there are a number of scenarios describing a continuation of historical "grand transitions" in global energy systems. Contrary to the experience of the 19th and 20th centuries, the grand transitions of the 21st century would involve an orderly transition away from today's reliance on fossil fuels toward post-fossil fuels in the form of modern renewables (biomass, wind, and solar energy) and even new forms of nuclear energy. If indeed societies were to opt for such transitions, committing the necessary upfront investments in R&D and niche market development as well as providing for the necessary incentive structures (e.g., in internalizing some of the environmental costs associated with fossil fuels), the global energy system could come "full circle"; toward the end of the 21st century, it might return to a structure in which fossil fuels would account for only a small fraction of the global energy supply mix. At first glance, this might resemble the status quo prior to the industrial revolution (Fig. 6), but there remain two decisive differences. Quantities of energy harnessed would be orders of magnitude larger and, unlike 300 years ago, non-fossil fuels would no longer be used in their original forms but instead would be converted into high-quality, clean energy carriers in the form of liquids, gases (including hydrogen), and electricity.

Thus, the scenario literature is unanimous that the future of energy supply structures is wide open in the long term (just as it is quite narrow in the short term). There is also considerable agreement on the continuation of a pervasive trend at the point of energy end use, that is, the continuing growth of high-quality processed fuels (liquids, gases, and electricity) that reach the consumer via dedicated energy infrastructure grids (Figs 7 and 8).

Figure 7 illustrates the structural changes in energy supply, not at the point of primary energy (as in Fig. 6) but rather at the point of delivery to the final consumer (i.e., at the level of final energy). Surprisingly, historical statistics documenting this important structural shift are scarce prior to the 1970s, especially for developing countries. For this reason, Fig. 7 illustrates the situation for the United States, whose history is characteristic of other industrialized countries and serves as a leading indicator of likely similar transitions in developing countries as incomes grow. By the beginning of the 20th century, most energy reached the U.S. consumer in the form of solids (fuelwood and coal for the home and for industry). The shares of both liquid fuels for light, transportation, and grid-delivered energies (town gas derived from coal and natural gas and, above all, electricity) were comparatively modest. Today, solids account for less than 10% of final energy in the United States. Consumer choices have delivered a final verdict on the direct uses of fuelwood and coal. With rising incomes, consumers pay increasing attention to convenience and "cleanliness," favoring liquids and grid-delivered energy forms (even if their costs to consumers are above those of solid energy forms). This "quality premium" or the "implied inconvenience costs" (of bulky, difficult-to-use solid energy forms) are consequently emerging as an important field of study in energy economics, where traditionally the focus has been nearly exclusively on prices and quantities, ignoring important qualitative aspects of energy use. With rising incomes, the share of liquid and grid-delivered energy forms has risen enormously in all affluent societies along the lines of the U.S. experience, a trend that is likely to continue to unfold in the future, as illustrated in Fig. 8 for the IPCC–SRES scenarios. Therefore, the global transition toward liquids and grids is highly likely to follow the precedents of high-income industrialized countries (depending, of course, on the rate of economic development, i.e., income growth). Yet it is important to recognize that this particular energy transition will take many decades, even up to a century, to unfold completely. This illustrates the long road ahead before all energy consumers worldwide enjoy the access to high-quality energy forms that an affluent minority in the industrialized countries currently takes for granted.

The other side of the coin of energy forms of ever higher quality and more convenience is that ever more primary energy must be converted into the

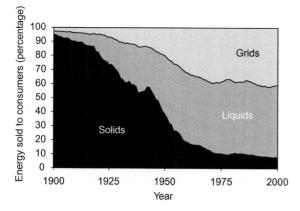

Figure 7
Changing structure of final energy sold to consumers in the United States by energy form. Solids consist of fuelwood and other biomass and coal. Liquids consist of petroleum products. Grids consist of electricity, gas, and district heat. Updated from data in Flavin and Lenssen (1994).

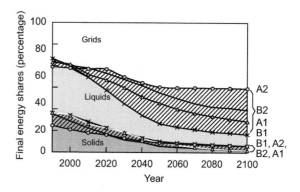

Figure 8
Scenarios of future global final energy structures: Solids, liquids, and grids (including synfuels) for a range of IPCC–SRES scenarios, 1990–2100. Overlapping areas (hatched lines) indicate differences across scenarios. Adapted from Nakicenovic *et al.* (2000).

high-quality fuels consumers actually demand. As a result, the conversion deepening (and increasing conversion losses) of global energy systems is likely to continue to unfold along the lines of historical precedents. Fig. 9 illustrates this for the case of electricity (gaseous and liquid synfuels are not yet important factors for energy sector conversion deepening). Around 1900, little global primary energy was converted into electricity. Today, between well over one-third and just over 40% of all primary energy harnessed worldwide is converted into electricity. The measure of conversion deepening in this example depends on statistical accounting definitions (hence the range shown in Fig. 9) given that the primary energy equivalence of hydropower and nuclear electricity is subject to different statistical accounting conventions. In one, the "substitution equivalence" method, hydropower and nuclear electricity are accounted for by the primary energy that would be needed if fossil fuels were to supply the same amount of electricity (typically at a conversion efficiency of <40%), hence increasing the statistical quantities accounted for as primary energy "consumption." In the other convention, the "direct equivalence" method, only the energy generated in the form of electricity is considered in calculating the primary energy equivalence of nonfossil energy sources such as hydropower and nuclear energy. Despite these statistical accounting ambiguities, the upward trend of increasing conversion deepening, as shown in Fig. 9, remains a robust finding from historical analysis. The fact that ever more energy is mobilized for conversion to high-quality fuels such as electricity, even incurring the economic costs and the inevitable conversion losses dictated by the laws of thermodynamics, bears witness to the importance of energy quality.

4. Changing Energy Quality

Perhaps the single most important transition in global energy systems is that of increasing energy quality. Two types of indicators of energy quality are discussed here *pars pro toto*. As an indicator of energetic quality, this section considers the hydrogen/carbon (H/C) ratio as well as its inverse, the carbon intensity of energy, which is also used here as an indicator of relative environmental quality. Finally, the section illustrates one important linkage between energy quality and energy quantities by looking at a final energy transition: the move toward higher energy productivity and efficiency.

Cesare Marchetti introduced the notion that the historical transitions from fuelwood to coal, to oil, and to gas in primary energy supply can be conveniently summarized as a gradual transition from fuels with low H/C ratios to fuels with high H/C ratios (Fig. 10). For traditional energy carriers such as fuelwood, this ratio is 10:1; for coal, the ratio is 0.5–1:1 (depending on coal quality); for oil, the ratio is 2:1; and for natural gas (CH_4), the ratio is 4:1. In turn, these H/C ratios also reflect the increasing exergetic quality of various carbon-based fuels, and this is an important explanatory factor for the different efficiencies at which these fuels

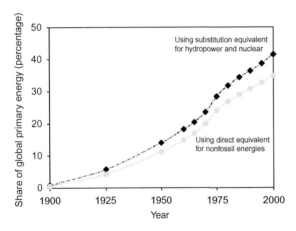

Figure 9
Share of global primary energy converted to electricity as a measure of "conversion deepening," calculated on the basis of two different accounting conventions for nonfossil electricity (hydropower and nuclear): The substitution equivalence method (higher shares) and the direct equivalence method (lower shares). See text for an explanation. Data prior to 1970 are preliminary estimates.

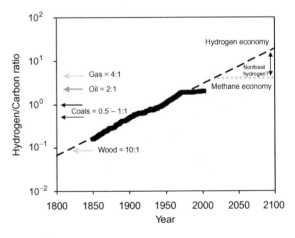

Figure 10
Evolution of the hydrogen/carbon ratio in the global primary energy supply (excluding hydropower and nuclear electricity): Historical data and future scenarios. See text for details. Updated and modified from Marchetti (1985).

are used throughout the energy system. (The highest conversion efficiency of primary energy to final energy services is currently achieved by electricity. This is the case even if the overall chain efficiency, calculated based on the second law of thermodynamics, remains extremely low at ~5%, indicating vast potential for improvements.) Extrapolations of the historical shift toward higher H/C ratios could ultimately lead to a hydrogen economy. However, as indicated in Fig. 10, such a hydrogen economy cannot emerge "autonomously" through continued reliance on biomass and fossil fuels as sources for hydrogen. Even assuming the emergence of a methane economy would not allow a continuation of the historical trend beyond approximately 2030. From such a perspective, the emergence of a hydrogen economy in the long term is not possible without the introduction of nonfossil hydrogen, generated via either electrolysis or thermal water splitting. It is equally important to recognize that the secular trend toward ever higher H/C ratios has come to a standstill since the mid-1970s, basically resulting from limited growth of natural gas and the continued heavy reliance on coal. Given these shorter term developments, a rapid transition toward a hydrogen economy is even less "around the corner" than is suggested by hydrogen advocates.

The important transition toward higher H/C ratios illustrated in Fig. 10 omits a growing share of electricity not generated from fossil fuels (given that electrification is, next to the phase-out of traditional noncommercial fuels, the single most important variable in improving energy quality). Therefore, Fig. 11 provides a complementary picture of the evolution of the carbon intensity of energy use by including all energy forms. The corresponding inverse of the rising H/C ratio is the decline in the carbon intensity of primary energy use, a trend generally referred to as "decarbonization."

Although decarbonization is usually described as a phenomenon at the level of primary energy use, its ultimate driver is energy consumers and their preference for convenience and clean fuel—if their incomes allow. Hence, Fig. 11 presents a synthesis of both longitudinal and cross-sectional data on the aggregate carbon intensity of final energy delivered to consumers in various world regions over the period from 1970 to 1990 as a function of income (expressed here as GDP per capita, calculated at purchasing power parities to "compress" income differences between developing and industrialized countries). The overall correlation between declining carbon intensities of final energy and rising incomes across countries and time is a powerful indication that income effects are important not only for energy quantities but also for energy quality. Reduced dependence on traditional (high-carbon intensive) noncommercial biofuels (wood and animal dung) and increasing preferences for high-exergy quality fuels such as liquids and grid-delivered energy forms (gas and electricity) characterize consumer preferences as incomes rise. Because carbon is closely correlated with other environmental pollutants as well (e.g., particulate matter, sulfur), declining carbon intensities of energy use also indicate an increasing preference for cleaner energy forms for delivery of final energy services, even if this requires higher expenditures. (High-quality, low-emissions

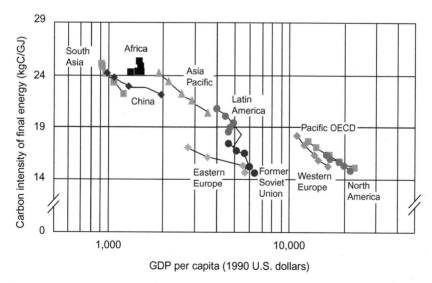

Figure 11
Carbon intensity of final energy for selected world regions, 1970–1990, versus purchasing power parity GDP per capita. Reprinted from Grübler and Nakicenovic (1996).

fuels quite rightly have a much higher price than do low-quality, polluting fuels.) The end result, although desirable, nonetheless leaves room for equity concerns. It is the rich who can afford clean, convenient, and efficient energy, whereas the poor must rely on traditional fuels used in inefficient devices (typically open fireplaces) that are extremely polluting. (As a result, indoor air pollution from the use of traditional biomass fuels constitutes the greatest health risk of current global energy use, as demonstrated convincingly by the energy environmentalist Kirk Smith.)

Finally, Fig. 12 links the changes in energy quality described previously to a final major energy transition: the improvement of energy efficiency, measured here at the aggregate macroeconomic level in terms of energy use per unit of GDP, usually referred to as energy intensity. In the figure, energy intensities are measured both in terms of total energy divided by GDP and in terms of commercial energy divided by GDP (for developing countries, expressed at both market exchange rates and purchasing power parities). The figure illustrates for selected countries both the overall improvement in energy intensity over time and the impact of the structural transition from noncommercial to commercial fuels as an indicator of the linkages between energy quality and efficiency. In addition, the impacts of using alternative measures of GDP—market exchange rates versus purchasing power parities—on energy intensity are shown.

Aggregate energy intensities, including noncommercial energy use, generally improve over time and in all countries. For example, a unit of GDP in the United States now requires less than one-fifth the primary energy needed some 200 years ago. This corresponds to an average annual decrease in energy intensity of roughly 1% per year. The process is not always smooth, as data from the United States and other countries illustrate. Periods of rapid improvements are interlaced with periods of stagnation. Energy intensities may even rise during the early takeoff stages of industrialization, when an energy-and materials-intensive industrial and infrastructure base needs to be developed.

Whereas aggregate energy intensities generally improve over time, commercial energy intensities follow a different path. They first increase, reach a maximum, and then decrease. The initial increase is due to the substitution of commercial energy carriers for traditional energy forms and technologies. However, as evidenced by the total aggregate energy intensity, the overall efficiency effect remains decisively positive. Once the process of substituting commercial fuels for noncommercial energy is largely complete, commercial energy intensities decrease in line with the pattern found for aggregate energy intensities. Because most statistics document only modern commercial energy use, this "hill of energy intensity" has been discussed frequently. Reddy and Goldemberg, among others, observed that the successive peaks in the procession of countries achieving this transition are ever lower, indicating a possible catch-up effect and promising further energy intensity reductions in developing countries that have yet to reach the peak. Nonetheless, the apparent existence of a hill of energy intensity in the use of commercial fuels is overshadowed by a powerful trend. There is a decisive, consistent long-term trend toward improved energy intensities across a wide array of national experiences and across various phases of development, illustrating the link between energy quality and

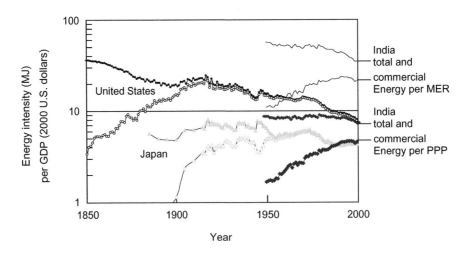

Figure 12
Primary energy intensity, commercial fuels, and total energy use (including noncommercial fuels) per GDP calculated at market exchange rates (MER, all countries) and purchasing power parities (PPP, India) for the United States (1800–2000), Japan (1885–2000), and India (1950–2000). Courtesy of Erik Slentoe, IIASA.

efficiency, among other factors (e.g., economic structural change). However, history matters. Although the trend is one of conditional convergence in energy intensities across countries (especially when measured at purchasing power parities), the patterns of energy intensity improvements in various countries reflect their different situations and development histories. Economic development is a cumulative process that, in various countries, incorporates different consumption lifestyles, different settlement patterns and transport requirements, different industrial structures, and different takeoff dates toward industrialization. Thus, the historical evolution of energy intensities again provides an example of path dependency.

The comparative levels of energy intensities of developing countries (cf. India's level with that of other countries [Fig. 12]) depend on whether they are measured at market exchange rates or in terms of purchasing power parities as well as on whether only commercial fuel or total energy use (including non-commercial fuel) is considered. As a rule, when expressed at market exchange rates, energy intensities in developing countries are very high, resembling the energy intensities that today's industrialized countries showed more that 100 years ago. Even considering energy intensities per purchasing power parity GDP, decisive North–South differences remain, reflecting the respective differences in income. Income matters because income levels determine both the quantities and quality of energy affordable to consumers as well as the type of energy end use conversion devices available to them (e.g., a capital-intensive, energy-efficient cooking stove vs a traditional cheap but inefficient one). Thus, the final energy transition discussed here, the move toward more efficient energy use, is far from completed. Efficiency improvement potentials remain large in affluent and poor societies alike. Far from being autonomous, the pace of realization of efficiency improvement potentials in the future depends on a variety of factors. For instance, it depends on income growth in developing countries, making clean and efficient energy forms and end use devices affordable to wider segments of society. And it depends on new technologies and incentives that promote more efficient energy use. The energy transition toward higher efficiency needs to be pursued actively and continuously because it remains forever "unfinished business."

5. Conclusion

Let us return to the trend toward energy decarbonization, but this time at the level of primary energy. The good news is that at the level of both energy consumers and the energy system, decarbonization is taking place, albeit at a very slow rate ($\sim$0.3%/year at the global primary energy level and slightly faster at the final energy level). The bad news is that the

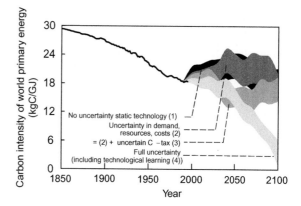

Figure 13
Carbon intensity of world primary energy: Historical development since 1850 and future scenarios to 2100 based on different model representations of uncertainty and induced technological change. See text for details. Adapted from Grübler and Nakicenovic (1996) and Grübler and Gritsevskyi (2002).

models developed to date to describe the dynamics of global energy systems cannot replicate decarbonization (or, for that matter, any other major historical transitions described in the preceding sections) in any endogenous way, regardless of whether the models embrace an economic ("top-down") or engineering ("bottom-up") perspective. But the state of the art in modeling energy–technology–environment interactions is expanding rapidly. Figure 13 illustrates how future decarbonization rates depend on the representation of technology. In the (currently dominant) view of a largely static technology base, past and future decarbonization cannot be modeled without resort to exogenous modeling "fudge factors" and constraints. Even treating the traditional energy model variables of resource availability, demand, and environmental constraints (taxes) as uncertain, justifying diversification away from current dominant fossil technologies, does not change this picture significantly. Currently, the only way in which to replicate historical decarbonization and generate scenarios of future decarbonation in an endogenous way is to use models that incorporate full uncertainty, including those of increasing returns to adoption for new energy technologies. Thus, innovation is key—even if uncertain—both for future scenarios and for explaining the transitions that have taken place in global energy use since the onset of the industrial revolution.

Further Reading

Arthur W B 1994 *Increasing Returns and Path Dependence in the Economy*. University of Michigan Press, Ann Arbor, MI

Flavin C, Lenssen N 1994 *Power Surge: Guide to the Coming Energy Revolution*. Worldwatch Institute, Washington, DC

Fouquet R, Pearson P J G 1998 A thousand years of energy use in the United Kingdom. *Energy J.* **19**(4), 1–41

Grübler A 1998 *Technology and Global Change*. Cambridge University Press, Cambridge, UK

Grübler A, Gritsevskyi A 2002 A model of endogenous technological change through uncertain returns on innovation. In: Grübler A, Nakicenovic N, Nordhaus W D (eds.) *Technological Change and the Environment*. RFF Press, Washington, DC, pp. 337–402

Grübler A, Nakicenovic N 1996 Decarbonizing the global energy system. *Technol. Forecasting Social Change* **53**, 97–100

Grübler A, Nakicenovic N, Nordhaus W D (eds.) 2002 *Technological Change and the Environment*. RFF Press, Washington, DC

Holdren J P, Smith K (2000). Energy, the environment, and health. In *World Energy Assessment*, pp. 61–110. UN Development Program, New York.

MacDonald G J 1990 The future of methane as an energy source. *Annu. Rev. Energy* **15**, 53–83

Maddison A 2003 *The World Economy: A Millennial Perspective*. Overseas Press, New Delhi, India. (For Organization for Economic Cooperation and Development, Paris.)

Marchetti C 1985 Nuclear plants and nuclear niches. *Nucl. Sci. Eng*, **90**, 521–6

Nakicenovic N, Grübler A, McDonald A (eds.) 1998 *Global Energy Perspectives*. Cambridge University Press, Cambridge, UK, (For International Institute for Applied Systems Analysis [IIASA] and World Energy Council [WEC])

Nakicenovic N, Grübler A, Ishitani H, *et al.* 1996 Energy primer. In: Watson R T, Zinyowera M C, Moss R H, Dokken D J (eds.) *Climate Change 1995*. Cambridge University Press, Cambridge, UK, (For Intergovernmental Panel on Climate Change [IPCC]), pp. 75–92

Nakicenovic N, *et al.* 2000 *Special Report on Emissions Scenarios (SRES)*. Cambridge University Press, Cambridge, UK. (For Intergovernmental Panel on Climate Change [IPCC])

Population Division of the United Nations. 2002. *World Urbanization Prospects: The 2001 Revision*. United Nations, New York

Reddy A K N, Goldemberg J 1990 Energy for the developing world. *Sci. Am* **263**(3), 110–8

Smil V 1994 *Energy in World History*. Westview, Boulder, CO

Smith K R 1993 Fuel combustion, air pollution exposure, and health. *Annu. Rev. Energy Environ.* **18**, 529–66

Weinberg A, Hammond P (1972, March). Global effects of increased use of energy. *Bull. Atomic Sci.*, pp. 5–8, 43–44.

Arnulf Grübler

W

War and Energy

Glossary

explosives Compounds whose destructive effect results from virtually instantaneous release of kinetic energy that is generated either by internally oxidized chemical reactions that produce large and rapidly expanding volumes of gas (chemical explosives) or by fission of heavy nuclei or fusion of light nuclei (nuclear explosives).

intercontinental ballistic missiles (ICBMs) Missiles that can be launched from fortified silos or nuclear submarines and can carry a single warhead or multiple independently targeted reentry vehicles (MIRVs).

military spending A category of expenses whose definition varies widely among countries and is often greatly undervalued in official accounts; for example, official Soviet military spending included merely operation and maintenance costs, and the real total (including weapons research and development and production) was roughly an order of magnitude higher.

nuclear weapons Both strategic and tactical warheads, bombs, and munitions, carried by intercontinental ballistic and medium- and short-range missiles or delivered by bombers or field artillery, whose destructive power is released either through nuclear fission or, in thermonuclear weapons, through fusion.

Wars demand an extraordinary mobilization of energy resources, they represent the most concentrated and the most devastating release of destructive power, and their common consequence is a major disruption of energy supplies in regions or countries that were affected by combat or subjected to prolonged bombing. Given these obvious realities, it is inexplicable that wars have received very little attention as energy phenomena. At the same time, there is a fairly common perception—one that has been greatly reinforced by the American conquest of Iraq in 2003—that energy is often the main reason why nations go to war. This article addresses all of these concerns. Because the destructiveness of war depends largely on the weapons used, it first outlines their brief history. The energy cost of individual armed conflicts, as well as the peacetime energy cost of preparing for war, is difficult to assess, and the article presents some representative calculations and cites some recent statistics using the major 20 th-century wars waged by the United States. Casualties are the most horrifying consequence of wars, and the greater destructive power led to their increase until the recent development of precision-guided munitions made it possible to minimize civilian deaths by careful targeting. Impacts of war on energy use during the conflict and in its aftermath, particularly in defeated countries that suffered a great deal of destruction, are clearly seen in consumption statistics. Finally, the article argues that the wish to be in actual physical control of energy resources has not been the sole reason, or even the primary reason, for any major modern armed conflict.

1. A Brief History of Weapons

Weapons are the prime movers of war. They are designed to inflict damage through a sudden release of kinetic energy (all handheld weapons, projectiles, and explosives), heat, or a combination of both. Nuclear weapons kill nearly instantaneously by the combination of blast and thermal radiation, and they also cause delayed deaths and sickness due to the exposure to ionizing radiation. Classification of wars based on these destructive prime movers divides history into four distinct periods. All prehistoric, classical, and early medieval warfare was powered only by human and animal muscles. Invention of gunpowder led to a rapid diffusion of initially clumsy front- and breach-loading rifles and to much more powerful field and ship guns.

By the late 1860s Nobel's combination of nitroglycerine and diatomaceous earth produced the first practical high explosive, and other variants were soon made into munitions for newly invented machine guns as well as into large-caliber gun shells and airplane bombs. Wars of the 20th century were dominated by the use of these high explosives delivered in shells (from land and ship-borne artillery and tanks), torpedoes, bombs, and missiles. Finally, strategic thinking and the global balance of power changed after World War II with the development of nuclear weapons.

301

The period of muscle-powered warfare lasted nearly until the very end of the Middle Ages. The daily range of advancing armies was limited less by the stamina of walking or running troops than by the speed of their supply trains, made up of animals ranging from oxen and horses in Europe to camels and elephants in Asia. Men used axes, daggers, swords, and lances in close combat and used spears and bows and arrows for attacking unprotected enemies as far as 200 m away. Much more powerful crossbows were used since the fourth century BC in both Greece and China. Inaccurate targeting and relatively low kinetic energy of these assaults (Table I) limited the magnitude and frequency of injuries that could be inflicted by these weapons. Massed human power was also used to wind winches of catapults and to move assorted siege machines designed to overcome urban fortifications.

Light horses were initially used to pull chariots, and only after the general adoption of stirrups (beginning in the third century AD) did mounted warriors become a particularly effective fighting force. Asian riders were unarmored, but they had small and extraordinarily hardy horses and could move with high speed and maneuverability. These abilities brought the Mongol invaders from the heart of Asia to the center of Europe between 1223 and 1241. In contrast, European knights relied on their heavy armor as they rode increasingly heavier animals. Their most spectacular series of long-distance forays brought them as Crusaders from many countries of Europe to the Eastern Mediterranean, where they established temporary rule (between 1096 and 1291) over fluctuating areas of the littoral. The only important nonanimate energies used in the

pre-gunpowder era were various incendiary materials (sulfur, asphalt, petroleum, and quicklime were used in their preparation) that were either fastened to arrowheads or hurled across moats and walls from catapults.

Gunpowder's origins can be traced to the experiments of medieval Chinese alchemists and metallurgists. Clear directions for preparing gunpowder were published in 1040, and eventually the proportions for its mixing settled at approximately 75% saltpeter (KNO_3), 15% charcoal, and 10% sulfur. Whereas ordinary combustion must draw oxygen from the surrounding air, the ignited potassium nitrate provides it internally and gunpowder undergoes a rapid expansion equal to roughly 3000 times its volume in gas. The first true guns were cast in China before the end of the 13th century, and Europe was just a few decades behind. Gunpowder raised the destructiveness of weapons and radically changed the conduct of both land and maritime battles.

When confined and directed in barrels of rifles, gunpowder could impart to bullets kinetic energy an order of magnitude higher than that of a heavy arrow shot from a crossbow gun, and larger charges in field artillery and ship guns could propel even heavy projectiles (Table I). Increasingly accurate gunfire from far beyond the range of archers eliminated the defensive value of moats and walls and did away with lengthy sieges of cities and castles. As a new defense, the fortified structures were built as low spreading polygons with massive earthen embankments and huge water ditches. French military engineer Sebastien Vauban became the most famous designer of these fortifications that embodied large amounts of energy. His largest project, at Longwy in northeastern France, required moving 640,000 m^3 of rock and earth (volume equivalent to about a quarter of Khufu's pyramid) and emplacing 120,000 m^3 of masonry.

The impact of guns was even greater in maritime engagements. Detailed historical accounts document how gunned ships (equipped with two other Chinese innovations, compass and good rudders, as well as with better sails) became the carriers of European technical superiority. By the beginning of the 16th century, they were the principal means of global empire building by the nations of the Atlantic Europe. By the late 17th century, the largest "men-of-war" carried up to 100 guns each. Dominance of these ships did not end until the introduction of naval steam engines during the 19th century.

The next weapons era began with the formulation of high explosives, which are prepared by the nitration of organic compounds such as cellulose, glycerine, phenol, and toluene. Ascanio Sobrero prepared nitroglycerin in 1846, but its practical use did not begin until Nobel mixed it with an inert porous substance (diatomaceous earth) to create dynamite and introduced a practical detonator (Nobel igniter)

Table I
Kinetic Energy of Projectiles.

Weapon	Projectile	Kinetic energy (J)
Bow and arrow	Arrow	20
Heavy crossbow	Arrow	100
Civil war musket	Bullet	1×10^3
Assault rifle (M16)	Bullet	2×10^3
Medieval cannon	Stone ball	50×10^3
18 th-century cannon	Iron ball	300×10^3
World War I artillery gun	Shrapnell shell	1×10^6
World War II heavy AA gun	High–explosive shell	6×10^6
M1A1 Abrams tank	Depleted U shell	6×10^6
Unguided World War II rocket	Missile with payload	18×10^6
Boeing 767 (September 11, 2001)	Hijacked plane	4×10^9

Table II
Detonation Velocity of Common Explosives.

Explosive	Density (g/cm^3)	Detonation velocity (m/s)
Gunpowder	1.0	1350
Dynamite	1.6	5000
TNT	1.6	6700
RDX	1.8	8800
ANFO	1.0	3200

without which the new explosive would be nearly impossible to use. A single comparison conveys the explosive power of dynamite: its velocity of detonation is as much as 6800 m/s compared with less than 400 m/s for gunpowder (Table II). Slower acting, and preferably smokeless, propellants needed for weapons were produced during the 1880s: *poudre B* (gelatinized and extruded nitrocellulose) by Paul Vieille in 1884, Nobel's own ballistite (using nitroglycerine instead of ether and alcohol) in 1887, and cordite, patented in England by Frederick Abel and James Dewar in 1889.

Ammonium picrate was prepared in 1886. Trinitrotoluene (TNT), which was synthesized by Joseph Wilbrand in 1863 and which must be detonated by a high-velocity initiator, was used as an explosive by the end of the 19th century. The most powerful of all prenuclear explosives, cyclonite (cyclotrimethylenetrinitramine, commonly known as RDX [royal demolition explosive] and now a favorite of some terrorists), was first made by Hans Henning in 1899 by treating a formaldehyde derivative with nitric acid (Table II). Better propellants and inexpensive, high-quality steels increased the power and range of field of naval guns from less than 2 km during the 1860s to more than 30 km by 1900. The combination of long-range guns, heavy armor, and steam turbines (a superior new prime mover invented by Charles Parsons during the 1880s that made much faster speeds possible) resulted in new heavy battleships, with *Dreadnought*, launched in 1906, being their prototype.

Other new highly destructive weapons whose use contributed to the unprecedented casualties of World War I included machine guns, tanks, submarines, and the first military planes (e.g., light bombers). The two decades between the world wars brought rapid development of battle tanks, fighter planes, and long-range bombers and aircraft carriers, all of which were the decisive weapons of World War II. The German defeat of France in 1940 and advances of Wehrmacht in Russia in 1941 and 1942 were made possible by rapid tank-led penetrations. Japan's surprising assault on Pearl Harbor on December 7, 1941, would have been impossible without a large carrier force that launched fighter planes and dive bombers.

The same types of weapons—Soviet T-42 tanks driving all the way to Berlin and American dive bombers deployed by the U.S. Navy in the Pacific—eventually led to the defeat of the Axis.

The closing months of World War II saw the deployment of two new prime movers and of an entirely new class of weapons. Gas turbines, independently invented during the 1930s by Frank Whittle and Hans Pabst von Ohain in the United Kingdom and Germany, respectively, were installed in the first jet fighters (the British Gloster Meteor and the German Messerschmitt 262) in 1944. During that same year, rocket engines were used in the German ballistic missile V-2 to terrorize England. And the first fission bomb was tested at Alamogordo, New Mexico, on July 11, 1945, with the second one destroying Hiroshima, Japan, on August 6, 1945, and the third one destroying Nagasaki, Japan, 4 days later.

Jet propulsion enabled the fastest fighter aircraft to surpass the speed of sound (in 1947) and eventually to reach maximum velocities in excess of mach 3. The postwar arms race between the United States and the Soviet Union began with the assembly of more powerful fission bombs to be carried by strategic bombers. The first fusion bombs were tested in 1952 and 1953, and by the early 1960s the two antagonists were engaged in a spiraling accumulation of intercontinental ballistic missiles (ICBMs). But these are not technically weapons of war given that their real purpose was to deter their use by the other side. However, to achieve this objective, the superpowers did not have to amass more than 20,000 nuclear warheads. The sudden end of this expensive arms race did not usher in an era of extended peace; ironically, it initiated an era of even more acute security threats.

In a complete reversal of dominant concerns, the most common—and generally the most feared—weapons in the new war of terror are both inexpensive and easily available. A few kilograms of high explosives fastened to the bodies of suicide bombers (often spiked with metal bits) can cause dozens of deaths and gruesome injuries and can create mass psychosis among the attacked population. Simple "ANFO" car bombs are much more devastating (Table III). These devices, some of which weigh hundreds of kilograms and are able to destroy massive buildings and kill hundreds of people, are made from the mixture of two readily available materials: ammonium nitrate (a common solid fertilizer that can be purchased at, or stolen from, thousands of locations around the world) and fuel oil (which is even more widely available).

The most shocking weapons were fashioned on September 11, 2001, by 19 Islamic hijackers simply by commandeering rapidly moving massive objects, Boeings 757 and 767, and steering two of them into the World Trade Center in New York and one into the Pentagon in Washington, D.C. The World Trade

Table III
Kinetic Energy of Explosives and Total Energy Released by Nuclear Weapons.

Explosive device	Explosive	Kinetic energy (J)
Hand grenade	TNT	2×10^6
Suicide bomber	RDX	100×10^6
World War II gun shrapnell	TNT	600×10^6
Truck bomb (500 kg)	ANFO	2×10^9

Nuclear bomb	Reaction	Total energy (J)
Hiroshima bomb (1945)	Fission	52×10^{12}
U.S. ICBM	Fusion	1×10^{15}
Novaya Zemlya bomb (1961)	Fusion	240×10^{15}

Center towers were designed to absorb an impact of a slow-flying Boeing 707 lost in the fog and searching for a landing at the JFK or Newark airport. Gross weight and fuel capacity of that plane are just slightly smaller (15 and 5%, respectively) than the specifications for the Boeing 767-200, and the structures performed as intended even though the impact velocity of hijacked planes as they rammed into the buildings was more than three times higher (262 vs 80 m/s) than that of a slowly flying plane close to the landing.

As a result, the kinetic energy at impact was approximately 11 times greater than envisaged in the original design (~ 4.3 GJ vs 390 MJ). But because each tower had the mass of more than 2500 times that of the impacting aircraft, the enormous concentrated kinetic energy of the planes acted much like a bullet hitting a massive tree. That is, it penetrated instead of pushing; it was absorbed by bending, tearing, and distortion of structural steel and concrete; and the perimeter tube design redistributed lost loads to nearby columns. Consequently, it was not this instantaneous massive kinetic insult but rather a more gradual flux of the ignited fuel (each 767 carried more than 50 metric tons (t) of kerosene whose heat content was more than 2 TJ) that weakened the columns of structural steel.

Unfortunately, an eventuality of such a fire was not considered in the original World Trade Center design. Moreover, no fireproofing systems to control such fires were available at that time. Once the jet fuel spilled into the building, it ignited an even larger mass of combustible materials (mainly paper and plastics) inside the structures, and the fires burned with a diffuse flame with low-power densities of less than 10 W/cm². This left enough time for most people to leave the buildings before the thermally weakened structural steel (the fuel-rich, open-air fire could not reach the 1500 °C needed to actually melt the metal), thermal gradients on outside columns, and nonuniform heating of long floor joists precipitated the staggered floor collapse that soon reached the free-fall speed as the towers fell in only approximately 10 s.

Although the preparation of the most feared viral and bacterial weapons of mass destruction requires a fair degree of scientific expertise and a high degree of technical expertise (particularly to prepare the cultured pathogens for widespread dispersal), the overall energy cost of their development is very low in comparison with their potential impact, be it measured in terms of actual casualties and ensuing population-wide fear or in terms of long-term economic losses. There is no doubt that even a largely failed large-scale smallpox or anthrax attack on a major urban area would send property values falling and would lead to the flight of residents and businesses to less crowded locations.

2. Energy Costs and Consequences of War

Preindustrial land wars, fought largely by foot soldiers with simple weapons, did not require large amounts of energy embodied in arms and equipment, but even so, that investment was often a significant part of annual energy use in many poor subsistence societies. Large armies in transit, or during periods of prolonged sieges, almost always had a deep, and often quite ruinous, effect on regional supplies of food and fuelwood as they commandeered their provisions from the surrounding countryside. In contrast, maritime forays far from friendly shores always required careful planning of supplies that had to be carried by the ships during months of self-sufficient sailing.

Modern wars are waged with weaponry whose construction requires some of the most energy-intensive materials and whose deployment relies on incessant flows of secondary fossil fuels (e.g., gasoline, kerosene) and electricity to energize the machines that carry them and to equip and provision the troops who operate them. Production of special steels in heavy armored equipment typically needs 40 to 50 MJ/kg, and obviously, the use of depleted uranium (for armor-piercing shells and enhanced armor protection) is much more energy intensive. Aluminum, titanium, and composite fibers, the principal construction materials of modern aircraft, typically embody 170 to 250 MJ/kg, as much as 450 MJ/kg, and 100 to 150 MJ/kg, respectively.

The most powerful modern war machines are naturally designed for maximized performance and not for minimized energy consumption. For example, America's 60-tM1/A1 Abrams main battle tank, powered by a 1.1-MWAGT-1500 Honeywell gas turbine, needs (depending on mission, terrain, and weather) 400 to 800 L/100 km; in comparison, a large Mercedes S600 automobile consumes approximately 15 L/100 km, and a Honda Civic needs 8 L/100 km. Jet fuel requirements of highly maneuverable supersonic combat aircraft, such as the F-16 (Lockheed Falcon) and F-18 (McDonnell Douglas Hornet), are so high that no extended mission is possible without in-flight refueling from large tanker planes (the KC-10, KC-135, and Boeing 767).

Moreover, these highly energy-intensive weapons have been used in increasingly massive configurations; hence, the overall energy cost of a conflict can mount rapidly. The most concentrated tank attack during the final year of World War I involved nearly 600 machines, whereas nearly 8000 tanks, supported by 11,000 planes and by more than 50,000 guns and rocket launchers, took part in the final Soviet assault on Berlin in April 1945. During the Gulf War in 1991 ("Desert Storm," January–April 1991) and the months leading to it ("Desert Shield," August 1990–January 1991), some 1300 combat aircraft flew more than 116,000 sorties and the supporting transport and aerial refueling planes logged more than 18,000 deployment missions.

Another historically recurrent phenomenon is the necessity to expand the mass production of these energy-intensive machines in the shortest possible period of time. In August 1914, Britain had only 154 airplanes, but just 4 years later, the country's aircraft factories were sending out 30,000 planes per year. Similarly, when the United States declared war on Germany in April 1917, it had fewer than 300 second-rate planes, none of which could carry machine guns or bombs on a combat mission, but 3 months later Congress approved what was at that time an unprecedented appropriation ($640 million or ∼$8 billion in 2000 dollars) to build 22,500 Liberty engines for new fighters. The situation was reprised during World War II. American industries delivered just 514 aircraft to the U.S. forces during the final quarter of 1940, but the total wartime production reached more than 250,000 planes, exceeding the combined output of Germany and Britain.

There are no detailed reasoned studies of energy cost of modern wars. This is not surprising given that even their financial costs cannot be accounted for with a fair degree of accuracy. This has to do primarily with deciding what to include in such accounts. When the very physical survival of a society is at stake, it becomes impossible to separate the output of such a wartime economy into easily identifiable civilian and military sectors and then to assign approximate energy costs to these activities. Available aggregates show the total U.S. expenditures on major 20th-century conflicts as approximately $250 billion for World War I, $2.75 trillion for World War II, and $450 billion for the Vietnam war (all in constant 2000 dollars). By June 30, 2008 all congressional appropriations for the post-9/11 conflicts (wars in Afghanistan and Iraq and all other Global War on Terror operations) added up to about $775 billion (again, in constant 2000 dollars, including appropriations from the fiscal year 2001 through part of 2009). Expressing these costs in monies of the day and multiplying the totals by the adjusted averages of respective average energy intensities of the country's gross domestic product (GDP) during those periods sets the minimum energy costs of these conflicts.

Adjustments to average energy intensities are necessary because the war-related industrial production and transportation require considerably more energy per unit of output than does the rest of economic activity. Given the typical sectoral energy intensities of past GDPs, this article uses the conservative multiples of 1.5 for World War I, 2.0 for World War II, and 3.0 for the Vietnam war and for the post-9/11 conflicts. Although this procedure cannot yield any accurate figures, it conveys well the most likely magnitude of the energy burden. This burden was approximately 15% of the total U.S. energy consumption during the 1917–1918 period (World War I), averaged roughly 40% during the 1941–1945 period (World War II), but was definitely less than 4% during the 1964–1972 period (main combat action of Vietnam war) and around 3% for the post-9/11 conflicts. Naturally, these shares could be significantly higher during the years of peak war endeavor. For example, the peak World War II spending, expressed as a share of national economic product, ranged from 54% in the United States (in 1944) to 76% in the Soviet Union (in 1942) and in Germany (in 1943).

Peacetime expenditures directly attributable to the preparation for armed conflicts can also be significant. For decades, the highly militarized Soviet economy was spending on the order of 15% of its GDP on the development, procurement, and maintenance of weapons and on its large standing army and border guard and paramilitary forces. In contrast, the U.S. defense spending reached nearly 10% of GDP during the peak year of the Vietnam war, fell to below 5% by 1978, rose to 6.5% during President Reagan's mid-1980s buildup, and has stayed below 5% since 1991: it fell to just 3% in 2000 and, despite the conflicts in Iraq and Afghanistan, it was still only about 4% by 2008, well below the mean of the preceding half a century (about 5.5%). Given the enormous U.S. energy consumption (equivalent to ∼2.5 Gt of oil in 2000), the direct peacetime use of fuel and electricity by the U.S. military is a tiny fraction of the total. For example, the U.S. Department of Defense claimed less than 0.9% of the country's total primary energy supply in 2007.

But the comparison looks different in absolute terms. The average direct annual consumption of primary energy by the U.S. armed services was equivalent to approximately 25 Mt of oil since 2000. That is roughly the same amount of primary energy consumed annually by Switzerland or Austria; more remarkably, it is a total higher than the commercial energy consumption in nearly two-thirds of the world's countries. Naturally, operations such as the Gulf War, the war in Afghanistan, and the Iraq war boost the energy use by military not only due to the combat but also due to enormous long-distance logistic support. And of course, there are additional (and not easily quantifiable) expenditures of energy for the military research and development (R&D) and the procurement of weapons. And it is even more difficult to translate long-term costs of caring for wounded and traumatized soldiers into energy equivalents. Because of the advances in battlefield medicine many more US soldiers have been surviving their injuries (the ratio of wounded in combat to killed has been 7:1 in Iraq compared to 2.6:1 in Vietnam and 2:1 during WW II) the long-term financial cost of caring for the veterans of the Iraq War will at least match the cost of waging the war, or it may even exceed it.

The link between energy use and success in modern war (or in preventing war) is far from being a simple matter of strong positive correlations. Of course, there is no doubt that the possession of nuclear weapons (the MAD [mutually assured destruction] concept) was the main reason why the two superpowers did not fight a thermonuclear war, but the nuclear stockpiles, and hence their energy cost, went far beyond any rational deterrent level. Development and deployment of these weapons, beginning with the separation of the fissile isotope of uranium and ending with the construction of nuclear submarines to carry them with nearly complete invulnerability during the extended submerged missions, has been highly energy intensive.

A conservative estimate might be that at least 5% of all U.S. and Soviet commercial energy that was consumed between 1950 and 1990 was claimed by developing and amassing these weapons and the means of their delivery. And the burden of these activities continues with expensive safeguarding and cleanup of contaminated production sites. Estimated costs of these operations in the United States have been steadily rising, and much greater investment would be needed to clean up the more severely contaminated nuclear weapons assembly and testing sites in Russia and Kazakhstan. Even so, given the potentially horrific toll of a thermonuclear exchange—even when limited to targeting strategic facilities rather than cities, the direct effects of blast, fire, and ionizing radiation would have caused 27 million to 59 million deaths during the late 1980s—one could argue that the overall cost/benefit ratio of the nuclear arms race has been acceptable.

Of course, there is no doubt that the rapid mobilization of America's economic might, which was energized by a 46% increase in the total use of fuels and primary electricity between 1939 and 1944, was instrumental in winning the war against Japan and Germany. In contrast, the Vietnam war was a perfect illustration of the fact that to win a war, it is not enough to use an enormous amount of explosives (the total was nearly three times as much as all bombs dropped by the U.S. Air Force on Germany and Japan during World War II) and to deploy sophisticated weapons (state-of-the-art jet fighters, bombers, helicopters, aircraft carriers, defoliants, etc.). The attacks of September 11, 2001, on the World Trade Center and the Pentagon illustrate the perils and penalties of the aptly named asymmetrical threats. In those attacks, 19 Muslim *jihadis*, at the cost of their lives and an investment that might not have surpassed $100,000, caused approximately 3000 virtually instantaneous deaths (and the death toll could have easily surpassed 10,000 if the towers were not so structurally sound), created enormous direct and indirect economic dislocations, and led to costly deployments of military and covert power.

While the direct costs of 9/11 attacks can be estimated with reasonable accuracy, quantifying the overall burden to the US and to the world economy depends heavily on the choice of analytical boundaries and on assumptions regarding the duration (and even causation) of some of the included effects (the key items being insurance losses, reduced air travel and tourism, higher costs of providing transportation security, higher risk premiums, higher operating costs for businesses, spending on the new Department of Homeland Security and, above all, direct and indirect costs of military intervention in Afghanistan and Iraq). As a result, various cost estimates have ranged from several hundred billion to more than $ 2 trillion. Unfortunately, there is no easy military solution for this predicament. Both the classical powerful weapons and the new "smart" machines are of very limited use in this new, highly asymmetric global war, as are most of the security checks and precautions that have been taken so far at the airports.

The two most important and unprecedented consequences of warfare waged by modern societies are the just noted extent of economic mobilization for major armed conflicts and the growing magnitude of destructive power. The latter trend has brought increased casualties, including larger numbers of noncombatants, and huge material losses. World War I was the unmistakable divide. Enormous resources had to be rapidly diverted to war production, and the death toll, in both absolute and relative terms, surpassed that of any previous experience. The increasing destructiveness of modern wars is best illustrated by comparing the overall conflict casualties. Battle deaths, expressed as fatalities per 1000 men of armed forces fielded at the beginning of a conflict, were less than 200 during the Crimean War of 1853–1856 and the Franco-Prussian War of 1870-1871. They surpassed 1500 during World War I and 2000 during World War II (when they were more than 4000 for Russia).

Civilian casualties of modern warfare grew even faster. During World War II, they reached approximately 40 million, more than 70% of the 55 million total. Roughly 100,000 people died during nighttime raids by B-29 bombers using incendiary bombs that leveled approximately 83 square kilometers of Japan's four principal cities between March 10 and March 20, 1945. Five months later, the explosion of two nuclear bombs, which killed at least 100,000 people, released energy of 52.5 and 92.4 TJ, respectively (Table III). At that time, only a few people envisioned how much more powerful these weapons would get. Expressed in common units of TNT equivalents (1 t TNT = 4.184 GJ), the two bombs dropped on Japan rated 12.5 and 22 kt, respectively. The most powerful thermonuclear bomb tested by the Soviet Union, over the Novaya Zemlya on October 30, 1961, rated 58 Mt, equal to 4600 Hiroshima bombs, and by 1990 the total power of U.S. and Soviet nuclear warheads surpassed 10 Gt, the equivalent of 800,000 Hiroshima bombs.

Effects of major wars on energy consumption, be it in defeated countries or in the aftermath of major civil conflicts, are usually quite pronounced. The Soviet Union after the civil war of 1918–1921 and Germany and Japan after their defeat during the late 1940s are the best documented examples. Japan's accurate historical statistics reveal the magnitude of this impact. In 1940, the country's primary energy supply was an equivalent of approximately 63 Mt of oil; by 1945, the total had fallen nearly exactly by half; and in 1946, the aggregate declined by another 10%. The 1940 level was not surpassed until 1955, 10 years after the surrender.

3. Energy as the Cause of War

Finally, this article concludes with a few paragraphs on energy resources as the *cassus belli*. Many historians single out Japan's decision to attack the United States in December 1941 as a classic case of such causation. In January 1940, President Roosevelt's administration, determined to aid the United Kingdom that was under Nazi attack, abrogated the 1911 Treaty of Commerce and Navigation with Japan. In July 1940, Roosevelt terminated the licenses for exports of aviation gasoline and machine tools to Japan, and in September the ban was extended to scrap iron and steel. Already by July 1940, some of Japan's top military officers warned that the navy was running out of oil and that they wanted a quick decision to act. An attack on the United States was to clear the way for the assault on Southeast Asia with its Sumatran and Burmese oilfields.

Although it would be wrong to deny the proximate role that Japan's declining oil supplies might have played in launching the attack on Pearl Harbor, it is indefensible to see Japan's aggression as an energy-driven quest. The attack on Pearl Harbor was preceded by nearly a decade of expansive Japanese militarism (so clearly demonstrated by the 1933 conquest of Manchuria and by the attack on China in 1937), and Marius Jansen, one of the leading historians of modern Japan, wrote about a peculiarly self-inflicted nature of the entire confrontation with the United States. No convincing arguments can made to explain either Hitler's serial aggression—against Czechoslovakia (in 1938 and 1939), Poland (in 1939), Western Europe (beginning in 1939 and 1940), and the Soviet Union (in 1941)—or his genocidal war against Jews by concerns about energy supplies.

The same is true about the geneses of the Korean War (North Korea is the coal-rich part of the peninsula), the Vietnam war (waged by France until 1954 and by the United States after 1964), the Soviet occupation of Afghanistan (1979–1989), and the U.S. war against the Taliban (launched in October 2001) as well as about nearly all cross-border wars (e.g., Sino–Indian, Indo–Pakistani, Eritrean–Ethiopian) and civil wars (e.g., Sri Lanka, Uganda, Angola, Colombia) that have taken place (or are still unfolding) during the past two generations in Asia, Africa, and Latin America. And although it could be argued that Nigeria's war with the secessionist Biafra (1967–1970) and Sudan's endless civil war had clear oil components, both were undeniably precipitated by ethnic differences and the latter one began decades before any oil was discovered in the central Sudan.

On the other hand, there have been various indirect foreign interventions in the Middle Eastern countries—through arms sales, military training, and covert action—that aimed at either stabilizing or subverting governments in the oil-rich region. Their most obvious manifestation during the cold war was the sales (or simply transfers) of Soviet arms to Egypt, Syria, Libya, and Iraq and the concurrent American arms shipments to Iran (before 1979), Saudi Arabia, and the Gulf states. During the 1980s, these actions included strong and, as it turned out, highly regrettable Western support of Iraq during its long war with Iran (1980–1988). This brings us to the two wars where energy resources have been widely seen as the real cause of the conflicts.

By invading Kuwait in August 1990, Iraq not only doubled crude oil reserves under its control, raising them to approximately 20% of the world total, but it also directly threatened the nearby Saudi oilfields (including the four giants—Safania, Zuluf, Marjan, and Manifa—that are on- and offshore just south of Kuwait) and, hence, the very survival of the monarchy that controls 25% of the world's oil reserves. Yet even in this seemingly clear-cut case, there were other compelling reasons to roll back the Iraqi expansion. At that time, the United States was importing a much smaller share of its oil from the Middle East than were Western Europe and Japan (25% vs more than 40% and 67%, respectively, in 1990), but the Iraqi quest for nuclear and other nonconventional weapons with which the country could dominate and destabilize the entire region, implications of this shift for the security of U.S. allies, and risks of another Iraqi–Iranian or Arab–Israeli war (recall Saddam Hussein's missile attacks on Israel designed to provoke such a conflict) certainly mattered a great deal.

There is one very obvious question to ask: if the control of oil resources was the sole objective, or at least the primary objective, of the 1991 Gulf War, why was the victorious army ordered to stop its uncheckable progress as the routed Iraqi divisions were fleeing north, and why did it not occupy at least Iraq's southern oilfields? Similarly, more complex considerations were behind the decision to conquer Iraq in March 2003. Although no stockpiles of weapons of mass destruction were discovered, it could be argued that Hussein's past actions and the best available prewar evidence did not leave any responsible post-September 11 leadership in the United States with the luxury of waiting indefinitely for what might happen next. Such a wait-and-see attitude was already assumed once, with fatal consequences, after the World Trade Center bombing in February 1993.

Of course, a no less important reason for the Iraq war has been the grand strategic objective of toppling the Baathist regime and to eventually replace it with a representative government that might serve as a powerful and stabilizing political example in a very unsettled region. Once again, after the events of September 11, one could argue that no responsible U.S. leadership could have ignored such a risky but potentially immensely rewarding opportunity to redefine the prospects for a new Middle East. Any naïve claims about oil rewards of the Iraqi invasion are convincingly dispelled by indisputable realities. Iraq continues to sell its oil as a member of OPEC, it gives no discounts to any buyers, imports of the Iraqi oil supplied less than 5% of all US oil purchases (dominated by Canadian, Saudi, Mexican and Venezuelan sales), no US-based multinational oil companies have received any preferential contracts for the development of Iraqi oil, and the first oil deal signed by a new Iraqi government in 2008 (and inaugurated in March 2009) was actually with the Chinese National Petroleum Company. Those who see the US invasion of Iraq as a quest for oil have perhaps never thought about its benefit/cost ratio: in order to buy annually 25–30 Mt of Iraqi oil — oil that

could have been easily imported from other producers and for which the American consumers have been paying the world price (over $ 8 billion/year at March 2009 prices, nearly $ 25 billion during the year 2008) – the US has spent so far (adding up 2003–2009 congressional appropriations for the Operation Iraqi Freedom) nearly $ 700 billion, with a large part of it borrowed from its overseas creditors. And so even this instance, which so many commentators portray so simplistically as a clear-cut case of energy-driven war, is anything but that—yet again confirming the conclusion that in modern wars, resource-related objectives have been generally determined by broader strategic aims and not vice versa.

Further Reading

Belasco A 2008 *The Cost of Iraq, Afghanistan, and Other Global War on Terror Operations Since 9/11.* Congressional Research Service, Washington, DC

Clark G B 1981 Basic properties of ammonium nitrate fuel oil explosives (ANFO). *Colorado School Mines Q* **76**, 1–32

Committee for the Compilation of Materials on Damage Caused by the Atomic Bombs in Hiroshima and Nagasaki 1981 Hiroshima and Nagasaki. Basic Books, New York

Eagar T W, Musso C 2001 Why did the World Trade Center collapse? Science, engineering, and speculation. *J. Metals* **53**, 8–11. www.tms.org/pubs/journals/jom/0112/eagar/eagar-0112.html

Federation of American Scientists 1998. Operation Desert Storm. FAS, Washington, DC. www.fas.org/man/dod-101/ops/desert_storm.htm

Federation of American Scientists 2003. America's War on Terrorism. FAS, Washington, DC. www.fas.org/terrorism/index.html

Jansen M B 2000 *The Making of Modern Japan.* Belknap Press, Cambridge, MA

Lesser I O 1991 *Oil, the Persian Gulf, and Grand Strategy.* RAND, Santa Monica, CA

McNeill W H 1989 *The Age of Gunpowder Empires, 1450–1800.* American Historical Association, Washington, DC

MoveOn.org. (2002). Energy and War. www.moveon.org/moveonbulletin/bulletin8.html#section-9

Robertson L E 2002 Reflections on the world trade center. *The Bridge* **32**(1), 5–10

Singer J D, Small M 1972 *The Wages of War 1816–1965: A Statistical Handbook.* John Wiley, New York

Smil V 1994 *Energy in World History.* Westview, Boulder, CO

Smil V 2008 *Energy in Nature and Society: General Energetics of Complex Systems.* MIT Press, Cambridge, MA

Stiglitz J E, Bilmes L J 2008 *The Three Trillion Dollar War: The True Cost of the Iraq Conflict.* W.W. Norton, New York

Urbanski T 1967 *Chemistry and Technology of Explosives.* Pergamon, Oxford, UK

Vaclav Smil

Wind Energy, History of

By the first decade of the 21st century, wind power had become the best hope for the future of alternative energy. Of all the possible "new" sources of electricity, it had rather unexpectedly found itself the most touted and fastest growing alternative energy resource in the world, generating significant amounts of electricity in several countries, such as the United States, Denmark, Germany, and Spain, as well as impressive amounts in many other countries. The new era of wind development was led by the United States during the 1980s, but Europe has overtaken that ranking, accounting for two-thirds of total worldwide wind development by itself in 2001. Now a global multi-billion-dollar industry, wind energy is regaining its once prominent place in the energy firmament. For millennia, wind power has been used for everything from kite flying and propulsion to grinding and pumping, but never before has it been so important for the generation of electricity. It is an old resource applied to a new mission with unprecedented success. This article describes how wind power came to achieve this status.

1. Harnessing the Wind

1.1 Prehistory Use

For longer than one can grasp, winds have been churning seas, carrying silt, eroding continents, creating dunes, rustling leaves, and filling sails. They are responsible for loess soils in China, sand seas in Libya, and dust storms in Arizona. They have created, destroyed, altered, eroded, and changed the shape of land, the vegetation that would cover it, and the lives of millions of people.

Wind energy exists without end, yet at no time have humans ever employed more than a tiny fraction of its kinetic energy. Its power is so beyond human control that cultures throughout the world have paid it homage in their legends and mythologies. Creation stories of nearly all cultures involve the power of the wind.

Like the mystery of who invented the wheel, we do not know when humans first employed the force of the wind to do work. However, in terms of utility, sailing would seem to be one of the most likely early uses, perhaps 40,000 years ago when Asians migrated to greater Australia. Although we know little of these voyages, the far later water journeys of Polynesians in their double-hulled sailing craft are unquestioned. For these early sailors, wind provided the kinetic energy for the exploration and settlement of the Pacific Ocean islands, including the Hawaiian Islands, where it later led to the invention of surfing, the earliest use of wind-powered recreational activity.

1.2 Historic Uses of the Wind

As in prehistory, the first historic use of wind power involved its use in transportation. Egyptians plied the Nile River as early as 3100 BC using craft equipped with sails of linen and papyrus. In *The Odyssey*, Homer wrote of Odysseus, who sailed the Ionian and Aegean seas, eventually making a crucial error when he angered the god of the winds. Aeolus stopped the wind for 6 days, forcing Odysseus and his crew to rely on rowing, a very poor substitute. "No breeze, no help in sight, by our own folly—six indistinguishable nights and days," lamented Odysseus. The story of Odysseus reminds one of the capricious nature of the wind. Although it can be used, it cannot be put under human control any more than ocean currents can be put under human control. For this reason, wind has been successfully employed only sparingly, with inventors and engineers preferring power sources that they can manipulate as needed. Yet the importance of wind energy to human history is undeniable. From the time of Columbus until the middle of the 19th century, nations depended on sailing vessels. Advances in navigation and sail design combined to make the great age of the sailing craft feasible. In the billowing sails of thousands of ships for thousands of years, the following paragraphs trace the paths of commerce and conquest.

Whereas once the longest history of using wind power was on the sea and not the land, today the reverse is true. While the importance of sailing has faded to recreational use, the terrestrial applications of wind have multiplied. Where and when did people first turn to the wind to help in their harvests? We do not know exactly, but we do know that in the 10th century AD, windmills were turning in the blustery Seistan region of Persia (now Iran). For this, we have both artifact proof and written evidence. These were primitive windmills by modern standards, using vertical sails of reed bundles, and their owners built them to grind grain and lift water from streams to irrigate gardens. For centuries, these rudimentary yet ingenious machines were carried to other parts of the world, including India and China, where farmers employed them to pump water, grind grain, and crush sugarcane.

Independently, the post mill arose in England. Unlike the Persian sails, which follow a carousel's path around a vertical axis, the sails of the traditional European windmill follow a path around a horizontal axis. The whole assembly of the post mill, including the blades, axle, and milling cabin, rests on a massive vertical post. The miller orients the post mill into the wind by swinging the entire windmill around the post, hence the name post mill. In 1137, William of Almoner designed and constructed the first such mill in Leicester, England. For many years, historians of technology assumed that the English post mill idea was a by-product of the Christian Crusades, with the English post mill being inspired by the Persian horizontal windmills of Seistan. However, the designs of the two mills are very different, and the evidence points to multiple inventions rather than diffusion from a single source.

Environmental and social reasons help to explain why the post mill took root where it did in England. Much of the land in Leicester is level, limiting the use of water wheels. However, perhaps just as significant was social class. In England, the kings had vested the water rights in the nobility and in the established church. These rights were jealously guarded, and in a practical sense they gave control of energy production to the nobility. In this earlier world of vested and hereditary privilege came a new source of energy, one that the upper class could not control. In such an atmosphere, the wind became the great leveler of class. As the middle-class entrepreneur Herbert of Bury, Suffolk, stated in 1180, "The free benefit of the wind ought not be denied to any man."

Spreading throughout England, the use of the post mill eventually made its way to the European continent. By 1300, one saw horizontal axis windmills similar to those in England in Spain, France, Belgium, The Netherlands, Denmark, the German principalities, and the Italian states. Subsequently, the more powerful and more elaborate tower mill replaced the simple post mill, especially in The Netherlands.

As the name implies, tower mills reached skyward by using wood framing, brick, and stone. The tower, like those of the modern wind turbine, was immovable. The cap with the rotor, horizontal axle, and gearing rested on a curb at the top of the tower. The miller pointed the rotor into the wind by swinging the cap on its curb. During the "golden age" of the European windmill, the Dutch alone operated some 10,000 tower mills, and they did more than pump water and grind grain. For example, they reduced pepper and other spices, cocoa, dyes, chalk, and paint pigments. Lumber companies employed them as primary power for sawmills. Paper companies used windmills to reduce wood pulp to paper. One authority estimated that the mills provided as much as 25% of Europe's industrial energy from 1300 to the coming of the steam engine and cheap coal during the 1800s. The remainder came from hydro-power, assisted by human and animal labor.

1.3 Wind Energy in the New World

Although European countries explored and colonized the New World, there is disagreement as to whether the windmill accompanied them. The seal of New York City contains a windmill, early woodcuts of "Nieuw Amsterdaam" depicted windmills, and there are other examples of early windmills as well. But the European windmill was simply too large, cumbersome, and expensive to be easily adapted to New World conditions. Also, it was a labor-intensive device in a new land where labor was scarce and valuable. Any apparatus that required continual attention would have found limited use on North American soil. Furthermore, the climate and topography of the New England region favored water wheels, and by 1840 more than 50,000 water wheels were in operation. It was the energy source of choice.

The windmill did eventually find a home in the United States, albeit of a design different from that used in Europe, in the vast western part of the country. Inventors there came up with a blueprint for what would become known as the "American farm windmill." Adapted to a land of little rain but abundant wind, the American farm windmill was small, light, movable, self-regulating, inexpensive, and easy to maintain. It liberated groundwater in an arid area, bringing it to the surface and making agriculture possible. Between 1850 and 1900, an impressive and bewildering number of styles and brands of farm windmills appeared on the market. There is no way in which to estimate accurately how many such windmills once dotted the Great Plains, but some authorities have suggested a number as high as 6 million. These windmills added a vertical dimension to the otherwise horizontal space of the Great Plains and once were the most common landscape feature between the Mississippi and the Rockies. Today, the sight of an American windmill evokes a certain nostalgia for a simpler bucolic age. The windmill became an American icon.

1.4 The Marriage of Wind and Electricity

The development of wind power to grind grain and pump water was historically significant, but during the 21 st century it has been the marriage of wind and electricity that offers the greatest promise. However, it was not a union that took place quickly. In 1748, Benjamin Franklin wrote a scientific friend that, in regard to electricity, he was "chagrined a little that we have hitherto been able to produce nothing in the way of use to mankind." It would be many years before the first practical use of electricity appeared in the form of the telegraph in 1844.

Although the telegraph was used extensively for more than half a century, little more was done with electricity until the first generating plants began operating during the final quarter of the 19 th century. After that time, the use of electricity caught on quickly, and by 1900 it was becoming an essential element of the urban lifestyle. In this atmosphere of discovery, inventors contemplated the coupling of wind power and electricity. Journals such as *Scientific American* challenged American ingenuity to use "the natural forces which are in play about us." The most pressing problem was how to store this elusive form of energy. How could one use the power of the wind if one could not store electricity, especially given that the wind did not necessarily blow when one needed it? By the early 1880s, the French inventor Camille Faure had devised what he termed a "box of electricity" or storage battery. It seemed possible to generate electricity from the wind, store it, and use it when needed.

The first to build a practical large-scale wind turbine was Charles Brush, a scientist from Cleveland, Ohio, who had made a fortune through his electric arc lighting system. The basement of his Euclid Avenue mansion became his laboratory, but his experiments were limited by the lack of electricity. To resolve the problem, Brush in 1886 designed and constructed a wind turbine that was of immense proportions. Visitors to his 5-acre backyard saw a 40-ton (46,300-kg) tower rising 60 feet (18.3 m) into the air. From the tower turned a rotor 56 feet (17.1 m) in diameter with 144 slender blades. Within the tower, he located his dynamo and the necessary gearing to drive it. In his basement, Brush installed 12 batteries. The turbine and batteries infallibly provided electricity for his 100 incandescent lights, three arc lights, and a number of electric motors. The Brush wind dynamo worked incredibly well for

approximately 15 years, using "one of nature's most unruly motive agents," as *Scientific American* reported. The inventor used his machine only occasionally after 1900, which was the year Cleveland began offering the convenience of centrally generated electricity. He abandoned his machine in 1908.

Despite its success, the Brush wind dynamo was never duplicated. Although fuel was free, the machinery to capture it was not. Furthermore, Brush's extensive knowledge of both electricity and the working of the wind dynamo could not be easily replicated elsewhere. Although a smaller similar design was employed in New England by the Lewis Electrical Company (the first commercial wind turbine), it did not prove to be popular. Brush offered the public a concept, but not one that could be mass produced.

1.5 The Search for a Practical Wind Turbine

Experimentation continued sporadically between 1890 and 1920, but these decades represented the pause between invention and application. It was only after World War I that some mechanically minded Americans applied advances in aeronautics to the design of a practical inexpensive wind turbine. In 1920, few North American farmers had electricity, and those who did produced it with gasoline-powered generators such as Delco "light plants." By the late 1920s, wind turbines began to compete in the effort to provide the American farm family with the convenience of electricity. A number of companies produced wind turbines, but the most successful were Jacobs's Windelectric and Wincharger. Joe and Marcellus Jacobs were not engineers but rather farm boys with an inventive flair. They experimented on their parents' eastern Montana ranch. At first, they converted a multibladed farm windmill from pumping water to generating electricity, but it turned too slowly. Marcellus had learned to fly, and he soon realized that a three-bladed propeller might be a better solution to the problem of wind-powered electrical generation. The brothers eventually perfected the rotor blades, the blade feathering system, and a powerful generator that would characterize their wind turbine. Soon their ranch neighbors were purchasing the Jacobs's turbines. Uniquely, all of their experimentation and manufacturing took place on the family ranch. It was a perfect wind laboratory, but soon the brothers realized that truly building turbines in quantity would require relocation to a midwestern industrial center. They chose Minneapolis, Minnesota, in 1927, and between that year and 1957 the Jacobs brothers turned out approximately 30,000 small wind turbines. The Jacobs machines became legendary for their reliability. A Christian missionary in Ethiopia installed a Jacobs plant in 1938. In 1968, the missionary sent for a replacement set of generator brushes—the first repair in 30 years of operation. In 1933, Admiral Byrd installed a Jacobs plant on a 70-foot (21.3-m) tower at Little America. When Byrd abandoned his Antarctica outpost, he left the wind turbine. When his son, Richard Byrd, Jr., visited the site in 1947, the Jacobs was "still turning in the breeze," although only 15 feet (4.6 m) of the tower was free of ice. In 1955, one of the 1933 veterans returned and removed the blades from the still operating turbine because they were threatened by the advancing ice.

Although Jacobs marketed its wind turbine as the "Cadillac" of windchargers, many ranchers and farmers found that a "Chevrolet" would be sufficient. In general, they purchased a Wincharger brand from a company that produced models from 6 to 110 V and from 200 to 3000 W R. E. Weinig, the general manager of Wincharger, testified in 1945 that some 400,000 of the company's wind plants operated worldwide. Other brands included the Miller Airlite, Universal AeroElectric, Paris–Dunn, Airline, Wind Kind, and Winpower. In 1946, the Sears Roebuck catalog marketed a Silvertone Aircharger for $32.50. Most of these turbines were small, powering a radio and perhaps a couple of 40-W lights, but nevertheless they filled the desire for electricity in rural regions of North America.

Meanwhile, active research on larger wind turbines was under way in Europe. In 1900, the world's expert on electricity-generating wind turbines was not Charles Brush but rather Poul La Cour, a Danish scientist who spent most of his professional life working on wind electric systems. By 1906, with the support of the Danish government, 40 wind turbines were generating electricity in the small nation, beginning a tradition that has lasted to this day. To the south, Germany capitalized on La Cour's work and moved forward under the leadership of engineer Hermann Honnef and others. The editor of *Scientific American* noted that wind power in Germany "has been developed to a point that is surprising." Across the channel, the British showed a similar interest. In 1924, Oxford University's Institute of Agricultural Engineering tested seven turbines from five manu-facturers. Two years later, the institute reported that the "cost of windmill-generated electricity for small lighting and small power purposes was found to be quite reasonable and, as such, justifies its wider use in rural districts where there is no general supply."

1.6 The Victory of Centralism over Individualism

By the 1930s, wind turbines had found a particular place in the rural landscape in both Europe and the Americas. However, as noted in the Oxford report, their use was recommended only "where there is no general supply" of electricity. In other words, engineers and entrepreneurs had become committed to central power systems and the creation of a grid of transmission lines to carry the electricity. The idea of individual power units was anathema to engineers who thought in terms of electrical systems that served large populations from a central generating source. Of course, private utility companies complained that getting power to rural America was costly. Stringing wires was expensive, and utility companies refused to serve rural homes unless they were in close proximity to each other and to urban centers. In 1930, only approximately 10% of American farm families could boast electricity despite a much higher percentage in urban areas. It was at this time that the Rural Electrification Act (REA) came into existence.

The REA, passed in 1936, would change these statistics dramatically in the United States. The act called for local farmers to establish cooperatives with the authority to make loans to bring electrical power to farms within their designated region. It was enormously successful. No one would deny the ac-complishment of the REA, but the federally backed cooperatives sounded the death knell for the wind-charger industry. Centralization and government subsidies for the cooperatives killed the industry. One wind energy executive attempted to persuade Con-gress to include windchargers as part of the REA in more isolated areas, but neither Congress nor REA administrators would consider such a deviation from their commitment to a centralized power system. By 1957, every American wind electric company had closed its doors.

1.7 Large Wind Machine Experiments

While federal legislation stilled the small, farm-oriented wind turbines, one engineer was thinking on a much grander scale. A young engineer named Palmer Cosslett Putnam first became interested in wind energy in 1934 when he built a house on Cape Cod and, as he put it, "found both the winds and the electric rates surprisingly high." Putnam had in mind hooking up to utility company power, using it but sparingly, and selling back the surplus from his wind turbine. However, no mechanism existed at that time to convert direct current from a windcharger to alternating current that central station power plants delivered. Putnam researched the problem, and as an engineer committed to "economies of scale," he decided to build a huge experimental wind turbine that would generate alternating current identical to that from conventional power plants.

Putnam stripped bare a rounded knoll near Rutland, Vermont. There he erected a massive two-bladed turbine with S. Morgan Smith Company, a manufacturer of large hydroelectric turbines. Truckers struggled to bring some 500 tons (90,700 kg) of material and parts to the crest of Grandpa's Knob. The 70-foot (21.3-m) length of the two 7.5-ton

(6800-kg) blades posed a particular challenge. However, on October 19, 1941, the rotor began to turn. It operated for approximately 16 months, producing 298,240 kWh in 695 h of online production. On February 20, 1943, a bearing failed, resulting in a 2-year hiatus because both parts and labor were difficult to obtain during World War II. Production finally resumed, but not for long. On March 26, 1945, one of the blades separated from the rotor and crashed to the ground and on down the side of the mountain. Although Putnam promised that he and his partners would repair the turbine, the expense of doing so was prohibitive. The great turbine never turned again.

What is most remarkable about the Smith–Putnam project was not its successful operation but rather the fact that it happened at all. Putnam had to rely on private industry for machinery, expertise, and (above all) funding. Fortunately, S. Morgan Smith bankrolled his effort. A number of distinguished scientists from prominent universities participated, as did some of the leading engineering firms of the day. Vannevar Bush, then the dean of engineering at the Massachusetts Institute of Technology, put it well when he wrote that the project was "conceived and carried through free enterprisers who were willing to accept the risks involved in exploring the frontiers of knowledge, in the hope of financial gain." Obviously, there was no financial gain, but in its brief life the turbine certainly opened new frontiers of knowledge for the wind energy field.

2. The First U.S. Boom

2.1 The Search for Alternatives

For less than 22 years, Americans held hope in the promise that nuclear power was going to provide all of the electricity they could ever need, but it was not to be. The idea that blossomed with such promise at the Shippingport Atomic Power Station west of Pittsburgh, Pennsylvania, in December 1957 ironically shriveled in the same state with the accident at the Three Mile Island nuclear power plant near Harrisburg in April 1979. The event, including the chaotic public and governmental responses it triggered, punctured the nuclear balloon, resulting in the cancellation of dozens of nuclear power plants. None has been ordered in the United States since then, and after the 1986 catastrophe at the Chernobyl plant in Ukraine, few have been built anywhere in the world. These events, coupled with the embargoes on oil and the constraints on coal burning during the 1970s, stimulated a more serious evaluation of alternative energy resources.

Although several resources did garner more attention, there was never a wholehearted effort. Nascent public and private alternative energy programs existed in scattered, small, and even persistent

forms in many areas, but they received little more than perfunctory support from government agencies and large energy companies. Even the high-profile advocacy of the Carter administration did not have much staying power in terms of national policy, although it helped to build a foundation for later advances in some locations, especially California.

California was famously short on energy resources, and these deficiencies were to become painfully obvious over the years. During the second half of the 19th century, people relied on wood, but it became scarce rather quickly. Various forms of hydrocarbons were used for lighting, and when electricity became popular, hydropower was quickly put to work. However, its practical limits were also soon reached as California continued to grow. Soon the state ranked first in economic might, and feeding its appetite required huge amounts of electricity. Having virtually no coal, insufficient natural gas, and oil that was too expensive to use in power plants, and facing rising opposition to nuclear energy, Californians turned to alternative resources that have always been available in the state. At first, geothermal energy was the resource of choice, and it grew quickly. Yet ultimately its luster dulled, and by the mid-1980s serious attention was shifting to another resource—wind.

The attention given to wind power in California resulted not from the fact that the state was a dominantly windy place but rather from the convergence of local demand, state and federal financial subsidies, a favorable political climate, and the geographical convenience that found three sites with great promise virtually in the backyards of millions of people. Anyone who has driven the stretch of Interstate 10 through San Gorgonio Pass 100 miles east of Los Angeles, or the stretch of Interstate 580 through Altamont Pass 50 miles east of San Francisco, knows the strength and persistence of winds there. Even windier, albeit less traveled, Tehachapi Pass 50 miles east of Bakersfield was the third target of early development (Fig. 1).

With all of the favorable conditions for wind development finally in place by the mid-1980s, thousands of wind energy conversion systems (WECS) sprouted with a suddenness that startled the public. In unexpected numbers and with unexpected vigor, people began to complain how the wind turbines changed aesthetics, increased bird mortality, produced electronic interference, and created noise. However, there was no turning back; California's experiment in large-scale development of wind power was just getting started.

2.2 Reactions and Early Lessons

The early stages of modern wind development in California faced some persistent barriers. Wind turbines are unavoidably visible and impossible to hide,

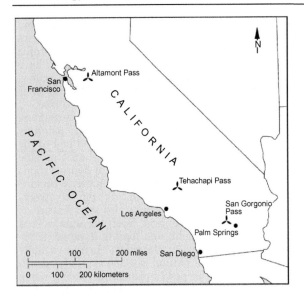

Figure 1
Principal developed wind resource areas of California.

especially in those locations where they were being concentrated in the state. Complaints about these new landscape additions continued, although they diminished in intensity as the technology improved and regulatory controls were imposed. Once financial subsidies ceased during the mid-1980s, wind power was left to survive on its own, and this was not possible for all installations or designs. A weeding out would soon take place as wind developers continued to face public complaints, technical difficulties, and financial challenges. Poorly sited and inoperable equipment was slowly removed, and regulators began flexing their bureaucratic muscles. The authorities who governed development in places like San Gorgonio Pass imposed controls on height, color, and reflectivity, instigated operational standards on design and noise, and required decommissioning bonds and protection of migrating birds and endangered species.

Despite the multiple regulations and engineering improvements that are now part of wind energy development in the United States, nothing can render them invisible and, at the same time, still economical to run. Simply put, the only way in which to hide wind turbines is by using the buffer of space. But in California, most wind turbines are along busy highways, always available for public scrutiny. Because of the slipshod way in which many developments were first installed, wind power received an early black eye that has taken years to heal. In the meantime, development activities picked up elsewhere, especially in Europe.

On the heels of the California experience and benefiting from improvements prompted by that experience, European wind developers began moving quickly to install their own projects. Their ambitions were encouraged by shortages of sufficient local resources, concerns about nuclear power and global warming, the political weight of "green" parties, and ample subsidies that became available to alternative energy developers. All of these factors taken together were to push wind development in Europe ahead of those in the United States during the 1990s, but Europeans' attraction to wind power had actually started years earlier.

3. The European Emphasis

3.1 The Early Years

Denmark has assumed a leadership role in the development of wind power in Europe, but it took a path far different from that taken by the United States. Unlike American wind turbine design, which was concentrated in the hands of the aerospace industry, Danish wind technology grew out of the agricultural sector as a natural by-product of the Danish economy and the long history that Danes have with the wind. They used it to power the ships of their early conquests and to carry Norsemen as far as the New World. So-called "Dutch" windmills eventually became a common sight in Denmark, with a some 3000 of them providing the equivalent of half again as much energy as all the animal power then supporting Danish agriculture by 1890. In 1891, La Cour, the "Danish Edison," began experimenting with wind-generated electricity, and as early as 1903, La Cour's Danish Wind Power Society was fostering the idea of wind-generated electricity.

By the end of World War I, more than one-fourth of all rural power stations in Denmark used wind turbines. During the long wartime blockade, the 3 MW provided by these crude wind generators and the widespread use of small farm windmills for grinding grain were invaluable to the impoverished rural population. Although most windmills were used for mechanical power, it has been estimated that wind turbines were providing the equivalent of 120 to 150 MW in Denmark by 1920. Nearly 90 turbines were installed during World War II, including the 30-kW Lykkegaard wind turbine patterned after La Cour's *klapsejlsmølle*, and F. L. Smidth's more modern "Aeromotors." This design, like those of La Cour, used four or five wide blades resembling those of traditional windmills except that the blades were covered with metal shutters instead of cloth sails. These wind turbines were controlled by opening and closing the shutters. In contrast, the Smidth company became one of the world's first firms to marry the rapidly advancing field of aerodynamics to wind turbine design. The Smidth company's wind turbines incorporated both technologies and used modern airfoils upwind of a concrete tower.

After the war, interest in wind energy again waned, although Johannes Juul remained resolute. In 1950, he began testing a prototype wind turbine for the Danish utility SEAS, and subsequently he modified a Smidth turbine used on the island of Bogø. With the experience gained from developing these two machines, Juul began work on his crowning achievement, the three-bladed, stall-regulated, upwind rotor at Gedser that spanned 24 m (79 feet) in diameter. Installed in 1956, the Gedser mill operated in regular service from 1959 through 1967. By the late 1970s, modern wind turbine manufacturers and experimenters erected their first prototypes. By 1980, when the European wind energy conference was held north of Copenhagen, there were several designs in operation.

Meanwhile, Ulrich Hütter was designing and building wind turbines for the Nazi-owned Ventimotor company outside of Weimar, Germany, in the hope of reducing reliance on foreign sources of energy. Hütter's team experimented with several designs, including one of the first uses of a wind turbine to drive an asynchronous or induction generator directly coupled to the electric utility network. After the war, Hütter went on to become a prominent aeronautical designer who continued experimenting with wind energy. During the late 1950s, he installed a novel turbine that operated at high rotor speeds with only two slender blades.

At the opposite technical pole from Hütter, Juul was a traditionally trained engineer working outside of academia. Juul's design grew out of the Danish craft school tradition, and he built on the experience gained by Smidth during the war. Juul built a turbine for the Danish utility SEAS at Gedser incorporating the lessons learned. Meanwhile, both the British and the French installed prototype wind turbines during the 1950s and 1960s.

The availability of cheap oil from the Middle East doomed most of these programs, but Juul's turbine at Gedser and Hütter's turbine in the Schwabian Alps continued in operation for the next decade. Hütter's design and his argument for it would captivate German and American engineers for years to come. However, it was Juul's design and its adherents in Denmark that would come to dominate the field of wind energy some decades later.

Juul's simple robust design was adopted by the Danish wind revival of the late 1970s following the oil crises of the period. When the California "wind rush" began during the early 1980s, budding Danish wind turbine manufacturers were poised to seize the opportunity to export technology and expertise. Europeans of several nationalities flocked to California to seek their fortunes in wind energy, but none more so than the Danes. The trickle of Danish machines entering California in 1981 soon became a torrent. At the peak of the rush in 1985, Europeans shipped more than 2000 wind turbines, mostly Danish ones, to the United States. Later, after the collapse of the California market in 1985, Denmark looked elsewhere and helped Germany to become the leading wind power country in the world. Indeed, in 2001, Germans installed a record 2600 MW, and all of these wind turbines were technologically derived from Juul's work at Gedser.

3.2 Dependable Home Market

In 1981, Denmark set a national goal of installing 1000 MW of wind power by the year 2000. Denmark became self-sufficient in oil a decade later following discoveries in the North Sea, but the country continued its wind development program as a means of reducing greenhouse gas emissions. Denmark later increased wind's expected contribution to 10% of the nation's electricity supply by the year 2000, a target it exceeded. In 2001, Denmark was generating 16% of its electricity with wind energy (Fig. 2).

The late 1970s brought a change that helped to further wind power development in Denmark when the European market for Danish farm equipment slackened, forcing manufacturers to seek new products for their rural customers. With generous incentives from the Danish government stimulating demand, Danish manufacturers quickly adapted their

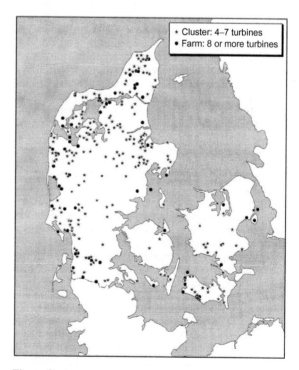

Figure 2
Wind-generating installations in Denmark as of February 2002.

surplus capacity to the new wind turbine market. The demand for modern wind turbines by a population spread across the Danish landscape, a good wind resource off the North Sea, and a manufacturing sector accustomed to building heavy machinery for a discerning rural market all contributed to launching the world's most successful domestic wind industry. In addition, the small size of the country allowed manufacturers to service their own turbines, often directly from the factory. This spatial proximity enabled companies to learn quickly from their mistakes and to keep their turbines in operation as physical proof to potential buyers that the companies' machines were good investments.

Danes pride themselves on their unusual ability to act individually while working cooperatively. Even though most farms are owner operated, nearly all Danish farmers are members of farm cooperatives. These cooperatives process the Danish foods found on shelves worldwide. Danish wind turbine cooperatives and an association of wind turbine owners have had a profound effect on the development of wind energy in the country. Some 100,000 Danish households ($\sim 5\%$ of the population) own shares in wind cooperatives.

Another contributor to Danish success was the Danish Windmill Owners Association (*Danske Vindkraftværker*). Formed in 1978, the group grew from a series of informal quarterly meetings of backyard experimenters, hobbyists, and environmentalists. Because many early wind turbines were unreliable, the group demanded minimum design standards. The most important element in the future success of wind energy was the requirement for a failsafe redundant braking system such as the tip brakes invented by Juul during the 1950s for use on his Gedser turbine. This single provision, more than any other, furthered Danish wind technology because it succeeded in ensuring the survival of the wind turbine when something went wrong. The owners association also compiled statistics on the reliability of Danish turbines and compelled Danish manufacturers to fulfill their product guarantees when the turbines underperformed. Collective action by Danish wind turbine owners was so successful that German wind turbine owners replicated the format south of the Danish–German border.

Wind energy grew rapidly in Denmark, not only due to early government incentives but also because Danish utilities were required to pay a premium price for wind-generated electricity. By agreement with the government, utilities paid 85% of the retail rate for electricity from privately owned wind turbines. Coupled with credits against energy taxes, Danish utilities were paying the equivalent of U.S. \$0.10/kWh for wind-generated electricity during the mid 1990s in comparison with retail rates of nearly \$0.17/kWh.

Many factors have contributed to the success of wind power in Denmark. One is two decades of consistent national policies supporting a strong domestic market for Danish wind machines, mostly for farmers and cooperatives. Another is the often-overlooked backing of Danish financial institutions; Danish banks and finance societies during the 1990s provided 10- to 12-year loans for 60 to 80% of the installed cost. Banks competed for wind projects by advertising themselves as "the wind banks" in local newspapers.

3.3 Nationwide Land Use Planning

Although thousands of wind turbines have been installed in Denmark by individuals without much notoriety or difficulty, Danish utilities encountered public opposition when they proposed wind projects of their own. These companies were viewed as outsiders, and the new neighbors of their proposed wind turbines considered them to be no different from conventional power plants. In response, and at the utilities' request, the government appointed a special wind turbine siting committee in 1991 to identify locations for wind turbines with a broad base of acceptability. Land use planning became necessary as wind turbines increased in size dramatically since the early 1980s and as their numbers continued to increase. The government committee estimated that the Danish landscape could absorb 1000 to 2800 MW of wind capacity, taking into account local objections and the preservation of scenic areas. Regional Danish jurisdictions now include wind energy in their land use plans. These plans designate zones where wind turbines are prohibited, where single turbines may be installed, where wind farms or wind power plants are permitted, and where clusters or single turbines may be erected with special approval. Many municipalities, including the capital city, also include wind turbines in their local plans.

There are wind turbines in urban areas throughout Denmark and three projects within the environs of Copenhagen. One project, Lynetten, is visible from the Christianborg Palace, the seat of Denmark's parliament, the *Folketing*. The turbines are also visible from the Little Mermaid, the most visited tourist attraction in the city. Today, wind turbines in Denmark are dispersed across the landscape, in contrast to the giant wind farms typically seen in North America, India, and Spain. In fact, many Danish installations include only one turbine. In total, after 16 years of development, Denmark reached 2000 MW of installed wind capacity and paved the way for other countries to have similar success.

3.4 Germany Becomes the Leader

Wind developments in Germany differ somewhat from those in Denmark. The German market clearly demonstrates the effectiveness of their incentives to stimulate the growth of wind energy. In 1991, Germany's conservative government introduced the *Stromeinspeisungsgesetz* (electricity feed law), requiring utilities to pay 90% of their average retail rates for purchases from renewable energy sources such as wind turbines. The law, which encompassed only a few paragraphs, resulted in extensive wind development. By the early 21st century, wind turbines were in every part of Germany, from polders on the North Sea to hilltops of the central highlands (Fig. 3).

The parliament (*Bundestag*) modified the feed law in 1999, stabilizing both the price and the period during which the price would be paid. The measure ensured a stable and continually expanding domestic market. In 2001, wind turbines were producing 3.5% of the electricity in Germany, a nation of 80 million and the world's third largest economy. As a result of its precedent-setting feed law, Germany had the most dynamic and transparent markets for wind turbines in the world during the first decade of the new millennium. Like Denmark, Germany's large domestic market spawned the growth of some of the world's largest wind turbine manufacturers. Also like

Denmark, many of the wind turbines in Germany were installed as single turbines or small clusters.

3.5 Other Markets

There are several other wind markets in Europe outside of Denmark and Germany. For example, development began in earnest in Spain during the mid-1990 s when the government introduced its own electricity feed law. Under the law, Spanish projects can choose a fixed price per kilowatt-hour of wind-generated electricity or the wholesale rate. Most have chosen the fixed price tariff. As a result of the combination of the country's electricity feed law, good wind resources, and supportive regional governments that see wind energy as a means of economic development, Spain has a bustling wind industry that rivals that of Germany. In 2001, Spain was the world's second largest manufacturer of wind turbines.

Great Britain and The Netherlands have decent wind energy resources, although they have had different experiences. Both countries launched wind development programs during the early 1980s but have lagged behind their European neighbors. Both misdirected research and development funds toward electric utilities and large industrial concerns in centrally directed programs that were incapable of supplying competitive products for a limited domestic market. Furthermore, programs designed to stimulate a market instead often antagonized prospective neighbors by forcing wind development onto the most sensitive sites. Although Great Britain has the best wind resource in Europe (Fig. 4), organized opposition nearly brought the industry to collapse during the late 1990s. Despite an early beginning and an excellent wind resource, total installed wind capacity in both Great Britain and The Netherlands represented only 5% of the total installed in Germany in 2002.

France, with its longtime emphasis on nuclear power, has been one of the last European countries to turn their attention to the possibilities of wind energy. After watching how wind development was stymied in Great Britain and being familiar with the opposition to new nuclear construction, France sought to adapt the German feed law to French conditions. The French electricity feed offers three fixed tariffs: one price for windy sites, a second price for modest wind resources, and a third price for the least windy areas. It also limits the size of the wind farms that can be built under the program. These provisions encourage a greater number of smaller projects than those found in North America and spread wind development beyond just a few windy provinces such as Brittany in the west and Languedoc in the south.

Several other countries, inside Europe as well as elsewhere, are installing wind generators. Within Europe, these include Italy, Greece, and Sweden. Outside of Europe, wind development is prominent

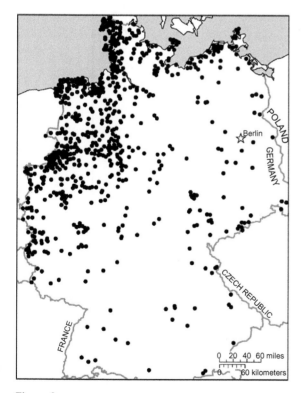

Figure 3
Wind-generating installations in Germany as of September 2001.

Figure 4
Wind-generating installations in the United Kingdom as of January 2002. Courtesy of British Wind Energy Association.

in India, particularly in the states of Gujarat and Tamil Nadu. Several large projects were also built in North Africa and China during the late 1990s. By the turn of the 21st century, wind turbines in commercial projects were operating on every continent.

4. Rejuvenated North America

4.1 The Great Plains

After the burden of bad press coming out of California, progress in wind power in the United States dwindled virtually to a standstill during the late 1990s, just as it was picking up momentum in Europe. Ironically, support and success in Europe rekindled the American market. Texas has been one of the leaders of this resurgence, followed by several other states such as Minnesota, Iowa, Oregon, Washington, and Pennsylvania (and soon perhaps the Dakotas and Massachusetts). Many reasons explain this renewed interest. First, wind technology has matured, becoming more efficient and reliable and less threatening to wildlife, especially to birds. Second, certain economic incentives now exist for utility companies to promote the resource, most specifically their ability to capture "credits" as a part of renewable portfolio standards. For example, these standards in Texas require utilities to provide a percentage of their electricity from renewable energy. Once this step was taken, Texas surged to nearly 1100 MW by 2001, second only to California's 1700 MW Most of this new Texas capacity is in West Texas, in the vicinity of Iraan and McCamey. The quick growth in Texas is due in part to a policy that allows companies who exceed the minimum required percentage to sell their additional credits to other providers.

A third reason behind the renewed growth of wind power in the United States is that farmers have learned that wind power can make them money and help them to keep their land. For this reason, the Great Plains seem destined to assume the leadership position in U.S. wind power development and perhaps the world, a result of a combination of natural and cultural conditions. Most notably, the U.S. wind resource is concentrated there as the result of favorable climatic and topographic conditions. It was estimated that the potential in the Dakotas would match 60% of the electrical demand for the entire country in 1990 (Fig. 5 and Table 1).

Another important reason for the increased activity in the Great Plains is found in the region's history of land use. More than in any other part of the country, agriculture is the economic mainstay of the region, and it has fallen on hard times, particularly at the expense of the small farm. Word has spread quickly across the Great Plains that wind turbines can substantially improve the economic return from property in the region. In fact, farmers in many areas today generate more income from electricity than from traditional farm products while at the same time using only a small portion of their land for the physical wind equipment. Ironically, the very area of the country that several decades ago was home to the greatest concentration of windmills in the world will possibly become so again, this time having them pumping out electrons instead of water.

In addition to Texas, development on the Great Plains has been most noticeable in Iowa and Minnesota, where 324 and 319 MW of generating capacity, respectively, were installed at the time of this writing. Unlike Texas, where most of the wind installations are on marginal grazing land or in spent oilfields, the wind turbines in Iowa and Minnesota are on working farms.

Also on the Great Plains, albeit farther west, wind developments have been appearing in Wyoming ranching country, where at the end of 2001 there was about 140 MW of installed generating capacity. In some locations in that state, such as near Interstate 80 west of Laramie, the wind is ferocious, and for the first time it is being seen not as a burden of life in the area but rather as an asset. In the Great Plains, NIMBY (not in my backyard) is being supplanted by a new acronym, PIMBY (please in my backyard).

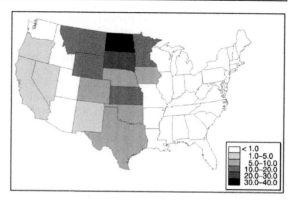

Figure 5
Wind energy potential as a percentage of total electric consumption. Reprinted from Pacific Northwest National Laboratory. (1991). "An Assessment of the Available Wind Land Area and Wind Energy Potential in the Contiguous United States." PNNL, Richland, WA.

Table 1
Wind Potential in Selected States.

Rank	State	Billions of kilowatt-hours
1	North Dakota	1210
2	Texas	1190
3	Kansas	1070
4	South Dakota	1030
5	Montana	1020
6	Nebraska	868
7	Wyoming	747
8	Oklahoma	725
9	Minnesota	657
10	Iowa	551
11	Colorado	481
12	New Mexico	435
13	Idaho	73
14	Michigan	65
15	New York	62
17	California	59[a]

[a]Illustrates that numerous states have greater wind potential than does California, where the majority (~ 90%) of U.S. wind development has occurred to date.
Source. Data from Pacific Northwest National Laboratory. (1991). "An Assessment of the Available Windy Land Area and Wind Energy Potential in the Contiguous United States." PNNL, Richland, WA.

4.2 The Pacific Northwest

The other location of substantially increased wind power installations has been the Pacific Northwest, an area that recently has witnessed the second fastest growth after Texas. Several sites in the windy Columbia River Gorge east of Portland are being monitored, and the Stateline Project, along the eastern border of Washington and Oregon, amounts to approximately 180 MW of capacity. Here too it is making money for local people; according to an article in the March 25, 2002, issue of the *Register Guard* of Eugene, Oregon, Northwestern Wind Power is paying 60 farmers in four counties alone to build wind turbines on at least 150,000 acres. Another developer, SeaWest Wind Power Inc., has leased tens of thousands of acres from at least 20 landowners in Oregon and Washington. In Sherman County, some farmers can expect a $10,000 sign-up payment, $15,000 for each turbine installed, and up to $5000 per turbine per year. That does not include the money that developers pay to reserve the land—approximately $2000 per month—while they conduct wind studies. Here, as in the Great Plains, wind development is becoming a profitable business.

4.3 Nantucket Sound

Objections to wind power, infrequent in the Northwest and Great Plains, still exist in some areas where population is concentrated, especially in those areas that are considered attractive tourist destinations. During recent years, such public opposition, which began in California, has been concentrated on a site off the coast of Massachusetts, where 170 turbines have been proposed to spread across the sea on a 28-acre patch on Horseshoe Shoal in Nantucket Sound, 9 miles off Martha's Vineyard and as close as 4 miles to Cape Cod. Objections, particularly from an organization called the Alliance to Protect Nantucket Sound, have focused on the visual intrusion of these turbines and to the potential disruption of

traditionally productive fishing grounds. At least 10 other wind power projects are being considered elsewhere in New England, but because the $700 million Cape Wind project came first, the discussion about offshore wind power is concentrating here. Regardless of how the controversy off Cape Cod is resolved, it has underscored a key point in the future potential for this renewable energy source. In some places wind power is welcomed warmly, whereas elsewhere it is opposed with vehemence.

5. Trends

Wind power has become the best hope for the future of alternative energy. Of all the possible "new" sources of electricity, it has rather unexpectedly found itself the most touted and fastest growing alternative energy resource in the world, generating significant amounts of electricity in several countries, such as the United States, Denmark, Germany, and Spain, as well as impressive amounts in many other countries (Fig. 6).

Development of wind power has been following several trends. One has been the effort toward greater economies of scale that results through the use of multimegawatt turbines. Most of the early turbines of this size were abandoned by the mid-1980s after failing in three areas: hours of operation, energy generated, and contribution toward making wind technology more competitive with conventional fuels. Another round of large turbines during the late 1980s and early 1990s performed somewhat better than their forebears, renewing hope among government-sponsored laboratories that large wind turbines did have a future after all, but they too were not very

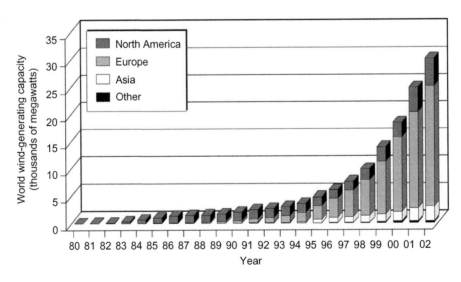

Figure 6
World generating capacity, illustrating the steep and recent rise in installed wind power.

successful. In 1990, the European Commission intended to chart the way for a third round of funding for large wind turbine development. In contrast to the past, funds were directed toward wind turbine manufacturers rather than the aerospace industry or other large industries. European governments, like their North American counterparts, had promoted megawatt-sized wind turbines for more than a decade with little success. The European Community worried that the progression toward larger wind turbines produced by Danish and German manufacturers would take too long to reach the size that program managers in Brussels, Belgium, believed were the optimum for Europe.

Manufacturers participating in a more aggressive program erected their prototype megawatt-class turbines during the mid-1990s. Commercial production began during the late 1990s, and by early 2000 increasing numbers of the turbines were being installed in Germany. In 2000, the average wind turbine installed in Germany was exceeding 1 MW for the first time, a milestone that took much longer to achieve in the United States because development rested not with wind manufacturers but rather with the National Aeronautics and Space Administration (NASA). However, circumstances eventually improved. By 2002, megawatt-scale wind turbines, some with more than 100 turbines of 1.5 MW each, were becoming common in Europe and on large wind farms in North America.

Another trend has been to install new turbines not on land but rather offshore. This strategy has been practiced with special vigor in Europe in the hope that offshore locations would reduce siting conflicts. Because offshore turbines require more expensive infrastructure, projects have longer lead times and must be larger than those on land. For this reason, such projects have been slow to develop, but the Danes and Swedes completed small near-shore projects during the late 1990s. One large project of 202-MW turbines, Middlegrunden, was installed outside of Copenhagen's harbor in 2000. In 2002, Denmark began installing 300 MW of offshore capacity, including the 160-MW Horns Rev project in the North Sea and the 158-MW R.0dsand project in the Baltic Sea. To try to visualize the impact of the turbines in advance, planners have taken to using computer visualization, for example, at Samsø (Fig. 7).

6. Conclusion

In most ways, wind power has come of age. The technology has improved, the economics has become more appealing, and substantial progress has been achieved in reducing environmental impacts. Planting wind turbines as a kind of semipermanent new "crop" on the Great Plains is becoming more common, making wind power more of a welcomed intrusion than a stimulus for citizen opposition.

Where public perceptions continue to slow growth in wind power, attitudes are largely influenced by benefits versus costs, and this has been most apparent off Cape Cod. Although one might argue that the sight of wind turbines helps to remind the public of the costs of their energy requirements, the proposed installation of fields of wind turbines in view of a heavily frequented tourist destinations is perhaps more than can reasonably be supported. In this case and probably others, the wind industry would be well advised to concentrate their developments where they are wanted and leave other environmentally sensitive areas alone. If one can agree that wind power has come of age, it will make a valuable contribution to providing us with the electricity we need in a world with increasing sensitivity to its environmental costs.

Figure 7
Visualization of a suggested offshore wind farm near Samsø, Denmark. The original idea was to place 10 turbines in a circular array. The project was expected to be built in 2003 in a linear array. Courtesy of Frode Birk Nielsen.

Further Reading

Ackermann T, Söder L 2002 An overview of wind energy—Status 2002. *Renewable Sustainable Energy Rev.* **6**, 67–128

Asmus P 2000 *Reaping the Wind: How Mechanical Wizards, Visionaries, and Profiteers Helped Shape Our Energy Future.* Island Press, Washington, DC

Baker T L 1985 *A Field Guide to American Windmills.* University of Oklahoma Press, Norman

Gipe P 1995 *Wind Energy Comes of Age.* John Wiley, New York

Golding E W 1955 *The Generation of Electricity by Wind Power.* E. & F. N. Spon, London

Pasqualetti M J, Gipe P, Righter R W (eds.) 2002 *Wind Power in View: Energy Landscapes in a Crowded World*. Academic Press, San Diego

Putnam P C 1948 *Power from the Wind*. Van Nostrand Reinhold, New York

Righter R W 1996 *Wind Energy in America: A History*. University of Oklahoma Press, Norman

van Est R 1999 *Winds of Change: A Comparative Study of the Politics of Wind Energy Innovation in California and Denmark*. International Books, Utrecht, Netherlands

Martin Pasqualetti, Robert Righter, and Paul Gipe

Wood Energy, History of

Glossary

Bronze Age A period between the Stone and Iron Ages characterized by the manufacture and use of bronze tools and weapons. This smelting required huge volumes of wood.

Mycenaean Age in Greece (600–1100 BC) A period of high cultural achievement, forming the backdrop and basis for subsequent myths of the heroes. It was named for the kingdom of Mycenae and the archaeological site where fabulous works in gold were unearthed. The Mycenaean Age was cut short by widespread destruction ushering in the Greek Dark Age.

smelting To melt or fuse a metal-bearing ore for the purpose of separating and refining the metal.

Wood was the primary fuel for the entire world from the discovery of fire to the age of fossil fuels. It remains the primary fuel to this day for those living in much of the developing world.

1. The Importance of Wood in Human Evolution

It may seem bold to assert wood's crucial place in human evolution. However, consider that trees have provided the material to make fire, the key to humanity's domination over all other animals. It allowed *Homo erectus,* from which *Homo sapiens* evolved, to migrate from their warm niche in equatorial Africa to lands throughout the Old World, where heat from fire made these colder climates habitable. Increasing the range of habitat gave humanity better odds for survival. Light from wood fires permitted *H. erectus* to work after nightfall. Hence, the discovery freed humanity from the strictures of the diurnal cycle. Wood fires also enhanced humanity's access to food. With torches, *H. erectus* hunted at night. This gave them the element of surprise in killing diurnal animals and added nocturnal animals to their diet. Fire also widened the types of available food, freeing them from the constraints of the hunt. Vegetable matter such as tubers, when cooked over fire, became eatable. A diet based on local plants promoted population stability because humans no longer had to follow migrating herds to eat. A wider choice of food also enhanced survival and proliferation (Fig. 1). Fire also provided humans with greater security. They could safely live indoors out of harms way because they had the means of illumination. Lighting also warded off nocturnal predators. It lessened the chance of stepping on

venomous snakes or tripping over rocks by improving vision at night. Wood fires changed not only the social evolution of humanity but also its physical development. Eating food softened by cooking led to smaller molars. Although many animals, such as chimpanzees and otters, have used tools as aids, no other animal except the human genus beginning with *H. erectus* has ever built fires.

2. The Importance of Wood in the Development of Civilization

Only by cooking grains over wood-fueled fires do grains become edible, thus enabling the agricultural revolution of the past 10,000 years (Figs. 2 and 3). In charcoal-fueled kilns, potters turned earth into durable ceramics in which goods could be stored and shipped. Providing an efficient means of storage stimulated trade on land and sea. Charcoal-fueled fires also gave humanity the capability to extract metal from ore, resulting in tools and weaponry so revolutionary that historians categorize society's evolution according to the dominant metal in use at the time. Wood fuel, in fact, is the unsung hero of the technological developments that brought us from a bone and stone culture to the industrial revolution.

3. The Importance of Wood Appreciated by Our Antecedents

Lucretius, probably the most scientifically oriented thinker of Antiquity, believed that the discovery of fire made civilization possible. He conjectured that the technique of metallurgy was born when great fires "devoured the high forests and thoroughly heated the earth," melting metal embedded in rock.

Figure 1
Native Americans using wood for cooking. From Perlin Forestry Archives.

Figure 2
Medieval people carry firewood from the forest to their
hearths. From Perlin Forestry Archives.

Figure 3
Wood fires turned grains into eatable bread. From
Perlin Forestry Archives.

When people saw the hardened metal, "the thought
then came to them," according to Lucretius, "that
these pieces could be made liquid by heat and
cast into the form and shape of anything, and then
by hammering, could be drawn into the form of
blades as sharp and thin as one pleased, so they might
equip themselves with tools." Lucretius remarked
that tools, in turn, made forestry and carpentry
possible, enabling humans "to cut forests, hew timber, smooth, and even fashion it with auger, chisel,
and gouge." By this process, in Lucretius' opinion,
civilization emerged.

Ibn Khaldun, writing in the 14th century ad, discussed the crucial role that wood fuel played in the
world of Islam. In "The Muqaddimah," his major
work, he noted that "God made all created things
useful for man so as to supply his necessities and
needs. Trees belong among these things. They give
humanity its fuel to make fires, which it needs to
survive."

The English of the 16th and 17th centuries
also recognized the crucial role of wood fuel in their
lives. Gabriel Plattes, writing in 1639, observed that
all tools and instruments are made of wood and iron.
Upon weighing the relative importance of the two
materials, he chose wood over iron because without
wood for fuel "no iron can be provided."

4. Examples of Wood Used for Fuel Over the Millennium

4.1 Bronze Age Greece and Cyprus

Mycenaean Greece attained an unprecedented level
of material growth during the late Bronze Age.
However, it did not have much copper to make sufficient quantities of bronze, the essential metal of
the time. To sustain the booming economy of the
Mycenaean states, abundant and reliable sources of
copper had to be found. Fortunately, there was one
area close by, Cyprus, that had lots of copper ore and
plenty of wood with which to smelt and refine it. In
response to this need, the Cypriots smelted as much
copper ore as they possibly could for the overseas
market. However, the growing demand put a great
burden on the island's woods because charcoal was
the fuel for smelting and refining copper. One hundred and twenty pine trees were required to prepare
6 tons of charcoal needed to produce 1 ingot of

copper, deforesting approximately 4 acres. Underwater archaeologists found on board a Bronze Age shipwreck 200 ingots of copper that had been mined and smelted in Cyprus. The production of just this shipload cost the island approximately 24,000 pine trees. The lively commerce in ingots during the 14th and 13th centuries BC surely consisted of many such shipments and the concomitant deforestation of a large expanse of woodlands. The industry's consumption of wood deforested approximately 4 or 5 square miles of woods per year. Another 4 or 5 square miles of forest was cut to supply fuel for heating and cooking and for other industries, such as pottery works and lime kilns. The cumulative effect of deforestation on such a scale must have been felt quite soon on an island of only 3600 square miles.

Cutting trees for fuel apparently made significant changes in the flora of Cyprus during the late Bronze Age. Bronze axes found dating to this period suggest that logging was an important occupation. Those wielding the axes cleared the majority of forest along the coast and hillsides near the larger late Bronze Age sites in Cyprus. As a result, pigs, which thrive in a moist, woody habitat, could no longer survive, giving way to sheep and goat tending, which flourish in a relatively barren environment.

With the forest cover gone, natural calamities struck. Silting of major harbors and an increase in flooding and mudslides in urban areas ensued. The Tremithos River, for example, transported tons of soil from the deforested hills below the Troodos Massif and deposited them into the Mediterranean. Currents moved these deposits of sediment toward the important late Bronze Age town of Hala Sultan Tekke. So much alluvium accumulated near the mouth of its harbor that during the 12th century BC. Hala Sultan Tekke was sealed off from the sea and could no longer function as a port.

Similarly, excessive amounts of alluvium carried by the Pedieos River, which flowed through Enkomi, a major Bronze Age city that faced the Near East, formed a delta as it drained into the ocean, changing Enkomi from a coastal to inland city. The number and severity of floods and mudslides also increased as a consequence of deforestation, periodically transforming the streets of the city into raging torrents full of mud and other debris.

With wood difficult to come by, interest heightened in the search for ways to stretch limited supplies, resulting in technological advances and recycling. The adoption of hydrometallurgy to prepare copper ore for smelting appears to be one of the strategies chosen by late Bronze Age Cypriot metalworkers to conserve fuel. The process required the exposure of mined ore to the elements so that the ambient moisture would leach impurities. The leached ore could be directly smelted, circumventing an initial roast and reducing the number of smelts needed to reduce the ore. Hence, the amount of fuel expended

in this phase was decreased by one-third. In another move to save fuel, metallurgists collected old and broken tools in order to resmelt them. Recycling bronze became a major source of the metal during the period of wood depletion on Cyprus.

Despite such highly innovative ways to save energy, metallurgists could not sustain the high productivity of the previous two centuries. Copper production peaked in approximately 1200 BC, and the last copper furnaces were shut down in 1050 BC. During the same time period, 90% of the island's settlements were abandoned and the population withered away, as did the economy and material culture.

The depressed state of the Cypriot copper industry apparently affected the entire eastern Mediterranean region. Bronze was in such short supply that the smiths throughout the area bought whatever pieces they could find, whether virgin ingots, ingots composed of recycled bronze, or scrap.

Often, smiths in southern Europe could obtain only minute quantities of bronze because of copper shortages. Such was the case in Greek Messenia in approximately 1200 BC. With so little bronze available, everyone suffered. Without metal hoes, plowshares, and scythes, farmers were far less productive. Without axes, adzes, and saws, few ships could be built for commerce or war. Without enough arrowheads, spearheads, and blades for swords, soldiers were no match for a well-armed foe.

Taking advantage of Messenia's vulnerability, a group of insurgents overpowered local forces and destroyed the palace at its capital, Pylos. After the catastrophe, the populace did not attempt to rebuild. Events at Messenia presaged trouble that other societies in the eastern Mediterranean would soon face. Just as bronze gave these civilizations the material to expand to heights never before attained, conversely, the lack of bronze played a major role in their demise.

In this sea of troubles, new hope for future generations arose. The fuel crisis that had caused copper production to decline served as the incentive for metallurgists to begin working with iron. Because Bronze Age Cypriots smelted copper ore that contained 10 times more iron than copper, metallurgists could obtain more useable iron than copper with the same investment of fuel. Hence, common sense dictated switching to iron smelting when fuel was at a premium. Furthermore, the refuse from Cypriot copper smelting contained significant amounts of iron. As long as there had been plenty of fuel, metallurgists smelted virgin ore and ignored the slag that had accumulated. When fuel became scarce and forced production cuts, metallurgists began to mine their industrial garbage. They discovered that the slag contained a great amount of iron, which could be removed simply by hammering. Taking iron out of the slag manually permitted metallurgists to bypass any processes requiring fuel and still obtain usable metal. The labor-intensive nature of working in this manner drastically reduced the overall output of metal on Cyprus and yielded relatively small amounts of iron. However, the success metallurgists had in working with iron at this early stage laid the foundation for the coming of the Iron Age in the Mediterranean as well as the rest of Europe.

4.2 Classical Greece

In classical Greece, the silver smelted at Laurion, southeast of Athens, near the sea, paid for the Athenian fleet that defeated the Persians at Salamis, turning the tide in the war against the Persians. Coins minted with Laurion's silver were accepted as the currency of the Greek world. With its treasury full of bullion, the Athenian economy could well afford to spend lavishly. Miners did not recover pure silver at Laurion but galena ore. Once mined, metallurgists had to heat the ore to very high temperatures to extract the silver. Charcoal was the fuel metallurgists at Laurion used in the smelting process and trees, or course, had to be cut down to produce this charcoal (Fig. 4). To extract the silver from ore at Laurion required the burning of more than 24 million pines or more than 52 million oaks. The largest outlays of fuel occurred during the two most active periods of mining, from 482 to 404 BC and from the second decade of the fourth century BC to its end. The surrounding areas could supply only a fraction of Laurion's fuel, forcing metallurgists to locate their furnaces on the coast so they could easily receive fuel imported from Thrace by ship.

4.3 Classical Rome

Rome financed its growth largely with silver extracted from Spanish ore. Production increased considerably during the end of the republic and the first years of the empire. However, this was accomplished only by great expense to the Iberian woodlands since silver smelting consumed more than 500 million trees during the 400 years of operation. Woodsmen had to deforest more than 7000 square miles to provide fuel for the furnaces. Near the end of the period of peak production, the need to sustain high output so strained the area's fuel supplies that it merited intervention by the Roman state. Under the reign of the emperor Vespasian, the Roman government directed all mining areas of southwestern Spain to prohibit the sale of burnable wood by those who ran bathhouses in the region (Fig. 5).

To produce enough silver to support the habits of a succession of rulers who spent as extravagantly as Caligula and Nero, a time had to come when the tree supply in Spain would dwindle and production in silver would decline accordingly. Conservation laws could only temporarily stave off wood shortages when silver was spent so wastefully. At approximately the end of the second century AD, the inevitable occurred: Silver production declined. Further output was limited not by the supply of ore, which remained abundant, but by the accessibility of fuel.

The decline in silver production offered later emperors two choices: cut expenditures or find alternative financing. They unanimously chose the latter but differed in methodology. The emperor Commodus "stretched" silver money by adding base metal, which comprised 30% of the coin. He also went on a killing spree, enraged that the empire's revenues could not meet his expenditures. When he finally calmed down, he decided to auction off whatever he could, offering provincial and administrative offices to the highest bidder.

Septimius Severus, who ruled a few years after Commodus, added 20% more alloy to the silver

Figure 4
Ancient Greek urn depicts a worker stoking a metallurgical furnace with charcoal. From Perlin Forestry Archives.

Figure 5
Longshoremen loading a log destined to fuel the baths of Rome. From Perlin Forestry Archives.

coinage, thus reducing the silver content to a mere 50%. Because Roman money was now so badly debased, Severus instituted the requisitioning of commodities rather than collecting worthless currency through taxation. Further debasements forced the government to search for creative ways of staying afloat. Most of the methods chosen circumscribed the freedom of Rome's citizens. Providing the government with the provisions it needed became compulsory. The government also established guilds, expecting them to produce according to obligations it set but rewarding members with monopolies in their respective trades.

By the end of the third century AD, Rome's currency had lost 98% of its silver content, and the public placed as little value on it as did the government. People increasingly took to trading in commodities and services so that by the first part of the fourth century AD barter had become institutionalized.

To keep those in Rome from becoming too anxious over the declining economy, the rulers of Rome had to constantly find ways to keep the population placated. The later emperors were well aware of the Romans' love of bathing and added many new baths to the city, eventually bringing the total to more than 900. The largest held as many as 2000 bathers at a time. Bathwater had to be hot if the Romans were to remain happy. Because keeping the Roman populace satisfied was paramount in the minds of those in power, the authorities were willing to go to great lengths to ensure a constant flow of fuel to the bathing establishments.

In the third century ad, the emperor Severus Alexander saw to it that entire woods were cut down to keep the baths in Rome well heated. When these forests gave out a century later, the authorities founded a guild with 60 ships at its disposal that was solely responsible for supplying the baths with wood. Sometimes, wood could be obtained as close to home as the Campania region. Usually, though, the guild had to make its wood runs primarily to the forests in North Africa. That Romans would travel such great distances for fuel indicates just how little wood was left in its vicinity and how dependent the Romans had become on foreign supplies.

Interestingly, the rise and decline of fuel supplies in Rome closely parallel the fortunes of the empire. The pioneering ecologist George Perkins Marsh demonstrated the fact by describing the changes in brick and masonry work in Rome over the centuries. Bricks in early buildings were extremely thin, well fired, and held together by liberally applied quantities of lime mortar. In contrast, as the imperial period progressed, the opposite proved true: Bricks were very thick, usually poorly fired, and held together by a minimum of mortar. Marsh hypothesized that the difference was "due to the abundance and cheapness of fuel in early times, and its growing scarceness

and dearness in later ages." He elaborated on his observation:

> When wood cost little, constructors could afford to burn their brick thoroughly and burn and use a great quantity of lime. As the price of firewood advanced, they were able to consume less fuel in brick and lime kilns and the quality and quantity of brick and lime used in building were gradually reversed in proportion.

4.4 England, 1500s–1700s

At the beginning of the 16th century, England depended on the Continent for its shot and artillery. However, Henry VIII's apostasy resulted in an arms embargo and a threat of invasion. Self-sufficiency seemed the only recourse. Henry therefore saw to the development of a local arms industry in Sussex, where rich veins of high-phosphorous iron were particularly advantageous for the casting of guns and where oak and beech would provide ample fuel for the gun founders. They did not smelt their iron in simple hearths but used blast furnaces and an accompanying forge. They chose the blast furnace and forge over the simpler hearths because the former could produce approximately 20 times more iron than the latter. The blast furnace and forge also consumed much more charcoal (Figures 6 and 7). By the late 1540s, the new English arms industry annually consumed approximately 117,000 cords of wood, causing great destruction of nearby woods. People living in the vicinity viewed these blast furnaces and forges as menacing since the wood they consumed was essential to the locals' survival, who used it to heat their homes, cook their meals, make their tools and fishing boats, and build and repair their houses. By the 1700s, the iron industry had deforested such a large area of southeast England that the iron masters had to ration the amount of iron ore they could smelt. To increase iron production required learning to smelt ore with coal without adding its impurities to the finished metal. By discovering how to make coal as close to charcoal as possible, the English freed themselves from the constraints of its dwindling forests and began the industrial revolution, an age that has qualitatively separated people living since the mid-19th century from those throughout the rest of history.

4.5 Madeira, West Indies, and Brazil, 1400s–1600s

Sugar brought great wealth to those who raised the commodity. Hence, when the Portuguese landed on the warm, greatly forested, fertile, and well-watered island of Madeira, they immediately started to plant cane. Once the juice was forced out of the stalks, workers poured the extracted liquid into kettles. A fire, rarely extinguished, burned underneath each pot. The wood from linden trees furnished much of

Figure 6
Wood cutters chop off tree limbs to be used as charcoal.
From Perlin Forestry Archives.

Figure 7
Charcoal fires also fueled English glassworks. From
Perlin Forestry Archives.

the fuel. No doubt cedar was also used, being the
most common wood on Madeira. Sugar workers
aptly called the room in which the cane juice
was boiled the "sweet inferno." The juice continued

boiling until judged ready for removal to an area
where it would solidify into sugar. By the end of the
15th century, the island's sugar industry required
approximately 60,000 tons of wood just for boiling
the cane. Four of the 16 mills operating on Madeira
consumed 80,000 pack animal loads of wood
per year.

Almost immediately after the discovery of the
West Indies and South America, Europeans recognized that the land offered the same ideal conditions
for growing and processing sugar as existed on
Madeira. The Spanish planted sugarcane on Espanola soon after Columbus's first journey to the New
World. As expected, the cane flourished, and by the
end of the 16th century 40 sugar mills were operating
on the island. It was no different in other parts of the
Indies, and sugarcane proliferated and became the
area's main source of revenue. Nor did the Portuguese, after their success on Madeira, wait long
to establish sugar plantations after taking possession
of Brazil. Near the end of the 16th century, a visitor
reported 70 sugar mills at work in the Pernambuco
region and 40 in the area of Bahia. Experts in sugar
production in the New World estimated that six to
eight slaves had to be constantly employed in cutting
fuel in the forest and transporting it to the mills for
optimum efficiency. Providing fuel for 1 mill stripped
approximately 90 acres of forestland each year.

Such large-scale consumption of wood took its toll
on the forests of the New World. Two hundred and
forty years after the Portuguese had found Madeira
so thickly wooded that they called it "isola de Madeira" (island of timber), it had become the island of
wood in name only, so deforested that those passing
it by ship could find absolutely no greenery in view.
Likewise, the very thickly forested island of Espanola
had become an open land. The pace of deforestation
on Barbados exceeded that on Espanola and Madeira. In little more than 20 years, the representatives
of the planters admitted to having used and destroyed all the timber formerly growing on the island.

4.6 American Colonies

America's great forests offered the pilgrims a higher
standard of living than that of their peers in England
since they had upon arrival more wood fuel with
which to heat than did most noblemen in England.
Families also used the enormous fireplace around
which they huddled for lighting. The fireplace was so
huge as to require logs of dimensions that could only
be dragged into the house by a horse or oxen.

America's plethora of wood gave rise to a large
and prosperous iron industry. Despite the fact that
American woodcutters earned three times more than
their counterparts in England and Americans who
coaled the wood made approximately twice as much
as those who did the same work in England, because

a cord of wood cost 14 times more in England than in America, American ironmasters could manufacture pig iron in the Colonies and export it to England, underselling English iron.

Iron furnaces and forges first began operating in the backwoods of Virginia and Maryland and then spread to Pennsylvania and New Jersey. They not only produced iron for the English market but also produced cast iron goods for local use. The proliferation of home-produced iron goods frightened the English, whose mercantile economy was based on buying raw materials from the Colonies at bargain prices and selling back finished products at much higher prices. The English therefore passed a law in 1749 that encouraged the importation of pig iron from the American Colonies, because wood shortages had hindered England's iron production, while forbidding the Americans from developing forges to turn out finished iron products.

The act stirred already growing feelings for independence among many of the colonists. It hardly seemed fair or rational to Americans that they had to send to England the iron they produced in order to buy it back again as manufactured items, costing them almost twice as much compared to if they done the entire process at home. The intent of the act reminded many of the servile conditions forced upon the Children of Israel by the Philistines when the Jews were not allowed any smiths of their own. The prohibitory nature of the act aroused revolutionary rhetoric among the more radical Americans such as James Otis, credited by John Adams as the father of the idea of independence, and John Dickinson, who later authored revolutionary America's "Declaration of the Causes of Taking Up Arms." Both feared that prohibiting the manufacture of iron goods was the first step in stopping the construction of any machinery that would further develop the Colonies' economy.

Despite the anger the act aroused in America, it did increase American exports of pig iron to England. By 1769, approximately half of the pig iron worked in English forges came from America. By 1776, as much iron was produced in the Colonies as in Britain. Even more encouraging to those favoring revolution, and frightening to those wishing to hold onto the American Colonies, was the fact that not only did America's innumerable iron mines and endless forests give it the very sinews of power but also a certain type of iron produced in America, called best principio, was judged as good as any in the world for making shot and cannon.

4.7 America after Independence

Alexander Hamilton, probably the greatest advocate of industrialization during the early republic, predicted that America was very fortunate to have vast supplies of iron ore and wood with which to make charcoal. Trench Coxe, a colleague of Thomas Jefferson and James Madison, predicted that America's rich store of wood fuel would provide invaluable to the rise of industry and turn America into the most powerful country in the world.

Breweries, distilleries, salt and potash works, and casting and steel furnaces all needed heat to produce finished products. Francois Michaux, a French botanist who studied America's trees at the beginning of the 18th century, found that the bakers and brick makers of New York, Philadelphia, and Baltimore commonly consumed prodigious quantities of pitch pine. Hatters of Pittsburgh, on the other hand, preferred charcoal made from white maple. Boats sailed along the Erie Canal picking up wood to fuel the nation's largest salt works, located in upstate New York. Boiling rooms, in which the salt water was evaporated over charcoal fires, produced 2 million bushels of salt per year. The salt went to Canada, Michigan, Chicago, and all points west. Farmers were the largest purchasers, using the salt for preserving meat that they marketed. Steam engines, which in the 1830s began to replace factories run on water power, usually burned wood as their fuel.

Americans could see the Hand of Providence at work in providing great amounts of wood and iron next to each other. Wood- and iron-rich western Pennsylvania, northern Michigan, and southeast Ohio became the principal producers of iron in 19th-century America. Throughout the nation, charcoal-burning iron mills produced 19 million tons of iron between 1830 and 1890. To truly appreciate the magnitude of the output of America's wood-fueled iron furnaces, comparison to the amount of iron produced during the heyday of the British charcoal-burning irons industry, which dated from the 1640s to the 1780s, is in order. In its most productive years, England's charcoal-run ironworks produced approximately 1 million tons of iron in 60 years.

The steamboat opened the old American west (lands east of the Mississippi River, north of the Ohio River, and west of the Allegheny Mountains) to settlement (Fig. 8). How else could farmers in places such as Ohio, Indiana, and Illinois expect to move their goods to distant markets except by navigation? Had farmers had to rely on land transport instead, they would have needed 5000 miles of good road, which did not exist at the time, and 250,000 wagons to transport the same amount of goods annually carried by steamboat. Steamboats were lauded by mid-19th-century pundits as the most important technological development to have occurred in the United States because they contributed more than any other single development to advance the prosperity of those living in the Mississippi and Ohio River basins.

Great quantities of wood were needed for fuel. The large steamboat Eclipse's 15 boilers consumed wood

Figure 8
Hauling timber to fuel steamboats. From Perlin Forestry Archives.

by the carload. Thousands of wood yards were located along the banks of every navigable river simply to provide steamers with fuel. Backwoodsmen brought the timber they had cut to these depots and hacked it into the proper size for the ships' furnaces. At night, the owners of these wood lots kept gigantic fires blazing so those onboard the boats could see the yards and fuel up.

Railroads liberated Americans from their dependence on waterways for shipping freight and personal travel. They slashed the cost of and the time required to travel. With plenty of timber growing along the right-of-way of most railroads, or at least close by, locomotives used wood as their only fuel from when the first tracks were laid in 1830 to the beginning of the Civil War. Fuel needs of engines owned by the New York Central Line required the railroad to build 115 woodsheds along its track. If the woodsheds were stacked side by side, they would have stretched approximately 5 miles. A wood yard for trains at Columbus, Nebraska, was half a mile long.

Between 1810 and 1867, approximately 5 billion cords of wood was consumed for fuel in fireplaces, industrial furnaces, steamboats, and for railroads. To obtain such a quantity of wood meant the destruction of approximately 200,000 square miles of forest land, an area nearly equal to all the land that comprises the states of Illinois, Michigan, Ohio, and Wisconsin.

4.8 Use of Wood for Fuel, the Present

Today, half of the wood cut throughout the world is used for fuel. Eighty percent of all wood used for fuel is consumed in the developing world. In fact, the majority of people in the developing world depend on wood as their primary energy source. Charcoal and firewood in the Cameroons, for example, account for 80% of all energy consumption. The demand for firewood and charcoal is also increasing as the population increases. In Africa, the amount of wood consumed for energy purposes increased from 250 million m³ in 1970 to 502 million m³ in 1994. In Latin America, the vast majority of the rural population uses wood as its primary fuel. The true danger to the forests of the developing world is that growing numbers of people exploit local forests, threatening their viability and forcing people to travel greater distances for wood fuel. One person can ably supply himself with firewood from 0.5 ha of forestland; today, however, usually approximately 15 people try to supply themselves with firewood from 1 ha of woods. Such overexploitation will surely cause local supplies to quickly diminish, forcing people to seek wood in more distant forests. Demand for wood and charcoal in growing urban areas of the developing world has introduced a new and growing problem—the industrialization of acquiring firewood and charcoal. Instead of individuals fanning neighboring woods to cut down trees for fuel, well-capitalized charcoal dealers search throughout the countryside for supplies. Bangkok's 5 million citizens obtain a large amount of their charcoal in this manner from forests throughout Thailand. There is no question that as the population in the developing world grows, so too will the demand for firewood and charcoal and deforestation will continue to accelerate.

Bibliography

Bryant D, Nielsen D, Tangley L 1997 *The Last Frontier Forests: Ecosystems and Economies on the Edge.* World Resources Institute, Washington, DC

Myers N 1980 The present status and future prospects of tropical moist forests. *Environ. Conserv.* **7**(2), 101–14

Perlin J 1991 *A Forest Journey: The Role of Wood in the Development of Civilization.* Harvard Univ. Press, Cambridge, MA

Richards J, Tucker R (eds.) 1988 *World Deforestation in the Twentieth Century.* Duke Univ Press, Durham, NC

John Perlin

World History and Energy

Glossary

antiquity   The era of ancient (from the Western perspective, mostly the Middle Eastern and Mediterranean) civilizations extant between prehistory and the Middle Ages.

determinism   A doctrine claiming that human actions are determined by external factors; its prominent varieties include environmental and geographic determinism.

early modern world   The period immediately following the Middle Ages, variously dated as 1493–1800, 1550–1850, or the 16th to 18th centuries.

energy transition   A period of passing from one configuration of prime movers and dominant fuels to a new setup.

middle ages   The period between antiquity and the modern era, often circumscribed by the years 500–1500 CE.

prehistory   The period of human evolution predating recorded history.

A strict thermodynamic perspective must see energy—its overall use, quality, intensity, and conversion efficiency—as the key factor in the history of the human species. Energy flows and conversions sustain and delimit the lives of all organisms and hence also of superorganisms such as societies and civilizations. No action—be it a better crop harvest that ends a famine or the defeat of an aggressive neighbor—can take place without harnessing and transforming energies through management, innovation, or daring. Inevitably, the availability and quality of particular prime movers and sources of heat and the modes of their conversions must have left deep imprints on history. But no energetic perspective can explain why complex entities such as cultures and civilizations arise and no thermodynamic interpretation can reveal the reasons for either their remarkable history or their astounding diversity of beliefs, habits, and attitudes from which their actions spring. This article examines both of these contrasting views of energy and world history.

1. A Deterministic View of History

Countless energy imperatives—ranging from the solar flux reaching the earth to minimum temperatures required for the functioning of thousands of enzymes—have always shaped life on Earth by controlling the environment and by setting the limits on the performance of organisms. Deterministic interpretations of energy's role in world history seems to be a natural proposition, with history seen as a quest for increased complexity made possible by mastering higher energy flows. Periodization of this quest on the basis of prevailing prime movers and dominant sources of heat is another obvious proposition. This approach divides the evolution of the human species into distinct energy eras and brings out the importance of energy transitions that usher in more powerful, and more flexible, prime movers and more efficient ways of energy conversion. Perhaps the most intriguing conclusion arising from this grand view of history is the shrinking duration of successive energy eras and the accelerating pace of grand energy transitions.

The first energy era started more than 300,000 years ago when the human species, *Homo sapiens*, became differentiated from *Homo erectus*, and the era continued until the beginning of settled societies some 10,000 years ago. Throughout prehistory, all efforts to control greater energy flows were capped by the limited power of human metabolism and by the inefficient use of fire. Domestication of draft animals and harnessing of fire for producing metals and other durable materials constituted the first great energy transition: reliance on these extrasomatic energies had raised energy throughput of preindustrial societies by more than an order of magnitude. The second transition got under way only several millennia later; it was not as universal as the first one and its effects made a profound, and relatively early, difference only in some places: it came as some traditional societies substituted large shares of their muscular exertions by waterwheels and windmills, simple but ingenious inanimate prime movers that were designed to convert the two common renewable energy flows with increasing power and efficiency.

The third great energy transition—substitution of animate prime movers by engines and of biomass energies by fossil fuels—began only several centuries ago in a few European countries and it was accomplished by all industrialized nations during the 20th century. That transition is yet to run its course in most low-income economies, particularly in Africa. The latest energy transition has been under way since 1882 when the world's first electricity-generating stations were commissioned in London and New York (both Edison's coal-fired plants) and in Appleton, Wisconsin (the first hydroelectric station). Since that time, all modernizing economies have been consuming increasing shares of their fossil fuels indirectly as electricity and introducing new modes of primary electricity generation—nuclear fission starting in the mid-1950s, and later also wind turbines and photovoltaic cells—to boost the overall output of this most flexible and most convenient form of energy. The second key attribute of this transition has been a steady relative retreat of coal mirrored by the rise of hydrocarbons, first crude oil and later natural gas.

Improving the quality of life has been the principal individual benefit of this quest for higher energy use that has brought increased food harvests, greater accumulation of personal possessions, abundance of educational and leisure opportunities, and vastly enhanced personal mobility. The growth of the world's population, the rising economic might of nations, the extension of empires and military capabilities, the expansion of world trade, and the globalization of human affairs have been the key collective consequences of the quest. These advances are discussed in this article and the limits of prime movers and heat sources that were available during the successive eras of energy use and the major accomplishments that were achieved through ingenuity and better organization are noted.

2. The Earliest Energy Eras

During the long span of prehistory, the human species relied only on its somatic energy, using muscles to secure a basic food supply and then to improve shelters and acquire meager material possessions. Organismic imperatives (above all, the basal metabolism scaling as the body mass raised to 0.75 power) and the mechanical efficiency of muscles (able to convert no more than 20–25% of ingested food to kinetic energy) governed these exertions: healthy adults of smaller statures cannot sustain useful work at rates of more than 50–90 W and can develop power of 10^2 W only during brief spells of concentrated exertion. The former performance sufficed for all but a few extreme forms of food gathering and the latter exertions were called on for some forms of hunting. Simple tools made some foraging and processing tasks more efficient and extended the reach of human muscles.

Energy returns in foraging (energy in food/energy spent in collecting and hunting) ranged from barely positive (particularly for some types of hunting) to seasonally fairly high (up to 40-fold for digging up tubers). The choice of the collected plants was determined above all by their accessibility, nutritional density, and palatability, with grasslands offering generally a better selection of such species than did dense forests. Collective hunting of large mammals brought the highest net energy returns (because of their high fat content) and it also contributed to the emergence of social complexity. Only a few coastal societies collecting and hunting marine species had sufficiently high and secure energy returns (due to seasonal migrations of fish or whales) such that they were able to live in permanent settlements and devote surplus energy to elaborate rituals and impressive artistic creations (for example, the tall

ornate wooden totems of the Indian tribes of the Pacific Northwest).

The only extrasomatic energy conversion mastered by prehistoric societies was the use of fire for warmth and cooking, which can be indisputably dated to approximately 250,000 years ago. Eventual shifts from foraging to shifting cultivation and then to sedentary farming were gradual processes driven by a number of energy-related, nutritional, and social factors: there was no short and sharp agricultural revolution. These changes were accompanied by declining net energy returns in food production, but these declines had a rewarding corollary as the higher investment of metabolic energy in clearing land and in planting, weeding, fertilizing, harvesting, and processing crops, as well as storing grains or tubers, made it possible to support much higher population densities. Whereas the most affluent coastal foraging societies had densities less than 1 person/km² (and most foraging societies had carrying capacities well below 0.1 person/km²), shifting agricultures would easily support 20–30 people/km² and even the earliest extensive forms of settled farming (ancient Mesopotamia, Egypt and China's Huanghe Valley) could feed 100–200 people/km², that is, 1–2 people/ha of cultivated land (Fig. 1).

The increasing size of fields could not be managed by slow and laborious hoeing but plowing is either exceedingly taxing or, in heavy soils, outright impossible without draft animals. Farming intensification thus led to harnessing the first important extrasomatic source of mechanical energy by domesticating draft animals throughout the Old World (the pre-Columbian Americas had only pack animals). Continuous energy investment was then needed for animal breeding and feeding, as well as for producing more complex implements.

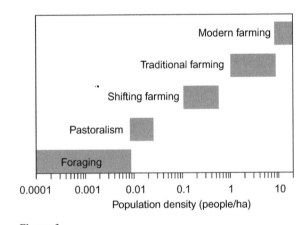

Figure 1

Successive stages of sedentary agricultures have required higher energy inputs but have been able to support 10^3 to 10^4 more people per unit of land than did foraging.

Small bovines would rate less than 200 W, stronger animals could sustain more than 300 W, and the best oxen and good early draft horses could surpass 500 W, equal to the labor of 6-8 adult men (Fig. 2). Draft animals thus speeded up field, transportation, and crop processing tasks and boosted agricultural productivity. Their numbers were governed by an obvious energetic imperative: no society could afford to cultivate feed crops where harvests were barely adequate to provide subsistence diets. Those agroecosystems where grazing land was also limited (the rice regions of Asia) could support only relatively small numbers of draft animals.

Limited unit power of muscles could be overcome by massing people, or draft animals, and the combination of tools and organized deployment of massed labor made it possible to build impressive structures solely with human labor. Massed forces of 20–100 adults could deliver sustained power of 1.5–8 kW and could support brief exertions of up to

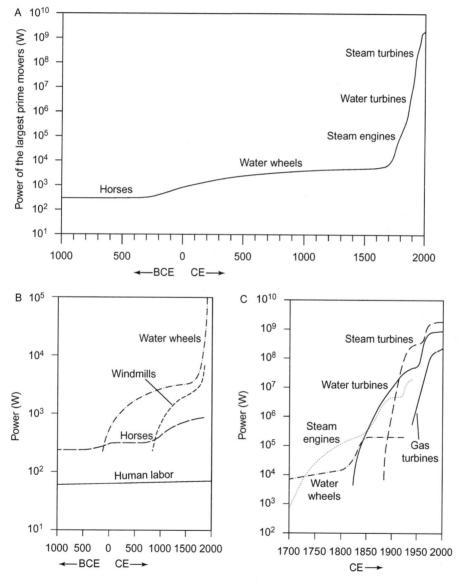

Figure 2
The maximum power of prime movers shown as the sequence of the highest capacity converters for the span of the past 3000 years (A) and shown in detail for the periods 1000 BCE to 1700 CE and 1700 CE to 2000 CE (B and C).

100 kW, enough to transport and erect (with the help of simple devices) megaliths and to build impressive stone structures on all continents except Australia. In contrast to this massed deployment of human labor in construction, no Old High culture took steps to a truly large-scale manufacture of goods and the atomization of production remained the norm. In addition, violent conflict powered solely by an individual's muscles could take place only as hand-to-hand combat or by an attack with an arrow launched from less than a couple hundred meters away, a limit ordained by the maximum distance between one extended and one flexed arm when drawing a bow. Eventually catapults, tensioned by many hands, increased the mass of projectiles, but did not substantially lengthen the maximum distance of attack.

Shifting agriculturalists extended the use of fire to the regular removal of vegetation, and early settled societies also adopted fire to produce bricks and containers and to smelt metals, beginning with copper (before 4000 BCE) and progressing to iron (common in some parts of the Old World after 1400 BCE). Charcoaling was used to convert wood to a fuel of higher energy density (29 MJ/kg compared to no more than 20 MJ/kg for wood and less than 15 MJ/kg for crop residues) and superior quality (essentially smokeless and hence suitable for burning indoors in fixed or portable hearths). But open fireplaces and braziers converted less than 10% of the supplied energy into useful heat and traditional charcoaling turned less than one-fifth of the charged wood energy into the smokeless fuel. Primitive furnaces used for metal smelting were also very inefficient, requiring as much as 8–10 units of charcoal for a unit of pig iron. The resulting high demand for wood was a leading cause of extensive deforestation, but a nearly complete disappearance of forests from parts of the Mediterranean (Spain, Cyprus, and Syria) and the Near East (Iran, Afghanistan) was caused by smelting copper rather than iron.

Small-scale and highly energy-intensive metallurgy meant that no early societies could smelt enough metal to make it the dominant material in daily use and simple machines, farming implements, and household utensils of the antiquity remained overwhelmingly wooden. This changed radically only when coke-based smelting began producing inexpensive iron after 1750. Similarly, the inherently limited power of oxen, the most affordable draft animals, which were rarely fed any concentrates, meant a ponderous pace of field operations: plowing a hectare of a loamy field with a pair of these animals was easily four times faster than hoeing the same land, but a pair of well-fed horses would have accomplished the same task in less than one-half the time required by oxen. And the combination of weak animals, inefficient harnessing, and poor (unpaved) roads greatly restricted the size of maximum loads and the greatest distance of daily travel.

European antiquity also saw the first uses of water-driven prime movers. Their origins are obscure, with the first reference to their existence, by Antipater of Thessalonica during the first century BCE, describing their use in grain milling. The earliest wheels were horizontal, with water directed through a sloping wooden trough onto wooden paddles fitted to a shaft that was directly attached to a millstone above. More efficient vertical water wheels, first mentioned by Vitruvius in 27 BCE, turned the millstones by right-angle gears and operated with overhead, breast, or undershot flows. Although there were some multiple installations of Roman water wheels—perhaps most notably the set of 16 wheels at Barbegal near Arles amounting to over 30 kW of capacity—cheap slave labor clearly limited the adoption of these machines.

3. Medieval and Early Modern Advances

The dominance of animate prime movers extended throughout the Middle Ages but their efficiency had improved and they were increasingly joined by gradually more powerful converters of flowing water and wind. Human statures did not show any notable growth during the medieval era but better designs of some man-powered machines were able to harness muscle power more efficiently. Man- and animal-powered tread-wheels were deployed in the construction of tall buildings and in loading and unloading ship cargoes. The combination of breeding, better feeds, more efficient harnessing, and shoeing eventually raised the performance of the best draft animals as much as 50% above the capacities that prevailed during antiquity.

The collar harness, optimizing the deployment of powerful breast and shoulder muscles, had its origins in China of the 5th century of the CE and its improved version became common in Europe five centuries later. Iron horseshoes, preventing excessive wear of hooves and improving traction, became common at approximately the same time. But it took centuries before the intensification of cropping, with more widespread rotation of food and feed (particularly leguminous) species, increased the availability of concentrate feeds and allowed for harder working draft animals. During the 18th century, a good horse was considered to be equivalent to 10 men, or at least 700 W, and the best horses would eventually surpass power equivalent to 1 kW. Whereas a pair of early medieval oxen could sustain no more than 600 W, a pair of good horses in early modern Europe delivered 1.2 kW and large horse teams (up to 40 animals) deployed in the United States after 1870 to pull gang plows or the first grain combines had draft power of at least 8 kW and up to 30 kW.

Some medieval societies began to rely on inanimate prime movers for a number of demanding tasks including grain milling, oil pressing, wood sawing, powering of furnace bellows and forge hammers, and the mechanization of manufacturing processes ranging from wire pulling to tile glazing. Waterwheels were the first machines to spread widely and the famous Domesday Book attests how common they were in England of the late 11th century: it lists 5624 water mills in southern and eastern England, one for every 350 people. A subsequent increase in the highest capacities of waterwheels was slow: it took nearly 800 years to boost the performance by an order of magnitude. Early modern Europe developed some relatively very large water-driven wheels, and although typical unit capacities of these wooden machines remained limited (by 1700 they averaged less than 4 kW), they were the most powerful prime movers of the early modern era (Fig. 2).

Ships with simple square sails were used by the earliest Old World civilizations, but the first written record of wind-driven machines comes only approximately 1000 years after the first mention of water wheels. In 947 CE, al-Masudi's report described their use to raise water for irrigating gardens in what is now eastern Iran, and the first European record dates only from the closing decades of the 12th century. Subsequently, inefficient windmills continued to be used infrequently throughout the Mediterranean and the Middle East, and even less so in India and in East Asia, and they had undergone a great deal of development in only a small number of European regions. The earliest European windmills pivoted on a massive central post that was supported usually by four diagonal quarter-bars, had to be turned to face the wind, and were unstable in high winds and their low height limited their efficiency.

However, they were widely used in grain milling and water pumping (the Dutch drainage mills being the best known example), as well as in some industrial operations. Post mills were gradually replaced by tower mills and smock mills, and during the early 17th century the Dutch millers introduced first relatively efficient blade designs (however, true airfoils, aerodynamically contoured blades with thick leading edges, originated only just before the end of the 19th century), and after 1745 the English invention of using a fantail to power a winding gear turned the sails into the wind automatically. Even with these innovations, the average power of the 18th century windmills remained below 5 kW.

The Late Middle Ages and the early modern era were also the time when wind energy was harnessed more effectively for Europe's great seafaring voyages. The rise of the West clearly owes a great deal to an unprecedented combination of harnessing two different kinds of energy: better, and larger, sailships equipped with newly developed heavy guns. Once the medieval ships became rigged with a larger number of loftier and better adjustable sails, increased in size, and acquired stern-post rudders and magnetic compasses (both invented in China), they became much more efficient, and much more dirigible, converters of wind energy. These ships carried first the Portuguese and then other European sailors on increasingly more audacious voyages. The equator was crossed in 1472, Columbus led three Spanish ships to the Caribbean in 1492, Vasco da Gama rounded the Cape of Good Hope and crossed the Indian Ocean to India in 1497, and in 1519 Magellan's *Victoria* completed the first circumnavigation of the earth. The inexorable trend of globalization was launched with these sailings.

Gunpowder was another Chinese invention (during the 11th century) that was better exploited by the Europeans. The Chinese cast their first guns before the end of the 13th century, but Europeans were only a few decades behind. Within a century, superior European gun designs transformed the medieval art of war on land and gave an offensive superiority to large sailships. Better gun-making was obviously predicated on major medieval advances in ore-mining and metal-smelting techniques that are exhaustively described in such classics as Biringuccio's and Agricola's volumes from 1540 and 1556, respectively.

These innovations reduced the need for energy inputs, particularly in the iron-making blast furnaces that appeared first in the lower Rhine valley before the end of the 15th century. As these furnaces grew in volume, charcoal's fragility limited their height and the annual volume of individual smelting operations. Larger operations also required the use of waterpower (for blasting and subsequent metal forging) and this demand restricted the location to mountainous areas. But the main challenge was to keep them supplied with charcoal, and the English predicament is the best illustration.

By the early 18th century, a typical English furnace produced only approximately 300 tons of pig iron per year, but with at least 8 kg of charcoal per kilogram of iron and 5 kg of wood per kilogram of charcoal, its annual demand was approximately 12,000 tons of wood. With nearly all natural forests gone, the wood was cut in 10- to 20-year rotations from coppicing hardwoods yielding between 5 and 10 tons/ha. This means that approximately 2000 ha of coppicing hardwoods, a circle with a radius of 2.5 km, were needed for perpetual operation. Nationwide (with nearly 20,000 tons of pig iron produced annually during the 1720s), this translated (together with charcoal needed for forges) to at least 1100 km^2 of coppiced growth. To produce 1 million tons with the same process would have required putting at least one-quarter of the British Isles under coppiced wood, an obvious impossibility. Yet starting in the mid-1830s, Great Britain began smelting more than 1 million tons of iron per year and yet some the country's forests began regrowing; coke and steam engines made that possible.

4. Transitions to modernity

Millennia of dependence on animate power and biomass fuels came to an end only gradually and the great transition to fossil fuels and fuel-consuming engines had highly country-specific onsets and durations. Differences in accessibility and afford-ability explain why traditional energy sources were used for so long after the introduction of new fuels and prime movers. For example, four kinds of distinct prime movers, in addition to human labor, coexisted in parts of Europe for more than 150 years between the late 18th and the mid-20th centuries before internal combustion engines and electric motors became totally dominant: draft animals (both in agriculture and in city traffic), water wheels (and, since the 1830s, water turbines), windmills, and steam engines. In the wood-rich United States, coal surpassed fuelwood combustion and coke became more important than charcoal only during the 1880s.

Moreover, the epochal energy transition from animate to inanimate prime movers and from biomass to fossil fuels has yet to run its global course. By 1900, several European countries were almost completely energized by coal—but energy use in rural China during the last year of the Qing dynasty (1911) differed little from the state that prevailed in the Chinese countryside 100 or 500 years earlier. Even in the early 1950s, more than one-half of China's total primary energy supply was derived from woody biomass and crop residues. The share of these fuels had been reduced to 15% of China's total energy use by the year 2000, but it remains above 70%, or even 80%, for most of the countries of sub-Saharan Africa (in the year 2000, India's share of traditional biomass fuels was approximately 30% and that of Brazil was approximately 25%), and globally it still accounts for nearly 10%.

Industrialization of the British Isles is, of course, the best known case of an early transition from wood to coal and England was the first country to accomplish the shift from wood to coal during the 16th and 17th centuries. Much less known is the fact that the Dutch Republic completed a transition from wood to peat during its Golden Age of the 17th century when it also replaced a large share of its mechanical energy needs by sailships, which moved goods through inland canals and on the high seas, and by windmills. In England and Wales, the process started as a straightforward fuel substitution in a society where the combined demand for charcoaling, ship- and housebuilding, heating, and cooking led to extensive deforestation. The use of coal as the fuel for a new mechanical prime mover began only after 1700 with Newcomen's inefficient steam engine. James Watt's separate condenser and other improvements (patented in 1769) transformed the existing engine from a machine of limited utility (mostly water pumping in coal mines) into a prime mover of unprecedented power suitable for many different tasks.

Watt's improved machines still had low conversion efficiency (less than 5%) but his engines averaged approximately 20 kW, more than 5 times that of the typical contemporary watermills, nearly 3 times that of windmills, and 25 times the performance of a good horse. One hundred years later, the largest stationary steam engines were 10 times as efficient as Watt's machines and rated ≈ 1 MW After the expiration of Watt's patent, the development of high-pressure steam engines progressed rapidly, radically transforming both land and maritime travel. For centuries, horse-drawn carriages averaged less than 10 km/h, but by 1900, trains (the first scheduled services began during the 1830s) could go easily 10 times faster and carry passengers in much greater comfort (Fig. 3). Railways had also drastically lowered the cost of moving voluminous loads in areas where inexpensive canal transport was not possible. Steamships cut the length of intercontinental travel and greatly expanded and speeded longdistance trade. For example, trans-Atlantic crossing, which took more than 1 week with the sailships of the 1830s, was cut to less than 6 days by 1890.

Steam engines will be always seen as the quintessential energizers of the Industrial Revolution, that great divide between the traditional and the modern world. But this notion is far from correct, and so is the very idea of the Industrial Revolution. Coal did power the expansion of iron-making, but the textile industry, the key component of that productive transformation, was commonly energized, both in Europe and in North America, by water power rather than by coal combustion. Innovations to meet this demand included Benoit Fourneyron's reaction water turbine (in 1832) and an inward–flow turbine of James B. Francis (in 1847). Before the century's

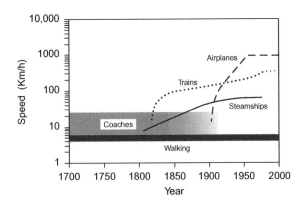

Figure 3

Maximum speeds of transportation during the years 1700–2000.

end, Francis' turbines and also Pelton machines (jet–driven turbines introduced in 1889) helped to launch modern water-powered electricity generation (only Viktor Kaplan's axial flow turbines are a 20th century innovation). As for the Industrial Revolution, the author sides with those historians who see it much more as a complex, multifaceted, and protracted process rather than as a clearly definable, intensive transformation powered by coal and coal-generated steam. After all, in 1850, England still had more shoemakers than coal miners and more blacksmiths than ironworkers, and in the United States, coal began providing more energy than wood only during the early 1880s.

Another little appreciated fact is that the harnessing of wind energy reached its peak at the same time that the steam engine did. America's westward territorial progression across the windy Great Plains created a demand for windmills to pump water for steam locomotives, households, and cattle. These machines were made generally of large numbers of fairly narrow slats that were fastened to solid or sectional wheels and usually equipped with either the centrifugal or the side-vane governor and independent rudders. Total sales of these relatively small (less than 1 kW) machines amounted to millions of units after 1860, whereas the capacity of approximately 30,000 larger windmills in countries around the North Sea peaked at approximately 100 MW by the year 1900.

From the vantage point of the early 21st century, it is clear that it was the next energy transition—the process of electrification and the rising global dependence on hydrocarbons—that has left the greatest mark on individual lives as well as on the fortunes of economies and nations during the 20th century. The invention of a commercially viable system of electricity generation, transmission, and use (beginning only with incandescent lighting) by Thomas Λ. Edison and his associates was compressed into a remarkably short period during the early 1880s. This was followed by a no less intensive period of innovation with fundamental contributions by George Westinghouse [who correctly insisted on alternate current (AC) for transmission], Charles Parsons (who patented the first steam turbine in 1884), William Stanley (who introduced an efficient transformer in 1885), and Nikola Tesla (who invented the electric motor in 1888).

As a result, by the late 1890s, the entire electric system was basically perfected and standardized in the form that is still relied on today; the challenge ahead was to keep enlarging its unit sizes and improving efficiencies and this challenge has been met in many impressive ways. Since 1900, the maximum sizes of turbogenerators grew from 10 to ≈ 1.5 GW, AC transmission voltages rose from less than 30 kV to more than 700 kV, and in 2003, the best efficiencies of thermal generation surpassed 40% (and with

cogeneration 60%), compared to as little as 5% in 1900. An inexpensive and reliable supply of electricity transformed every aspect of everyday activities by bringing bright and affordable light to both interiors and streets, by powering a still-growing array of time-saving and leisure-enhancing gadgets, and by energizing urban and intercity trains.

But the most revolutionary consequence of electrification was in industrial production. The reciprocating motion of steam engines had to be transmitted by rotating shafts and belts, resulting in friction and a great deal of lost time with accidents and allowing only limited control of power at individual workplaces. Inexpensive electric motors of all sizes changed all that: no shafts and belts, no noise and dangerous accidents, only precise and flexible individual power controls. American manufacturing was the first to make the transition in just three decades: by 1929, the capacities of industrial electrical motors accounted for over 80% of all installed mechanical power. Highly productive assembly lines (Ford pioneered them in 1913) were an obvious product of this transformation, as were many new specialized industries. And although the experimental beginnings of radio and television predate World War I, it was only after World War II that electricity began powering the new computer age with its immense flow of information and entertainment options.

Electricity has been also the principal means of easing the burden of female household labor as a growing variety of machines and gadgets took over common chores. Another great change brought about by inexpensive electricity has been the global spread of air conditioning (first patented by William Carrier in 1902). Its availability opened up the American Sunbelt to mass migration from the Snowbelt, and since the 1980s, room units have also spread rapidly among more affluent households of subtropical and tropical countries.

The age of crude oil was launched during the same decade as that of electricity, and the three key ingredients of a modern automobile—Gottlieb Daimler's gasoline-fueled engine, Karl Benz's electrical ignition, and Wilhelm Maybach's float-feed carburetor—came together during the 1890s when Rudolf Diesel also introduced a different type of internal combustion engine. Subsequent decades have seen a great deal of improvement but no fundamental change of the prime mover's essentials. Only in the United States and Canada did car ownership reach high levels before World War II; Western Europe and Japan became nearly saturated only during the 1980s. By 1904, the Wright brothers built their own four-cylinder engine with an aluminum body and a steel crankshaft to power the first flight of a machine heavier than air, and the first 50 years of commercial and military flight were dominated by airplanes powered by reciprocating engines. Jet engines powered the first warplanes by 1944; the age of commercial jet flights began during the 1950s and was elevated to a new level by the Boeing 747, introduced in 1969.

Transportation has been the principal reason for the higher demand for crude oil, but liquid fuels, and later natural gas, also became very important for heating, and both hydrocarbons are excellent feedstocks for many chemical syntheses. By 1950, crude oil and natural gas claimed approximately 35% of the world's primary energy supply, and by 2000, their combined share was just over 60% compared to coal's approximately 25% (Fig. 4). Fossil fuels thus provided approximately 90% of all commercial primary energy supply, with the rest coming from primary (hydro and nuclear) electricity. Despite a great deal of research and public interest, new renewable conversions (above all, wind turbines and photovoltaics) still have only a negligible role, as do fuel cells whose high conversion efficiencies and pollution-free operation offer a much better way of converting gaseous and liquid fossil fuels to kinetic energy than does air-polluting combustion.

5. High-energy Civilization and Its Attributes

Fossil fuels and electricity have helped to create the modern world by driving up farm productivity and hence drastically reducing agricultural populations, by mechanizing industrial production and letting the labor force move into the service sector, by making megacities and conurbations a reality, by globalizing trade and culture, and by imposing many structural uniformities onto the diverse world. Inevitably, all of these developments had enormous personal and collective consequences as they released hundreds of millions of people from hard physical labor, improved health and longevity, spread literacy, allowed

for rising material affluence, broke traditional social and economic confines, and made the Western ideas of personal freedom and democracy into a powerfully appealing (as well as fanatically resented) global force.

But these benefits are fully, or largely, enjoyed only by a minority (only $\approx 15\%$) of the world's population. The great energy transitions of the past century raised standard of living everywhere but it has not been accompanied by any impressive decline of disparities between rich and poor societies. In the year 2000, the richest 10% of the world's population consumed more than 40% of all commercial primary energy (Fig. 5). In addition to being liberating and constructive, modern high-energy civilization is an enormous source of environmental pollution and degradation of ecosystems (perhaps even endangering the very maintenance of a habitable biosphere), is prone to many social ills accentuated by urban living, has acquired weapons of mass destruction, and is highly vulnerable to asymmetrical threats of terrorism.

People who live in affluent societies take the levels of energy that individuals and collectives control for granted, but the claims now made on energy resources are still astounding no matter if they are compared across the entire span of human evolution or just across the 20th century. Peak unit capacities of prime movers rose from less than 100 W of sustained human labor for the late Neolithic foragers to

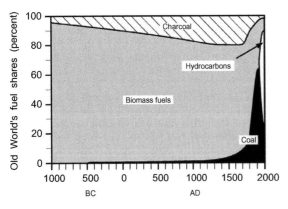

Figure 4
Approximate shares of global fuel consumption in the period from 1000 BCE to 2000 CE.

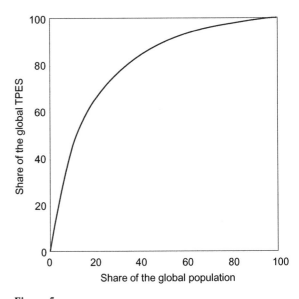

Figure 5
Inequality of global commercial energy use in the year 2000 is indicated by a highly convex shape of the Lorenz curve: the richest 10% of the world's population claimed approximately 45% of all energy, whereas the poorest 50% had access to just 10% of the total.

approximately 300 W for a draft ox of the early antiquity and to 2 kW for the largest Roman water-wheels. Improved versions of those machines rated approximately 5 kW by the end of the first millennium of the CE and still no more than 8 kW by 1700. A century later, Watt's steam engines pushed the peak ratings to 100 kW, by 1900 the largest steam and water turbines had capacities 100 times higher (10 MW), and steam turbines reached eventually their maximum at 1.5 GW (Fig. 2). Peak unit capacities of prime movers that deliver sustained power thus rose approximately 15 million times in 10,000 years, with more than 99% of the rise taking place during the 20th century. Increases in the destructive discharge of energies have been even more stunning: the largest tested thermonuclear weapon (the Soviet Union's 100-megaton bomb in 1961) had power 16 orders of magnitude higher than the kinetic energy of a forager's arrow.

Because of the rapid growth of the global population, per capita comparisons yield naturally smaller multiples. Despite the near quadrupling of the global population—from 1.6 billion in 1900 to 6.1 billion in 2000—the average gross annual per capita supply of commercial energy more than quadrupled from just 14 GJ to approximately 60 GJ. In the United States, per capita energy use more than tripled to approximately 340 GJ/year, Japan's more than quadrupled to just over 170 GJ/ year, and China's per capita fossil fuel use, exceptionally low in 1900, rose 13-fold between 1950 and 2000, from just over 2 to $\approx$ 30 GJ/year. These gains are far more impressive when expressed in more meaningful terms as useful energy services. Conservative calculations indicate that because of better conversion efficiencies, the world in the year 2000 had at its disposal at least 25 times more useful commercial energy than in 1900 and the corresponding multiples exceed 30 in some rapidly industrializing countries.

Perhaps the best way to compare the secular energy gains at a household level is to compare installed electric power. In 1900, a typical urban U.S. household had only a few low-power light bulbs, adding up to less than 500 W. In the year 2000, an all-electric, air-conditioned suburban house with some 400 m^2 of living area had more than 80 switches and outlets ready to power every imaginable household appliance and drawing upward of 30 kW, at least a 60-fold jump in one century. Three vehicles owned by that household would boost the total power under the household's control to close to half a megawatt! Equivalent power—though nothing like the convenience, versatility, flexibility, and reliability of delivered energy services—would have been available only to a Roman *latifundia* owner of $\approx$ 6000 strong slaves or to a 19th century landlord employing 3000 workers and 400 large draft horses.

Because the adoption of new energy sources and new prime movers needs substantial investment, it is not surprising that this process broadly correlates with the upswings of business cycles. The first wave, well documented by Schumpeter, corresponds to the rising extraction of coal and the introduction of stationary steam engines (1787–1814). The second wave (1843–1869) was stimulated by railroads, steamships, and iron metallurgy, and the third wave (1898–1924) was stimulated by the rise of electricity generation and the replacement of steam-driven machinery by electric motors in manufacturing (Fig. 6).

The initial stages of energy transitions also correlate significantly with the starts of major innovation waves that appear to be triggered by economic depressions. The first innovation cluster, peaking in 1828, was associated mainly with mobile steam engines; the second, peaking in 1880, was associated with the introduction of electricity generation and internal combustion engines, and the third cluster, peaking in 1937, included gas turbines, fluorescent lights, and nuclear energy. Post-World War II extension of these waves would include the global substitution of hydrocarbons for coal and mass car ownership; this wave was checked in 1973 by the sudden increase in the price of oil initiated by the Organization of Petroleum Exporting Countries.

6. Limits of Energetic Determinism

When seen from a strictly biophysical point of view, energy may have an unchallenged primacy among the variables determining the course of history, but when it is considered from broader cultural and social perspectives, it may not even rank as *primus inter pares*. In addition to all of those indisputable energy imperatives, there are a multitude of none-nergy factors that initiate, control, shape, and direct human decisions to harness and use energies in a myriad of specific ways. Only if one were to equate the quality of life, or the accomplishments of a civilization, with the mindless accumulation of material possessions, would the rising consumption of energy be an inevitable precondition. But such a primitive perspective excludes the multitude of moral, intellectual, and esthetic values whose inculcation, pursuit, and upholding have no links to any particular level of energy use.

To begin with, timeless artistic expressions show no correlation with levels or kinds of energy consumption: the bison in the famous cave paintings of Altamira are not less elegant than Picasso's bulls drawn nearly 15,000 years later. It should also noted that all universal and durable ethical precepts, be they of freedom and democracy or compassion and charity, originated in antiquity, when an inadequate and inefficient energy supply was but a small fraction of today's usage. And to cite some more recent examples, the United States adopted a visionary constitution while the country was still a subsistence

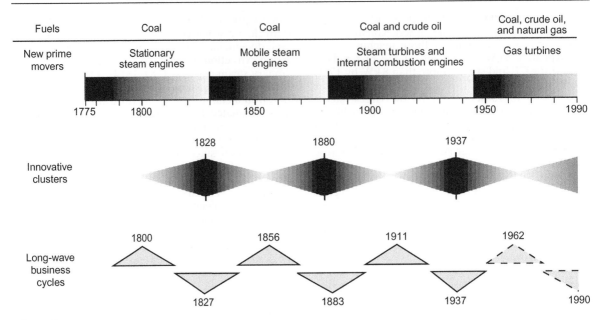

Figure 6
Timelines of major energy eras, innovative clusters (according to G. Mensch), and long-wave business cycles (according to J. A. Schumpeter) in the years 1775–1990.

society running on wood; in contrast, before its collapse, the Soviet Union was the world's largest producer of crude oil and natural gas—yet all that country could offer its citizens was an impoverished life in fear, in a cage they were not allowed to leave. That political freedoms have little to do with energy use can be clearly seen by inspecting the list of the world's least free countries: it includes not only energy-poor Afghanistan, Vietnam, and Sudan, but also oil-rich Libya and Saudi Arabia.

Long-term trends in population growth are another key historic variable that is hard to relate to changes in the energy base and to levels of energy use. Improved nutrition could be seen as the principal cause of tripling the European population between 1750 and 1900, but such a claim cannot be reconciled with careful reconstructions of average food energy intakes. China's example is even more persuasive: between 1700 and 1900, the Qing dynasty did not see any major change in energy sources and prime movers, and no rise in the average per capita use of wood and straw, but the country's population tripled to approximately 475 million people.

Even the links between economic output and energy use are not that simple. When seen from a physical (thermodynamic) perspective, economies are complex systems that incessantly acquire and transform enormous quantities of fossil fuels and electricity, and some very high correlations between the rate of energy use and the level of economic performance suggest that the latter may be a direct function of the former. There is little doubt that the successive positions of economic preeminence and international influence wielded by the Dutch Republic in the 17th century, Great Britain in the 19th century, and the United States in the 20th century had their material genesis in the early exploitation of fuels that yielded higher net energy returns and allowed for higher conversion efficiencies (peat, coal, and crude oil, respectively).

A closer analysis, however, reveals that the link between energy use and economy cannot be encompassed by any easily quantifiable function as national specificities preclude any normative conclusions and invalidate many intuitive expectations. Possession of abundant energy resources has been no guarantee of national economic success and their virtual absence has been no obstacle to achieving enviable economic prosperity. A long list of energy-rich nations that have nevertheless mismanaged their fortunes must include, to name just the three most prominent cases, the Soviet Union, Iran, and Nigeria. The list of energy-poor nations that have done very well by any global standard must be headed by Japan, South Korea, and Taiwan. And countries do not have to attain specific levels of energy use in order to enjoy a comparably high quality of life.

Although it is obvious that a decent quality of life is predicated on certain minima of energy use, those countries that focus on correct public policies may realize fairly large rewards at levels not far above such minima, whereas ostentatious overconsumption wastes energy without enhancing the real quality of life. A society concerned about equity and willing to channel its resources into the provision of adequate diets, availability of good health care, and accessibility to basic schooling could guarantee decent physical well-being, high life expectancy, varied nutrition, and fairly good educational opportunities with an annual per capita use of as little as 40–50 GJ of primary energy converted with efficiencies prevailing during the 1990s.

A better performance, pushing infant mortalities below 20/1000, raising female life expectancies above 75 years, and elevating the UNDP's Human Development Index (HDI) above 0.8 appears to require at least 1400–1500 kgoe (kilograms of oil equivalent) of energy per capita, and in the year 2000, the best global rates (infant mortalities below 10/1000, female life expectancies above 80 years, HDI above 0.9) needed no less than approximately 2600 kgoe/capita (Fig. 7). All of the quality-of-life variables relate to average per capita energy use in a nonlinear manner,

with clear inflections evident at between 40 and 70 GJ/capita, with diminishing returns afterward and with basically *no* additional gains accompanying consumption above 110 GJ/capita or ≈ 2.6 metric tons of crude oil equivalent. The United States consumes exactly twice as much primary energy per capita as does Japan or the richest countries of the European Union (340 GJ/year versus 170 GJ/year), but it would be ludicrous to suggest that American lives are twice as good. In reality, the United States falls behind Europe and Japan in a broad range of quality-of-life variables, including higher infant mortality rates, more homicides, lower scientific literacy, and less leisure time.

Finally, energy use is of little help in explaining the demise of the established order. The long decline of the Western Roman Empire cannot be tied to any loss of energy supplies or a drastic weakening of energy conversion capabilities and neither can the fall of the French monarchy during the 1780s, the collapse of the Czarist empire in 1917, or the Nationalist

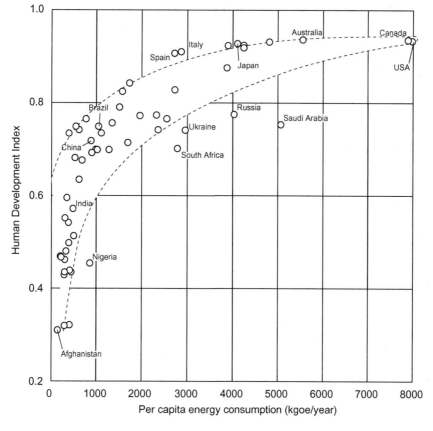

Figure 7
Plot of the Human Development Index against the average annual per capita use of commercial energy in the year 2000, which shows that virtually no quality-of-life gains accrue with consumption above 2.6 metric tons of oil equivalent.

retreat from mainland China during the late 1940s. Conversely, many historically far-reaching consolidations (including the gradual rise of Egypt's Old Kingdom, the rise of the Roman Republic, or the rise of the United States) and lightning expansions of power (including the spread of Islam during the 7th and 8th centuries or the Mongolian conquest of the 13th century) cannot be linked to any new prime movers or to better uses of prevailing fuel.

There is clearly an ambivalent link between energy and history. Energy sources and prime movers delimit the options of human history and determine the tempo of life, and, everything else being equal, thermodynamics requires that higher socioeconomic complexity must be supported by more intensive flows of energy. And yet, neither the possession of abundant energy sources nor a high rate of energy consumption guarantees the security of a nation, economic comfort, or personal happiness. Access to energies and the modes of their use constrain the options for human beings' actions but do not explain the sources of people's aspirations and the reasons for the choices they make and do not preordain the success or failure of individual societies at a particular time in history. Indeed, the only guaranteed outcome of higher energy use is greater environmental burdens whose global impacts may imperil the very habit-ability of the biosphere. To prevent this from happening, humanity's most important future choice may be to limit the use of energy and thus to embark on an entirely new chapter of history.

Further Reading

Cowan C W, Watson P J (eds.) 1992 *The Origins of Agriculture*. Smithsonian Institution Press, Washington, DC

Finniston M, *et al.* (eds.) 1992 *Oxford Illustrated Encyclopedia of Invention and Technology*. Oxford University Press, Oxford, UK

Forbes R J 1972 *Studies in Ancient Technology*. E.J. Brill, Leiden, the Netherlands

Goudsblom J 1992 *Fire and Civilization*. Allen Lane, London

Langdon J 1986 *Horses, Oxen and Technological Innovation*. Cambridge University Press, Cambridge, UK

McNeill W H 1989 *The Age of Gunpowder Empires, 1450–1800*. American Historical Association, Washington, DC

Nef J U 1932 *The Rise of the British Coal Industry*. Routledge, London

Pacey A 1990 *Technology in World Civilization*. The MIT Press, Cambridge, MA

Schumpeter J A 1939 *Business Cycles: A Theoretical and Statistical Analysis of the Capitalist Processes*. McGraw-Hill, New York

Schurr S H 1984 *Energy use, technological change, and productive efficiency: An economic-historical interpretation*. Annu. Rev. Energy 9, 409–25

Singer C, *et al.* (eds.) (1954–1984). *A History of Technology*. Clarendon Press, Oxford, UK.

Smil V 1991 *General Energetics: Energy in the Biosphere and Civilization*. Wiley, New York

Smil V 1994 *Energy in World History*. Westview Press, Boulder, CO

Smil V 2003 *Energy at the Crossroads: Global Perspectives and Uncertainties*. MIT Press, Cambridge, MA

Vaclav Smil

LIST OF CONTRIBUTORS

Contributors are listed in alphabetical order together with their addresses. Titles of articles that they have authored follow in alphabetical order. Where articles are coauthored, this has been indicated by an asterisk preceding the title.

Alhajji, A. F.
Ohio Northern University
Ada, Ohio
United States
OPEC Market Behavior, 1973–2003

Arndt, Roger E. A.
University of Minnesota
Minneapolis, Minnesota
United States
Hydropower, History and Technology of

Berck, Peter
University of California
Berkeley, California
United States
Prices of Energy, History of

Buckley, Geoffrey L.
Ohio University
Athens, Ohio
United States
Coal Mining in Appalachia, History of

Castaneda, Christopher J.
California State University
Sacramento, California
United States
Natural Gas, History of

Chalabi, Fadhil J.
Centre for Global Energy Studies
London
United Kingdom
OPEC, History of

Christensen, Paul
Hofstra University
Hempstead, New York
United States
Economic Thought, History of Energy in

Crease, Robert P.
Stony Brook University
Stony Brook, New York
United States
Energy in the History and Philosophy of Science

Daemen, Jaak J. K.
University of Nevada
Reno, Nevada
United States
Coal Industry, History of

Duffy, Robert J.
Colorado State University
Fort Collins, Colorado
United States
Nuclear Power, History of

Dunn, Seth
Worldwatch Institute
New Haven, Connecticut
United States
Hydrogen, History of

Garber, Elizabeth
State University of New York at Stony Brook
Stony Brook, New York
United States
Conservation of Energy Concept, History of

Gibbons, John H.
Resource Strategies
The Plains, Virginia
United States
Conservation Measures for Energy, History of

Giebelhaus, August W.
Georgia Institute of Technology
Atlanta, Georgia
United States
Oil Industry, History of

Gipe, Paul
Paul Gipe & Associates
Tehachapi, California
United States
Wind Energy, History of

Goudsblom, Johan
University of Amsterdam
Amsterdam
The Netherlands
Fire: A Socioecological and Historical Survey

Grübler, Arnulf
International Institute for Applied Systems Analysis
Laxenburg
Austria
Transitions in Energy Use

Gulliver, John S.
University of Minnesota
Minneapolis, Minnesota
United States
Hydropower, History and Technology of

Gwin, Holly L.
Resource Strategies
The Plains, Virginia
United States
Conservation Measures for Energy, History of

Hall, Charles A. S.
State University of New York
College of Environmental Science and Forestry
Syracuse
New York
United States
Ecosystems and Energy: History and Overview

Hutchison, Keith
University of Melbourne
Melbourne
Australia
Thermodynamic Sciences, History of

Nye, David E.
University of Southern Denmark
Odense
Denmark
Electricity Use, History of

Pasqualetti, Martin J.
Arizona State University
Tempe, Arizona
United States
*Geographic Thought, History of Energy in
Wind Energy, History of

Periman, Richard D.
Rocky Mountain Research Station
USDA Forest Service, Albuquerque
New Mexico
United States
Early Industrial World, Energy Flow in

Perlin, John
Rahus Institute Martinez
California
United States
Solar Energy, History of
Wood Energy, History of

Righter, Robert
Southern Methodist University
Dallas, Texas
United States
Wind Energy, History of

Roberts, Michael J.
Agricultural and Resource Economics
North Carolina State University, NC
USA
Prices of Energy, History of

Salameh, Mamdouh G.
Oil Market Consultancy Service
Haslemere, Surrey
United Kingdom
Oil Crises, Historical Perspective

Simmons, I. G.
University of Durham
Durham
United Kingdom
Environmental Change and Energy

Smil, Vaclav
University of Manitoba
Winnipeg, Manitoba
Canada
War and Energy
World History and Energy

Solomon, Barry D.
Michigan Technological University
Houghton, Michigan
United States
Geographic Thought, History of Energy in

Tainter, Joseph A.
Rocky Mountain Research Station
Albuquerque
New Mexico
Sociopolitical Collapse, Energy and

Tarr, Joel A.
Carnegie Mellon University
Pittsburgh, Pennsylvania
United States
Manufactured Gas, History of

SUBJECT INDEX

The Subject Index has been compiled to assist the reader in locating all references to a particular topic in the Encyclopedia. Index entries are filed in letter-by-letter sequence, and arranged in set-out style with up to three levels of heading. Major coverage of a subject is indicated by ***bold italic*** page numbers. Page numbers followed by *T* (or *F*) refer to Tables (or Figures). *See* references point the user to preferred terms; *see also* references indicate terms of related interest. Every effort has been made to make the index as comprehensive as possible and to standardize the terms used.

Printed and bound by CPI Group (UK) Ltd, Croydon, CR0 4YY

08/05/2025

01864827-0009